Bioremediation for Sustainable Environmental Cleanup

Editors

Anju Malik

Department of Energy and Environmental Sciences
Chaudhary Devilal University
Sirsa, Haryana, India

Vinod Kumar Garg

Department of Environmental Science and Technology
School of Environment and Earth Sciences
Central University of Punjab
Bathinda, Punjab, India

CRC Press
Taylor & Francis Group
Boca Raton London New York

CRC Press is an imprint of the
Taylor & Francis Group, an **informa** business

A SCIENCE PUBLISHERS BOOK

First edition published 2024
by CRC Press
2385 NW Executive Center Drive, Suite 320, Boca Raton FL 33431

and by CRC Press
4 Park Square, Milton Park, Abingdon, Oxon, OX14 4RN

© 2024 Anju Malik and Vinod Kumar Garg

CRC Press is an imprint of Taylor & Francis Group, LLC

Library of Congress Cataloging-in-Publication Data (applied for)

ISBN: 978-1-032-23491-5 (hbk)
ISBN: 978-1-032-23492-2 (pbk)
ISBN: 978-1-003-27794-1 (ebk)

DOI: 10.1201/9781003277941

Typeset in Times New Roman
by Radiant Productions

Preface

In the last century, rapid industrialization and urbanization have left a large quantity of a wide range of pollutants emitted from both natural and anthropogenic sources. These pollutants have detrimental effects on the flora, fauna and natural environment. The persistence and prevalence of these pollutants virtually in all the ecosystems of Earth are of a major concern. Although a large number of conventional physical and chemical methods are available for cleaning up the environment, they are not sustainable and have inherent limitations. Hence, an alternate eco-sustainable, bio-based and environment-friendly approach recognized as *Bioremediation* has emerged. In bioremediation, biodiversity acts as a toolbox providing various processes/mechanisms to eliminate, immobilize/stabilize, degrade or transform various hazardous contaminants into innocuous and value-added products. This approach uses a vast array of biological agents, especially bacteria (microbial remediation), fungi (mycoremediation), algae (phycoremediation), higher plants (phytoremediation), biochar, nano-biomaterials, etc.

This book *Bioremediation for Sustainable Environmental Cleanup* has a compilation of seventeen chapters contributed by eminent researchers from various countries located in diverse geographical areas including Argentina, Canada, Germany, India, Pakistan, South Africa, the United Kingdom and the United States of America. The book starts with a brief overview of various bioremediation approaches used to clean up the polluted environment. It is followed by the chapters on bioremediation strategies like Bioprecipitation, Bioaccumulation and Biosorption in Chapters 2, 3 and 4. Further, the remediation of Polycyclic Aromatic Hydrocarbons using phytoremediation and mycoremediation has been described in Chapters 5 and 6. The current status and prospects of microbe-assisted bioremediation of pesticides have also been included in Chapter 7. Chapter 8 focuses on the degradation of different pesticides by the bacterium *Pseudomonas putida*. Chapters 9 to 13 extensively discuss the remediation of metal(loid)s. Chapter 14 presents an overview of the potential of Water Hyacinth (*Eichhornia crassipes*) in remediating wastewater discharged from the paper and pulp, textile and dairy industries. Furthermore, the applications of novel materials viz. biohybrids, nano-biomaterials and graphitic Carbon Nitride (g-C_3N_4), in environmental clean up of heavy metals, pesticides, dyes and other organic pollutants have been addressed/covered in Chapters 15, 16 and 17.

Conclusively, this book comprehensively describes the state-of-the-art and potential of emerging bioremediation approaches, which have been employed for sustainable environmental clean up of various environmental pollutants such as metal(loid)s, polycyclic aromatic hydrocarbons, dyes, pesticides, petroleum hydrocarbons, etc., by using bacteria, fungi, algae, higher plants and novel materials like biohybrids, nano-biomaterials and graphitic Carbon Nitride (g-C_3N_4). The emphasis throughout, however, is on sustainable environmental clean up. Each chapter of the book can stand alone, contains updated information and is clearly illustrated with figures, pictures and tables in a scientific way.

We firmly believe that this book will cater to the interest of researchers, academicians, environmentalists, agriculturalists, extension workers, industrialists, students at undergraduate and postgraduate levels, practising engineers, policymakers and other enthusiastic people who are unreservedly devoted to sustainable environmental clean up of pollutants.

The editors have sincerely worked for about two years to carefully select, edit, revise and compile the contributed manuscripts to the scope of the book. We have strived hard to ensure that this book is free from any erroneous or deceptive information and any such mistake is completely inadvertent. Further, we wholeheartedly thank all our contributors for readily accepting our invitation, timely submitting their innovative, high quality and valuable chapters, and their cooperation at all stages in making this voluminous and high-quality outcome a successful attempt. Last but not the least, we also express our deep sense of gratitude to our family members for their unconditional support during this long journey.

Anju Malik
Sirsa, India
Vinod Kumar Garg
Bathinda, India

Contents

CHAPTER 1

Bioremediation
A Sustainable Approach for Environmental Cleanup

Bharti Singh,[1] *Anju Malik*[1,]* and *Vinod Kumar Garg*[2]

1.1 Introduction

Both urbanization and industrialization have resulted in a rise of contaminated land and water. As both these activities are growing at a rapid rate, they are creating stress on renewable as well as non-renewable resources. As a result, environmental pollution has increased globally during the last several decades. The expansion of the manufacturing and agricultural industries has led to an increase in the release of a wide variety of xenobiotic substances into the environment. Excessive release of hazardous waste has resulted in a lack of clean water and soil disturbances, restricting agricultural output (Kamaludeen et al. 2003). Heavy metals, petroleum oil, pesticides, hydrocarbons such as aliphatic, aromatic and polycyclic aromatic hydrocarbons, nitroaromatic compounds, chlorinated hydrocarbons such as polychlorinated biphenyls (PCBs) and perchloroethylene, organophosphorus compounds, organic solvents (phenolic compounds) and phthalates are all possible contaminants in hazardous waste (Megharaj et al. 2011). The economically and environmentally viable management of these wastes is a serious challenge around the world. According to third-world network data, more than 450 million kg of hazardous waste is released into the air, water and land on a global scale (Singh 2014). If these pollutants are discharged into the environment without being correctly treated, they may cause serious health concerns as well as the extinction of many species of fauna and flora. Various kinds of contaminants are carried into the ground via air and water, resulting in a slew of major environmental and health issues around the world (Boopathy 2000). Different types of technologies have been developed for waste disposal that use high-temperature incineration, soil washing, adsorption, flocculation, landfilling, pyrolysis and chemical decomposition etc. Despite their high potential for effectiveness, these techniques are also inconvenient, expensive and undesirable.

Many researchers have increased their efforts to identify more environmentally friendly and cost-effective alternatives for the use of hazardous chemicals and treatments to eliminate current dangerous contaminants for sustainable cleanup of the environment. Several underdeveloped countries cannot afford to build environmentally friendly and economically viable machinery for

[1] Department of Energy and Environmental Sciences, Chaudhary Devilal University, Sirsa, Haryana, India.
[2] Department of Environmental Science and Technology, School of Environment and Earth Sciences, Central University of Punjab, Bhatinda, Punjab, India.
* Corresponding author: anjumalik@cdlu.ac.in, anjumalik27@yahoo.com

waste handling. In such desperate situations, a few environmentally beneficial techniques can overcome the restrictions associated with safe and cost-effective waste management technology. Bioremediation approaches are increasingly being used as preferred methods for the detoxification and biodegradation of pollutants in waste management and cleanup programs to revert contaminated sites to their natural state, as they are cost-effective, natural environment friendly and long-term solutions (Kumar et al. 2018).

Bioremediation is described as a process that uses living organisms, primarily microorganisms, green plants and their enzymes, to remove, degrade, mineralize, transform and detoxify environmental pollutants and hazardous components of waste into harmless or less toxic forms during the treatment of contaminated sites in order to restore them to their original state (Azubuike et al. 2016). Pesticides, polyaromatic hydrocarbons, halogenated petroleum hydrocarbons, nitroaromatic chemicals, metals and industrial solvents have all been reduced in concentration and toxicity using the bioremediation technique (Dua et al. 2002).

Microorganisms are used in bioremediation to immobilize or change the chemical structure of pollutants found in soils, sediments, water and air, resulting in partial degradation, mineralization or transformation of the molecule. However, most of the pollutants exhibit resistance to degradation, which causes them to persist in the environment. Such accumulation may sometimes pose a serious risk. Bioremediation techniques may be effectively set in polluted fields with the right use of natural and engineered microorganisms. The treatment of polluted groundwater, soil, wetlands, industrial wastes and sludge can all benefit from bioremediation approaches. Phytoremediation is the process in which different types of plants are used in the bioremediation of contaminants. It is a non-conventional, cost-effective and eco-friendly technology that utilizes plants to remove, transform or stabilize a variety of contaminants located in water, sediments or soils (Prasad et al. 2001, Ladislas et al. 2012). This chapter aims to provide a brief overview of bioremediation and its guiding principles, as well as the numerous bioremediation approaches, specifically the *in-situ* and *ex-situ* remediation strategies, their benefits and drawbacks.

1.2 Bioremediation

Bioremediation is the "use of living organisms to clean up toxins from soil, water, or wastewater" (EPA 2016). The utilization of living organisms to ameliorate the contaminated environment viz. water, soil, etc., is referred to as bioremediation (Brar et al. 2006, Antizar-Ladislao 2010, Latha and Reddy 2013). In other words, it is a technique for eliminating environmental contaminants and restoring the natural ecosystem while also avoiding additional pollution. Various species and their products, such as bacteria (bacterial bioremediation), fungi (mycoremediation), plants (phytoremediation) and biomolecules generated from organisms, are engaged in bioremediation processes (derivative bioremediation). Bioremediation can be aerobic (Wiegel and Wu 2000, Bedard and May 1995) or anaerobic (Komancová et al. 2003). Bioremediation can occur naturally, which is referred to as "natural attenuation" or it can take place artificially, which is referred to as "biostimulation". Note that not all toxins can be effectively removed with bioremediation. Bioremediation aids humanity and society in dealing with harmful chemical pollutants that, if not removed or rendered safe, may be damaging to human health and the environment.

1.3 Principles of Bioremediation

Recent research in ecology has revealed a variety of possibilities for improving biological systems. The remediation of contaminated water and land regions is one of these studies' notable successes. Living organisms including bacteria, fungi and algae, as well as plants, are utilized in bioremediation to break down and detoxify dangerous contaminants in the environment, converting them to CO_2, H_2O, microbial biomass and metabolites. These microorganisms might be native to the polluted location or they could have been isolated elsewhere and transported to the contaminated site. Different types of organisms are involved in the biodegradation of a substance. A process known as bioaugmentation involves the use of different cultured microorganisms to degrade soil and water contaminants. Microorganisms must enzymatically attack contaminants and transform them into harmless compounds for bioremediation to be efficient (Vidali 2001). Since environmental conditions must be favorable for microbial development for bioremediation to be effective, environmental parameters are regulated/monitored to accelerate microbial growth and deterioration. Bioremediation procedures are often less expensive than traditional methods like incineration, and certain contaminants may be treated on-site, lowering exposure hazards for cleanup workers and possibly broader exposure due to transportation problems. Bioremediation is more widely accepted than other approaches since it is based on natural attenuation. Most bioremediation systems are operated under aerobic circumstances; however, operating one under anaerobic conditions may allow microbial organisms to degrade compounds that are typically resistant to degradation (Colberg and Young 1995).

1.4 Types of Bioremediation

The bioremediation process is divided into *in-situ* and *ex-situ* based on the origin and removal of pollutants, as shown in Figure 1.1.

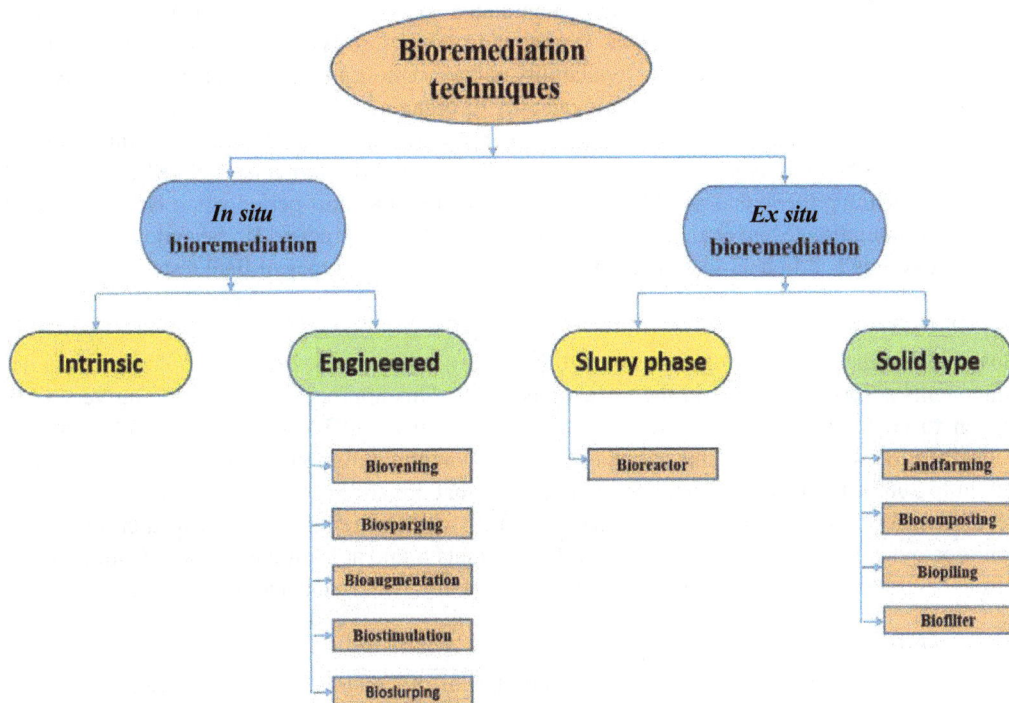

Figure 1.1. Types of bioremediation techniques (Tyagi and Kumar 2021).

1.4.1 In-situ Bioremediation

In-situ bioremediation is defined as the process of degrading the contaminants in a naturalistic environment to produce carbon dioxide and water. It is a low-cost, low-maintenance, environmentally beneficial and long-term solution for contaminated site cleaning (Aggarwal et al. 1990, Jørgensen, 2007, Megharaj et al. 2011, Latha and Reddy 2013). *In-situ* bioremediation is most commonly used to degrade pollutants in saturated soils and groundwater. This technique is basically applied to polluted material on-site. Some of the examples of *in-situ* bioremediation strategies as shown in Figure 1.1 are described next.

1.4.1.1 Bioaugmentation

Bioaugmentation is an effective method to increase the biodegradation of chemical compounds which are added to the contaminated soil or water. When bioattenuation or biostimulation is ineffective against resistant compounds, this strategy is thought to be effective. This approach is primarily used to remediate aromatic and chlorinated hydrocarbon-contaminated municipal wastewater and soil. Microorganisms from different cultures are brought together to enhance bioaugmentation efficiency. The proper utilization of microbes which are naturally occurring can help to degrade and entirely mineralize organic pollutants (Bender and Phillips 2004). Bioaugmentation-aided phytoextraction with Plant Growth-Promoting Rhizobacteria (PGPR) or Arbuscular Mycorrhizal Fungi (AMF) is also a potential option for metal-contaminated soil cleanup (Lebeau et al. 2008).

1.4.1.2 Biostimulation

The process of stimulating microbial enzymes for the bioremediation of diverse xenobiotic chemicals is known as biostimulation, which involves adding soil nutrients, trace minerals, electron acceptors or donors to a contaminated site until the optimum pH is reached (Li et al. 2010). If natural degradation does not occur or occurs at a slow rate, the environment must be manipulated in order to stimulate biodegradation and increase reaction rates. This method is primarily intended to remediate fuel, hydrocarbon, metal contaminated sites, non-halogenated Volatile Organic Compounds and Semi-Volatile Organic Compounds (VOCs, SVOCs) pesticides and herbicide pollution of soil and groundwater. The availability of carbon, nutrients such as N and P, temperature, available oxygen, soil pH and the kind and quantity of organic pollutant itself all influence the rate of microbial turnover of chemical pollutants (Carberry and Wik 2001). There are several examples of contaminants being biostimulated by indigenous microbes. Microorganisms have been observed to convert trichloroethene and perchloroethylene to ethane in a short period when lactate is added during biostimulation (Shan et al. 2010). *In-situ* biostimulation is predicted to become a dependable and safe cleaning solution as scientific data accumulates via these types of field tests.

1.4.1.3 Bioventing

Bioventing is an *in-situ* approach in which different microorganisms are used to degrade organic components adsorbed on soils in the unsaturated zone. This approach involves adding very little oxygen to contaminated soil at low airflow rates, which is sufficient for successful microbial biodegradation while minimizing pollutant volatilization and emission into the atmosphere (Atlas and Philp 2005). In this technique, organic materials that are added to soil in the unsaturated zone are biodegraded by the native microorganisms. TCE (trichloroethane), ethylene dibromide and dichloroethylene are the most common volatile pollutants found at locations containing mid-weight petroleum products (such as diesel and jet fuel) (Lee et al. 2006, Latha and Reddy 2013).

1.4.1.4 Bioslurping

Bioslurping is a technology that combines a number of procedures, including vacuum-enhanced pumping, bioventing and soil vapor extraction, to remove contaminants from soil and groundwater while also accelerating microbial biodegradation. One disadvantage of this strategy is the soil's low permeability, which reduces the pace at which oxygen is transferred and further suppresses microbial activity. These methods are often used to remove volatile and semivolatile organic pollutants from soil and liquids (Vidali 2001).

1.4.1.5 Biosparging

Biosparging also known as air sparging is an *in-situ* treatment method that involves injecting air under pressure below the water table to clean polluted groundwater and raising subsurface oxygen concentrations to enhance biodegradation in saturated and unsaturated soils. VOCs that have polluted groundwater or soils in the saturated zone are treated using this method. Vapors that have been volatilized pass into the unsaturated zone, where they are vacuum-extracted by a soil vapor extraction system (Hardisty and Ozdemiroglu 2005, Neilson and Allard 2008).

1.4.2 Ex-situ Bioremediation

Ex-situ bioremediation procedures treat pollutants elsewhere than where they were initially discovered. *Ex-situ* bioremediation techniques are further divided into other categories, such as landfarming, composting, biopiling, bioreactors and biofilters, based on the type of pollutant, depth and degree of pollution, treatment cost and geographical and geological characteristics of the contaminated location (Atlas and Philp 2005). Some of the *ex-situ* bioremediation strategies as shown in Figure 1.1 are discussed below:

1.4.2.1 Biopiling

A biopile is one of the several bioremediation strategies for treating hydrocarbon-contaminated soil that involves piling the dirt on top of an air distribution system and aerating it. Landfarming and composting processes are combined in the biopiling approach and are mostly used in the treatment of areas polluted with petroleum hydrocarbons. This whole setup includes a treatment bed and nutrients, an aeration system, an underground irrigation system and a leachate collecting system (Azubuike et al. 2016). For its cost-effectiveness and the ability to manage temperature, pH, and nutrient conditions, this approach is increasingly being employed for bioremediation (Whelan et al. 2015).

1.4.2.2 Landfarming

Landfarming is a technique that includes dumping polluted material onto the soil surface and tilling it to mix and aerate it (Harmsen et al. 2007, Maciel et al. 2009). Fuels, non-halogenated volatile organic carbons, pesticides and herbicides are among the contaminants that this technology is meant to handle. The idea is to encourage indigenous biodegradative bacteria and make it easier for them to degrade pollutants aerobically. Landfarming appears to be limited to the treatment of 10–35 cm of surface soil (Kumar et al. 2018). This technology is commonly used to clean up polluted areas with aliphatic and polycyclic aromatic hydrocarbons, as well as PCBs (Silva-Castro et al. 2012).

1.4.2.3 Bioreactors

In this process, the pollutant is degraded in a reactor or container under controlled conditions. The bioreactor technique is used to remediate soil or water that has been polluted by volatile organic pollutants such as BTEX (benzene, toluene, ethylbenzene and xylene). A slurry bioreactor is a vessel and apparatus which is used to increase the bioremediation rate of soil-bound and water-soluble pollutants as a contaminated soil and biomass slurry (Kumar et al. 2011). Batch, continuous, sequential batch biofilm, membrane, fluidized bed, biofilm and airlift bioreactors are among the many types of bioreactors available globally. Before being placed in a bioreactor, the contaminated soil must be pretreated or the contamination can be eliminated from the soil in different ways (EPA 2000).

1.4.2.4 Biofilters

Biofilters are most commonly used to remove gaseous contaminants. The removal of gaseous contaminants is accomplished using columns packed with microorganisms by converting unwanted elements, such as CO_2, H_2O and cell mass, into harmless objects (Boopathy 2000). Biofilters use a combination of basic techniques such as assimilation, adsorption, desorption and debasement of gaseous phase contaminants. Microorganisms form a biofilm that adheres to the surface of a solid-packed medium. The filter bed medium is often composed of natural materials with a wide surface area in the zone and some supplement supplies. Biofilters are commonly used in water supply systems to manage moisture and add nutrients (Gopinath et al. 2018).

1.5 Different types of Pollutants in the Environment and their Bioremediation

Many organic pollutants, such as polychlorinated biphenyls (PCBs), Polycyclic Aromatic Hydrocarbons (PAHs), pesticides, oil spills and dyes have been found to be toxic in the environment. They can be detrimental to other species and human health due to their large-scale manufacturing and use, toxicity, bioaccumulation and persistence in the environment (Figure 1.2). Inorganic pollution by Heavy Metal (HM) is less visible and direct than other types of pollutants, but its effects on marine ecosystems and humans are concentrated and very extensive (Necibi and Mzoughi 2017).

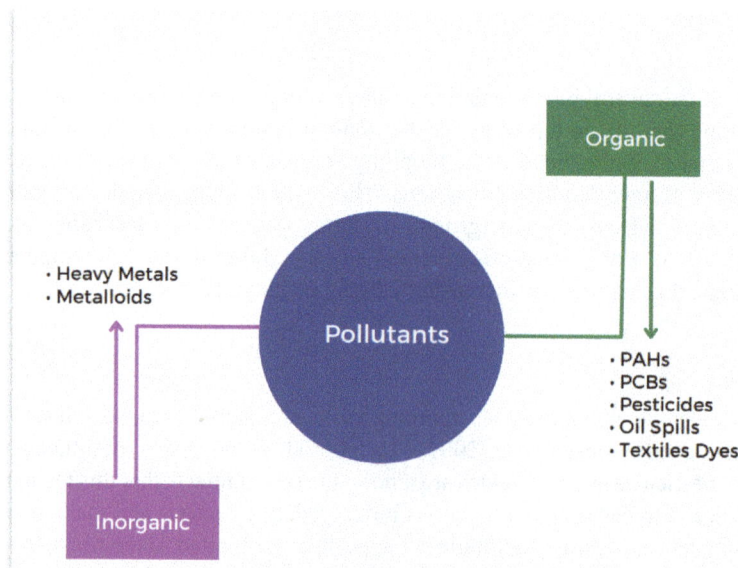

Figure 1.2. Different types of pollutants in the environment.

1.5.1 Bioremediation of Polycyclic Aromatic Hydrocarbons (PAHs)

Polycyclic Aromatic Hydrocarbons (PAHs) consist of a large and diverse class of organic compounds consisting of fused aromatic rings in various forms. They are colorless, white or pale yellow-green solids, have low water solubility and vapor pressure, and a high octanol/water partitioning coefficient and low microbial bioavailability which are produced as a result of incomplete combustion of organic compounds (wood, coal, oil and gasoline). US Environmental Protection Agency (USEPA) has identified a collection of 16 different PAHs as priority pollutants. PAHs are used to make pesticides, dyes, plastics and medicines. PAHs are widely distributed in environments which are toxic and persistent for a long period (Cerniglia 1984, Wang 2009). Due to their mutagenic and carcinogenic toxicity, PAH's outcome in nature is of major environmental concern.

Bioremediation is a natural process that recovers PAH-polluted environments by converting hazardous PAHs into non-toxic products. The number and kind of microorganisms present, as well as the nature and chemical structure of the chemical component being degraded, all influence PAH breakdown. Some of the microorganisms that are used in the degradation of PAHs are given in Table 1.1. During biodegradation, microorganisms either convert or mineralize PAHs and other xenobiotic chemicals to CO_2 and H_2O. The rate of biodegradation is influenced by factors like pH, temperature, oxygen, microbial population, degree of acclimation, nutrient accessibility, the chemical structure of the substance, cellular transport characteristics and chemical partitioning in growing media (Haritash and Kaushik 2009). Benzo[a]pyrene (BaP) is a PAH that consists of five fused benzene rings and is one of the most potent carcinogenic PAHs. In mammals and

Table 1.1. Polycyclic Aromatic Hydrocarbons (PAHs) degrading microorganisms.

PAHs	Microorganism	References
Phenanthrene	*Arabidopsis thaliana* ATCG5600	Hernández-Vega et al. 2017
Pyrene	*Achromobacter xylosxidans* PY4	Nazifa et al. 2018
Benzo (a) pyrene	*Serratia marcescens*	Kotoky and Pandey 2020
Phenanthrene, Pyrene, fluoranthene	*Pseudomonas aeruginosa,Ralstonia* sp.	Sangkharak et al. 2020
Phenanthrene	*Bacillus thuringiensis, Pleusotus cornucopiae, Pseudomonas*	Jiang et al. 2015
Pyrene	*Roseobacter clade*	Zhou et al. 2020
Benzo (a) pyrene	*Aspergillus nidulans*	Ostrem Loss et al. 2019
Benzo (a) pyrene	*Lasiodiplodia theobromae*	Cao et al. 2020
Benzo (a) pyrene	*Megasporoporia* sp. S47	De Lima Souza et al. 2016
Phenanthrene	*Coriolopsis byrsinaRyvarden* strain APC5	Agrawal et al. 2021

aquatic species, Benzo(a)Pyrene (BaP), is known for its mutagenic, carcinogenic and teratogenic characteristics (IARC 1983, Juhasz and Naidu 2000, Jennings 2012). In literature, ligninolytic and non-ligninolytic strains of fungi with the ability to breakdown PAH have been documented. The degradation of BaP by white-rot fungus has been the subject of recent research (Hadibarataa and Kristanti 2012, Bhattacharya et al. 2014). White rot fungi such as *Phanerochaete chrysosporium*, *Trametes versicolor*, *Cirnipellis stipitaria* and *Pleurotus ostreatus* can breakdown most PAHs efficiently as a carbon source. The white rot fungus *Phanerochaete chrysosporium* has a remarkable ability to degrade and/or mineralize high-molecular-weight PAHs, and its genome has around 150 Polymorphic Cytochrome P450 Enzymes (CYPs) (Yadav et al. 2006) and has the ability to oxidize BaP to 3-hydroxybenzo[a]pyrene (Syed et al. 2010). These CYPs were inducible by naphthalene, phenanthrene, pyrene and BaP. *Aspergillus*, the most prevalent species of soil-dwelling fungi, may metabolize some PAHs and is frequently found in contaminated areas (Cerniglia and Sutherland 2010).

1.5.2 Bioremediation of Polychlorinated Biphenyls (PCBs)

Polychlorinated biphenyls (PCBs) are combinations of 209 types of synthetic organic chemicals called congeners (US EPA 2000). This substantial number of different chemical forms results from the binding of 110 chlorine atoms to the carbon atoms of the biphenyl core. The level of chlorination has a significant impact on the physical and chemical characteristics of PCBs. As the level of chlorination rises, PCBs become more viscous and waxy, while their solubility in water tends to decrease. The Agency for Toxic Substances and Disease Registry (ATSDR) states that PCBs are "oily liquids or solids, colourless to light yellow, and have no recognized odour". The unusual properties of PCBs include high thermal stability, chemical inertness, non-flammability and high electrical resistivity, that are relatively used in hydraulic fluids, capacitor dielectrics and electrical transformers. Other potential sources of PCBs include leaks, spills and slow release from PCB-contaminated areas (Van Aken and Bhalla 2011). PCBs are hazardous xenobiotics, mostly found in soils and sediments and are widely dispersed in the environment. It is well known that biphenyl dioxygenase is essential for the breakdown of PCBs. Due to the presence of a large number of congeners; efficient microbial breakdown of PCBs requires a number of metabolic processes. The microorganisms responsible for PCB transformation are unable to grow on PCBs as the only carbon source (Boyle et al. 1992) and require a co-substrate for microbial growth and degradation activity, demonstrating that PCB degradation occurs predominantly via co-metabolism. PCBs are hazardous chemicals that can affect hormones and cause cancer. As a result, PCB poisoning in the environment is becoming increasingly problematic and is of great concern (Seeger et al. 2010). Several approaches have been proposed for PCB degradation in the environment. The study is now concentrated on the creation of a less expensive bioremediation strategy, however, physical and chemical treatments are often costly. Bioremediation, which uses microorganisms capable of digesting harmful substances, is regarded as a viable, ecologically beneficent and economically advantageous technique for getting rid of PCBs (Dercova et al. 2015). PCB bioremediation relies primarily on bacterial aerobic utilization of the pollutant molecules, while benzoic acids that can be further degraded by specialized strains of bacteria or substances with lower toxicity are typically the products of biphenyl dioxygenase-initiated degradation (Murínová et al. 2014). The microorganisms used in the remediation/degradation of PCBs are given in Table 1.2.

Table 1.2. Polychlorinated biphenyls (PCBs) degrading microorganisms.

PCBs	Microorganism	References
PCB (Delor 103)	*Achromobacter xylosoxidans, Stenotrophomonas maltophilia, Rhodococcus ruber*	Horváthová et al. 2018
PCB	*Paraburkholderia xenovorans* LB400	Bako et al. 2021
PCBs (18, 52, 77)	*Streptococcus* sp. SPco	Lin et al. 2022
bph- gene	Bacterial gene (*Ralstonia, Cupriavidus*)	Jiang et al. 2018
PCB	*Rhodococcus biphenylivorans* strain TG9T	Ye et al. 2020
PCB	*Klebsiella* Lw3	Zhu et al. 2022
PCB	*Eisenia fetida*	Eslami et al. 2022
Trichlorobiphenyl	*Rhodococcus* strains KT112-7, CH628, P25	Gorbunova et al. 2021
PCB	*Ligninolytic* fungi	Šrédlová et al. 2021
PCB	*Ascomycetes* strain	Germain et al. 2021

1.5.3 Bioremediation of Pesticides

Pesticides are organic chemical compounds which are frequently used in agricultural applications to control or eradicate pests and boost crop yields. These are inexpensive, simple to make and easily accessible. Globally more than half of the pesticides are used in Asia. India is third in Asia behind China and Turkey in terms of pesticide-usage, ranking 12th globally (Nayak and Solanki 2021). Pesticides can be categorized into the following groups: insecticides, fungicides, herbicides, rodenticides and fumigants, etc. Pesticides that dissolve quickly are known as non-persistent, whereas those that resist degradation are known as persistent. They accumulate across the food chain and cause severe hazards as a result of their excessive use and persistence. Different types of degradation procedures are involved in the remediation of pesticides such as biodegradation, chemical degradation, hydrolysis, oxidation-reduction (redox), ionization and photo-degradation (Zhang and Qiao 2002). The metabolic activity of different types of microorganisms plays a crucial part in the degradation process. Some of the pesticide-degrading microorganisms are given in Table 1.3.

Different types of enzymes play a crucial role in the biodegradation of pesticides. The biodegradation of pesticides involves a series of enzyme-catalyzed processes which includes oxidation, reduction, dechlorination, dehydration and hydrolysis. The compound enters the body of the microorganism in a specific way, and after a series of physiological and biochemical reactions involving various enzymes, the pesticide could be completely degraded or broken down into smaller molecular compounds (CO_2 and H_2O) with no or low toxicity (Chen et al. 2011, Tang 2018).

A majority of the genes encoding for the degradation of pesticides have been reported to be on catabolic plasmids of *Pseudomonas* sp., *Micrococcus* sp., *Actinobacter* sp., *Rhodoccus* sp., *Fusarium* sp., and *Arthrobacter* sp., *Actinomycetes* could also be used for the biotransformation as well as biodegradation of pesticides.

Table 1.3. Pesticides degrading microorganisms.

Pesticides	Microorganism	References
Endosulfan	*Micrococcus* sp. strain, 2385	Pathak et al. 2016
DDT	*Ochrobacterum* sp.	Pan et al. 2017
Nitrophenol	*Rhodoccus* sp.	Sengupta et al. 2019
Cypermethrin chlordecone	*Bacillus subtilis*	Gangola et al. 2018
	Citrobacter	Chaussonnerie et al. 2016
(Lindane) hexachloro-cyclohexane	*Microbacterium*	Zhang et al. 2020
Chloropyrifos, diazinon	*Streptomyces* sp.	Briceño et al. 2018
S- triazine (atrazine)	*Arthrobacter* sp. Strain Ak-YNIO	Sagarkar et al. 2016
Hexachloro-cyclohexane	*Paenibacillus dendritiformis* SJPS-4	Jaiswal et al. 2022
Glyphosate, 2-4D, Atrazine	*Bradyrhizobium* sp. BR 3901	Barroso et al. 2020
Endosulfan	*Aspergillus trichoderma* spp.	Gangola et al. 2015
Atrazine	*Fusarium* sp.	Esparza-Naranjo et al. 2021

Degrading enzymes are relatively more resistant to anomalous environmental circumstances than microbial cells capable of producing such enzymes, and their degradable efficiency is substantially higher than that of microorganisms, especially at low pesticide doses. As a result, using degrading enzymes to detoxify the environment that has been contaminated by pesticides would be a more effective method.

1.5.4 Bioremediation of Textile Dyes

Various industrial effluents, including those from textile, printing, pharmaceutical and other industries, cause wastewater contamination (Uday et al. 2016). Dyes industries were established during the same period of rapid industrialization and urbanization as other sectors. Many different kinds of dyes are present in textile effluents. Based on their chemical makeup, dyes are divided into several categories, such as Anthraquinone bases, metal complex dyes, di-azo and basic dyes. These dyes might be cationic, anionic or neutral (Vikrant et al. 2018). According to Daneshvar et al. (2007), artificial dyes, such as azo, xanthenes and anthraquinone dyes, are extremely toxic to living things. When the dye forms compounds with other contaminants and deteriorates materials in the environment, its toxicity increases. Additionally, these dyes can contribute to hereditary illnesses that are incurable (Lellis et al. 2019).

Dyes can come from natural or synthetic sources or they can be generated via an inorganic technique, in the wastewater produced by textile manufacturers (Varjani et al. 2021). Synthetic dyes are inexpensive ingredients that come in large quantities, as well as being commonly used (Rossi et al. 2017). Common dyes are anthraquinone dyes, that have polycondensed ring structures and hydroxyl or amino functionalities. Different kinds of microorganisms used in the biodegradation of textile dyes are given in Table 1.4.

Various enzymes are used in the decolorization of textile dyes, including formate dehydrogenase, oxidases based on glyoxal and aryl-alcohol compounds and peroxidase (Chen et al. 2016, Sarkar et al. 2017). Laccases are valuable enzymes in the decolorizing process because aromatic compounds with azo and anthraquinone linkages are some of their substrates. Laccase is a multicooper oxidase enzyme that may be found in a variety of microorganisms (e.g., microalgae, fungi and bacteria) (Motamedi et al. 2021). A biodegradation pathway for Congo Red Dye utilizing azoreductase and laccase was proposed by Lade et al. (2015). By breaking azo bonds in the presence of azoreductase, the biodegradation of Congo Red Dye produces biphenyl-4, 4′-diamine and an unexplained intermediate. As a result of this microbial degradation phase, the dye is decolored and aromatic amines are formed as an end product. Furthermore, the presence of laccase causes the production

Table 1.4. Different kinds of dye-degrading microorganisms.

Dye	Microorganism	References
Reactive Orange 16 Dye	*Bacillus stratosphericus* SCA 1007	Akansha et al. 2022
Methylene blue	*Bacillus subtilis* MTCC441	Upendar et al. 2017
Indigo blue RBBR, Sulfur Black	*Cyanobacteria* (Anabena flosaquae UTCC 64) *Phormidium autumnale* UTEX B 1580 *Synechococcus* sp. PCC7942	Dellamatrice et al. 2017
Sumifex Tourgi blue dye	*Alishewanella* sp. CBL -2	Ajaz et al. 2018
Azo dye	*Enterobacter cloacae* and *Bacillus subtilis*	Priyanka et al. 2022
Congo Red Dye	*Aspergillus* sp.	Singh and Dwivedi 2022
Methylene Blue Dye	*Penicillium* P1	Liu et al. 2022
Congo Red Dye	*Penicillium oxalicum* and *Aspergillus tubingensis*	Thakor et al. 2022
Azo dye	*Pichia occidentalis* A2 (yeast)	Wang et al. 2020

of biphenyl and naphthalene by deamination of biphenyl-4, 4' - diamine and desulfonation of an unexplained intermediate product (Lade et al. 2015).

1.5.5 Bioremediation of Oil Spills

Bioremediation is considered one of the most effective oil spill remediation methods. Aliphatic (straight chain) and aromatic hydrocarbons are organic molecules largely composed of hydrogen and carbon (cyclic). Any combination of hydrocarbons contained in crude oil is referred to as petroleum hydrocarbons or Total Petroleum Hydrocarbons (TPHs). Oil spills, whether accidental or deliberate, have a significant influence on environmental contamination. Oil spills from oil tankers and far-off oil spills are known to pose a serious threat to the environment. By using microorganisms to remove hydrocarbon pollution from soil and water, bioremediation of oil spills makes these environments safe for both aquatic and terrestrial animals. Bacterial species, fungal species (a technique known as mycoremediation) and plant species can all employ bioremediation (by a process called phytoremediation) (Table 1.5). Different factors influencing hydrocarbon degradation have been reported by Cooney et al. (1985). Various types of aquatic plant species are capable of accumulating HMs. Petroleum hydrocarbon molecules link to soil components, making removal and degradation difficult (Barathi and Vasudevan 2001). The bulk of organic contaminants degrade most quickly and completely when exposed to aerobic conditions.

In literature, toluene and other monocyclic aromatic hydrocarbons, such as benzene, toluene and xylene, can be degraded and eliminated by *Pseudomonas putida* (Saptakee 2011, Sarang et al. 2013). When aromatic hydrocarbons are broken down by bacteria, a diol first forms, then the aromatic ring is broken and a diacid, such as cis-cis muconic acid, is produced. *Penicillium chrysosporium*, a white rot fungus, can degrade compounds such as biphenyl and triphenylmethane (Wolskm et al. 2012).

Table 1.5. Petroleum hydrocarbons (PHCs) degrading microorganisms.

PHCs	Microorganism	References
n-alkane (C_{14}-C_{30})	*Pseudomonas* sp.	Zheng et al. 2018 Varjani and Upasani 2017
n-alkane	*Bacillus* sp.	Dellagnezze et al. 2016
tetradecane	*Bacillus cerus*	Li et al. 2017
hexadecane	*Alcanivorax borkumensis*	Omarova et al.2018
n-alkane	*Aspergillus flavus*	Maruthi et al. 2013
n-alkane (C_{11}-C_{25})	*Penicillium* sp.	Govarthanan et al. 2017
n-alkane	*Trichoderma* sp.	Nazifa et al. 2018

1.6 Phytoremediation

Phytoremediation is an effective biotechnology for dealing with metal and metalloid contamination and has a distinct environmental application area (Gu 2018). Green technology for environmental remediation plays a competitive role in transferring the soluble and bioavailable fractions of hazardous metals and metalloids from the solution or adsorption phase into green plants for accumulation, detoxification and stabilization (Yu and Gu 2007a,b, 2008a,b).

Phytoremediation is a non-conventional, cost-effective and eco-friendly technology that utilizes plants to remove, transform or stabilize a variety of contaminants present in water, sediments or soils (Prasad et al. 2001, Ladislas et al. 2012). Phytoremediation is the application of plants for *in-situ* or *ex-situ* treatment/removal of heavy metals from contaminated soils, sediments and water (Garbisu and Alkorta 2001, Shah and Daverey 2020) provides many benefits compared to traditional approaches of contaminated land remediation such as lowered cost, maintenance, exposure to workers, and is generally more aesthetically pleasing (Sharma and Yeh 2020). Plants that have more metal-removal capacity through accumulation are known as hyperaccumulators. These plants are used to remediate contaminants by the uptake or transpiration of contaminated water (Cho-Ruk et al. 2006, Smolyakov 2012). Table 1.6 shows various types of aquatic plant species capable of accumulating HMs. Plants absorb a significant number of toxic elements and nutrients, but only a small portion of them are damaging, affecting plants at higher concentrations. When the degree of pollution in plants rises, they are harmed or die. Various processes for treating the water have been designed, for example, biological, physical and chemical, but they are costly and only applicable to a small amount of wastewater (Rezania et al. 2015). As a result, an alternative wastewater treatment procedure, phytoremediation has been proposed, in which diverse plants are used to clean wastewater and eliminate hazardous contaminants.

Table 1.6. Various types of aquatic plant species capable of accumulating HMs.

Heavy Metals	Plant species	References
Cd, Zn, Pb, Cu	*Sedum plumbizincicola*	Li et al. 2018
Fe, Cu, Zn	*Eichhornia crassipes* *Typha latifolia*	Abbas et al. 2021
Ni	*Ricinus communis*	Çelik and Akdaş 2019
Al, Fe	*Ipomoea aquatica* *Centella asiatica*	Hanafiah et al. 2020
Pb, Cr, Cd	*Trifolium repens* L.	Lin et al. 2021
Cd, Zn, Cu	*Rhazya stricta* L.	Azab and Hegazy 2020
Zn, Cd, Cr, Ni	*Jatropha curcas*	Chang et al. 2014
Cd, Ni	*Phragmites australis*	Eid et al. 2020
Cu, Pb, Zn	*Polygonum hydropiperoides*	Rudin et al. 2017

1.6.1 Mechanisms of Phytoremediation

1.6.1.1 Phytoextraction

Phytoextraction or phytoaccumulation is the process in which metal contaminants are accumulated and stabilized in the upper part of plants. During this process, the root uptake the metals and transfer them to the upper portion. Selecting plants that absorb and concentrate harmful heavy metals in various areas of the plant, this method aids in the removal of heavy metal pollution from the soil.

1.6.1.2 Phytovolatilization

In this process, contaminants are uptaken by roots, translocated in upper plant parts and released through the leaves in the volatile form into the atmosphere (USEPA 2000). This process involves the diffusion of volatile pollutants through open stomata of the leaves in a less toxic form. This involves the removal of the pollutants in a gaseous form and the particular pollutant removal in safer forms.

1.6.1.3 Phytostabilization

Phytostabilization is a process in which the stabilization or fixation of heavy metals occurs so that proper absorption and precipitation take place mainly through the soil, sediment and sludge (USEPA 2000). In this process, contaminants are absorbed and collected by roots or precipitate inside the root zone of plants (rhizosphere).

1.6.1.4 Phytofiltration/Rhizofiltration

Precipitation and absorption by plants from soil and water are the main mechanisms in rhizofiltration. In this process, the contaminants are restricted only to the root system. Various heavy metals are retained by the root system in rhizofiltration (USEPA 2000). In rhizofiltration, plant roots grow very rapidly and require minimal time for decontamination (Sarkar et al. 2011).

1.6.1.5 Phytotransformation

Phytotransformation, also known as phytodegradation is the breakdown of organic pollutants sequestered by plants through (1) metabolic processes inside the plant; or (2) the influence of substances produced by the plant, such as enzymes (EPA 1998). It contributes to the removal of organic pollutants such as chlorinated solvents, herbicides and the breakdown of complex organic compounds into simple ones (EPA 2000). As plants do not contain active transporters, these organic contaminants are absorbed through passive uptake. When the degradation of contaminants occurs in the rhizosphere, the process is called rhizodegradation. Organics such as polyaromatic hydrocarbons and polychlorinated biphenyls can be mineralized by rhizospheric bacteria. Furthermore, enzymatic breakdown by enzymes released by specific plants and associated microbe species, such as dehalogenase, nitro-reductase, laccase peroxidase and others, acts on dangerous xenobiotics (Cherian and Oliveira 2005).

1.6.2 Advantages and Disadvantages of Bioremediation

Bioremediation is the best option over conventional methods for remediation such as incineration, landfilling, etc. Bioremediation can be done on-site, thereby reducing expenses and risks associated with transportation, can treat or eliminate a wide range of diffused contaminants permanently, and can be applied to large-scale operations (Van Aken 2009). Other advantages include being environmentally friendly, affordable due to its minimal installation and a higher level of public acceptance (Ali and Sajad 2013). Advantages of phytoremediation include the development of soil fertility, recovery of valuable metals and avoidance of metal erosion and leaching (Mench et al. 2009).

Several disadvantages also exist with the bioremediation technique. The use of bioremediation is restricted to biodegradable substances only. Substances like metals, chlorinated organic pollutants and radionuclides are not capable of complete and quick degradation. Sometimes, the by-products of biodegradation may be more toxic and persistent during the metabolism of contaminants. Some drawbacks of the phytoremediation technique include longer treatment time, higher pollutant concentrations and bioavailability to plants, the toxicity of pollutants to plants and the inability to treat organic contaminants due to the lack of enzymes for their degradation (Ali et al. 2013).

1.7 Conclusion

In the chapter, a brief account of the principle, types, applications, benefits and drawbacks of bioremediation techniques have been discussed. As a result, one can see that compared to other physical and chemical remediation approaches, it is an emerging, interdisciplinary, effective and environmentally friendly remediation approach that is currently paving the way to a more promising future. Approaches to bioremediation largely depend on the concepts of *in-situ* and *ex-situ* technology. In order to select the bioremediation method that would remove contaminants effectively, different factors such as the presence of a particular microbial community, the bioavailability of pollutants and environmental conditions have to be taken into consideration. As a result, improving our understanding of microbial populations and how they interact with their specific environment and pollutants, learning more about microbial genomics to increase their capacity for biodegradation and evaluating the efficacy of new bioremediation strategies in the field will allow one to develop more convenient bioremediation techniques.

References

Abbas, N., M. T. Butt, M. M. Ahmad, F. Deeba and N. Hussain. 2021. Phytoremediation potential of *Typha latifolia* and water hyacinth for removal of heavy metals from industrial wastewater. Chem. Int. 7(2): 103–111.

Agrawal, N., V. Kumar and S. K. Shahi. 2021. Biodegradation and detoxification of phenanthrene in *in-vitro* and *in vivo* conditions by a newly isolated ligninolytic fungus *Coriolopsis byrsina* strain APC5 and characterization of their metabolites for environmental safety. Environ. Sci. Pollut. Res., 1–16.

Aggarwal, P. K., J. L. Means, R. E. Hinchee, G. L. Headington and A. R. Gavaskar. 1990. Methods to select chemicals for *in situ* biodegradation of fuel hydrocarbons. Batt. Col. Div. Oh.

Ajaz, M., A. Elahi and A. Rehman. 2018. Degradation of azo dye by bacterium, *Alishewanella* sp. CBL-2 isolated from industrial effluent and its potential use in decontamination of wastewater. J. of Water Reuse Desalin. 8(4): 507–515.

Akansha, K., A. N. Yadav, M. Kumar, D. Chakraborty and S. Ghosh Sachan. 2022. Decolorization and degradation of reactive orange 16 by *Bacillus stratosphericus* SCA1007. Folia Microbiol. 67(1): 91–102.

Ali, H., E. Khan and M. A. Sajad. 2013. Phytoremediation of heavy metals—concepts and applications. Chemosphere. 91(7): 869–881.

Antizar-Ladislao, B. 2010. Bioremediation: working with bacteria. Elements. 6(6): 389–394.

Atlas, R. M. and J. Philp. 2005. Bioremediation. Applied Microbial Solutions for Real-world Environmental Cleanup. ASM Press.

Azab, E. and A. K. Hegazy. 2020. Monitoring the efficiency of *Rhazya stricta* L. plants in phytoremediation of heavy metal-contaminated soil. Plants. 9(9): 1057.

Azubuike, C. C., C. B. Chikere and G. C. Okpokwasili. 2016. Bioremediation techniques–classification based on site of application: principles, advantages, limitations and prospects. World J. Microbiol. Biotechnol. 32(11): 1–18.

Bako, C. M., T. E. Mattes, R. F. Marek, K. C. Hornbuckle and J. L. Schnoor. 2021. Dataset describing biodegradation of individual polychlorinated biphenyl congeners (PCBs) by *Paraburkholderia xenovorans* LB400 in presence and absence of sediment slurry. Data in Brief. 35: 106821.

Barathi, S. and N. Vasudevan. 2001. Utilization of petroleum hydrocarbons by *Pseudomonas fluorescens* isolated from a petroleum-contaminated soil. Environ. Int. 26(5-6): 413–416.

Barroso, G. M., J. B. dos Santos, I. T. de Oliveira, T. K. M. R. Nunes, E. A. Ferreira, I. M. Pereira, D. V. Silva and M. de Freitas Souza. 2020. Tolerance of *Bradyrhizobium* sp. BR 3901 to herbicides and their ability to use these pesticides as a nutritional source. Ecol. Indic. 119: 106783.

Bedard, D. L. and R. J. May. 1995. Characterization of the polychlorinated biphenyls in the sediments of Woods Pond: evidence for microbial dechlorination of Aroclor 1260 *in situ*. Environ. Sci. Technol. 30(1): 237–245.

Bender, J. and P. Phillips. 2004. Microbial mats for multiple applications in aquaculture and bioremediation. Bioresour. Technol. 94(3): 229–238.

Bhattacharya, S., A. Das, K. Prashanthi, M. Palaniswamy and J. Angayarkanni. 2014. Mycoremediation of Benzo [a] pyrene by *Pleurotus ostreatus* in the presence of heavy metals and mediators. 3 Biotech. 4(2): 205–211.

Boopathy, R. 2000. Factors limiting bioremediation technologies. Bioresour. Technol. 74(1): 63–67.

Boyle, A. W., C. J. Silvin, J. P. Hassett, J. P. Nakas and S. W. Tanenbaum. 1992. Bacterial PCB biodegradation. Biodegradation. 3(2): 285–298.

Brar, S. K., M. Verma, R. Y. Surampalli, K. Misra, R. D. Tyagi, N. Meunier and J. F. Blais. 2006. Bioremediation of hazardous wastes—a review. Practice Periodical of Hazardous. J. Hazard. Toxic Radioact. Waste. 10(2): 59–72.

Briceño, G., K. Vergara, H. Schalchli, G. Palma, G. Tortella, M. S. Fuentes and M. C. Diez. 2018. Organophosphorus pesticide mixture removal from environmental matrices by a soil Streptomyces mixed culture. Environ. Sci. Pollut. Res. 25(22): 21296–21307.

Cao, H., C. Wang, H. Liu, W. Jia and H. Sun. 2020. Enzyme activities during Benzo [a] pyrene degradation by the fungus *Lasiodiplodia theobromae* isolated from a polluted soil. Sci. Rep. 10(1): 1–11.

Carberry, J. B. and J. Wik. 2001. Comparison of ex situ and *in situ* bioremediation of unsaturated soils contaminated by petroleum. J. Environ. Sci. Health, Part A. 36(8): 1491–1503.

Çelik, Ö. and E. Y. Akdaş. 2019. Tissue-specific transcriptional regulation of seven heavy metal stress-responsive miRNAs and their putative targets in nickel indicator castor bean (*R. communis* L.) plants. Ecotoxicol. Environ. Saf. 170: 682–690.

Cerniglia, C. E. 1984. Microbial metabolism of polycyclic aromatic hydrocarbons. Adv. Appl. Microbiol. 30: 31–71.

Cerniglia, C. E. and J. B. Sutherland. 2010. Degradation of polycyclic aromatic hydrocarbons by fungi. In Handbook of Hydrocarbon and Lipid Microbiology.

Chang, F. C., C. H. Ko, M. J. Tsai, Y. N. Wang and C. Y. Chung. 2014. Phytoremediation of heavy metal contaminated soil by *Jatropha curcas*. Ecotoxicol. 23: 1969–1978.

Chaussonnerie, S., P. L. Saaidi, E. Ugarte, A. Barbance, A. Fossey, V. Barbe, G. Gyapay, T. Bruls, M. Chevallier, L. Couturat, S. Fouteau, D. Muselet, E. Pateau, G. N. Cohen, N. Fonknechten, J. Weissenbach and D. Le Paslier. 2016. Microbial degradation of a recalcitrant pesticide: chlordecone. Front. Microbial. 7: 2025.

Cherian, S. and M. M. Oliveira. 2005. Transgenic plants in phytoremediation: recent advances and new possibilities. Environ. Sci. Technol. 39(24): 9377–9390.

Chen, B. Y., C. M. Ma, K. Han, P. L. Yueh, L. J. Qin and C. C. Hsueh. 2016. Influence of textile dye and decolorized metabolites on microbial fuel cell-assisted bioremediation. Bioresour. Technol. 200: 1033–1038.

Chen, S., Q. Hu, M. Hu, J. Luo, Q. Weng and K. Lai. 2011. Isolation and characterization of a fungus able to degrade pyrethroids and 3-phenoxybenzaldehyde. Bioresour. Technol. 102(17): 8110–8116.

Cho-Ruk, K., J. Kurukote, P. Supprung and S. Vetayasuporn. 2006. Perennial plants in the phytoremediation of lead contaminated soils. Biotechnol. 5: 1–4.

Colberg, P. J. S. and L. Y. Young. 1995. Anaerobic degradation of nonhalogenated homocyclic aromatic compounds coupled with nitrate, iron, or sulfate reduction. Microbial Transformation and Degradation of Toxic Organic Chemicals. 307330.

Cooney, J. J., S. A. Silver and E. A. Beck. 1985. Factors influencing hydrocarbon degradation in three freshwater lakes. Microb. Ecol. 11(2): 127–137.

Daneshvar, N., M. Ayazloo, A. R. Khataee and M. Pourhassan. 2007. Biological decolorization of dye solution containing Malachite Green by microalgae *Cosmarium* sp. Bioresour. Technol. 98(6): 1176–1182.

Dellagnezze, B. M., S. P. Vasconcellos, A. L. Angelim, V. M. M. Melo, S. Santisi, S. Cappello and V. M. Oliveira. 2016. Bioaugmentation strategy employing a microbial consortium immobilized in chitosan beads for oil degradation in mesocosm scale. Mar. Pollut. Bull. 107(1): 107–117.

Dellamatrice, P. M., M. E. Silva-Stenico, L. A. B. D. Moraes, M. F. Fiore and R. T. R. Monteiro. 2017. Degradation of textile dyes by cyanobacteria. Braz. J. Microbiol. 48: 25–31.

Dercova, K., K. Laszlova, H. Dudášová, S. Murinova, M. Balaščáková and J. Škarba. 2015. The hierarchy in selection of bioremediation techniques: the potentials of utilizing bacterial degraders. Chemické Listy 109: 279–288.

Dua, M., A. Singh, N. Sethunathan and A. Johri. 2002. Biotechnology and bioremediation: successes and limitations. Appl. Microbiol. Biotechnol. 59(2): 143–152.

de Lima Souza, H. M., L. D. Sette, A. J. Da Mota, J. F. do Nascimento Neto, A. Rodrigues, T. B. de Oliveira, L. A. de Oliveira, H. D. Santos Barroso and S. P. Zanott. 2016. Filamentous fungi isolates of contaminated sediment in the Amazon region with the potential for benzo (a) pyrene degradation. Water Air Soil Pollut. 227(12): 1–13.

Eid, E. M., T. M. Galal, N. A. Sewelam, N. I. Talha, S. M. and Abdallah. 2020. Phytoremediation of heavy metals by four aquatic macrophytes and their potential use as contamination indicators: a comparative assessment. Environ Sci. Pollut. Res. 27: 12138–12151.

EPA. 1998. A Citizen's Guide to Phytoremediation. EPA 542-F-98-011. U.S. Environmental Protection Agency, Washington.

EPA. 2000. A citizen guide to phytoremediation. EPA 542-F-98-011. United States Environmental Protection Agency, p.6. Available http//www.bugsatwork.com/XYCLONYX/EPA_GUIDES/PHYTO.PDF.

EPA. 2016. United States Environmental Protection Agency. https://www3.epa.gov/ (Accessed May 2016).

Eslami, N., A. Takdastan and F. Atabi. 2022. Biological Remediation of Polychlorinated Biphenyl (PCB)-Contaminated soil using the vermicomposting technology for the management of sewage sludge containing *Eisenia fetida* earthworms. Soil Sediment Contamin. An International Journal, 1–17.

Esparza-Naranjo, S. B., G. F. da Silva, D. C. Duque-Castaño, W. L. Araújo, C. K. Peres, M. Boroski and R. C. Bonugli-Santos. 2021. Potential for the biodegradation of atrazine using leaf litter fungi from a subtropical protection area. Curr. Microbiol. 78(1): 358–368.

Gangola, S., G. Negi, A. Srivastava and A. Sharma. 2015. Enhanced biodegradation of endosulfan by *Aspergillus* and *Trichoderma* spp. isolated from an agricultural field of tarai region of Uttarakhand. Pestic. Res. J. 27(2): 223–230.

Gangola, S., A. Sharma, P. Bhatt, P. Khati and P. Chaudhary. 2018. Presence of esterase and laccase in *Bacillus subtilis* facilitates biodegradation and detoxification of cypermethrin. Sci. Rep. 8(1): 1–11.

Garbisu, C. and I. Alkorta. 2001. Phytoextraction: a cost-effective plant-based technology for the removal of metals from the environment. Bioresour. Technol. 77(3): 229–236.

Germain, J., M. Raveton, M. N. Binet and B. Mouhamadou. 2021. Potentiality of Native *Ascomycete* strains in bioremediation of highly polychlorinated biphenyl contaminated soils. Microorganisms. 9(3): 612.

Gopinath, M., R. H. Pulla, K. S. Rajmohan, P. Vijay, C. Muthukumaran and B. Gurunathan. 2018. Bioremediation of volatile organic compounds in biofilters. In Bioremediation: Applications for Environmental Protection and Management. Springer, Singapore, 301–330.

Gorbunova, T. I., D. O. Egorova, M. G. Pervova, T. D. Kyrianova, V. A. Demakov, V. I. Saloutin and O. N. Chupakhin. 2021. Biodegradation of trichlorobiphenyls and their hydroxylated derivatives by *Rhodococcus*-strains. J. Hazard. Mater. 409: 124471.

Govarthanan, M., S. Fuzisawa, T. Hosogai and Y. C. Chang. 2017. Biodegradation of aliphatic and aromatic hydrocarbons using the filamentous fungus *Penicillium* sp. CHY-2 and characterization of its manganese peroxidase activity. RSC Adv. 7(34): 20716–20723.

Gu, J. D. 2018. Bioremediation of toxic metals and metalloids for cleaning up from soils and sediments. Appl. Environ. Biotechnol. 3(2): 48–51.

Hadibarata, T. and R. A. Kristanti. 2012. Fate and cometabolic degradation of benzo [a] pyrene by white-rot fungus *Armillaria* sp. F022. Bioresour. Technol. 107: 314–318.

Hanafiah, M. M., M. F. Zainuddin, N. U. Mohd Nizam, A. A. Halim and A. Rasool 2020. Phytoremediation of aluminum and iron from industrial wastewater using *Ipomoea aquatica* and *Centella asiatica*. Appl. Sci. 10(9): 3064.

Hardisty, P. E. and E. Ozdemiroglu. 2005. The economics of groundwater remediation and Protection, CRC Press, Boca Raton, FL.

Haritash, A. K. and C. P. Kaushik. 2009. Biodegradation aspects of polycyclic aromatic hydrocarbons (PAHs): a review. J. Hazard. Mater. 169(1-3): 1–15.

Harmsen, J., W. H. Rulkens, R. C. Sims, P. E. Rijtema and A. J. Zweers. 2007. Theory and application of landfarming to remediate polycyclic aromatic hydrocarbons and mineral oil contaminated sediments; beneficial reuse. J. Environ. Qual. 36: 1112–1122.

Hernández-Vega, J. C., B. Cady, G. Kayanja, A. Mauriello, N., Cervantes, A., Gillespie, L. Lavia, J. Trujillo, M. Alkio and A. Colón-Carmona. 2017. Detoxification of polycyclic aromatic hydrocarbons (PAHs) in *Arabidopsis thaliana* involves a putative flavonol synthase. J. Hazard. Mater. 321: 268–280.

Horváthová, H., K. Lászlová and K. Dercová. 2018. Bioremediation of PCB-contaminated shallow river sediments: the efficacy of biodegradation using individual bacterial strains and their consortia. Chemosphere. 193: 270–277.

IARC (International Agency for Research on Cancer) 1983. Polynuclear aromatic compounds, part 1, chemical, environmental, and experimental data. IARC Monographs on the Evaluation of the Carcinogenic Risk of Chemicals to Man. IARC Sci. Pub. 32: 33–451.

Jaiswal, S., D. K., Singh and P. Shukla. 2022. Lindane bioremediation by *Paenibacillus dendritiformis* SJPS-4, its metabolic pathway analysis and functional gene annotation. Environ. Technol. Innov. 27: 102433.

Jennings, A. A. 2012. Worldwide regulatory guidance values for surface soil exposure to carcinogenic or mutagenic polycyclic aromatic hydrocarbons. J. Environ. Manage. 110: 82–102.

Jiang, J., H. Liu, Q. Li, N. Gao, Y. Yao and H. Xu. 2015. Combined remediation of Cd–phenanthrene co-contaminated soil by *Pleurotus cornucopiae* and *Bacillus thuringiensis* FQ1 and the antioxidant responses in *Pleurotus cornucopiae*. Ecotoxicol. Environ. Saf. 120: 386–393.

Jiang, L., C. Luo, D. Zhang, M. Song, Y. Sun and G. Zhang. 2018. Biphenyl-metabolizing microbial community and a functional operon revealed in e-waste-contaminated soil. Environ. Sci. Technol. 52(15): 8558–8567.

Jørgensen, K. S. 2007. *In situ* bioremediation. Adv. Appl. Microbiol. 61: 285–305.

Juhasz, A. L. and R. Naidu. 2000. Bioremediation of high molecular weight polycyclic aromatic hydrocarbons: a review of the microbial degradation of benzo [a] pyrene. Int. Biodeterior. Biodegradation. 45(1-2): 57–88.

Kamaludeen, S. P. B., K. R. Arunkumar and K. Ramasamy. 2003. Bioremediation of chromium contaminated environments. Indian J. Exp. Biol. 41: 972–985.

Komancová, M., I. Jurčová, L. Kochánková and J. Burkhard. 2003. Metabolic pathways of polychlorinated biphenyls degradation by *Pseudomonas* sp. 2. Chemosphere. 50(4): 537–543.

Kotoky, R. and P. Pandey. 2020. Rhizosphere assisted biodegradation of benzo (a) pyrene by cadmium resistant plant-probiotic *Serratia marcescens* S2I7, and its genomic traits. Sci. Rep. 10(1): 1–15.

Kumar, A., B. S. Bisht, V. D. Joshi and T. Dhewa. 2011. Review on bioremediation of polluted environment: a management tool. Int. J. Environ. Sci. Technol. 1(6): 1079.

Kumar, V., S. K. Shahi and S. Singh. 2018. Bioremediation: an eco-sustainable approach for restoration of contaminated sites. In Microbial bioprospecting for sustainable development (pp. 115–136). Springer, Singapore https://doi.org/10.1007/978-981-13-0053-0_6.

Lade, H., S. Govindwar and D. Paul. 2015. Mineralization and detoxification of the carcinogenic Azo dye congo red and real textile effluent by a polyurethane foam immobilized microbial consortium in an upflow column bioreactor. Int. J. Environ. Res. Public Health. 12: 6894–6918.

Ladislas, S., A. El-Mufleh, C. Gérente, F. Chazarenc, Y. Andrès and B. Béchet. 2012. Potential of aquatic macrophytes as bioindicators of heavy metal pollution in urban stormwater runoff. Water Air Soil Pollut. 223(2): 877–888.

Latha, A. P. and S. S Reddy. 2013. Review on bioremediation—Potential tool for removing environmental pollution. Int. J. Basic Appl. Chem. Sci. 3(3): 21–33.

Lebeau, T., A. Braud and K. Jézéquel. 2008. Performance of bioaugmentation-assisted phytoextraction applied to metal contaminated soils: a review. Environ. Pollut. 153(3): 497–522.

Lee, T. H., I. G. Byun, Y. O. Kim, I. S. Hwang and T. J. Park. 2006. Monitoring biodegradation of diesel fuel in Bioventing processes using *in situ* respiration rate. Water Sci. Technol. 53(4-5): 263–272.

Lellis, B., C. Z. Fávaro-Polonio, J. A. Pamphile and J. C. Polonio. 2019. Effects of textile dyes on health and the environment and bioremediation potential of living organisms. Biotechnol. Res. Innov. 3(2): 275–290.

Li, C. H., Y. S., Wong and N. F. Y. Tam. 2010. Anaerobic biodegradation of polycyclic aromatic hydrocarbons with amendment of iron (III) in mangrove sediment slurry. Bioresour. Technol. 101(21): 8083–8092.

Li, Y., H. Gong, H. Cheng, L. Wang and M. Bao. 2017. Individually immobilized and surface-modified hydrocarbon-degrading bacteria for oil emulsification and biodegradation. Mar. Pollut. Bull. 125(1-2): 433–439.

Li, Z., L. Wu, Y. Luo and P. Christie. 2018. Changes in metal mobility assessed by EDTA kinetic extraction in three polluted soils after repeated phytoremediation using a cadmium/zinc hyperaccumulator. Chemosphere 194: 432–440.

Lin, H., C. Liu, B. Li and Y. Dong. 2021. *Trifolium repens* L. regulated phytoremediation of heavy metal contaminated soil by promoting soil enzyme activities and beneficial rhizosphere associated microorganisms. J. Hazard. Mater. 402: 123829.

Lin, Q., X. Zhou, S. Zhang, J. Gao, M. Xie, L. Tao, F. Sun, M. Z. Hashmi and X. Su. 2022. Oxidative dehalogenation and mineralization of polychlorinated biphenyls by a resuscitated strain *Streptococcus* sp. SPC0. Environ. Res. 207: 112648.

Liu, J., H. Su, J. Xue and X. Wei. 2022. Optimization of decoloration conditions of methylene blue wastewater by *Penicillium* P1. Indian J. Microbiol. 62(1): 103–111.

Maciel, B. M., A. C. F. Santos, J. C. T. Dias, R. O. Vidal, R. J. C. Dias, E. Gross and R. P. Rezende. 2009. Simple DNA extraction protocol for a 16S rDNA study of bacterial diversity in tropical landfarm soil used for bioremediation of oil waste. Genet. Mol. Res. 8(1): 375–388.

Maruthi, Y. A., K. Hossain and S. Thakre. 2013. *Aspergillus flavus*: A potential Bioremediator for oil contaminated soils. Eur. J. Sustain. Dev. 2(1): 57–57.

Megharaj, M., B. Ramakrishnan, K. Venkateswarlu, N. Sethunathan and R. Naidu. 2011. Bioremediation approaches for organic pollutants: a critical perspective. Environ. Int. 37(8): 1362–1375.

Mench, M., J. P. Schwitzguébel, P. Schroeder, V. Bert, S. Gawronski and S. Gupta. 2009. Assessment of successful experiments and limitations of phytotechnologies: contaminant uptake, detoxification and sequestration, and consequences for food safety. Environ. Sci. Pollut. Res.16(7): 876–900.

Motamedi, E., K. Kavousi, S. F. S. Motahar, M. R. Ghaffari, A. S. A. Mamaghani, G. H. Salekdeh and S. Ariaeenejad. 2021. Efficient removal of various textile dyes from wastewater by novel thermo-halotolerant laccase. Bioresour. Technol. 337: 125468.

Murínová, S., K. Dercová and H. Dudášová. 2014. Degradation of polychlorinated biphenyls (PCBs) by four bacterial isolates obtained from the PCB-contaminated soil and PCB-contaminated sediment. Int. Biodeterior. Biodegradation. 91: 52–59.

Nayak, P. and H. Solanki. 2021. Pesticides and Indian agriculture—a review. Int. J. Res. Granthaalayah. 9: 250–263.

Nazifa, T. H., M. A. Ahmad, T. Hadibarata and A. Aris. 2018. Bioremediation of diesel oil spill by filamentous fungus *Trichoderma reesei* H002 in aquatic environment. Int. J. Integr. Eng. 10(9).

Necibi, M. and N. Mzoughi. 2017. The distribution of organic and inorganic pollutants in marine environment. Micropollutants: sources, Ecotoxicological Effects and Control Strategies. 129: 1–43.

Neilson, A. H. and A. S. Allard. 2008. Environmental degradation and transformation of organic chemicals, CRC Press, Boca Raton, FL.

Omarova, M., L. T. Swientoniewski, I. K. M. Tsengam, A. Panchal, T. Yu, D. A. Blake, Y. M. Lvov and V. John. 2018. Engineered clays as sustainable oil dispersants in the presence of model hydrocarbon degrading bacteria: the role of bacterial sequestration and biofilm formation. ACS Sustain. Chem. Eng. 6(11): 14143–14153.

Ostrem Loss, E. M., M. K. Lee, M. Y. Wu, J. Martien, W. Chen, D. Amador-Noguez, C. Jefcoate, C. Remucal, S. Jung, S. C. Kim and J. H. Yu. 2019. Cytochrome P450 monooxygenase-mediated metabolic utilization of benzo [a] pyrene by *Aspergillus* species. MBio 10(3): e00558–19.

Pan, X., T. Xu, H. Xu, H. Fang and Y. Yu. 2017. Characterization and genome functional analysis of the DDT-degrading bacterium *Ochrobactrum* sp. DDT-2. Sci. Total Environ. 592: 593–599.

Pathak, A., A. Chauhan, A. Y. Ewida and P. Stothard. 2016. Whole genome sequence analysis of an Alachlor and Endosulfan degrading *Micrococcus* sp. strain 2385 isolated from Ochlockonee River, Florida. Journal of Genomics. 4: 42.

Prasad, M. N. V., P. Malec, A. Waloszek, M. Bojka and K. Strzallka. 2001. Physiological responses of *Lemna trisulca* L. (duckweed) to cadmium and copper bioaccumulation. Plant Sci. 161: 881.

Priyanka, J. V., S. Rajalakshmi, P. S. Kumar, V. G. Krishnaswamy, D. A. Al Farraj, M. S. Elshikh and M. R. A. Gawwad. 2022. Bioremediation of soil contaminated with toxic mixed reactive azo dyes by co-cultured cells of *Enterobacter cloacae* and *Bacillus subtilis*. Environ. Res. 204: 112136.

Rezania, S., M. Ponraj, M. F. M. Din, A. R. Songip, F. M. Sairan and S. Chelliapan. 2015. The diverse applications of water hyacinth with main focus on sustainable energy and production for new era: an overview. Renew. Sustain. Energy Rev. 41: 943–954.

Rossi, T., P. M. S. Silva, L. F. De Moura, M. C. Araújo, J. O. Brito and H. S. Freeman. 2017. Waste from eucalyptus wood steaming as a natural dye source for textile fibers. J. Clean. Prod. 143: 303–310.

Rudin, S. M., D. W. Murray and T. J. Whitfeld. 2017. Retrospective analysis of heavy metal contamination in Rhode Island based on old and new herbarium specimens. Appl. Plant Sci. 5(1): 1600108.

Sagarkar, S., P. Bhardwaj, V. Storck, M. Devers-Lamrani, F. Martin-Laurent and A. Kapley. 2016. s-triazine degrading bacterial isolate *Arthrobacter* sp. AK-YN10, a candidate for bioaugmentation of atrazine contaminated soil. Appl. Microbial. Biotechnol. 100(2): 903–913.

Sangkharak, K., A. Choonut, T. Rakkan and P. Prasertsan. 2020. The degradation of phenanthrene, pyrene, and fluoranthene and its conversion into medium-chain-length polyhydroxyalkanoate by novel polycyclic aromatic hydrocarbon-degrading bacteria. Curr. Microbiol. 77(6): 897–909.

Saptakee, S. 2011. Journal on Bioremediation for Oil spills.www.buzzle.com/articles.

Sarang, B., K. Richa and C. Ram. 2013. Comparative study of bioremediation of hydrocarbon fuel. Int. J. Biotechnol. Bioeng. Res. 4: 677–686.

Sarkar, D., R. Datta and R. Hannigan. 2011. Geochemical cycling of trace and rare earth elements in Lake Tanganyika and its major tributaries. pp. 135–171. *In*: Elsevier [ed.]. Concepts and Applications in Environmental Geochemistry, 135.

Sarkar, S., A. Banerjee, U. Halder, R. Biswas and R. Bandopadhyay. 2017. Degradation of synthetic Azo Dyes of textile industry: a sustainable approach using microbial enzymes. Water Conserv. Sci. Eng. 24(2): 121–131.

Seeger, M., M. Hernández, V. Méndez, B. Ponce, M. Córdova and M. González. 2010. Bacterial degradation and bioremediation of chlorinated herbicides and biphenyls. J. Plant. Nutr. Soil Sci. 10(3): 320–332.

Sengupta, K., M. T. Swain, P. G. Livingstone, D. E. Whitworth and P. Saha. 2019. Genome sequencing and comparative transcriptomics provide a holistic view of 4-nitrophenol degradation and concurrent fatty acid catabolism by *Rhodococcus* sp. strain BUPNP1. Front. Microbial. 9: 3209.

Shan, H., H. D. Kurtz Jr. and D. L. Freedman. 2010. Evaluation of strategies for anaerobic bioremediation of high concentrations of halomethanes. Water Res. 44(5): 1317–1328.

Shah, V. and A. Daverey. 2020. Phytoremediation: A multidisciplinary approach to clean up heavy metal contaminated soil. Environ. Technol. Innov. 18: 100774.

Sharma, R. and K. C. Yeh. 2020. The dual benefit of a dominant mutation in Arabidopsis IRON DEFICIENCY TOLERANT1 for iron biofortification and heavy metal phytoremediation. Plant Biotechnol. J. 18(5): 1200–1210.

Silva-Castro, G. A., I. Uad, J. Go´nzalez-Lo´pez, C. G. Fandino´, F. L. Toledo and C. Calvo. 2012. Application of selected microbial consortia combined with inorganic and oleophilic fertilizers to recuperate oil-polluted soil using land farming technology. Clean Technol. Environ. Pol. 14: 719–726.

Singh, S. P. 2014. Application of bioremediation on solid waste management: a review. J. Bioremed. Biodegr. 5(06).

Singh, G. and S. K. Dwivedi. 2022. Biosorptive and biodegradative mechanistic approach for the decolorization of congo red dye by *Aspergillus* species. Bull. Environ. Contam. Toxicol. 108(3): 457–467.

Smolyakov, B. S. 2012. Uptake of Zn, Cu, Pb, and Cd by water hyacinth in the initial stage of water system remediation. Appl. Geochem. 27(6): 1214–1219.

Šrédlová, K., K. Šírová, T. Stella and T. Cajthaml. 2021. Degradation products of polychlorinated biphenyls and their *in vitro* transformation by ligninolytic fungi. Toxics. 9(4): 81.

Syed, K., H. Doddapaneni, V. Subramanian, Y. W. Lam and J. S. Yadav. 2010. Genome-to-function characterization of novel fungal P450 monooxygenases oxidizing polycyclic aromatic hydrocarbons (PAHs). Biochem. Biophys. Res. Commun. 399(4): 492–497.

Tang, W. 2018. Research progress of microbial degradation of organophosphorus pesticides. Prog. Appl. Microbiol. 1: 29–35.

Thakor, R., H. Mistry, K. Tapodhan and H. Bariya. 2022. Efficient biodegradation of Congo red dye using fungal consortium incorporated with *Penicillium oxalicum* and *Aspergillus tubingensis*. Folia Microbiol. 67(1): 33–43.

Tyagi, B. and N. Kumar. 2021. Bioremediation: principles and applications in environmental management. In Bioremediation for Environmental Sustainability. 3–28. Elsevier.

Uday, U. S. P., T. K. Bandyopadhyay and B. Bhunia. 2016. Bioremediation and detoxification technology for treatment of dye (s) from textile effluent. Textile Wastewater Treatment, 75–92.

United States Environmental Protection Agency (USEPA). 2000. Introduction Phytoremediation. EPA 600/R-99/107, U.S. Environmental Protection Agency, Office of Research and Development, Cincinnati, OH.

Upendar, G., S. Dutta, P. Bhattacharya and A. Dutta. 2017. Bioremediation of methylene blue dye using *Bacillus subtilis* MTCC 441. Water Sci. Technol. 75(7): 1572–1583.

U.S. EPA Seminars. Bioremediation of Hazardous Waste Sites: Practical Approach to Implementation, EPA/625/ K96/ 001.

Van Aken, B. 2009. Transgenic plants for enhanced phytoremediation of toxic explosives. Curr. Opin. Biotechnol. 20(2): 231–236.

Van Aken, B. and R. Bhalla. 2011. Microbial degradation of polychlorinated biphenyls. Ind. Toxic Wastes 152166.

Varjani, S. J. and V. N. Upasani. 2017. Crude oil degradation by *Pseudomonas aeruginosa* NCIM 5514: influence of process parameters Indian J. Exp. Biol. 55: 493–497.

Varjani, S., P. Rakholiya, T. Shindhal, A. V. Shah and H. H. Ngo. 2021. Trends in dye industry effluent treatment and recovery of value added products. J. Water Process. Eng. 39: 101734.

Vidali, M. 2001. Bioremediation. an overview. Pure Appl. Chem. 73(7): 1163–1172.

Vikrant, K., B. S. Giri, N. Raza, K. Roy, K. H. Kim, B. N. Rai and R. S. Singh. 2018. Recent advancements in bioremediation of dye: current status and challenges. Bioresour. Technol. 253: 355–367.

Wagh, M. S., W. J. Osborne and S. Sivarajan. 2022. *Bacillus xiamenensis* and earthworm *Eisenia fetida* in bio removal of lead, nickel and cadmium: a combined bioremediation approach. Appl. Soil Ecol. 176: 104459.

Wang, D. G., M. Yang, H. L. Jia, L. Zhou and Y. F. Li. 2009. Polycyclic aromatic hydrocarbons in urban street dust and surface soil: comparisons of concentration, profile, and source. Arch. Environ. Contam. Toxicol. 56(2): 173–180.

Wang, X., Y. Wang, S. Ning, S. Shi and L. Tan. 2020. Improving azo dye decolorization performance and halotolerance of *Pichia occidentalis* A2 by static magnetic field and possible mechanisms through comparative transcriptome analysis. Front. Microbiol. 11: 1–10.

Whelan, M. J., F. Coulon, G. Hince, J. Rayner, R. McWatters, T. Spedding and I. Snape. 2015. Fate and transport of petroleum hydrocarbons in engineered biopiles in polar regions. Chemosphere. 131: 232240.

Wiegel, J. and Q. Wu. 2000. Microbial reductive dehalogenation of polychlorinated biphenyls. FEMS Microb. Ecol. 32(1): 1–15.

Wolskm, E. A., V. Barrera, C. Castellari and J. F. González. 2012. Biodegradation of phenol in static cultures by *Penicillium chrysogenum* ERK1: catalytic abilities and residual phototoxicity. Rev. Argent. Microbiol. 44(2): 113–121.

Yadav, J. S., H. Doddapaneni and V. Subramanian. 2006. P450ome of the white rot fungus *Phanerochaete chrysosporium*: structure, evolution and regulation of expression of genomic P450 clusters.

Ye, Z., H. Li, Y. Jia, J. Fan, J. Wan, L. Guo, X. Su, Y Zhang, W. M. Wu and C. Shen. 2020. Supplementing resuscitation-promoting factor (Rpf) enhanced biodegradation of polychlorinated biphenyls (PCBs) by *Rhodococcus biphenylivorans* strain TG9T. Environ. Pollut. 263: 114488.

Yu, X. Z. and J. D. Gu. 2007a. Accumulation and distribution of trivalent chromium and effects on hybrid willow (*Salix matsudana Koidz × alba* L.) metabolism. Arch. Environ. Contam. Toxicol. 52: 503–511.

Yu, X. Z. and J.D. Gu. 2007b. Metabolic responses of weeping willows to selenate and selenite. Env. Sci. Pollut. Res. 14: 510–517.

Yu, X. Z. and J. D. Gu. 2008a. The role of EDTA in phytoextraction of hexavalent and trivalent chromium by two willow trees. Ecotoxicol. 17:143-152.

Yu, X. Z. and J. D. Gu. 2008b. Effect of available nitrogen on phytoavailability and bioaccumulation of hexavalent and trivalent chromium in hankow willows (*Salix matsudana Koidz*) Ecotoxicol. Environ. Saf. 70: 216–222.

Zhang, J. L. and C. L. Qiao. 2002. Novel approaches for remediation of pesticide pollutants. Int. J. Environ. Pollut. 18(5): 423–433.

Zhang, W., Z. Lin, S. Pang, P. Bhatt and S. Chen. 2020. Insights into the biodegradation of lindane (γ-hexachlorocyclohexane) using a microbial system. Frontiers in Microbiology. 11: 522.

Zheng, M., W. Wang, M. Hayes, A. Nydell, M. A. Tarr, S. A. Van Bael and K. Papadopoulos. 2018. Degradation of Macondo 252 oil by endophytic *Pseudomonas putida*. J. Environ. Chem. Eng. 6(1): 643–648.

Zhou, H., S. Zhang, J. Xie, H. Wei, Z. Hu and H. Wang. 2020. Pyrene biodegradation and its potential pathway involving *Roseobacter clade* bacteria. Int. Biodeterior. Biodegrad. 150: 104961.

Zhu, D. H., F. H. Nie, Q. L. Song, W. Wei, M. Zhang, Y. Hu, H. Y. Lin, D. J. Kang, Z. B. Chen and J. J. Chen. 2022. Isolation and genomic characterization of *Klebsiella* Lw3 with polychlorinated biphenyl degradability. Environ. Technol., 1–11.

CHAPTER 2

Bioprecipitation as a Bioremediation Strategy for Environmental Cleanup

Samantha M. Wilcox,[1] *Catherine N. Mulligan*[1,]* and
Carmen Mihaela Neculita[2]

2.1 Introduction

The term 'sustainable' is generally used in today's culture and its true significance is often not clear. While typically used synonymously to environmentally friendly behavior, the term represents much more. According to the United Nations (UN) 17 Sustainable Development Goals, 'sustainability' refers to social, economic and environmental action (UNDP 2022). The term 'sustainability', used throughout this chapter, aligns itself with the United Nations definition.

Imperative to the achievement of 'sustainability' is its implementation during decision-making processes. The introduction of ISO Lifecycle Assessment 14040 and the field of Environmental Accounting has made it easier to adhere to sustainable practices in government and industry (ICF Incorporated 1995, ISO 2006). These approaches aim to minimize the environmental, social and economic impact at all stages of a product, a process or a service.

Remediation strategies, especially bio-based processes, are often considered inherently sustainable. However, it is important to apply these principles to improve longstanding engineering frameworks. An effective engineered clean-up strategy will appease the sustainability requirements and offer solutions that meet both environmental and governmental regulations. This chapter focuses on biological precipitation, also referred to as bioprecipitation as a remediation strategy for soil and groundwater contamination. The process enhances the already naturally occurring processes. It is therefore considered a cost effective, socially accepted and environmentally friendly technique. This chapter aims to document and justify the necessity and highlight the sustainable nature of bioprecipitation as a technique for environmental clean-up.

[1] Department of Building, Civil and Environmental Engineering, Concordia University, Montreal, Canada.
[2] Research Institute on Mines and the Environment (RIME), University of Quebec in Abitibi-Témiscamingue, Rouyn-Noranda, Canada.
* Corresponding author: mulligan@civil.concordia.ca

2.2 Overview of Contamination and Remediation

2.2.1 Soil and Groundwater Contamination

Metal and metalloid (metal(loid)s) contamination is prevalent in soil, sediment, groundwater and surface water. Contaminant release (from a point source or continuous) can be from a variety of industrial operations. The fate and transport of contamination depends largely on the metal(loid) present and the transport mechanisms. The characteristics of the metal(loid), i.e., speciation, solubility, vapor pressure and partitioning coefficients dictate the fate of the pollutant in media, while transport can be evaluated by advection, convection, diffusion or dispersion mechanisms.

The fate and transport of metal(loid) contamination can influence the area of impact and the concentration. This can affect its impact on humans, animals and ecosystems. Figure 2.1 shows the likely human exposure to metal(loid) contamination and the resulting consequences on bodily systems. This highlights the potentially severe impacts contaminant release can have on humans, but also ecosystems in general.

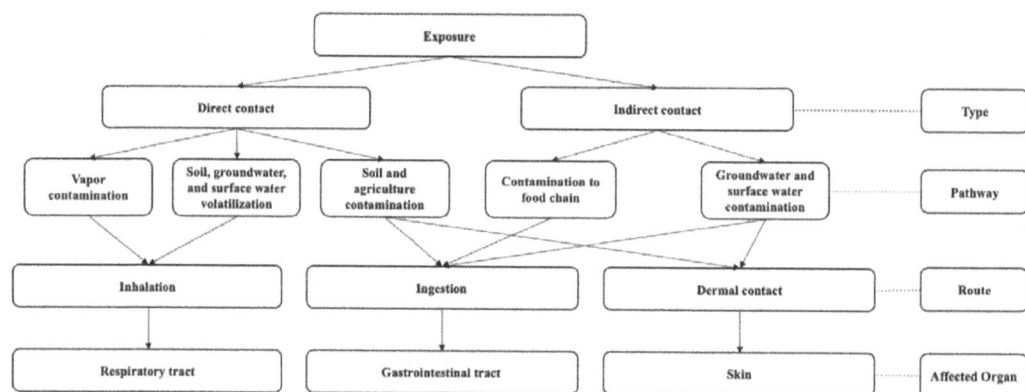

Figure 2.1. Human exposure to metal(loid) contamination including contact type, exposure pathways, exposure routes, and impacted organs (Adapted from Kuppusamy et al. 2020).

2.2.2 Clean-up Strategies for Soil and Groundwater Contamination

Soil and groundwater contamination can be mitigated with various clean-up strategies. Decontamination strategies aim to mitigate pollution via metal(loid) transformation (to a less toxic and/or less mobile form), immobilization, or removal. Table 2.1 lists numerous soil, groundwater and sediment clean-up strategies identifying the principal process and the overall goal of the treatment option. Often, multiple types of remediation strategies can be applied simultaneously or in series to achieve high efficacies of reclamation outlined by environmental and/or regulatory standards.

At present, biological remediation strategies are a topic of intensive research. These treatment methods often provide green and sustainable solutions that adhere to environmental concerns compared to other treatment methods (i.e., physical, thermal and chemical). Comparatively, they emit relatively low greenhouse gas emissions and are low water and energy consuming processes. Bioremediation strategies are typically of low cost with moderate to minimal maintenance. Due to their biological nature, they gain public acceptance more readily and therefore, are more easily implemented at a site.

Table 2.1. Clean-up strategies (Adapted from Government of Canada 2017, LaGrega et al. 1994, Ossai et al. 2020).

Treatment Type	Goal	Process	Method
Physical Treatment	Immobilization	Containment	Containment Booms
			Skimmers
			Physical Barriers
			Impermeable/Slurry Walls
			Frozen Walls
			Surface Capping
			Hydraulic Containment, Hydraulic Control
			Pump and Treat
		Sorption	Sequestration
	Removal	Removal	Dredging and Offsite Disposal
		Volatilization	Soil Vapor Extraction
			Air Sparging
			Air Stripping and Steam Stripping, Steam Distillation
		Desorption	Soil Washing, Soil Flushing
			Soil Extraction, Solvent Extraction
Chemical Treatment	Immobilization	Containment	Solidification/Stabilization
			Encapsulation
		Sorption	Immobilization
			Activated Carbon Treatment
	Transformation	Biodegradation	Dispersion
		Emulsify	Emulsification
		Hydrolysis	Dechlorination, Reductive Dehalogenation
		Redox Reactions	Ultraviolet Oxidation, Photocatalytic Oxidation
			Chemical Oxidation-Reduction
	Removal	Desorption	Supercritical Fluid Extraction and Supercritical Fluid Oxidation
Thermal/ Heat Treatment	Removal	Desorption	Thermal Desorption
			Hot Water Injection
		Volatilization	Smoldering Combustion
			Incineration
			Hot Air Injection
			Steam Injection
			Hot Gas Decontamination
			Vitrification
	Transformation	Degradation	Pyrolysis
Electric and Electromagnetic Treatment	Removal	Volatilization	Electrical Resistance Heating
		Desorption	Radio Frequency Heating
		Volatilization	Microwave Heating
		Electro-migration	Electrokinetic Remediation, Electrochemical Soil Process
	Transformation	Degradation	Photocatalytic Degradation

Table 2.1 contd. ...

...Table 2.1 contd.

Treatment Type	Goal	Process	Method
Acoustic and Ultrasonic Treatment	Removal	Desorption	Ultrasonic Extraction
	Transformation	Degradation	Sonochemical Degradation, Sonochemical Oxidation
Biological Treatment	Transformation	Biodegradation	Bioremediation
			Bioattenuation
			Biostimulation
			Bioaugmentation
			Bioventing
			Biosparging
			Bioslurry
			Biopiling
			Biotransformation
			Bioreactor
			Landfarming
			Composting
			Windrows
			Monitored Natural Attenuation
			Reductive Dechlorination
			Vermicomposting, Vermiremediation
			Trichoremediation
			Mycoremediation, Mycodegradation
			Phycoremediation
			Phytodegradation, Phytotransformation
			Rhizodegradation
		Degradation	Biodegradation
	Removal	Filter	Constructed Wetlands
		Translocation	Phytoextraction, Phytoaccumulation, Phytoabsorption, Phytosequestration
			Phytohydraulics, Hydraulic Plume Control
		Volatilization	Phytovolatilization
			Vegetative Cover Systems, Evaporatranspiration Cover Systems
		Redox Reactions	Bioelectrical System, Electrobioremediation
			Nanobioremediation, Nanoremediation
	Immobilization	Sorption	Phytostabilization, Phytoimmobilization
			Rhizofiltration

2.3 Bioprecipitation

2.3.1 Overview: Principles and Applications

2.3.1.1 Chemical Precipitation

Precipitation is an intricate phenomenon involving thermodynamic and kinetic processes. The process is governed by its thermodynamic properties, i.e., supersaturation state. A solution is considered supersaturated when the solute and solvent are no longer in an equilibrium (saturated) state (Karpiński and Bałdyga 2019, Lewis 2017). Equation 2.1 shows the supersaturation calculation, whereby the supersaturation (σ) is calculated from the actual molar chemical potential (μ), the molecule in equilibrium state (μ_{eq}), the universal gas constant (R) and the absolute temperature (T). Based on the equation:

- The solution is in equilibrium when $\Delta\mu$ ($\Delta\mu = \mu - \mu_{eq}$) is equal to 0
- If $\Delta\mu > 0$ the solution is supersaturated, spontaneous precipitation will occur
- If $\Delta\mu < 0$ the solution is below the saturated state, spontaneous dissolution will occur (Karpiński and Bałdyga 2019).

Supersaturation (Davey and Garside 2000, Karpiński and Bałdyga 2019)

$$\sigma = \frac{(\mu - \mu_{eq})}{RT}$$

Eq. 2.1

Phase diagrams can demonstrate the liquid or solid phase of a compound of interest (Karpiński and Bałdyga 2019). The thermodynamic component of precipitation can be better understood by the Gibbs phase rule, which identifies the number of possible phases and the degree of freedom of the multiphase system in equilibrium (Faghri and Zhang 2006). The Gibbs phase rule both identifies and inhibits solid precipitates formed from a solution (Yong et al. 2014).

Furthermore, the thermodynamic potential is described by Gibbs free energy. It explains the maximum energy transfer of a closed system. The change in Gibbs free energy (ΔG) can be used to define primary homogenous nucleation. Equation 2.2 shows the calculation, where the number of molecules (N), the reaction affinity (Φ), the crystal surface area (A) and the surface tension (σ) are used to explain the work to produce crystals during precipitation of a supersaturated solution. From Eq. 2.1, one can consider R as the molar chemical potential change or the Gibbs free energy (Karpiński and Bałdyga 2019).

Gibbs Free Energy Change (Karpiński and Bałdyga 2019, Nielsen 1964)

$$\Delta G = -N\Phi + A\sigma$$

Eq. 2.2

While supersaturation state is the driving force of precipitation, the kinetic process illustrates the rate of precipitation. Nucleation, crystal growth and agglomeration represent the primary kinetic process involved in precipitation (Lewis 2017), these terms are further defined in Table 2.2. Again from Eq. 2.1, the term $\frac{\Delta\mu}{RT}$ represents the rate of nucleation and crystal growth (Karpiński and Bałdyga 2019).

Precipitation can be applied to soil, groundwater, surface water and wastewater treatment. For soil and groundwater remediation, metal(loid)s can chemically precipitate out of the pore water (Yong et al. 2014). These precipitates can adsorb onto soil particle surfaces (Yong et al. 2014) or can form a cement matrix clogging the pore spaces (Mitchell and Soga 2005). For water and wastewater

Table 2.2. Kinetic Mechanisms of Precipitation (Adapted from Lewis 2017).

Kinetic Process	Definition	Impact on Precipitation

Nucleation	Creation of crystal particles occurring from enlargement of ions or molecules in a supersaturated state. Nucleation can be:
	- Primary: spontaneous precipitation from solution
	  - Homogenous nucleation (spontaneous precipitation)
	  - Heterogeneous nucleation (prompted by foreign particles)
	- Secondary: prompted by the presence of existing crystals
	  - Contact nucleation (via crystal-crystal contact)
	  - Shear nucleation (via fluid flow)
	  - Fracture nucleation (via particle impact)
	  - Attrition nucleation (via particle impact and disruption)
		High supersaturation → Primary nucleation
		Low saturation → Secondary nucleation

Crystal Growth	Growth of crystal particles due to adsorption on substance surfaces	High supersaturation → Rough crystals Low saturation → Spiral or smooth crystals High nucleation and low growth rate → small but many crystals
Agglomeration	Enlargement of crystals due to prolonged contact and formation of crystalline bridges between particles	High nucleation → many crystals → high agglomeration High agglomeration → Low crystal purity

treatment, chemical precipitation is applied through coagulation and flocculation, whereby the use of a reagent is used to promote precipitation and aggregation, respectively (Davis 2010, Mihelcic and Zimmerman 2014). Figure 2.2 demonstrates the various factors that affect precipitation and potential impacts to its use as a remediation strategy. Of the factors identified, pH is the most significant, whereby alkaline solutions tend to precipitate metal(loid)s and acidic solutions cause dissolution of metal(loid) precipitates (Yong et al. 2014).

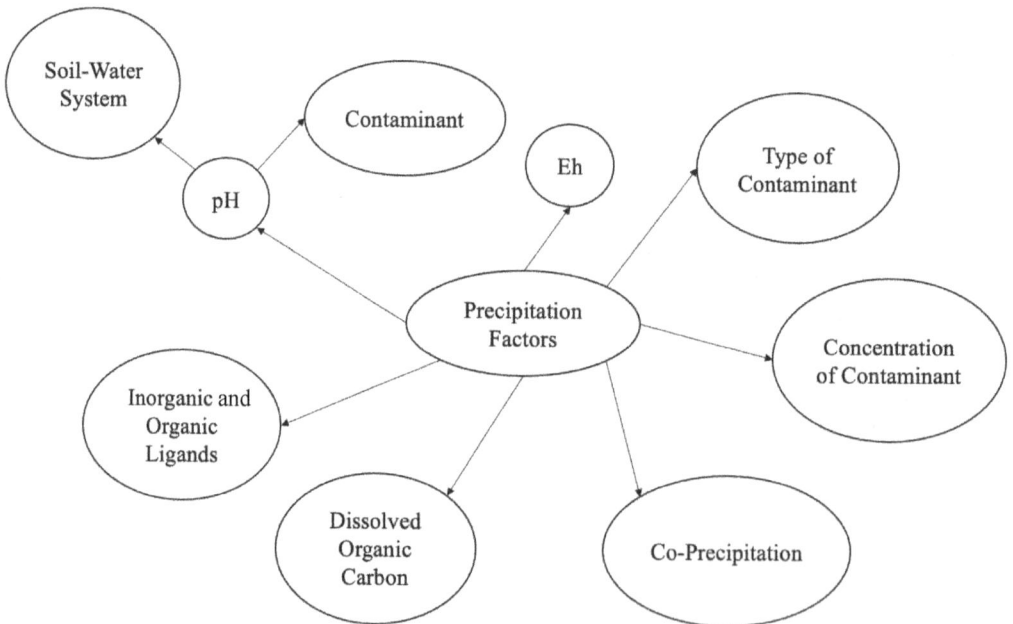

Figure 2.2. Factors influencing precipitation (Adapted from Yong et al. 2014).

2.3.1.2 Biological Precipitation

Bioprecipitation is an emerging bioremediation strategy. The process applies chemical precipitation with the presence of microorganisms to catalyze the reaction. The microorganism activity typically acts as the catalyst for the oxidative and reductive reactions. The reactions aim to generate alkalinity or consume acidity to precipitate metal(loid)s (Johnson and Santos 2020). Through local alkalinization or enzyme generated ligands, reactions can precipitate carbonate, hydroxide, sulfide, phosphate, oxalate, etc. compounds (Kumar et al. 2013).

The metal(loid) in focus drastically impacts the type and efficacy of precipitation. Metal(loid)s can react differently to the oxidative and reductive reactions. For example, some metals (Fe, Mn) increase solubility under reductive conditions, while others (U^{6+}, Cr^{6+}) decrease solubility (Gadd 2004). Further, the phase of the metal(loid) can impact its pH. Ferric iron (Fe^{3+}) in its amorphous form can generate alkalinity, while in soluble form it cannot (Johnson and Santos 2020). Therefore, the metal(loid) of focus should be analyzed to assess its chemical speciation and establish chemical stability.

The most common form of bioprecipitation is Biological Sulfate Reduction (BSR). Sulfate Reducing Bacteria (SRB) catalyze the dissimilatory sulfur reduction process, whereby oxidation of an electron donor facilitates the reduction of sulfate (SO_4^{2-}) to soluble sulfides (S^{2-}) (Sánchez-Andrea et al. 2014). The process aims to increase pH to precipitate metal(loid)s and reduce sulfate (Willis and Donati 2017). It is a common remediation strategy for acidic wastewaters with metal(loid) contamination (Sánchez-Andrea et al. 2014). This type of contamination is common among mining industries, electroplating industries and tannery industries (Sahinkaya et al. 2017). Equations 2.3 and 2.4 demonstrate how chemically BSR can induce bioprecipitation to remediate contaminated groundwater. Equation 2.3 uses formaldehyde (CH_2O) as an electron donor to reduce sulfate from acidic industrial groundwater to hydrogen sulfide. The hydrogen sulfide then reacts with other metal divalent cations (M^{2+}) to precipitate low solubility metal sulfides (MS), Eq. 2.4.

Reduction of Sulfate to Hydrogen Sulfide with the use of an Electron Donor (Sahinkaya et al. 2017, Willis and Donati 2017)

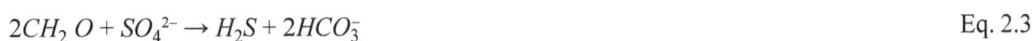

$$2CH_2O + SO_4^{2-} \rightarrow H_2S + 2HCO_3^- \qquad\qquad \text{Eq. 2.3}$$

Metal Precipitation from Hydrogen Sulfide (Sahinkaya et al. 2017, Willis and Donati 2017)

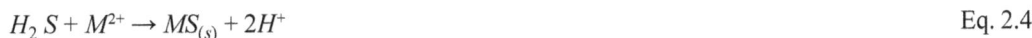

$$H_2S + M^{2+} \rightarrow MS_{(s)} + 2H^+ \qquad\qquad \text{Eq. 2.4}$$

SRB species can be both heterotrophs and autotrophs (chemolithotrophs) found in anaerobic environments (Barton et al. 2015, Hao et al. 2014). As of 2015, there were 59 genera and 220 species reported (Barton et al. 2015). However, the genera *Desulfovibrio* is the most highly reported bacteria from bioreactor studies (Kiran et al. 2017). The microorganisms are classified as complete oxidizers, incomplete oxidizers or both (Hao et al. 2014). There are over 75 energy sources that promote SRB growth (Barton et al. 2015). Although trace metals, selenium (Se) and molybdenum (Mo), are attributed to bacterial growth, the mechanisms linked to metal(loid) reduction are not currently considered as related (Barton et al. 2003).

To assure microorganisms can facilitate bioprecipitation, they require some degree of metal tolerance. The metal tolerance allows the microorganisms to flourish under normally toxic metal(loid) contamination. This is achieved via metabolism, cell wall structure, Extra-cellular Polymeric Substances (EPS), methylation, alkylation/dealkylation (Diels et al. 2006), intra-cellular and extra-cellular sequestration, active transport effluent pumps, enzymatic detoxification, reduction in metal sensitivity of cellular targets (Bruins et al. 2000, Diels et al. 2006) and exclusion by permeability barrier (Bruins et al. 2000). Metal resistant genes have been found in both gram-positive and gram-negative microorganisms (Abou-Shanab et al. 2007) and SRB species have also reported gram-positive and gram-negative microorganisms (Barton et al. 2015). To improve culture resistance a mixture of species is often selected for bioprecipitation processes (Kiran et al. 2017).

In addition to BSR, SRB have also been found to reduce other metal(loid)s with direct or indirect microorganism activity. SRB can directly reduce metal(loid)s (i.e., Cr, As, Al, Te and Sb) to a less toxic, insoluble form or can indirectly reduce metal(loid)s via hydrogen sulfide produced during BSR (Sánchez-Andrea et al. 2016, Willis and Donati 2017). After metal(loid) reduction, many species are capable of precipitation. For example, uranium (U^{6+}) can precipitate uraninite (UO_2), chromium (Cr^{6+}) in the presence of ferric iron (Fe^{3+}) can precipitate chromium hydroxide oxide (CrO (OH)) and chromium sulfide (Cr_2S_3) and arsenic can precipitate arsenic sulfide (As_2S_3) and arsenopyrite (FeAsS). Metal(loid) reduction using SRB activity is presented in Table 2.3. The following electron donors were used as an energy source in some of the noted reactions: acetate (CH_3COOH), formate (CH_2O_2), lactate ($C_3H_6O_3$) and pyruvate ($C_3H_4O_3$).

Table 2.3. Reduction of Metal(loid)s with SRB (Adapted from/Barton et al. 2015, Lovley 1993, Sahinkaya et al. 2017).

Metal(loid)	Electron Acceptor	Formula
Iron	Fe^{3+}	$CH_3COO^- + 8Fe^{3+} + 4H_2O \rightarrow 2HCO_3^- + 8Fe^{2+} + 9H^+$ $CO_2 H^- + 2Fe^{3+} + H_2O \rightarrow HCO_3^- + 2Fe^{2+} + 2H^+$ $C_3O_3H_5^- + 4Fe^{3+} + 2H_2O \rightarrow CH_3COO^- + HCO_3^- + 4Fe^{2+} + 5H^+$ $C_3O_3H_3^- + 2Fe^{3+} + 2H_2O \rightarrow CH_3COO^- + HCO_3^- + 2Fe^{2+} + 3H^+$
	Fe^0	$Fe^0 + 2H^+ \rightarrow Fe^{2+} + H_2$
Manganese	Mn^{4+}	$MnO_2 \rightarrow MnCO_3*$ $2H_2S + MnO_2 \rightarrow MnS + S^0 + 2H_2O$
	Mn^{6+}	$Mn^{6+} \rightarrow Mn^{2+}*$
Uranium	U^{6+}	$CH_3COO^- + 4U^{6+} + 4H_2O \rightarrow 2HCO_3^- + 4U^{4+} + 9H^+$ $H_2 + U^{6+} \rightarrow 2H^+ + U^{4+}$
Selenium	SeO_4^{2-}	$4CH_3COO^- + 3SeO_4^{2-} \rightarrow 3Se^0 + 8CO_2 + 4H_2O + 4H^+$ $CH_3COO^- + H^+ + 4SeO_4^{2-} \rightarrow 4SeO_3^{2-} + 2CO_2 + 2H_2O$
	SeO_3^{2-}	$2H_2S + SeO_3^{2-} + 2H^+ \rightarrow SeS_2 + 3H_2O$
Chromium	Cr^{6+}	$2Cr^{6+} + 3HS^- \rightarrow 2S^0 + 2Cr^{3+} + 3H^+$
Mercury	Hg^{2+}	$Hg^{2+} \rightarrow Hg^0*$
Cobalt	Co^{3+}	$Co^{3+} + H_2S \rightarrow CoS + 2H^+$
Palladium	Pd^{2+}	$Pd^{2+} \rightarrow Pd^0*$
Nickel	Ni^{3+}	$Ni^{3+} \rightarrow Ni^{2+}*$
Technetium	Tc^{7+}	$Tc^{7+} \rightarrow Tc^{4+}*$ $Tc^{7+} \rightarrow Tc^{5+}*$
Vanadium	V^{5+}	$V^{5+} \rightarrow V^{3+}*$
Molybdenum	Mo^{6+}	$Mo^{6+} \rightarrow Mo^{4+}*$
Arsenic	Ar^{5+}	$3Ar_2S_3 + 3H_2S \rightarrow 2H_2As_3S_6^- + 2H^+$ $As^{5+} \rightarrow As^{3+}*$
Gold	Au^{3+}	$Au^{3+} \rightarrow Au^0*$
	Au^+	$Au^+ \rightarrow Au^0*$

* Not a full formula. Showing the redox couple.

Electron donors provide a carbon and energy source for the reaction. The selection of an appropriate electron donor is the ratio of Chemical Oxygen Demand (COD) and sulfate ion concentration (SO_4^{2-}), i.e., COD/SO_4^{2-} is important. There is a correlation between the interaction of SRB with carbon source and electron acceptor (Barbosa et al. 2014, Kiran et al. 2017). The COD denotes oxygen content required to oxidize the organic material and the electron acceptor is SO_4^{2-} (Barbosa et al. 2014). However, COD is measured under aerobic conditions, and therefore does not accurately portray the organic carbon available to anerobic microorganisms (Neculita and Zagury 2008). The minimum theoretical value acceptable for organic degradation and BSR is 0.67 (Hao et al. 1996, Kiran et al. 2017, Neculita and Zagury 2008).

Other noteworthy indicators for an electron donors' performance are the carbon (C) to nitrogen (N) ratio (C/N) and the Dissolved Organic Carbon (DOC) to SO_4^{2-} ratio (DOC/SO_4^{2-}) (Neculita and Zagury 2008). The C/N ratio gives information about the capacity of biological degradation of an electron donor (Reinertsen et al. 1984, Zagury et al. 2006). While the ratio does not necessarily indicate the actual C and N available to the microorganisms (Reinertsen et al. 1984), a value of about 10 is generally accepted as a suitable substrate (Béchard et al. 1994, Neculita and Zagury 2008, Reinertsen et al. 1984, Zagury et al. 2006). The DOC/SO_4^{2-} ratio is like the COD/SO_4^{2-} ratio but is more easily quantified (Neculita and Zagury 2008).

An electron donor can be organic or synthetic. In keeping with the sustainable nature of remediation, an organic energy is a natural, economic and socially acceptable additive to the operation. Organic waste can be considered sewage sludge, animal manure, leaf mulch, wood chips, sawdust and cellulose (Liamleam and Annachhatre 2007). The high carbon content of organic waste is advantageous to BSR demonstrating high rates of sulfate reduction, specifically when wastes are applied as mixtures (Hao et al. 2014, Liamleam and Annachhatre 2007). The COD/SO_4^{2-} ratio for organic wastes ranges from 1.6–5 (Kiran et al. 2017). In general, for all performance indicators (i.e., C/N, COD/SO_4^{2-}, DOC/SO_4^{2-}), a higher ratio is linked to superior BSR (Neculita and Zagury 2008).

Molasses is the most cost effective and widely available electron donor for BSR (Janssen and Temminghoff 2004, Liamleam and Annachhatre 2007), indicating its preferential use for sustainable engineering. In addition to its use as an electron donor, it contains nutrients (i.e., P, K, Cl, amino acids) for SRB growth (Janssen and Temminghoff 2004). The process aims to ferment molasses into lactate which is used as the electron donor or carbon source (Liamleam and Annachhatre 2007). However, SRB growth is inhibited at molasses concentrations greater than 5 g/L (Hao et al. 2014, Janssen and Temminghoff 2004), and there is seemingly a link between the quantity applied and the pH rise (Janssen and Temminghoff 2004). Further, partial decomposition creates high COD in effluent (Hao et al. 2014).

Lactate is another organic substrate used as an energy source for BSR. Its use as an electron donor is not temperature dependent, and does not impact its oxidizing ability (Liamleam and Annachhatre 2007). Both lactate and ethanol are considered optimal for SRB growth (Janssen and Temminghoff 2004), although, ethanol is not an organic substrate it is considered the most cost effective (Gibert et al. 2002, Liamleam and Annachhatre 2007). Further, acetate is the most used energy source for Fe^{3+} bioprecipitation (Lovley 1993). Other energy sources include hydrogen, formate, methanol, propionate, butyrate, sugar and hydrocarbons (Liamleam and Annachhatre 2007). Zero-valent iron has also been used to establish an anaerobic environment for SRB growth by consuming oxygen (O_2) while generating hydrogen (H_2) which acts as an electron donor (Pagnanelli et al. 2009). A high value for Gibbs free energy is preferable for BSR to assure that sulfidogenic reactions prevail over methanogenic reactions (Liamleam and Annachhatre 2007). However, the concentration of the electron donor should be monitored as high carbon concentration can lead to methanogenic conditions (Diels et al. 2005, 2006). A mixture of organic substrates is often used with success, that can produce high rates of sulfate reduction (Liamleam and Annachhatre 2007). The cost of the substrate should be kept in mind and adhere to the projects' economic sustainability.

Bioprecipitation can also occur as an oxidative reaction. Ferrous iron (Fe^{2+}) is transformed via microbial oxidation into ferric iron (Fe^{3+}) for precipitation. Microbial oxidation can take place as an anaerobic process (using phototropic and nitrate-reducing microorganisms) and aerobic process (using neutrophilic and acidophilic microorganisms). The phylum *Proteobacteria* is the most common bacteria responsible for Fe^{2+} reduction (Kiskira et al. 2017). Zero-valent iron can oxidize to Fe^{2+} and Fe^{3+} ions, help to facilitate the bioprecipitation process (Pagnanelli et al. 2009). From the oxidative reaction, Fe^{3+} can precipitate ferric hydroxide (Fe $(OH)_3$), jarosite ($MFe_3(SO_4)_2(OH)_6$, where M is a monovalent cation, i.e., K^+, Na^+, NH_4^+, Ag^+, or H_3O^+), schwertmannite ($Fe_{16}O_{16}(SO_4)_2(OH)_6 \cdot nH_2O$, where n is between 10 and 12), goethite (FeO (OH)), hematite (Fe_2O_3) and scorodite ($FeAsO_4 \cdot 2H_2O$). The pH, temperature, and overall chemistry of solution influence the precipitate formed (Sahinkaya et al. 2017).

2.3.1.3 Microbially Induced Calcite Precipitation (MICP)

MICP is another form of bioprecipitation. However, unlike the earlier discussed oxidative and reductive bioprecipitation processes, MICP is not easily influenced by redox reactions (Achal et al. 2011, Xiangliang 2009). This form of biological precipitation aims to immobilize the contaminant via cementation. Microorganisms are used to facilitate the hydrolysis of urea creating carbonate (CO_3^{2-}) and ammonium (NH_4^+), Eq. 2.5. The NH_4^+ ions increase pH to ameliorate precipitation (Achal et al. 2013a,b), and a cementation solution with calcium (Ca^{2+}) ions is introduced to precipitate calcium carbonate ($CaCO_3$), Eq. 2.6. Other metal divalent cations (M^{2+}) from solution are also able to precipitate metal carbonate (MCO_3) compounds, Eq. 2.7.

Urease Hydrolysis (Mwandira et al. 2022)

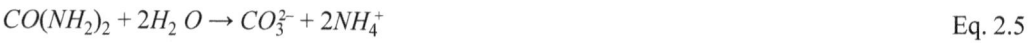

$$CO(NH_2)_2 + 2H_2O \rightarrow CO_3^{2-} + 2NH_4^+$$ Eq. 2.5

Calcium Carbonate Precipitation (Mwandira et al. 2022)

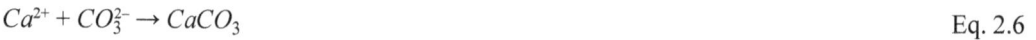

$$Ca^{2+} + CO_3^{2-} \rightarrow CaCO_3$$ Eq. 2.6

Metal Carbonate Precipitation (Mwandira et al. 2022)

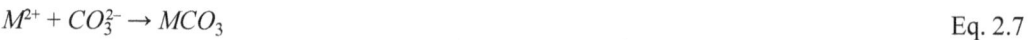

$$M^{2+} + CO_3^{2-} \rightarrow MCO_3$$ Eq. 2.7

The aim of this process is to immobilize the precipitates through the formation of a cement matrix, whereby precipitates form bridges between soil particles. Since the crystals precipitate out of the soil-groundwater system and clog the pore spaces, the soil properties are altered. The permeability, porosity, stiffness, shear strength, unconfined compressive strength, microstructure and shear wave velocity are all impacted by MICP (Mujah et al. 2016).

The main theory behind MICP remediation is solidification/stabilization (S/S). S/S is a strategy that immobilizes the soil and groundwater contaminants by using additives that alters the physical properties (i.e., solidifies/entraps the contaminant) and/or chemical properties (i.e., transforms the contaminant to a less toxic, less mobile form), respectively (LaGrega et al. 1994, Sharma and Reddy 2004). Historically, S/S used cements, pozzolans, thermoplastic materials or organic polymers to achieve contaminant entrapment (Sharma and Reddy 2004), however MICP offers a biological approach to reach S/S remediation. The mechanisms involved in S/S to remediate soil and groundwater include macroencapsulation, microencapsulation, adsorption, absorption, precipitation and detoxification (LaGrega et al. 1994). Through MICP, sorption can cause bioprecipitated $CaCO_3$ crystals and other MCO_3 compounds to bond to soil particle surfaces (LaGrega et al. 1994, Xiangliang 2009) via electrochemical bonds, such as van der Waal's forces or hydrogen bonds (LaGrega et al. 1994). This can aid the development of the cement matrix, which offers a more sustainable approach to S/S remediation.

The most common microorganism used for MICP is the *Bacillus* species (Achal and Pan 2011). To improve MICP performance a desirable microorganism should be high urease producing with high metal tolerance. Again, these organisms can be gram-positive or gram-negative, however gram-positive bacteria are more reactive (Beveridge and Fyfe 1985, Levett et al. 2020). The negative charge of the bacterial cell wall can attract Ca^{2+} ions causing MICP on the gram-positive and gram-negative cell walls (Achal and Pan 2011). Microorganism EPS can also affect MICP.

EPS is reported to impact the biofilm, cell adhesion, $CaCO_3$ capture (Achal and Pan 2011) and the $CaCO_3$ mineralogy (Kawaguchi and Decho 2002). The $CaCO_3$ mineralogy can be altered via various polymorphs (i.e., calcite, vaterite and aragonite). The type of polymorph produced during MICP can impact the stability of precipitates, where calcite is the most stable and desirable, while aragonite and vaterite are less stable.

Other factors influencing the efficacy of MICP as a remediation technique are temperature, bacterial concentration or density, pH, degree of saturation, concentration of cementation solution and field application (Mujah et al. 2016). This bioprecipitation method offers a promising remediation strategy. However, the long-term impact should be studied. Metal(loid) dissolution from redox and/or pH changes could release contaminants into the soil and groundwater. Additionally, over time metal(loid) leaching can occur from cracks, fissures or interstices formed in the cement matrix. These defects can be caused by wind, erosion, wetting and drying cycles, hot and cold cycles, rain and snow and/or other environmental elements.

2.3.2 Design

Bioprecipitation can be applied as a remediation solution in a variety of methods. The processes are classified as *in-situ* or *ex-situ* based on where the operation takes place. If clean-up occurs at the site, the operation is considered an *in-situ* process. However, if soil is extracted from the site and transported for treatment, the operation is an *ex-situ* process. In addition to these operation models, bioprecipitation often occurs either simultaneously or in sequence with other treatment methods. This is based on the requirements of the project, i.e., future land usage and the desired level of metal(loid) removal. This chapter will put emphasis on *in-situ* bioremediation strategies that are passive or rely on natural attenuation processes. These strategies are innately more sustainable and therefore, offer benefits as a clean-up protocol.

2.3.2.1 In-situ Bioprecipitation

In-situ bioprecipitation can take place as a natural or engineered process. These operations are typically passive, as they aim to enhance the natural processes that occur. They typically consume minimal energy, have little operation and maintenance requirements and are of low cost. Further, the reduction in transport and the recycle of materials negates environmental impacts developed with active processes (Hengen et al. 2014). However, these processes are harder to control and metal recovery is difficult (Kaksonen and Puhakka 2007, Kiran et al. 2017). Overall, *in-situ* bioprecipitation adheres to the social, economic and environmental demand expected by an eco-friendly, sustainable remediation strategy. Table 2.4 provides information on various case-studies demonstrating the efficacy of *in-situ* methods.

In-situ bioprecipitation can be achieved via injection wells. There are two ways to facilitate remediation with injection wells. In the first, the wells are constructed to provide the reactants to the contaminated zone. A mixture (including electron donors and microorganisms) is injected into the well, which follows the groundwater flow path with the aim to precipitate and immobilize the contaminant (Vanbroekhoven et al. 2008). The concentration of the electron donor during injection should be tested since a temporarily high carbon content at the well may produce methane without metal(loid) precipitation (Diels et al. 2005). The second, uses the pull-push-pull principle, in which groundwater is extracted from the well, is mixed with additives and then is reinjected into the same well (Janssen and Temminghoff 2004). In a pilot test using the latter method, BSR was achieved whereby zinc concentrations were significantly reduced (40 mg/L to < 0.001 mg/L), however more interesting was the longevity of BSR 5 wk post operation (Janssen and Temminghoff 2004). The long-term stability of the metal-sulfide precipitates can be of concern (Miao et al. 2012), requiring continuous monitoring.

Bioprecipitation can also be facilitated via permeable reactive barriers. A reactive barrier is implemented in soil to cut across the groundwater flow. The barrier is implemented downstream of a

Table 2.4. *In-situ* bioprecipitation case-studies.

Method	Location	Project Specifications	Method Efficacy	References
Reactive Barrier	Curilo mine district, Sophia	Microorganism: SRB Electron Donor: Leaves, compost, zero-valent iron, silica sand, perlite, limestone	25% SO_4^{2-} 6% Cd *Percentages are higher with sorption consideration	Pagnanelli et al. 2009
	Synthetic groundwater using contaminated sediment from Belgium	Microorganism: SRB Electron Donor: Zero-valent iron	47% As	Kumar et al. 2016
	Nickel Rim tailings impoundment	Microorganism: SRB Electron Donor: Compost, leaf mulch, wood chips	74% SO_4^{2-} >85% Fe	Benner et al. 1999
	Unknown	Microorganism: SRB Electron Donor: Composted leaf mulch, wood chips, sawdust, sewage sludge	98% SO_4^{2-} 26.67-99.99% Fe 75–99.17% Zn 98.75–99.92% Ni	Waybrant et al. 2002
Wetlands	Camborne, Cornwall	Microorganism: SRB Electron Donor: Sodium acetate, propionic acid, glycerol	3.1 and 4.0 $\mu mol \cdot l^{-1} \cdot h^{-1}$ Fe 1.31 and 2.44 $\mu mol \cdot l^{-1} \cdot h^{-1}$ Zn	Webb et al. 1998
Injection Wells	The Netherlands	Microorganism: SRB Electron Donor: Molasses	99.98% Zn	Janssen and Temminghoff 2004
	Laboratory tests for Umicore sites	Microorganism: SRB Electron Donor: Lactate, cheese whey, soy oil	> 99% Zn > 99% Co > 85% SO_4^{2-}	Vanbroekhoven et al. 2008
	Metal processing factory in Maasmechelen, Belgium	Microorganism: SRB Electron Donor: Lactate, glycerol, vegetable oil	96%–97% Zn	Lookman et al. 2013

metal(loid) contaminated plume, whereby the reactive barrier is designed to degrade or immobilize the contaminant via BSR. The two primary configurations of a reactive barrier are continuous (vertical barrier, perpendicular to the contaminant plume) and funnel-and-gate system (V-shaped funnel directing contaminant plume through the vertical reactive gate) (Sharma and Reddy 2004). The reactive barrier is designed for the specific site, such that reactive material (electron donor and microorganism consortium) is selected based on the desired type of bioprecipitation. Acid Mine Drainage (AMD) for example, which is highly acidic and heavily contaminated with sulfuric acid (SO_4^{2-} and H^+) and other heavy metals, could use a layered mixture of silica sand, organic waste and silica sand either in a horizontal or vertical sequence to decrease effluent AMD (Benner et al. 1999, Waybrant et al. 2002). Clogging of reactive material due to bioprecipitation may decrease the efficacy of the barrier over time (Kiran et al. 2017).

Wetlands and engineered wetlands aim to remove metal(loid) contaminants from water with degradation and bioprecipitation techniques. They can be in the form of aerobic wetlands, anaerobic wetlands (Johnson and Santos 2020) or anoxic ponds (Kiran et al. 2017). There are two main impacts to the redox potential of these designs: the hydraulic design and the mode of operation. The hydraulic designs can use a vertical flow (aerobic) treatment, horizontal subsurface flow (anoxic) treatment, subsurface flow treatment (anoxic) (Faulwetter et al. 2009) or a surface flow (aerobic) treatment (Kosolapov et al. 2004). The mode of operation refers to the feed mode (batch feed, intermittent flow feed or continuous flow feed), hydraulic load rate and Hydraulic Retention Time (HRT) (Faulwetter et al. 2009). The use of plant species to help mitigate pollution via the influence of the redox condition is specific to wetlands (Faulwetter et al. 2009) and phytoremediation (Stephen and Macnaughtont 1999). Phytoremediation can mitigate soil and groundwater pollution via five primary mechanisms: phytostabilization, phytoextraction, phytovolatilization, rhizodegradation and phytotransformation. Rhizodegradation, specifically, is applicable during wetland treatment, as the plant-soil-microorganism system in the rhizosphere can degrade contaminants (Shmaefsky 2020).

Biofilters offer an additional method to remediate soil and groundwater via *in-situ* bioprecipitation. The method consists of a filter media that allows microorganisms to attach and multiply. The biofilter essentially acts as a surface for microorganism immobilization leading to the development of a biofilm, which enables redox reactions necessary for bioprecipitation. There are numerous organic and inorganic filter materials that can be used to facilitate biofiltration. In keeping with the sustainable assessment of clean-up strategies, compost can be used as a viable, environmental filter media. The compost media has nutrients for microorganism growth, good water retention capacity for microorganism metabolism and good permeability for homogenous distribution that are beneficial to the biofilter process (Pachaiappan et al. 2022).

MICP is typically applied to the field via injection, surface percolation or pre-mixing. The latter two applications are *in-situ* treatment methods and therefore preferential. An issue experienced with MICP remediation is the uniform distribution of the mixture (cementation solution and microorganism) to adequately precipitate a $CaCO_3$ matrix. An injection can lead to clogging around the injection site, which inhibits its performance as a clean-up strategy. By reducing the injection rate, the mixture may increase its reach. Surface percolation, however, has a better uniform distribution. The mixture percolates through the soil via gravity. The depth of the required remediation should be previously assessed as the permeability of soil impacts the depth reached by percolation (Mujah et al. 2016).

Sorption and bioprecipitation are remediation mechanisms that typically occur simultaneously. Sorption includes adsorption (accumulation to surfaces) or absorption (penetration into substances). As mentioned earlier, sorption is part of the S/S technique aiming to immobilize metal(loid) contaminants *in-situ* (LaGrega et al. 1994). Once contaminants precipitate from groundwater or pore space, adsorption to soil particle surfaces is desirable. Again, considering sustainable remediation, the use of an organic matter energy source during bioprecipitation can enhance the sorption capacity. The carboxylic group of olive pomace, compost and leaves can facilitate sorption (Pagnanelli et al. 2009), improving immobilization.

These *in-situ* remediation methods offer a greener solution for industrial clean-up of soil and groundwater. Although some greenhouse gas emissions and energy will be consumed to either drill the injection wells or implement the reactive barrier, these strategies offer better sustainability to *ex-situ* methods. However, it is important to note that these strategies are not viable for all types of contamination. Often more intensive methods are required to achieve high degrees of metal(loid) contaminant removal required by environmental and government agencies.

2.3.2.2 Ex-situ Bioprecipitation

Ex-situ bioprecipitation uses engineered structures (i.e., bioreactors) to treat soil. The operation is considered an active treatment as it requires continuous chemical and/or biological additives to facilitate biological precipitation. The process is highly demanding, requiring constant labor requirements and monitoring. The operation has high capital and maintenance costs, where contaminated soil is excavated, transported and treated using reactor design. The operation is less sustainable than *in-situ* methods, however, it should be noted that the process is easily controlled with better predictability and metal(loid) recovery (Kaksonen and Puhakka 2007, Kiran et al. 2017).

An overview of reactor designs for *ex-situ* bioprecipitation is described in Table 2.5. In addition to the design itself, a reactor can operate under different modes or various stages. The operation mode refers to how the influent is added to the reactor, i.e., batch, continuous or semi-continuous application. The stages refer to the sequence of operations. In a one-stage process bioprecipitation occurs in one reactor, while a two-stage process will separate mechanisms (i.e., oxidative-reductive processes and precipitation) to occur in different reactors (Sánchez-Andrea et al. 2014).

2.3.3 Strategy Implementation

The implementation of a remediation strategy is developed based on the robust site assessment data. Samples of polluted soil, sediment and/or water are collected from the affected site to analyze the contaminant(s) and its potential hazard after release. Site characteristics (i.e., location, geology, weather, topography, etc.) are all documented or measured to evaluate the extent of potential contamination over time and brainstorm viable remediation options. The land use requirements are considered to establish project objects and outline environmental and regulatory standards. The desired future land use post contamination is evaluated and predicted to best suit the needs of nearby communities. A contaminated site should always be left in a state equal to or better than it was before contamination. Remediation strategies that adhere to the project objectives and site assessment are discussed, addressing the operation efficacy for each scenario, public acceptance and its green and sustainable nature. An optimal strategy is selected based on feasibility (i.e., cost, operation and maintenance and time to execute). If a solution is deemed unfeasible, alternative remediation solutions are established. After iterations of the process, a remediation strategy is selected and implemented at the site. Figure 2.3 outlines the process and various components that engineers take into consideration.

Laboratory testing is essential to the performance of a selected biological remediation strategy. Testing is required to ensure the microorganism mixture or species, energy source and design all functions to achieve high rates of immobilization, biotransformation or removal of the metal(loid) contaminant. During testing, the C/N, COD/SO$_4^{2-}$ and DOC/SO$_4^{2-}$ ratios can be measured and documented as indicators for performance. Further, laboratory testing can evaluate different modes of operation scenarios (i.e., feed mode, hydraulic load rate, HRT, etc.) to establish an optimal efficacy. Laboratory testing at initial stages of brainstorming can have beneficial impacts to the design, impacting its timeframe and cost. This can therefore affect the sustainability of a selected strategy.

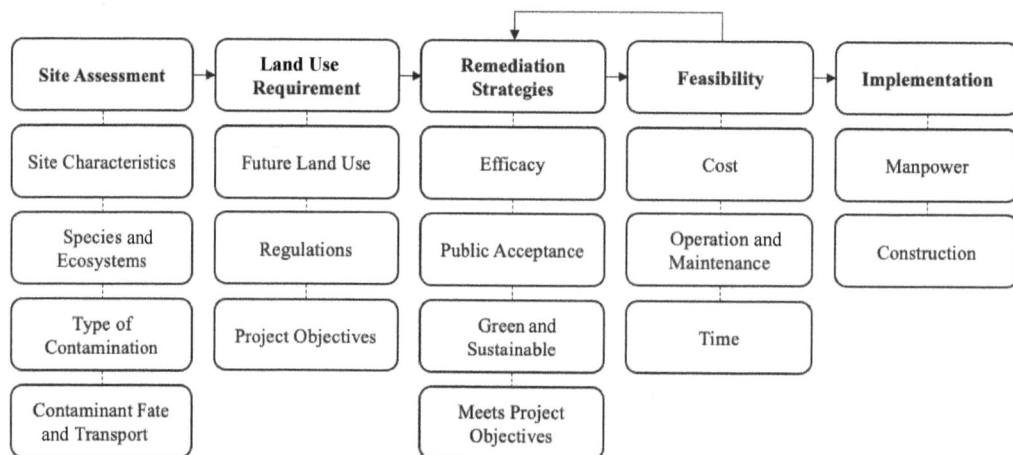

Figure 2.3. Selection process for clean-up strategy incorporating site assessment, land use requirements, feasibility and implementation (Adapted from Wilcox et al. 2022).

Table 2.5. *Ex-situ* reactor designs for bioprecipitation (Adapted from Kaksonen and Puhakka 2007, Kiran et al. 2017, Speece 1983).

Method	Design	Advantages & Disadvantages
Continuously Stirred Tank Reactor (CSTR)		Advantages: - High performance - Reliable Disadvantages: - Poor biomass retention
Anaerobic Contact Press (ACP)		Advantages: - Good biomass retention Disadvantages: - Breakdown of flocs and sludge due to high shear force
Upflow Anaerobic Filter Reactor (UAFR)		Advantages: - Good sludge retention - Low shear force Disadvantages: - Channeling due to shear force - Large pressure gradients
Downflow Anaerobic Filter Reactor (DAFR)		Advantages: - Utilizes gravitation force - Good sludge retention - Low shear force Disadvantages: - Channeling due to shear force - Large pressure gradients
Fluidized-Bed Reactor (FBR)		Advantages: - Large surface area for biofilm - Good biomass retention - Small pressure gradients - No channeling or clogging - Large mass transfer rates Disadvantages: - Biomass loss due to shear force - Energy requirement

Table 2.5 contd. ...

...Table 2.5 contd.

Method	Design	Advantages & Disadvantages
Downflow Fluidized-Bed Reactor (DFBR)		Advantages: - Solids recovery from reactor bottom - Large surface area for biofilm - Good biomass retention - Small pressure gradients - No channeling or clogging - Large mass transfer rates Disadvantages: - Biomass loss due to shear force - Energy requirement
Gas Lift Reactor (GLR)		Advantages: - Good mass transfer - Good mixing Disadvantages: - High pressure drop with gas substrate
Upflow Anaerobic Sludge Blanket (UASB)		Advantages: - No channeling, clogging, sludge compaction - No biomass carrier Disadvantages: - Biomass washout
Anaerobic Hybrid Reactor (AHR)		Advantages: - Less clogging - Easy sludge removal - Good biomass retention Disadvantages: - NA

Table 2.5 contd. ...

...Table 2.5 contd.

Method	Design	Advantages & Disadvantages
Anaerobic Baffled Reactor (ABR)		Advantages: - Good sludge retention times - No biomass carrier - High loading rates Disadvantages: - NA
Membrane Bioreactor (MBR)		Advantages: - Good biomass retention Disadvantages: - Fouling

* *Red arrow represents gas, blue arrow represents influent and effluent, and green arrow represents recycle process.*

2.4 Conclusions

Bioprecipitation offers a sustainable approach to soil and groundwater clean-up for metal(loid) contamination. The process offers social, economic and environmental benefits as a sustainable remediation strategy. As a biological treatment method using natural processes, public acceptance tends to be high in comparison to a more invasive treatment, such as physical, chemical, thermal or electrical operations.

While bioprecipitation can occur as an *in-situ* and *ex-situ* remediation technique, *in-situ* treatment offers a more sustainable approach. The operation occurs at the site with minimal soil disturbance, energy consumption and greenhouse gas emissions. The economic value is also high as capital, maintenance and transport costs are relatively low.

Bioprecipitation can occur as an oxidative-reductive reaction or a cementation technique. The desired mechanism of immobilization should be based on the type and degree of contamination with a focus on chemical stability. For oxidative-reductive reactions an energy source should be selected to facilitate the reaction. Organic waste is a viable electron donor, showing promise as a carbon source with high COD/SO_4^{2-} ratios for BSR. Organic waste can be repurposed to a valuable bioprecipitation additive, enhancing the sustainable nature of the operation.

A thorough site assessment should be conducted at each site prior to implementation as a remediation strategy. While bioprecipitation shows promise as a bioremediation strategy for metal(loid) clean-up, laboratory testing is required to better predict its applicability for each individual case . The strategy should be selected only if it meets the project objectives, the future land use requirements and can achieve high removal efficacies that adhere to environmental standards.

References

Abou-Shanab, R. A. I., P. van Berkum and J. S. Angle. 2007. Heavy metal resistance and genotypic analysis of metal resistance genes in gram-positive and gram-negative bacteria present in Ni-rich serpentine soil and in the rhizosphere of *Alyssum murale*. Chemosphere. 68(2): 360–367. https://doi.org/10.1016/j.chemosphere.2006.12.051.

Achal, V. and X. Pan. 2011. Characterization of urease and carbonic anhydrase producing bacteria and their role in calcite precipitation. Current Microbiol. 62(3): 894–902. http://dx.doi.org.qe2a-proxy.mun.ca/10.1007/s00284-010-9801-4.

Achal, V., X. Pan and D. Zhang. 2011. Remediation of copper-contaminated soil by *Kocuria flava* CR1, based on microbially induced calcite precipitation. Ecol. Eng. 37(10): 1601–1605. https://doi.org/10.1016/j.ecoleng.2011.06.008.

Achal, V., A. Mukerjee and M. Sudhakara Reddy. 2013a. Biogenic treatment improves the durability and remediates the cracks of concrete structures. Constr. Build. Mater. 48: 1–5. https://doi.org/10.1016/j.conbuildmat.2013.06.061.

Achal, V., X. Pan, D.-J. Lee, D. Kumari and D. Zhang. 2013b. Remediation of Cr(VI) from chromium slag by biocementation. Chemosphere. 93(7): 1352–1358. https://doi.org/10.1016/j.chemosphere.2013.08.008.

Barbosa, L. P., P. F. Costa, S. M. Bertolino, J. C. C. Silva, R. Guerra-Sá, V. A. Leão and M. C. Teixeira. 2014. Nickel, manganese and copper removal by a mixed consortium of sulfate reducing bacteria at a high COD/sulfate ratio. World J. Microbiol. Biotechnol. 30(8): 2171–2180. https://doi.org/10.1007/s11274-013-1592-x.

Barton, L. L., R. M. Plunkett and B. M. Thomson. 2003. Reduction of metals and nonessential elements by anaerobes. pp. 220–234. *In:* L. G. Ljungdahl, M. W. Adams, L. L. Barton, J. G. Ferry and M. K. Johnson [eds.]. Biochemistry and Physiology of Anaerobic Bacteria. Springer. https://doi.org/10.1007/0-387-22731-8_16.

Barton, L. L., F. A. Tomei-Torres, H. Xu and T. Zocco. 2015. Metabolism of metals and metalloids by the sulfate-reducing bacteria. pp. 57–83. *In:* D. Saffarini [ed.]. Bacteria-Metal Interactions. Springer International Publishing. https://doi.org/10.1007/978-3-319-18570-5_4.

Béchard, G., H. Yamazaki, W. D. Gould and P. Bédard. 1994. Use of cellulosic substrates for the microbial treatment of acid mine drainage. J. Environ. Qual. 23(1): 111–116. https://doi.org/10.2134/jeq1994.00472425002300010017x.

Benner, S. G., D. W. Blowes, W. D. Gould, R. B. Herbert and C. J. Ptacek. 1999. Geochemistry of a permeable reactive barrier for metals and acid mine drainage. Environ. Sci. Technol. 33(16): 2793–2799. https://doi.org/10.1021/es981040u.

Beveridge, T. J. and W. S. Fyfe. 1985. Metal fixation by bacterial cell walls. Can. J. Earth Sci. 22(12): 1893–1898. https://doi.org/10.1139/e85-204.

Bruins, M. R., S. Kapil and F. W. Oehme. 2000. Microbial resistance to metals in the environment. Ecotoxicol. Environ. Saf. 45(3): 198–207. https://doi.org/10.1006/eesa.1999.1860.

Davey, R. J. and J. Garside. 2000. From Molecules to Crystallizers: An Introduction to Crystallization. Oxford University Press.

Davis, M. L. 2010. Water and Wastewater Engineering: Design Principles and Practice. The McGraw-Hill Companies, Inc.

Diels, L., J. Geets, W. Dejonghe, S. van Roy, K. Vanbroekhoven and G. Malina. 2005. Heavy metal immobilization in groundwater by *in situ* bioprecipitation: comments and questions about carbon source use, efficiency and sustainability of the process. Proceedings of the 9th International FKZ/TNO Conference on Contaminated Soil (Consoil). France, 355–360.

Diels, L., J. Geets, W. Dejonghe, S. V. Roy, K. Vanbroekhoven, A. Szewczyk and G. Malina. 2006. Heavy metal immobilization in groundwater by in situ bioprecipitation: comments and questions about efficiency and sustainability of the process. Proceedings of the Annual International Conference on Soils, Sediments, Water and Energy. Amherst. 11: 99–112. https://scholarworks.umass.edu/soilsproceedings/vol11/iss1/7.

Faghri, A. and Y. Zhang. 2006. Chapter 2: thermodynamics of multiphase systems. pp. 107–176. *In:* Transport Phenomena in Multiphase Systems. Elsevier Academic Press.

Faulwetter, J. L., V. Gagnon, C. Sundberg, F. Chazarenc, M. D. Burr, J. Brisson, A. K. Camper and O. R. Stein. 2009. Microbial processes influencing performance of treatment wetlands: a review. Ecol. Eng. 35(6): 987–1004. https://doi.org/10.1016/j.ecoleng.2008.12.030.

Gadd, G. M. 2004. Microbial influence on metal mobility and application for bioremediation. Geoderma. 122(2): 109–119. https://doi.org/10.1016/j.geoderma.2004.01.002.

Gibert, O., J. de Pablo, J. L. Cortina and C. Ayora. 2002. Treatment of acid mine drainage by sulphate-reducing bacteria using permeable reactive barriers: a review from laboratory to full-scale experiments. Rev. Environ. Sci. Biotechnol. 1(4): 327–333. https://doi.org/10.1023/A:1023227616422.

Government of Canada. 2017. Compare decontamination technologies—Guidance and Orientation for the Selection of Technologies—Contaminated sites—Pollution and waste management—Environment and natural resources—Canada.ca. https://gost.tpsgc-pwgsc.gc.ca/Techlst.aspx?lang=eng#wb-auto-5.

Hao, O. J., J. M. Chen, L. Huang and R. L. Buglass. 1996. Sulfate-reducing bacteria. Crit. Rev. Environ. Sci. Technol. 26(2): 155–187. https://doi.org/10.1080/10643389609388489.

Hao, T., P. Xiang, H. R. Mackey, K. Chi, H. Lu, H. Chui, M. C. M. van Loosdrecht and G.-H. Chen. 2014. A review of biological sulfate conversions in wastewater treatment. Water Res. 65: 1–21. https://doi.org/10.1016/j.watres.2014.06.043.

Hengen, T. J., M. K. Squillace, A. D. O'Sullivan and J. J. Stone. 2014. Life cycle assessment analysis of active and passive acid mine drainage treatment technologies. Resour. Conserv. Recycl. 86: 160–167. https://doi.org/10.1016/j.resconrec.2014.01.003.

ICF Incorporated. 1995. An introduction to environmental accounting as a business management tool: key concepts and terms (EPA 742-R-95-001). https://www.epa.gov/sites/default/files/2014-01/documents/busmgt.pdf.

ISO. 2006. ISO 14040:2006(en), environmental management—life cycle assessment—principles and framework. International Organization for Standardization. https://www.iso.org/obp/ui/#iso:std:iso:14040:ed-2:v1:en.

Janssen, G. M. C. M. and E. J. M. Temminghoff. 2004. *In situ* metal precipitation in a zinc-contaminated, aerobic sandy aquifer by means of biological sulfate reduction. Environ. Sci. Technol. 38(14): 4002–4011. https://doi.org/10.1021/es030131a.

Johnson, D. B. and A. L. Santos. 2020. Biological removal of sulfurous compounds and metals from inorganic wastewaters. pp. 215–246. *In:* P. Lens [ed.]. Environmental Technologies to Treat Sulfur Pollution: Principles and Engineering (Second Edition). IWA Publishing.

Kaksonen, A. H. and J. A. Puhakka. 2007. Sulfate reduction based bioprocesses for the treatment of acid mine drainage and the recovery of metals. Eng. Life Sci. 7(6): 541–564. https://doi.org/10.1002/elsc.200720216.

Karpiński, P. H. and J. Bałdyga. 2019. Chapter 8: precipitation processes. pp. 216–265. *In:* A. S. Myerson, D. Erdemir and A. Y. Lee. [eds.]. Handbook of Industrial Crystallization (3rd Edition). Cambridge University Press. https://www-cambridge-org.qe2a-proxy.mun.ca/core/books/handbook-of-industrial-crystallization/precipitation-processes/011386C77A45C4AAAFD4DE1B2FE0D609.

Kawaguchi, T. and A. W. Decho. 2002. A laboratory investigation of cyanobacterial extracellular polymeric secretions (EPS) in influencing $CaCO_3$ polymorphism. J. Cryst. Growth. 240(1): 230–235. https://doi.org/10.1016/S0022-0248(02)00918-1.

Kiran, M. G., K. Pakshirajan and G. Das. 2017. An overview of sulfidogenic biological reactors for the simultaneous treatment of sulfate and heavy metal rich wastewater. Chem. Eng. Sci. 158: 606–620. https://doi.org/10.1016/j.ces.2016.11.002.

Kiskira, K., S. Papirio, E. D. van Hullebusch and G. Esposito. 2017. Fe(II)-mediated autotrophic denitrification: a new bioprocess for iron bioprecipitation/biorecovery and simultaneous treatment of nitrate-containing wastewaters. Int. Biodeterior. Biodegrad. 119: 631–648. https://doi.org/10.1016/j.ibiod.2016.09.020.

Kosolapov, D. B., P. Kuschk, M. B. Vainshtein, A. V. Vatsourina, A. Wießner, M. Kästner and R. A. Müller. 2004. Microbial processes of heavy metal removal from carbon-deficient effluents in constructed wetlands. Eng. Life Sci. 4(5): 403–411. https://doi.org/10.1002/elsc.200420048.

Kumar, N., R.-M. Couture, R. Millot, F. Battaglia-Brunet and J. Rose. 2016. Microbial sulfate reduction enhances arsenic mobility downstream of zerovalent-iron-based permeable reactive barrier. Environ. Sci. Technol. 50(14): 7610–7617. https://doi.org/10.1021/acs.est.6b00128.

Kumar, R., M. Nongkhlaw, C. Acharya and S. R. Joshi. 2013. Bacterial community structure from the perspective of the uranium ore deposits of domiasiat in India. Proceedings of the National Academy of Sciences, India Section B: Biological Sciences. 83(4): 485–497. https://doi.org/10.1007/s40011-013-0164-z.

Kuppusamy, S., N. R. Maddela, M. Megharaj and K. Venkateswarlu. 2020. Total Petroleum Hydrocarbons: Environmental Fate, Toxicity, and Remediation. Springer International Publishing. https://doi.org/10.1007/978-3-030-24035-6.

LaGrega, M. D., P. Buckingham and J. Evans. 1994. Hazardous Waste Management (Second Edition). Waveland Press, Inc.

Levett, A., E. J. Gagen, Y. Zhao, P. M. Vasconcelos and G. Southam. 2020. Biocement stabilization of an experimental-scale artificial slope and the reformation of iron-rich crusts. Proceedings of the National Academy of Sciences. 117(31): 18347–18354. https://doi.org/10.1073/pnas.2001740117.

Lewis, A. 2017. Precipitation of heavy metals. pp. 101–120. *In:* E. R. Rene, E. Sahinkaya, A. Lewis and P. N. L. Lens [eds.]. Sustainable Heavy Metal Remediation: Volume 1: Principles and Processes. Springer International Publishing. https://doi.org/10.1007/978-3-319-58622-9_4.

Liamleam, W. and A. P. Annachhatre. 2007. Electron donors for biological sulfate reduction. Biotechnol. Adv. 25(5): 452–463. https://doi.org/10.1016/j.biotechadv.2007.05.002.

Lookman, R., M. Verbeeck, J. Gemoets, S. Van Roy, J. Crynen and B. Lambié. 2013. *In-situ* zinc bioprecipitation by organic substrate injection in a high-flow, poorly reduced aquifer. J. Contam. Hydrol. 150: 25–34. https://doi. org/10.1016/j.jconhyd.2013.03.009.

Lovley, D. R. 1993. Dissimilatory metal reduction. Annu. Rev. Microbiol. 47(1): 263–290. https://doi.org/10.1146/ annurev.mi.47.100193.001403.

Miao, Z., M. L. Brusseau, K. C. Carroll, C. Carreón-Diazconti and B. Johnson. 2012. Sulfate reduction in groundwater: characterization and applications for remediation. Environ. Geochem. Health. 34(4): 539–550. https://doi. org/10.1007/s10653-011-9423-1.

Mihelcic, J. R. and J. B. Zimmerman. 2014. Environmental Engineering: Fundamentals, Sustainability, Design (Second Edition). John Wiley & Sons, Inc.

Mitchell, J. K. and K. Soga. 2005. Chapter 8: soil deposits—their formation, structure, geotechnical properties, and stability. pp. 195–250. *In:* Fundamentals of Soil Behavior (3rd Edition). John Wiley & Sons, Inc.

Mujah, D., M. Shahin and L. Cheng. 2016. State-of-the-art review of biocementation by microbially induced calcite precipitation (MICP) for soil stabilization. Geomicrobiol. 34: 524–537. https://doi.org/10.1080/01490451.20 16.1225866.

Mwandira, W., K. Nakashima and S. Kawasaki. 2022. Chapter 13: stabilization/solidification of mining waste via biocementation. pp. 201–209. *In:* D. C. W. Tsang and L. Wang [eds.]. Low Carbon Stabilization and Solidification of Hazardous Wastes. Elsevier. https://doi.org/10.1016/B978-0-12-824004-5.00014-1.

Neculita, C. M. and G. J. Zagury. 2008. Biological treatment of highly contaminated acid mine drainage in batch reactors: long-term treatment and reactive mixture characterization. J. Hazard Mater. 157(2): 358–366. https:// doi.org/10.1016/j.jhazmat.2008.01.002.

Nielsen, A. E. 1964. Kinetics of Precipitation. Pergamon Press.

Ossai, I. C., A. Ahmed, A. Hassan and F. S. Hamid. 2020. Remediation of soil and water contaminated with petroleum hydrocarbon: a review. Environ. Technol. Innov. 17: 100526. https://doi.org/10.1016/j.eti.2019.100526.

Pachaiappan, R., L. Cornejo-Ponce, R. Rajendran, K. Manavalan, V. Femilaa Rajan and F. Awad. 2022. A review on biofiltration techniques: recent advancements in the removal of volatile organic compounds and heavy metals in the treatment of polluted water. Bioeng. 13(4): 8432–8477. https://doi.org/10.1080/21655979.2022.2050538.

Pagnanelli, F., C. C. Viggi, S. Mainelli and L. Toro. 2009. Assessment of solid reactive mixtures for the development of biological permeable reactive barriers. J. Hazard. Mater. 170(2): 998–1005. https://doi.org/10.1016/j. jhazmat.2009.05.081.

Reinertsen, S. A., L. F. Elliott, V. L. Cochran and G. S. Campbell. 1984. Role of available carbon and nitrogen in determining the rate of wheat straw decomposition. Soil Biol. Biochem. 16(5): 459–464. https://doi. org/10.1016/0038-0717(84)90052-X.

Sahinkaya, E., D. Uçar and A. H. Kaksonen. 2017. Bioprecipitation of metals and metalloids. pp. 199–231. *In:* E. R. Rene, E. Sahinkaya, A. Lewis and P. N. L. Lens [eds.]. Sustainable Heavy Metal Remediation: Volume 1: Principles and Processes. Springer International Publishing. https://doi.org/10.1007/978-3-319-58622-9_7.

Sánchez-Andrea, I., J. L. Sanz, M. F. M. Bijmans and A. J. M. Stams. 2014. Sulfate reduction at low pH to remediate acid mine drainage. J. Hazard. Mater. 269: 98–109. https://doi.org/10.1016/j.jhazmat.2013.12.032.

Sánchez-Andrea, I., A. J. M. Stams, J. Weijma, P. Gonzalez Contreras, H. Dijkman, R. A. Rozendal and D. B. Johnson. 2016. A case in support of implementing innovative bio-processes in the metal mining industry. FEMS Microbiol. Lett. 363(11): 1–4. https://doi.org/10.1093/femsle/fnw106.

Sharma, H. and K. Reddy. 2004. Geoenvironmental Engineering: Site Remediation, Waste Containment and Emerging Waste Management Technologies. John Wiley & Sons.

Shmaefsky, B. R. 2020. Principles of phytoremediation. pp. 1–26. *In:* B. R. Shmaefsky [ed.]. Phytoremediation: *In-situ* Applications. Springer International Publishing. https://doi.org/10.1007/978-3-030-00099-8_1.

Speece, R. E. 1983. Anaerobic biotechnology for industrial wastewater treatment. Environ. Sci. Technol. 17(9): 416–427.

Stephen, J. R. and S. J. Macnaughtont. 1999. Developments in terrestrial bacterial remediation of metals. Curr. Opin. Biotechnol. 10(3): 230–233. https://doi.org/10.1016/S0958-1669(99)80040-8.

UNDP. 2022. Sustainable development goals: United Nations Development Programme. UNDP. https://www.undp. org/sustainable-development-goals.

Vanbroekhoven, K., S. Van Roy, L. Diels, J. Gemoets, P. Verkaeren, L. Zeuwts, K. Feyaerts and F. van den Broeck. 2008. Sustainable approach for the immobilization of metals in the saturated zone: *in situ* bioprecipitation. Hydrometall. 94(1–4): 110–115. https://doi.org/10.1016/j.hydromet.2008.05.048.

Waybrant, K. R., C. J. Ptacek and D. W. Blowes. 2002. Treatment of mine drainage using permeable reactive barriers: column experiments. Environ. Sci. Technol. 36(6): 1349–1356. https://doi.org/10.1021/es010751g.

Webb, J. S., S. McGinness and H. M. Lappin-Scott. 1998. Metal removal by sulphate-reducing bacteria from natural and constructed wetlands. J. Appl. Microbiol. 84(2): 240–248. https://doi.org/10.1046/j.1365-2672.1998.00337.x.

Wilcox, S. M., C. N. Mulligan and C. M. Neculita. 2022. Bioprecipitation as a remediation technique for metal(loid) contamination from mining activities. *In:* A. Malik, M. K. Kidwai and V. K. Garg [eds.]. Bioremediation of Toxic Metal(loid)s. CRC Press, Taylor and Francis group.

Willis, G. and E. R. Donati. 2017. Heavy metal bioprecipitation: use of sulfate-reducing microorganisms. pp. 114–130. *In:* E. R. Donati [ed.]. Heavy Metals in the Environment: Microorganisms and Bioremediation. CRC Press. https://doi.org/10.1201/b22013.

Xiangliang, P. 2009. Micrologically induced carbonate precipitation as a promising way to *in situ* immobilize heavy metals in groundwater and sediment. Res. J. Chem. Environ. 13(4): 3–4.

Yong, R. N., C. N. Mulligan and M. Fukue. 2014. Sustainable Practices in Geoenvironmental Engineering (2nd ed.). CRC Press. https://doi.org/10.1201/b17443.

Zagury, G. J., V. I. Kulnieks and C. M. Neculita. 2006. Characterization and reactivity assessment of organic substrates for sulphate-reducing bacteria in acid mine drainage treatment. Chemosphere. 64(6): 944–954. https://doi.org/10.1016/j.chemosphere.2006.01.001.

CHAPTER 3

Bio Adsorption
An Eco-friendly Alternative for Industrial Effluents Treatment

Andrea Saralegui,[1,*] *M. Natalia Piol,*[1]
Victoria Willson,[1] *Néstor Caracciolo,*[2] *Silvia Ramos*[3]
and *Susana Boeykens*[1]

3.1 Introduction

Industrial effluents are complex systems and therefore their treatment requires an integral approach. The composition and characteristics of each type of effluent, whether solid or liquid, determine the treatment to be required and its final disposal. In order to find a feasible solution for each specific case, it is necessary to recognize the system and look for alternatives that reduce costs.

Both waste and pollutants can be considered as misplaced or unexploited resources. That is to say, the lack of a process's holistic view means that an incorrect disposal of a material turns it into a waste or even a pollutant. The reuse of this material not only reduces the degree of contamination, but also reduces the operational costs of its disposal and/or treatment. Furthermore, it turns it into a valuable resource that could even have some added value in other activities. Thus, what was earlier called "waste" could be seen as a new raw material in a recycling and reuse process, becoming a loop within the circular economy (Saralegui et al. 2022). The study of these new sources of raw materials is a way forward to improve the sustainability of water and effluent treatment technologies.

Environmental contamination with metals can negatively affect ecosystems and the exposed population health; this contamination is mainly due to technological changes and the increasing use of metal-containing materials in industry. This has led to concern and has led to numerous investigations on treatment technologies for water, effluents and contaminated soils (Branzini and Zubillaga 2012, Cartaya et al. 2011, EPA 2000, Volke Sepúlveda and Velasco Trejo 2002).

[1] Universidad de Buenos Aires, Facultad de Ingeniería, Instituto de Química Aplicada a la Ingeniería (IQAI), Laboratorio de Química de Sistemas Heterogéneos (LaQuíSiHe). Av. Paseo Colón 850, C1063ACV, Buenos Aires, Argentina.
[2] Universidad de Buenos Aires, Facultad de Ingeniería, Instituto de Química Aplicada a la Ingeniería (IQAI), Laboratorio de Química Ambiental (LaQuíAmb). Av. Paseo Colón 850, C1063ACV, Buenos Aires, Argentina.
[3] Universidad de Buenos Aires, Facultad de Ingeniería, Grupo Modelos Aplicados a Gestión Industrial, Dpto. de Gestión-Cátedras de Modelos y Optimización I y III. Av. Las Heras 2214, C1126, Buenos Aires, Argentina.
* Corresponding author: laquisihe@fi.uba.ar

Aquatic systems are large bodies that receive many waste streams, mainly resulting from human activities (Acumar 2019, Liu et al. 2008). The diversity of industries that may discharge their effluents into a water body influences its degree and type of pollution. The assessment of the interferences that may exist between various types of pollutants present simultaneously in a water body should keep being studied as they may produce different effects on the ecosystem (Renaud et al. 2021). Prevention is the way forward. Avoiding the discharge of effluents with high pollutant content is necessary and finding systems that could selectively remove or decrease a specific pollutant concentration in the presence of other substances is very promising regarding their recovery and reuse.

Due to their high persistence in the environment, metals have the potential to get bioaccumulated in organisms of lower trophic levels and reach their biomagnification, with a direct implication on the ecosystem and on the health of populations not directly exposed to contaminated water (Di Giulio and Newman 2008). For these reasons, the Argentinian legislation considers them as hazardous substances and regulates their content both in effluent discharges and in drinking and irrigation waters (PLNRA 1992). However, this legislation as well as other water quality regulations and guidelines in the world do not consider the effect of the simultaneous presence of different metals (Renaud et al. 2021).

The search for low-cost processes for water treatment is fundamental to its applicability; waste reuse is a possible path to follow. When thinking about the low cost of the process, this implies that the waste should be sourced close to where it will be used in order to reduce transport expenses as well as the carbon footprint of the process. It also implies making just a basic conditioning to the waste, that is, treating the material as little as possible. Finally, the process must be adapted to the specific effluent conditions, so as not to require adjustments involving the use of large quantities of chemical reagents or energy. There are several established conventional processes for pollutants treatment and recovery from wastewater (Mihelcic and Zimmerman 2012). However, in recent years, studies on biosorption technologies development have increased due to their potential economic convenience. Numerous works using lignocellulosic materials indicate a high capacity to concentrate water pollutants in their structures (Boeykens et al. 2018, Boeykens et al. 2019, Piol et al. 2021, Saralegui et al. 2021, Saralegui et al. 2022). For this work, wastes with high availability in Argentina were selected: peanut shells (*Arachis hypogaea*), sugar cane bagasse (*Saccharum officinarum*), avocado stones (*Persea americana*), pecan nut shells (*Carya illinoinensis*), wheat bran (*Triticum aestivum*), banana shells (*Mussa paradisiaca*) and different parts of the moringa plant (*Moringa oleifera*) for the study of metal removal from water.

Selected wastes:

- **Peanut shells:** Argentina is the sixth-largest producer of peanuts in the world, accounting for 3% of global production, with 42.6 million tons produced during the 2016–2017 growing season (Pellegrino 2019). **Peanut shells**, the waste from the industrialization of the nuts, are often a drawback as they represent between a quarter and a fifth of the harvest and constitute a polluting waste which is usually incinerated in the open air, generating large amounts of smoke (BCC 2014).

- **Sugar cane bagasse:** another waste product is **sugar cane bagasse**, which is the fibrous material that remains after the sugar cane is crushed. Typically, for every 10 tons of crushed cane, 3 to 4 tons of wet bagasse (40–50% moisture) remains as residue. Increasingly, it is used as fuel for steam and electricity generation in sugar mills, but also as a source of organic matter when returned to the soil. According to statistics, in the last 10 yr' harvest, 18 to 20 million tons of sugar cane were milled in the country, producing a large volume of bagasse for disposal (Preciado Patiño 2015, Rios et al. 2017).

- **Avocado stone:** the **avocado stone** represents about 18% of the fruit weight. In Argentina it is currently a growing crop as different soils and climates are favorable for its growth and development (Carrere 2010).

- **Pecan nuts:** these agronomic conditions also promote the production of high quality, counter-seasonal **pecan nuts**. Commercial pecan cultivation in Argentina has grown exponentially in recent years, which in the medium term will make Argentina one of the world's top three producers of this nut and the world's leading exporter of high value-added pecan-based products (Frusso 2013).

- **Wheat:** On the other hand, **wheat** is one of the most widely grown cereals in the world. In Argentina, around 15.5 million tons of wheat are produced annually. Seventy-five per cent is exported and approximately 4.6 million tons are milled to produce 3.4 million tons of flour. As by-products of this process, approximately 500,000 tons of bran and germ are extracted annually and used almost entirely for animal feed (Apro et al. 2004).

- **Banana: Banana** cultivation is the most prominent of Argentina's tropical crops, with an annual production ranging from 180,000 to 205,000 tons, with bananas being the most consumed fruit in Argentina with an annual per capita average of 12 kg (Molina et al. 2015).

- **Moringa oleifera:** *Moringa oleifera*, a shrubby plant native to India, is little known in South America, although it is now abundant throughout the tropics. It is a hardy species that requires little horticultural attention and grows rapidly, recognized for its nutritional and medicinal characteristics, as well as being used in some water purification processes (Folkard and Sutherland 1996). Different parts of the plant are usable raw materials, the leaves, the undeveloped pods and seeds. Obtaining products and by-products generates around 80% by crop weight of waste, the husks during dehulling the seed or the green stalks are discarded during the obtaining of the leaf for infusions use. Currently, in Argentina there is a growing wave of *Moringa* cultivation with a large generation of new products that in the short term will result in a large amount of waste to be disposed of.

The waste from the agri-food industry proposed in this work will be used as biosorbents to study the retention of metal ions from aqueous effluents. In the case of liquid effluents, the aim is to reach the polluting discharge levels required by local legislation, considering industries such as tanning or electroplating. In the case of solid effluents, the goal is to achieve eco-friendly final disposal.

The design of the Reactor App software were intended to simplify the calculations and to give as a result the volume of the continuous reactor to be used in a first approximation from data obtained with few tests carried out on batch reactors. In this way, a first scale-up can be carried out in a simplified way. As a second objective, Reactor App aims to facilitate the processing of the data derived from the continuous tests.

3.2 Initial Considerations

The total adsorption capacity (mg adsorbate/g adsorbent) was calculated by Eq. 3.1 and the removal percentage (%R) by Eq. 3.2 (Boeykens et al. 2017):

$$q_0 = \frac{(C_0 - C_f)}{m} V \qquad \text{Eq. 3.1}$$

$$\%R = \frac{(C_0 - C_f)}{C_0} \qquad \text{Eq. 3.2}$$

where, q_a (mg g^{-1}) is the adsorption capacity, C_0 (mg L^{-1}) is the P or Cr initial concentration, C_f is the P or Cr final concentration (after adsorption), m (g) is the adsorbent mass and V (L) is the volume of adsorption solution.

3.3 Dosage Curves

Once the adsorbent has been selected and treated only by washing with water, drying and sieving (Figure 3.1), the working conditions should be selected, for which the so-called dosage curves are carried out to examine the adsorbent performance for the removal of each pollutant. This means performing two kinds of experiments: (i) varying the adsorbent mass in contact with a specified volume of a given concentration solution, and (ii) varying the concentration of a given solution volume with a fixed adsorbent mass. To ensure that equilibrium is reached, both types of experiments should be performed with the proper agitation, at a fixed pH and temperature, for a determined amount of time.

On the basis of these experiments, working conditions can be selected to allow the necessary kinetic and equilibrium studies to be carried out.

As an example of this procedure, the results found using peanut shells as adsorbent, and nickel as contaminant are presented.

Figure 3.1. Peanut shells before and after washing, drying and sieving.

These curves were developed at pH = 6, corresponding to the pH of the effluents to be treated, with a volume of 50 mL of solution and continuous stirring for 24 hr, verifying that equilibrium was reached after that time and that the contaminant was detectable in the remaining solution. The range of concentrations that can be measured in the atomic absorption spectroscopy equipment for nickel was in 0.1 to 8 mg L^{-1} so the concentrations used for the dosing test were between 10 and 50 mg L^{-1}. The adsorbent mass range was 0.30 to 1.0 g established based of the initial 24 hr experiment.

When the weight of the adsorbent increases, the percentage of nickel removal increases, but the adsorption capacity decreases as it can be observed in Figure 3.2a. Secondly, when the weight of the adsorbent is fixed and the nickel concentration increases, the removal percentage decreases, but the adsorption capacity increases as can be observed in Figure 3.2b. The maximum adsorption capacity was 30 mg $Ni^{2+} L^{-1}$.

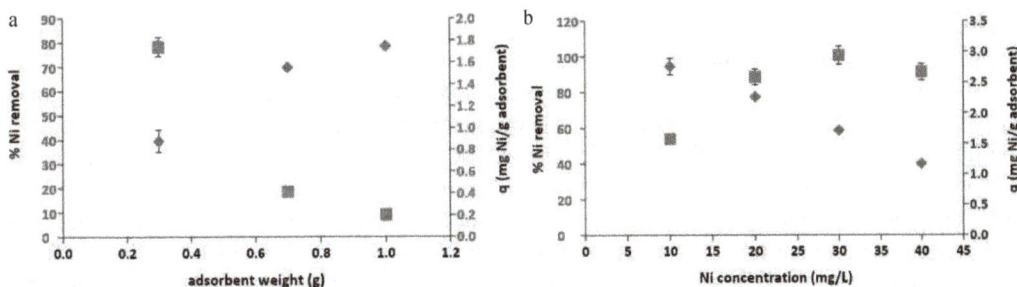

Figure 3.2. Peanut shell dosage curves. a. As a function of adsorbent mass. b. As a function of initial nickel solution concentration.

The compromise relation selected was 0.3 g of adsorbent and 20 mg $Ni^{2+} L^{-1}$ as the initial concentration. For this ratio the percentage of removal obtained was close to 80% and the adsorption capacity close to the maximum. This compromise relation also allows measurable results to be obtained within the concentration range suitable for the measuring instrument.

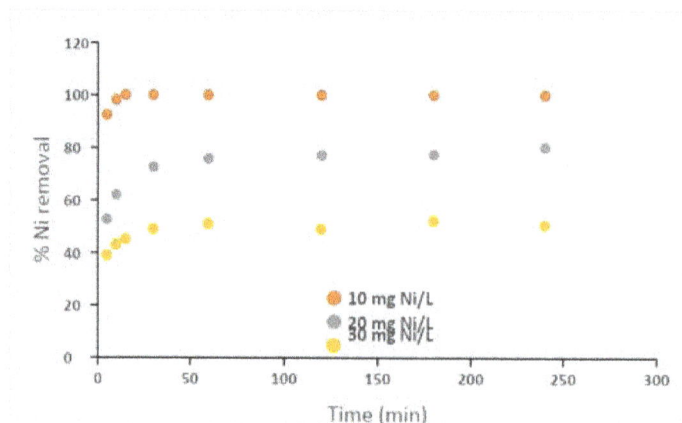

Figure 3.3. Kinetic curves obtained from the adsorption of nickel on peanut shells.

Once the ratio between the working concentration and the mass of the adsorbent to be used was established, the kinetic tests begin with the objective of obtaining the kinetic constant and the reaction order by applying different models.

As an example of the kinetic studies, the kinetic curves for 10, 20 and 30 mg Ni L^{-1}, using 0.3 g of adsorbent, are presented in Figure 3.3. The equilibrium time for all systems was 30 min.

By processing these studies one can determine the order of the reaction and the kinetic constant. For this, it is necessary to use pseudo-first order and pseudo-second order models. The mathematical expression for the pseudo-first order kinetics model (Lagergren 1898) is widely used for adsorption studies of liquids. The linear form is Eq. 3.3:

$$\ln(q_e - q_t) = \ln q_e - k_1 t \qquad \text{Eq. 3.3}$$

where q_t (mmol g^{-1}) is the amount of adsorbed solute at time t, and k (min^{-1}) is the constant velocity.

The pseudo-second order model was developed by Ho and McKay (1999). This model assumes that adsorption follows a second-rate kinetic mechanism. It indicates that one sorbate molecule is adsorbed on two active sites on the sorbent surface. The process can be expressed in a linear form by the Eq. 3.4:

$$\frac{t}{q_t} = \frac{1}{k_2 q_e^2} + \frac{t}{q_e} t \qquad \text{Eq. 3.4}$$

where, k_2 (mmol g^{-1} min^{-1}) is the velocity constant of pseudo-second order. The term $(k_2 q_e^2)$ represents the initial adsorption rate.

In the case of peanut shells, the best fit was obtained for the pseudo-second order model with regression coefficients higher than 0.998, while for the pseudo-first order fit the regression coefficient in all cases was less than 0.92. In general, the reaction order for lignocellulosic materials for metal removal responds to reaction of order 2. Several authors have demonstrated that the metal ions adsorption process in lignocellulosic materials occurs by complexation, it occurs where the carboxylic or carboxylate (R-COOH/R-COO-) and phenolic or phenolate (R'-OH / R'-O-) functional groups simultaneously chelate the metal ions (Mn$^+$), forming bidentate complexes R-COO-Me-O-R' (Neris et al. 2019, Guo et al. 2008, Piol et al. 2021, Boeykens et al. 2018).

Regarding the kinetic constant obtained, values between 600 and 15 g mmol^{-1} were obtained. The cyclic of these parameters (k and reaction order) is necessary to enter the Reactor App software.

3.4 Equilibrium Isotherms

Another extremely important data to work with the software and to be able to estimate the reactor's design dimensions is the q_{max} (maximum adsorption capacity that comes from the study of equilibrium isotherms).

For a given solid-liquid system, the dynamics of the adsorption equilibrium of the solute on the solid is expressed by the ratio between the amount adsorbed at equilibrium per unit mass of original adsorbent q_e (mmol g^{-1}) and the residual concentration at equilibrium in the liquid phase C_e (mmol L^{-1}), at a constant temperature. The graphical representation of these parameters is known as adsorption isotherm. The study of the isotherms serves to obtain parameters that characterize the adsorption phenomenon for a given adsorbent-adsorbate pair and to be able to establish the experimental conditions that maximize adsorption. In addition, the mathematical correlation of the process constitutes an important role for the analysis of the assumptions implied by the models and provide an idea of the adsorption mechanism, the surface properties, as well as the degree of affinity of the adsorbent (Foo and Hameed 2010).

The Langmuir (1918) and Freundlich (1906) models happen to be the most widely used isotherm models mentioned in literature. In Table 3.1 a comparison of the characteristic parameters and assumptions between each model is shown.

As an example, the nonlinear fit of the experimental data and the obtained parameters to the Langmuir and Freundlich adsorption isotherms for a biomass system of the aquatic macrophyte *Azolla* for the adsorption of Cu from an aqueous solution can be represented (Figure 3.4). The R^2 value closer to 1 is indicating that the experimental data fits better to the Langmuir model for the studied system, from which the model parameters are obtained (Figure 3.5).

For the proposed system, the model that best fits the data happens to be Langmuir. This model is the one that generally results best for lignocellulosic adsorbents as observed in other works (Guo et al. 2008, Boeykens et al. 2018).

Table 3.1. Comparison between Langmuir and Freundlich model with their respective parameters and assumptions.

Model	Langmuir 1918	Freundlich 1906
Mathematical expression	$q_e = \dfrac{q_{max} K_L C_e}{1 + K_L C_e}$	$q_e = K_f C_e^{1/n}$
Parameters	**Graph q_e vs C_e** • K_L (L mmol^{-1}) = constant. Langmuir, related to adsorption intensity • $q_{máx}$ (mmol g^{-1}) = maximum adsorption capacity	**Graph q_e vs C_e** • K_f (mmol g^{-1}) constant. Freundlich, related to the adsorption capacity • n = related to the intensity of adsorption or heterogeneity of the system • n > 1 → favorable adsorption • n < 1 → unfavorable adsorption
Assumptions	• Maximum adsorption in monolayer • Adsorption sites with constant energy • There is no lateral interaction between neighboring molecules	• Adsorption in multilayers • Adsorption sites with different affinities • Adsorption energy varies exponentially depending on the covered surface

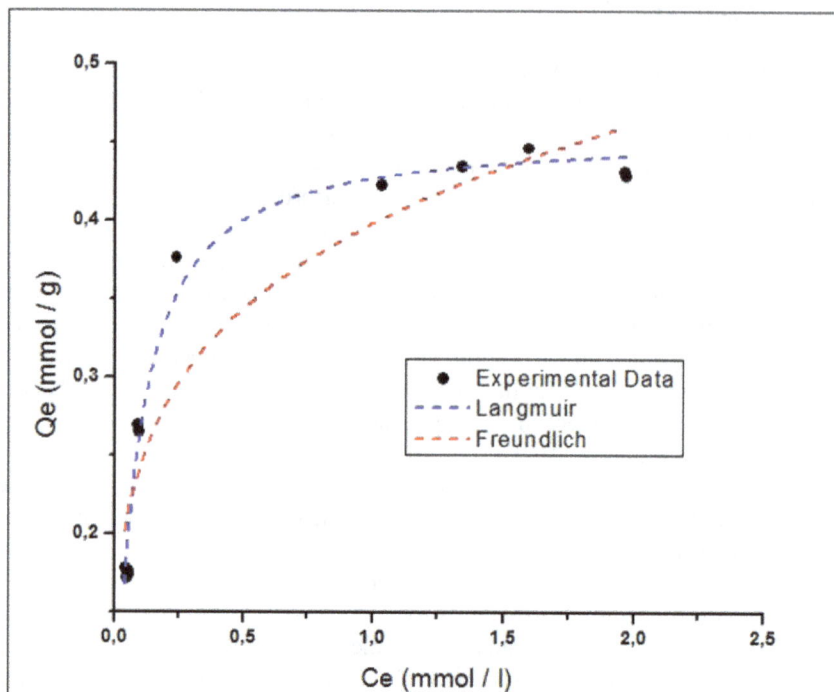

Figure 3.4. Experimental data and non-linear fit of the used models to the copper on *Azolla* adsorption isotherms.

Table 3.2. Obtained parameters from the adsorption isotherms fit to both models.

Adsorbent-adsorbate	Langmuir			Freundlich		
	$q_{máx}$	K_L	R^2	K_F	n	R^2
Azolla-Cu	0.457	14.19	0.9837	0.399	4.70	0.8895

3.5 Adsorption Column Design

In metal biosorption studies it is common to work in batch systems due to the speed of the process to obtain results and to be able to work on a very small scale in terms of the use of both the adsorbent and the adsorbate. The results obtained from these tests allow, from different modulizations, to attain parameters that permit quantifying the efficiency of adsorption in the elimination of specific adsorbates, as well as the maximum adsorption capacity. However, for larger scale processes, biosorption in fixed bed columns with continuous flow is preferable. These systems are in non-equilibrium and the concentration profiles in both the effluent and the fixed phase vary not only with time but also with space.

The fluid dynamic behavior of a fixed-bed column is described by means of concentration profiles of the adsorbate in the effluent versus time or volume. The curves obtained are a function of the adsorbent geometry, the operating conditions and also the adsorption data at equilibrium. The relative concentration curve (C/C_0) of the metal in the effluent versus total volume (or time) is known as breakthrough curve (BTC), and is obtained by passing a solution containing the solute to be adsorbed with an initial concentration C_0 through a column (fixed bed) packed with the adsorbent particles, measuring the output concentration C at different times.

The characteristics of the breakthrough curve are very important in the sense that the operation and dynamic response in a continuous reactor can be determined (Salamatinia et al. 2008). The breakthrough curve shows the performance of a fixed bed reactor from the point of view of the amount of metal that is possible to retain. In this way, a successful design of a continuous adsorption reactor requires being able to predict the characteristics of the breakthrough curve (Acheampong et al. 2013). The development of mathematical models describing the dynamic behavior of adsorption in fixed beds is difficult, because the adsorbate concentration, like the feed, moves through the column and a steady state cannot be assumed.

During the application of a model that approximates the experimental parameters, several factors should be considered, among them the adsorbate transport in fixed beds, which is described by the solid-liquid material balance equations. This description must account for the mass balance of the adsorbed solute, which in turn depends on the mechanism responsible for adsorption and may be controlled by mass transfer from within the solution to the adsorbent surface or by the chemical reaction between the adsorbent particles and the metal (Calero et al. 2009). In this sense, determining the parameters of models of varying complexity can be difficult, considering the large number of resources required and the complicated solution methods (Borba et al. 2008). This is why several simplified mathematical models have been developed, which provide parameters that qualitatively describe the effects during adsorption in the continuous reactor (Chu et al. 2007). Among them, the most widely applied models are Thomas (1944), Bohart-Adams (1920) and Yoon and Nelson (1984), which will be mentioned in the present investigation. Each model presents different mathematical equations so that from each, different parameters are obtained that provide different information about the processes under study. The assumptions and parameters of each model are shown in Table 3.3.

For the purpose of checking the applicability of the described models, the parameters provided by the different models (X_{cal}) should be compared with the experimentally obtained parameters

Table 3.3. Comparison between the used models for continuous reactors.

Model	Thomas	Bohart-Adams	Yoon-Nelson
Mathematic expression	$\dfrac{C}{C_0} = \dfrac{1}{1+\left[\exp\left(\dfrac{K_{TH}}{F}.(q_0.W - C_0.V_{ef})\right)\right]}$	$\dfrac{C}{C_0} = \exp\left(\dfrac{K_{AB}.C_0.V_{ef}}{F} - \dfrac{K_{AB}.N_0.Z}{U_0}\right)$	$\dfrac{C}{C_0} = \dfrac{\exp\left(K_{YN}.(\dfrac{V_{ef}}{F} - \tau)\right)}{1+\exp\left(K_{YN}.(\dfrac{V_{ef}}{F} - \tau)\right)}$
Parameters	**Graph C/C₀ Vs V_ef** K_{TH} is Thomas rate constant [mL min⁻1 g⁻¹], q_0 is the maximum solute concentration in the solid phase [mmol g⁻¹], V_{ef} is the effluent volume [L], F is the volumetric flow rate [mL min⁻¹], W is the amount of sorbent inside the reactor [g], C is the outlet concentration [mmol L⁻¹].	**Graph C/C₀ Vs V_ef** K_{AB} is the mass transfer coefficient [cm³ mmol⁻¹ min⁻¹], N_0 represents the maximum adsorption capacity [mmol cm⁻³], U is the linear liquid velocity [cm min⁻¹], Z is the bed height in the reactor [cm].	**Graph C/C₀ Vs V_ef** K_{YN} stands for the Yoon-Nelson rate constant [min⁻¹], τ [min] is the time required to retain 50% of the C_0.
Assumptions	• Adsorption behavior follows the Langmuir isotherm • Intraparticle diffusion and resistance to external mass transfer are negligible.	• Adsorption rate is proportional to the residual capacity of the solid and the adsorbate concentration • Intraparticle diffusion and resistance to external mass transfer are negligible.	• Adsorption rate decreases proportionally with the number of molecules adsorbed.

(X_{exp}) by means of goodness-of-fit analysis, which is evaluated by the regression coefficient R^2 and the average relative error (ARE) (Dhanasekaran et al. 2017) defined by the following expression:

$$ARE = \frac{100}{n} \sum_{i=1}^{n} \left| \frac{(X_{i,calc} - X_{i,exp})}{X_{i,exp}} \right| \qquad \text{Eq. 3.5}$$

The proximity of the value of R^2 to one (Tan and Hameed 2017) and ARE values lower than 35% (Tsai et al. 2016) indicate a good fit of the model to the experimental data.

To exemplify the application of these models, the work system was compounded by *Azolla* biomass with a contaminant solution of copper, for which a continuous reactor was assembled, whose dimensions were 15 cm in height, 15.9 cm³ in volume and an internal diameter of 1.161 cm, which was completed with 4.3 g of biomass with a particle size between 1.18–0.5 mm and then a Cu(II) solution of 50 mM concentration was passed upstream at a fixed flow rate of 0.5 mL min⁻¹ and the Cu(II) concentration at the reactor outlet was measured at different volumes by ultraviolet-visible spectrophotometry. From the breakthrough curves obtained (C/C_0 versus V graph) the nonlinear fitting models mentioned above were applied and the corresponding parameters listed in Table 3.4 were obtained.

Table 3.4. Obtained parameters with the Thomas–Bohart-Adams and Yoon-Nelson models for the adsorption of copper onto *Azolla* biomass in fixed bed reactor.

		Azolla-Cu
Thomas Model	K_{TH}	2.39
	q_0 calc	0.6222
	q_0 exp	0.679
	R^2	0.996
	ARE	8.39
Bohart-Adams Model	K_{AB}	1.637
	N_0 calc	0.1951
	N_0 exp	8.68E-2
	R^2	0.9770
	ARE	125
Yoon-Nelson Model	K_{yn}	0.1005
	τ_{calc}	136.3
	τ_{exp}	135.0
	R^2	0.996
	ARE	0.96

From the data obtained for the modeling of the experimental breakthrough curve it can be observed that both the Thomas and Yoon-Nelson models are the most appropriate to describe the behavior of the experimental data, since both have R^2 greater than 0.99 and an ARE less than 35%, which would indicate that the value of the parameters calculated by the model and the experimental one are very similar. These models adequately fit adsorption processes where external and internal diffusions are not the limiting step (Aksu and Gönen 2004). Both models have analogous mathematical equations, so they were expected to predict similar fits (Chu 2020), however, from each of them, different information about the system under study can be obtained. The Thomas model allows estimating the maximum amount of adsorbate retained in the solid phase, from the calculated q_0 which resulted to be 0.622 mmol g⁻¹. On the other hand, the Yoon-Nelson model allows knowing the value of the time required for the concentration in the effluent to be equal to 50% of the input, the latter, determined by the parameter τ, was 136 min.

The successful design of a fixed-bed column adsorption process requires the prediction of the effluent breakthrough curve (Chen et al. 2012), so the parameters obtained together with the adsorbent saturation times are very useful and represent a starting point for future studies. The description of these breakthrough curves was shown in the work presented by Saralegui et al. (2022). The optimization of a continuous reactor, varying parameters such as the amount of each adsorbent, flow rate, column size, among others are the subject of further studies.

3.6 Digital Application

The digital application called Reactor App was developed ad-hoc in order to manage the complex selection of the most efficient system for treatment. The application can be accessed by entering: https://laquisihereactorapp.fi.uba.ar/.

Reactor App (Figure 3.5) is a software designed with several objectives:

1. To assemble a database of the experimental data obtained for each of the sorbent-adsorbent systems studied, including data from the Argentinean legislation for effluent discharge.

2. To design a software that orders, analyses and adjusts theoretical or empirical models for the design of an adsorbate-adsorbent reactor, providing different mathematical models that assist in the design of reactors for business applications (tanneries, breweries, chrome works).

3. To create a platform for the laboratory to disseminate its work and to advise different organizations on different problems such as water treatment and effluent discharge.
   The system has three types of user profile (Figure 3.6):

   - External user, who will be able to view limited system information and will have reduced system functionality

   - Laboratory user, who will have all the information and functionality of the system, without being able to alter the system information.

   - Administrator user, who will also be able to modify the system information and moderate the users who have access to the application.

For reactor design (Saralegui et al. 2022), taking certain considerations into account, this application provides a first approximation of a continuous reactor using parameters estimated through the development of batch tests.

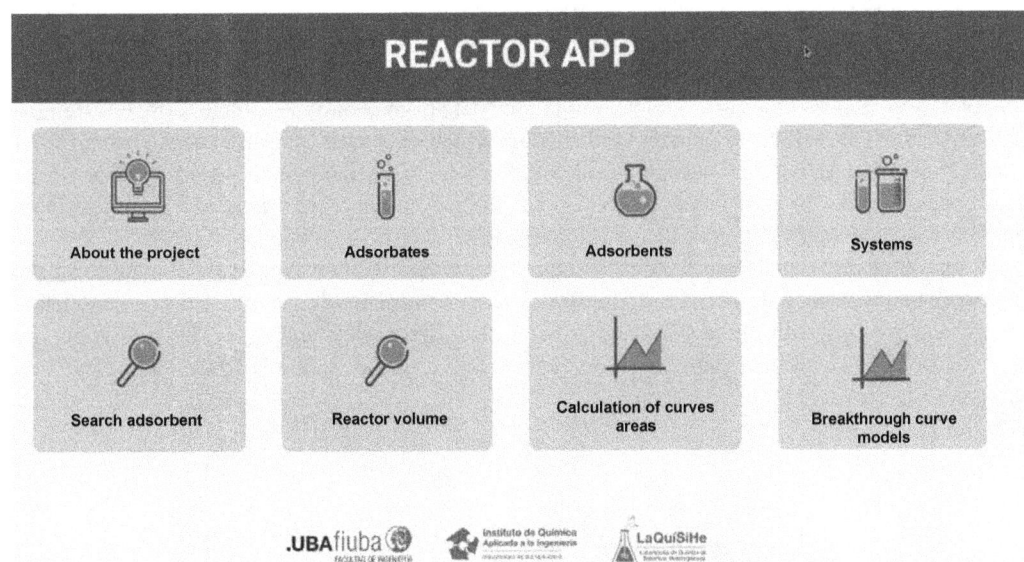

Figure 3.5. Reactor App software home screen.

Figure 3.6. Login screen of the Reactor App.

Data on adsorbents, adsorbates and pre-loaded adsorbent-adsorbate systems are needed to start the task. The software works through cards for each adsorbate (Figure 3.7) or adsorbent (Figure 3.8) that are pre-filled. These cards include the physicochemical characteristics of both the adsorbate and the adsorbent. These data can be obtained from literature (in the case of adsorbates, legislation or guide levels are included) or empirically and are used for the implementation of the models included in the software.

Figure 3.8 shows the cards corresponding to the systems or adsorbate-adsorbent pairs studied. In these cards, the data obtained in batch tests such as q_{max} and Langmuir constant, equilibrium time, reaction order and kinetic constant, among others are loaded.

Figure 3.9 shows the screen where the information to calculate the first approximation of the volume of a continuous reactor must be loaded.

Reactor App also allows uploading experimental data from continuous reactors (by uploading a CSV or XLSX file), both from fluid dynamic tests in which inert material is used as the reactor filler and additionally tests with the reactor filled with the adsorbent under study. By difference of the areas under the breakthrough curve (Saralegui et al. 2022), in both cases the amount of contaminant removed by the reactor can be calculated (Figure 3.10).

With the experimental data of the breakthrough curves, Reactor App is able to perform the estimation of different models used as approximations and return the characteristic fitting parameters of each one together with the regression coefficient R^2. The models that are possible to apply are Thomas, Bohart-Adams and Yoo-Nelson, explained above, and an example from the Reactor App return page is shown in Figure 3.11.

(a)

Adsorbates

| Name | Ionic charge | Search |

Arsenate (V) (arsenate)
AsO_4^{3-} Carga iónica: -3
Radio hidrodinámico: 0 Å
Límite de vertido (Ley 24.051): Sin datos
Ver procesos

Copper (II) (copper)
Cu^{2+} Carga iónica: 2
Radio hidrodinámico: 0.69 Å
Límite de vertido (Ley 24.051): Sin datos
Ver procesos

Chromate (VI) (chromate)
CrO_4^{2-} Carga iónica: -2
Radio hidrodinámico: 0 Å
Límite de vertido (Ley 24.051): Sin datos
Ver procesos

Ortho phosphate (VI) (phosphate)
PO_4^{3-} Carga iónica: -3
Radio hidrodinámico: 0 Å
Límite de vertido (Ley 24.051): Sin datos
Ver procesos

Nitrate (V) (nitrate)
NO_3^{-} Carga iónica: -1
Radio hidrodinámico: 0 Å
Límite de vertido (Ley 24.051): Sin datos
Ver procesos

Nickel (II) (nickel)
Ni^{2+} Carga iónica: 2
Radio hidrodinámico: 0.78 Å
Límite de vertido (Ley 24.051): Sin datos
Ver procesos

Silver (I) (silver)
Ag^{+} Carga iónica: 1
Radio hidrodinámico: 0.126 Å
Límite de vertido (Ley 24.051): Sin datos
Ver procesos

Lead (II) (lead)
Pb^{2+} Carga iónica: 2
Radio hidrodinámico: 1.2 Å
Límite de vertido (Ley 24.051): Sin datos
Ver procesos

Zinc (II) (zinc)
Zn^{2+} Carga iónica: 2
Radio hidrodinámico: 0.74 Å
Límite de vertido (Ley 24.051): Sin datos
Ver procesos

(b)

Adsorbents

| Name | Search |

Azolla pinnata
Tamaño de partícula: 1180-500 µm
sBet: 0.6463 m²/g   vBet: 0.0019 cm³/g
pH (carga cero): 1
Ver procesos

Activated carbon
Tamaño de partícula: 53-74 µm
sBet: 1.037 m²/g   vBet: 0.9 cm³/g
pH (carga cero): 14
Ver procesos

Avoacado stone
Tamaño de partícula: 500 µm
sBet: 0.0813 m²/g   vBet: 0.0001 cm³/g
pH (carga cero): 5
Ver procesos

Avocado stone activated with orthophosphoric acid
Tamaño de partícula: 500 µm
sBet: 0.1276 m²/g   vBet: 0.0001 cm³/g
pH (carga cero): 14
Ver procesos

Banana peel
Tamaño de partícula: 500 µm
sBet: 0.6463 m²/g   vBet: 0.0019 cm³/g
pH (carga cero): 4.64
Ver procesos

Peanut shell
Tamaño de partícula: 500 µm
sBet: 14 m²/g   vBet: 14 cm³/g
pH (carga cero): 14
Ver procesos

Dolomite
Tamaño de partícula: 53-74 µm
sBet: 1.83 m²/g   vBet: 0.003 cm³/g
pH (carga cero): 4.43
Ver procesos

Hydroxyapatite
Tamaño de partícula: 53-74 µm
sBet: 1.5 m²/g   vBet: 0.0044 cm³/g
pH (carga cero): 14
Ver procesos

Hydroxyapatite
Tamaño de partícula: piezas µm
sBet: 1.5 m²/g   vBet: 0.0044 cm³/g
pH (carga cero): 14
Ver procesos

Figure 3.7. Cards from the Reactor App software corresponding to the studied (a). Adsorbates (Text in Spanish indicating their different parameters such as ionic charge, hydrodynamic radio and discharge limit) and (b). Adsorbents (Text in Spanish indicating their particle size, BET area and pH).

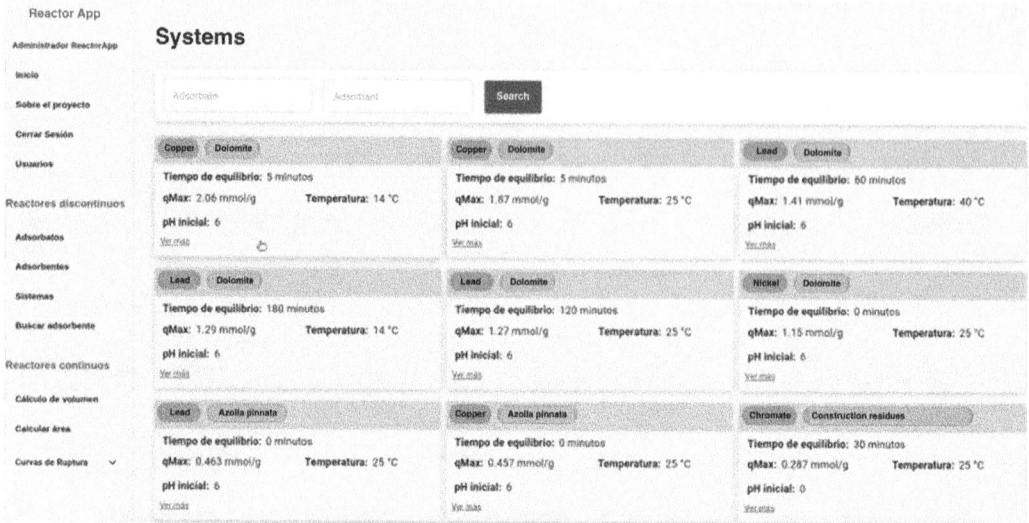

Figure 3.8. Cards of the Reactor App software corresponding to the systems (adsorbate/adsorbent pairs) studied (Text in Spanish indicating some parameters such as equilibrium time, qmax, temperature and initial pH).

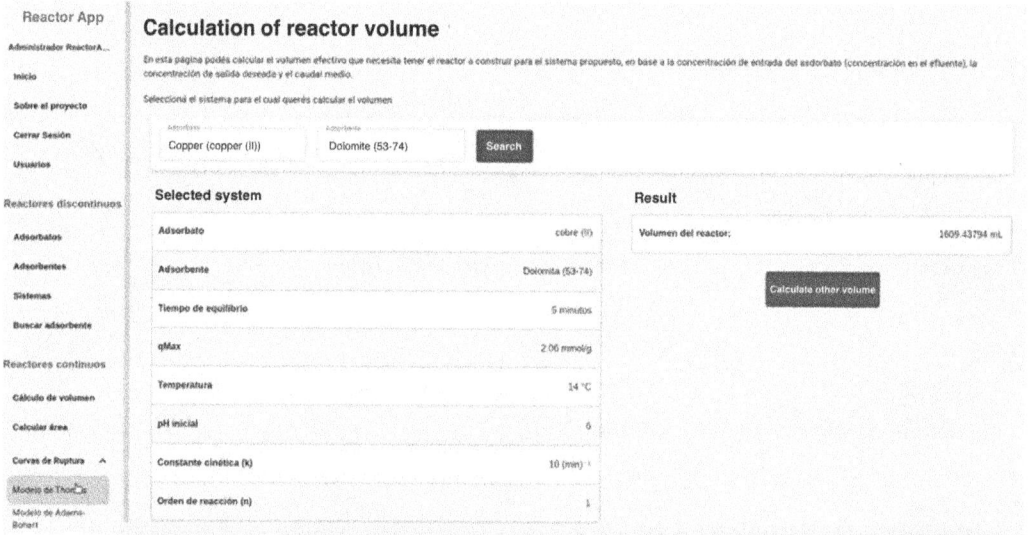

Figure 3.9. Data to be entered in Reactor App for the calculation of the first continuous reactors volume approximation. (Text in Spanish indicating the different variables for the selected system: adsorbate, adsorbent, equilibrium time, qmax, temperature, initial pH, kinetic constant and reaction order).

3.7 Conclusions

In conclusion, the choice of the low-cost adsorbent to be used should be based on the availability of the material in the same place where the effluent treatment is needed. In order to simplify the studies, first the dosage curve should be made and the working conditions should be established. Once the relation between the contaminant concentration and the quantity of adsorbent are fixed, the kinetic equation at a given temperature should be estimated. Then, the study of the equilibrium isotherm and the estimation of the maximum adsorption capacity could be accomplished. Finally, with this data the continuous reactor volume can be estimated using Reactor App and then a first

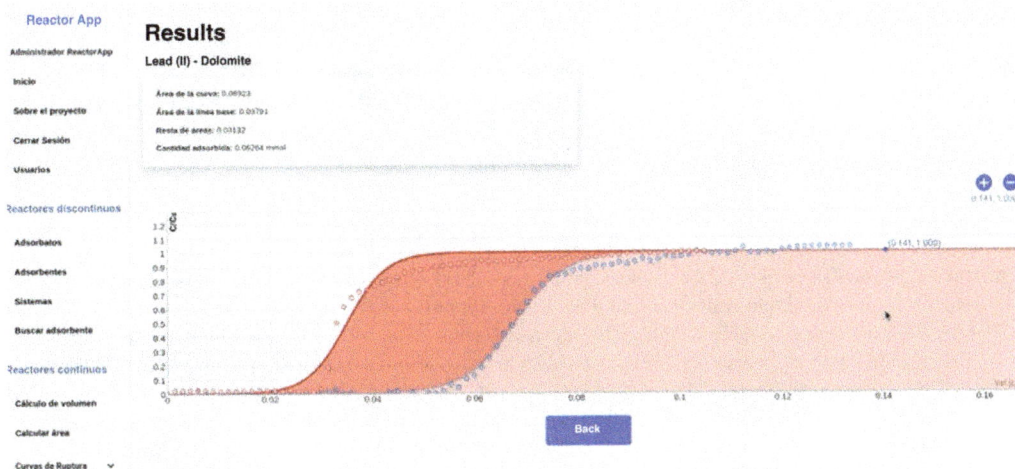

Figure 3.10. Graph obtained in Reactor App by entering the data of the breakthrough curves of the fluid dynamic test (pink dots) and the adsorption of Lead (II) on dolomite (light blue dots). The red area represents the amount of adsorbed lead, the text in Spanish additionally indicates the base line area, the difference between these both areas and the adsorbed amount in mmol.

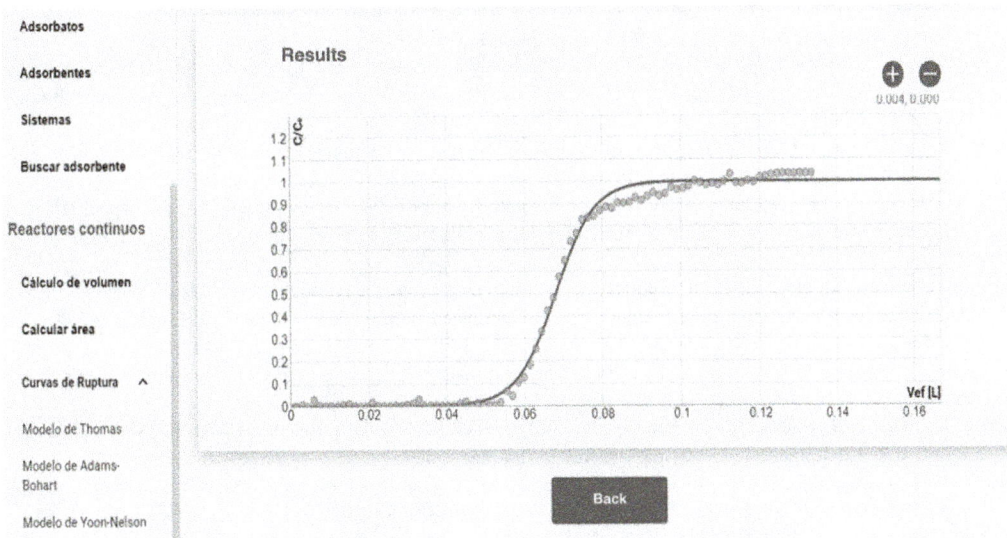

Figure 3.11. Example of the Reactor App return screen for the application of the Thomas model for the data obtained for copper (II) adsorption on *Azolla pinnata* in the continuous reactor.

approximation for the scale-up can be made. Once the continuous reactor is built, the results can be analyzed with Reactor App and the models can be used to parameterize the new reactor. In this way it is possible to consider a new scale jump.

In all the cases studied in our laboratory, the obtained results indicate that the use of lignocellulosic materials for the removal of ionic pollutants from wastewater is feasible, taking into account that the use of these materials is a contribution to the circular economy.

We are working on the mixing of contaminants and multiple adsorptions on these adsorbents, as well as on the safe final disposal of these solids when exhausted.

Acknowledgements

The authors acknowledge the financial support from Universidad de Buenos Aires (UBACyT N°20020190100323BA, N°20020190200302BA, PDE 032/2020) and would also like to thank the computer engineering students Matías Reimondo, Santiago Pinto and Lucas Lavandeira for their collaboration in the development of the application used in this work.

References

Acheampong, M., K. Pakshirajan, A. Annachhatre and P. Lens. 2013. Removal of Cu(II) by biosorption onto coconut shell in fixed-bed column systems. J. Ind. Eng. Chem. 19: 841–848.

ACUMAR. 2019. Records of establishments in the Matanza Riachuelo Basin 2019(in Spanish). Retrieved from http://datos.acumar.gob.ar/dataset/registro-de-establecimientos-de-la-cuenca-matanza-riachuelo-2019/archivo/dca9d704-8e4c-4bbb-bed1-2a7460788eb0 consulta 3/08/2019.

Aksu, Z. and F. Gönen. 2004. Biosorption of phenol by immobilized activated sludge in a continuous packed bed: prediction of breakthrough curves. Process Biochem. 39: 599–613.

Apro, N. J., C. J. Cuadrado and P. A. Secreto. 2004. Stabilization of wheat bran and germ, through the extrusion process as an input for the food industry (in Spanish). Development and Innovation Day. INTI-Cereals and Oilseeds.

BCC, Bolsa de comercio de Córdoba. 2014. Chapter 15: Peanut production chain (in Spanish). Retrieved from https://bolsacba.com.ar/buscador/?p=1354.

Boeykens, S. P., M. N. Piol, L. Samudio Legal, A. B. Saralegui and C. Vázquez. 2017. Eutrophication decrease: phosphate adsorption processes in presence of nitrates. J. Environ. Manage. 203: 888–895.

Boeykens, S. P., A. B. Saralegui, N. Caracciolo and M. N. Piol. 2018. Agroindustrial waste for lead and chromium biosorption J. Sustain. Dev. Energy Water Environ. Syst. 6: 341–350.

Boeykens, S. P., N. Redondo, R. A. Obeso, N. Caracciolo and C. Vázquez. 2019. Chromium and lead adsorption by avocado seed biomass study through the use of Total Reflection X-Ray Fluorescence analysis. Appl. Radiat. Isot. 153: 108809.

Bohart, G. and E. Adams. 1920. Some aspects of the behavior of charcoal with respect to chlorine. J. Am. Chem. Soc. 42: 523–544.

Borba, C., E. Da Silva, M. Fagundes-Klen, A. Kroumov and R. Guirardello. 2008. Prediction of the copper (II) ions dynamic removal from a medium by using mathematical models with analytical solution. J. Hazard. Mater. 152: 366–372.

Branzini, A. and M. Zubillaga. 2012. Remediation and monitoring of soils contaminated with heavy metals: Organic and inorganic amendments to remediate soils contaminated with metals (in Spanish). Editorial Académica Española.

Calero, M., F. Hernáinz, G. Blázquez, G. Tenorio and M. Martín-Lara. 2009. Study of Cr (III) biosorption in a fixed-bed column. J. Hazard. Mater. 171: 886–893.

Carrere, R. 2010. The avocado: a fruit tree for the family garden (in Spanish). Retrieved from https://docplayer.es/amp/14007991-La-palta-un-frutal-para-la-huerta-familiar.html.

Cartaya, O. E, I. Reynaldo, C. Peniche and M. L. Garrido. 2011. Use of natural polymers as an alternative for the remediation of soils contaminated by heavy metals (in Spanish). Rev. Int. de Contam. Ambient. 27: 41–46.

Chen, S., Q. Yue, B. Gao, Q. Li, X. Xu and K. Fu. 2012. Adsorption of hexavalent chromium from aqueous solution by modified corn stalk: a fixed-bed column study. Bioresource Technol. 113: 114–120. https://doi.org/10.1016/j.biortech.2011.11.110.

Chu, K. H. and M. A. Hashim. 2007.Copper biosorption on immobilized seaweed biomass: Column breakthrough Characteristics. J. Environ. Sci. 19: 928–932.

Chu, K. H. 2020. Breakthrough curve analysis by simplistic models of fixed bed adsorption: in defense of the century-old Bohart-Adams model. Chem. Eng. J. 380: 122513. https://doi.org/10.1016/j.cej.2019.122513.

Dhanasekaran P., P. Satya and K. I. Gnanasekar. 2017. Fixed bed adsorption of fluoride by *Artocarpus hirsutus* based adsorbent. J. Fluor. Chem. 195: 37–46.

Di Giulio, R. T. and M. C. Newman. 2008. Ecotoxicology. In Casarett and Doull's. Toxicology. The basic science of poisons, ed. C. D. Klaasen. Kansas: McGraw-Hill.

EPA, Environmental Protection Agency. 2000. *In situ* treatment of soil and groundwater contaminated with chromium-technical resource. 625/R-00/004. Retrieved from https://cfpub.epa.gov/si/si_public_record_report.cfm?Lab=NRMRL&dirEntryId=64150.

Folkard, G. and J. Sutherland. 1996. Moringa oleifera: A tree with enormous potential (in Spanish) Original in JAMA. 1931, 8(3): 211. Traducido por Ariadne Jiménez U.C.R., Turrialba, Costa Rica. Agroforestry Today, 8(3).

Foo, K. Y. and B. H. Hameed. 2010. Insights into the modeling of adsorption isotherm systems. Chem. Eng. J. 156: 2–10.

Freundlich, H. M. F. 1906. Über die Adsorption in Lösungen. Zeitschriftfür Physikalische Chemie (Leipzig) 57 A: 385–470.

Frusso, E. A. 2013. Influence of nitrogen, phosphorus and zinc on the chemical composition and yield of pecan nuts and their relationship with the variability of nutrients in the leaf (in Spanish) thesis for master's degree in Plant Production. Universidad de Buenos Aires.

Guo, X., S. Zhang and X. Shan. 2008. Adsorption of metal ions on lignin. J. Hazard. Mater. 151: 134–142. doi:https://doi.org/10.1016/j.jhazmat.2007.05.065.

Ho, Y. S. and G. McKay. 1999. Pseudo-second order model for sorption processes. Process. Biochem. 34: 451–465.

Lagergren, S. 1898. About the theory of so-called adsorption of soluble substances (in German: Zurtheorie der sogenannten adsorption gelosterstoffe), Supplement of the Royal Swedish Science Academiens Documents (in Swedish: Bihang Till Konglinga Svenska Vetenskaps Academiens Handlingar 24: 1–39.

Langmuir, I. 1918. The adsorption of gases on plane surfaces of glass, mica and platinum. J. Am. Chem. Soc. 40 (9): 1361–1403.

Liu, J., R. A. Goyer and M. P. Waalkes. 2008. Toxic effects of metals. In Casarett and Doull's. Toxicology. The Basic Science of Poisons., ed. C. D. Klaasen. New York, United State: McGraw - Hill.

Mihelcic, J. and J. Zimmerman. 2012. Environmental Engineering: Fundamentals, Sustainability, Design (in Spanish). Alfaomega. Mexico.

Molina, N. A., F. Scribano, G. Tenaglia and D. Rodríguez. 2015. Cost of banana production in Formosa (in Spanish). EEA Bella Vista. Serie Técnica, 50.

Neris, J. B., F. H. M. Luzardo, P. F. Santos, O. N. de Almeida and F. G. Velasco. 2019. Evaluation of single and tri-element adsorption of Pb(II), Ni(II) and Zn(II) ions in aqueous solution on modified water hyacinth (*Eichhornia crassipes*) fibers. J. Environ. Chem. Eng. 7(1): 102885. doi:https://doi.org/10.1016/j.jece.2019.102885.

Pellegrino, M. 2019. Peanut Chain - summary. Argentina: Presidency of the Nation (in Spanish) Retrieved from https://alimentosargentinos.magyp.gob.ar/HomeAlimentos/Cadenas%20de%20Valor%20de%20Alimentos%20y%20Bebidas/informes/Resumen_Cadena_2018_Mani_manies_crudos_y_Preparaciones_de_mani.pdf.

Piol, M. N., C. Dickerman, M. P. Ardanza, A. B. Saralegui and S. P. Boeykens. 2021. Simultaneous removal of chromate and phosphate using different operational combinations for their adsorption on dolomite and banana peel. J. Environ. Manage. 288: 112463.

PLNRA. Poder Legislativo Nacional. República Argentina. 1992. National Hazardous Waste Law No. 24.051 - Decree No. 831/93 - Resolution No. 242/93 Standards for Discharges from Industrial or Special Establishments reached by Decree 674/89 (in Spanish) C.F.R. 1992. http://servicios.infoleg.gob.ar/infolegInternet/anexos/10000-14999/12830/norma.htm.

Preciado Patiño, J. 2015. The Sugarcane Chain and its agro-industrial value (in Spanish). Council of Argentine Agro, Agrifood and Agroindustry Professionals, http://www.cpia.org.ar/agropost/201509/nota1.html.

Renaud, M., P. M. da Silva, T. Natal-da-Luz, S. D. Siciliano and J. P. Sousa. 2021. Community effect concentrations as a new concept to easily incorporate community data in environmental effect assessment of complex metal mixtures. J. Hazard. Mater. 411: 125088. doi:https://doi.org/10.1016/j.jhazmat.2021.125088.

Ríos, L., G. Pérez and A. Felipe. 2017. The Argentine sugar market and Economic analysis of the sugar harvest in Tucumán Campaign 2016–2017 (in Spanish). INTA.

Salamatinia, B., A. Kamaruddin and A. Abdullah. 2008. Modeling of the continuous copper and zinc removal by sorption onto sodium hydroxide-modified oil palm frond in a fixed-bed column. J. Chem. Eng. 145: 259–266.

Saralegui, A. B., V. Willson, N. Caracciolo, M. N. Piol and S. P. Boeykens. 2021. Macrophyte biomass productivity for heavy metal adsorption. J. Environ. Manage. 289: 112398.

Saralegui, A. B., M. N. Piol, V. Willson, N. Caracciolo and S. P. Boeykens. 2022. Lignocellulosic waste as adsorbent for water pollutants. A step towards sustainability and circular economy. pp. 168–182. *In*: A. Malik, M. K. Kidwai and V. K. Garg [eds.]. Bioremediation of Toxic Metal(loid)s : CRC press, Taylor and Francis group, Boca Raton.

Tan, K. L. and B. H. Hameed. 2017. Insight into the adsorption kinetics models for the removal of contaminants from aqueous solutions. J. Taiwan Inst. Chem. Eng. 74: 25–48.

Thomas, H. C. 1944. Heterogeneous ion exchange in a flowing system. J. Am. Chem. Soc. 66: 1664–1666.

Tsai, W., M. De Luna, H. Bermillo-Arriesgado, C. Futalan, J. Colades and Meng-WeiWan. 2016. Competitive fixed-bed adsorption of Pb(II), Cu(II), and Ni(II) from Aqueous Solution Using Chitosan-Coated Bentonite. Int. J. Polym. Sci. http://dx.doi.org/10.1155/2016/1608939.

Volke Sepúlveda, T. and J. Velasco Trejo. 2002. Tecnologías de remediación para suelos contaminados. Mexico: Instituto Nacional de Ecología (INE-SEMARNAT).

Yoon, Y. H. and J. H. Nelson. 1984. Application of gas adsorption kinetics I. A theoretical model for respirator cartridge service life. Am. Ind. Hyg. Assoc. J. 45: 509–516.

CHAPTER 4

Bioaccumulation and Biosorption
The Prospects and Future Applications

P.F. Steffi[1,*1] and *P.F. Mishel*[2]

4.1 Introduction

The rapid development of industry and modern technologies has shown the release of variable hazardous compounds into the surroundings, including heavy metals (Qiu et al. 2021). Heavy metals are utilized in the mining, metallurgical, electronics and electroplating sectors. Many modern operations produce trash that contains heavy metals which are harmful to both lower and higher living beings (Devanesan and AlSalhi 2021). There are a number of ways for removing heavy metal ions now in use. Precipitation, ion exchange, membrane processes, evaporation and filtration are the most significant (Danouche et al. 2021). The use of these approaches is frequently coupled with technological challenges, such as waste management (Pushkar et al. 2021). Biotechnological strategies that utilize biological materials such as bacteria and plants could be a viable alternative to the chemical and physical systems currently in use (Reddy et al. 2021).

Metals can be bound by biological material through biosorption and bioaccumulation processes (Abdul Jaffar et al. 2015). During the biosorption process, metal ions are adsorbed on the surface of a sorbent. Biosorption is a metabolically passive energy-producing process that uses dead biomass as a fuel source. Biosorption is a process that has a few distinct characteristics (Hansda et al. 2016). It has great effectiveness in sequestering dissolved metals from very dilute complicated solutions. As a result, biosorption is an excellent choice for treating high-volume, low-concentration complicated wastewaters (Joshi et al. 2011).

On the other hand, bioaccumulation can only occur in living beings due to the transit of pollutants into the cell and the accumulation of metals within the cell (Kumar et al. 2021). The initial phase in bioaccumulation is biosorption. Bioaccumulative chemicals are those that accumulate in living beings to the point where their concentrations in bodily tissues continue to rise (Kumar et al. 2007). Bioaccumulation is also known as bioconcentration in fish and other aquatic creatures. The

[1] PG and Research Department of Microbiology, Cauvery College for Women (Autonomous), Trichy.
[2] Department of Botany, Bharathidasan University, Trichy.
* Corresponding author: steffi.mb@cauverycollege.ac.in

general objective of this chapter is to look at the biosorption and bioaccumulation potentials for heavy metal removal from contaminated environments (Singh and Kumar 2020). The most utilized heavy metal removal technologies from water and wastewater are discussed.

4.1.1 Heavy Metals in the Environment

Recent advances in civilization have resulted in an overabundance of heavy metals entering the ecosystems. In both aqueous and terrestrial contexts, heavy metals can form indeterminate mineral and organometallic linkages (Yaashikaa 2020, Abdul Jaffar et al. 2015). Steel mills, plating facilities, tanning plants, chemical fertilizer, pesticide manufacturing plants, fabric dye plants, electroplating plants, motor and power engineering plants and battery and accumulator manufacturing plants produce wastewater (Ahmed et al. 2021). Heavy metals released into the environment have high mobility. Plants may absorb them, and they wind up in the digestive tracts of animals and humans (Al-Ansari et al. 2021). Due to their harmful effects on certain elements of the environment and bioaccumulation in the food chain, heavy metals constitute a significant hazard to living beings (Arroyo Herrera et al. 2021).

4.1.2 Heavy Metals in Surface Water

Industrial and municipal wastewater carries heavy metals into surface water. Leaching of chemical pollutants from landfills and deposition of atmospheric specific matter can also pollute surface water (Bai et al. 2019). Water contamination can also be caused by leaching of mineral fertilizers and pesticides from the soil. Heavy metals can be found in the following forms in the aquatic environment:

- Ionic form (the most harmful to living organisms);
- Ion linked with a variety of ligands (complex compounds);
- Precipitated molecules.

In surface water, many heavy metals bioaccumulate. Algae collect significant amounts of metals (particularly manganese and lead).

4.1.3 Heavy Metals in the Soil

Metal pollutants in the soil are caused by industrial and power engineering, as well as atmospheric pollutants and landfilling. Metals can be introduced to the soil through fertilizers with sewage sludge and insecticides. Soil contamination occurs because of transportation along with route systems (Bai et al. 2016). Metals in the soil can be dissolved, transported and taken up by various species depending on soil conditions, such as pH. Metals are present in the soluble form at lower pH—in acidic soils—and hence are more available to plants. High metal concentrations in soils always cause a disruption in chemical balance, which hinders the functioning of specific ecosystems (Baselga Cervera 2018). Metals accumulating in soils prevent soil microbes from growing. Microorganisms' basic physiological processes associated with the breakdown and modification of organic compounds are harmed.

4.1.4 Heavy Metals in Atmosphere

Heavy metals in the atmosphere are added mainly by emissions of specific matter from industry, transportation and power generation (Bindschedler et al. 2017). The pollution of the atmosphere has a worldwide scope. This is due to the small size of molecules, which allows them to float in the air for lengthy periods and travel to far-flung corners of the planet (Boriova et al. 2019). The quantity of heavy metals emitted into the atmosphere is determined by the economic development of a country as well as pollution control techniques. Airborne contamination by heavy metals has a negative impact on climate change, causes economic losses (mainly in agriculture and forestry) and offers a significant health risk to humans (Cachada et al. 2014).

4.2 Techniques for Heavy Metals Removal

Heavy alloys are usually deleted from water solutions employing physicochemical techniques. The following are the most prevalent methods:

- Infiltration
- Coagulation
- Chemical precipitation
- Ion exchange
- Membrane processes

Heavy metals can be removed using a single method or a combination of two or more methods. Many factors influence the method selection, including effluent kinds, content and forms and concentrations of elements to be removed, as well as the degree of removal required (Cachada et al. 2014). Precipitation of metals in the form of hydroxides, which are then removed from the solution during filtration or decantation, is the most common method for removing heavy metals from industrial effluent. The presence of organic and inorganic compounds, as well as temperature and pH, can all have a negative impact on the process's efficacy (Chamekh et al. 2021). Furthermore, handling moistened sludge raises costs. Another approach for removing heavy metals is ion exchange. Ions bound to the ion-exchanger are transferred for ions present in the surrounding solution in this process. Natural or synthetic ion-exchangers are available (Chen et al. 2019). The solutions should be pre-treated before the ion exchange since impurities in the water can disrupt the process. Membrane techniques can also remove heavy metals from wastewater. Contaminants can be separated using these methods. Membranes are essential for water treatment efficiency. A membrane is a thin partition that allows molecules to pass through it selectively. Natural and manmade materials can be used to make membranes (Chen et al. 2020). They should have excellent hydraulic efficiency, separation qualities, mechanical, chemical and heat resistance. Sorption on activated carbon or zeolite is used to remove heavy metals. Activated carbon is used in adsorption (present in granular of fine forms). Metals are removed from activated carbon's surface, where they are kept (Chojnacka 2010). The methods suggested are frequently time-consuming and costly to implement. As a result, new, more cost-effective and easier technologies for heavy metal removal are required.

4.3 Removal of Heavy Metals with Biosorption and Bioaccumulation

Since the turn of the century, there has been a growing awareness of the need to safeguard the natural environment. It is necessary to develop new technologies for eliminating toxins from the environment (Fabre et al. 2020). Biotechnological methods, which rely on the inherent characteristics of microbes to adsorb and accumulate heavy metals, may be a viable alternative to physicochemical methods. Heavy metals can be captured and accumulated by all microbial species in water solutions (Fargasova et al. 2010). Heavy metal absorption is linked to a microbial mechanism that allows the uptake of elements necessary for growth and metabolic processes. The ability of the biomass to bind and accumulate harmful metals can be utilized to build efficient and cost-effective wastewater treatment systems for the mining and electroplating industries (Flouty and Estephane 2012). Biosorption and bioaccumulation are terms used to describe processes that use biomass to remove metals (Figure 4.1).

Both approaches differ in terms of the mechanism that allows contaminants to be bound. Toxins stick largely to the surface of the microbial cell wall when biosorption occurs, whereas bioaccumulation allows pollutants to penetrate microbial cells (Gajda Meissner et al. 2020). As

**Heavy metals
in the environment**

|

Microorganisms

| |

Biosorption   **Bioaccumulation**

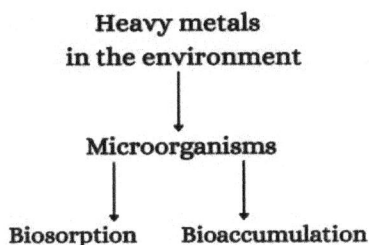

Figure 4.1. Biological methods for removal of heavy metals.

metal removal requires metabolic activity, bioaccumulation occurs only when microbial cells are alive. Biosorption is a technique that involves the utilization of dead biomass. Materials utilized in biological processes to remove metals include wasted biomass and organisms found naturally in the environment. This lowers the cost of biological approaches, making them more accessible to a wider range of applications (Garg et al. 2012).

4.4 Biosorption

The removal of heavy metals using live microbes is complicated. Heavy metals have a harmful influence on living organisms' cells, which is the main issue. Advanced research, on the other hand, revealed that dead cells could attach metal ions through a variety of physicochemical methods (Gerber et al. 2018). Biosorption and biosorbents of various origins have received a lot of interest as a result of this new finding. Several studies in this field have contributed to the spread of biosorption knowledge (Hansda et al. 2016). As a result, the number of new potential applications has increased. This process is known to be influenced not only by the kind and chemical composition of the biomass but also by external physicochemical factors. The processes behind the biosorption process have also been identified and elucidated. Many research centres across the world are working to better understand biosorption (Hlihor et al. 2015). Furthermore, the potential of various biological materials found in nature is being researched in order to improve their properties for heavy metal removal, learn about binding mechanisms and develop the most effective biosorbents for the removal of contaminants, including heavy metals (Ibuot et al. 2020). Metal ions (typically in the form of cations) are bound by cell membranes in a physicochemical procedure termed biosorption, i.e., by negatively charged substances present in cell membranes. Understanding the biosorption mechanisms that allow heavy metals to be removed is critical for process optimization (Irshad et al. 2021). During sorption, numerous distinct mechanisms take place. Due to the complexity of biological materials, several mechanisms may occur at different rates simultaneously. The following mechanisms are covered by biosorption:

- Ion exchange is a reversible chemical reaction that occurs on materials containing relevant functional groups and involves exchanging mobile ions for other ions of the same charge.
- Heavy metal ions bind to the cell membrane with functional elements, resulting in complexation.
- Physical adsorption is triggered by van der Waals forces, which are intermolecular interactions. There is no chemical bonding in the case of physisorption.

Heavy metals are bound by cells that have ceased to function metabolically, which is the major benefit of biosorption. As feeding living biomass necessitates an additional source of nutrients and energy, contaminants can be removed by dead organisms, simplifying and lowering the cost of the process (Jin et al. 2020). In this process, cell membranes play a crucial role. Before accessing the cell membrane and cytoplasm, all metal ions pass through the cell wall. As a cell wall's structure is made up of diverse polysaccharides and proteins, there are several active places for metal ion binding. Like commercially available resins, a cell wall can be treated as a complicated ion exchanger (Joshi et al. 2011). Different microbial cell wall compositions and intercellular changes have a major impact on the amount of adsorbed metal ions. Microbial cells have a surface made up of macromolecules containing a lot of charged functional elements, such as:

- Phosphate
- Amino
- Carboxylic
- Hydroxyl

Due to the lack of carboxylic and phosphate acid residues, the cell surface is frequently negatively charged. This enables passive cation binding on the cell surface. Positively charged metal ions are drawn to the cell and adsorbed on the cell's negatively charged surface (Kadukova 2016). The entire process is passive and takes place without the cell's metabolic processes. In Figure 4.2, the entire cycle of wastewater treatment with biosorption is shown.

First, heavy metal effluent is combined with biomass. Metal ions adsorb on the surface of bacteria as a result of this (biosorption). After that, biomass regeneration (desorption) is carried out, and metals can be collected from the leftover liquid portion (Leong and Chang 2020). The following are the phases of a typical biosorption laboratory procedure:

I.   A solution sample of volume V containing heavy metals at concentration C_i is created in the first phase.

II. In the second step, biomass M (g) is added to the sample under investigation.

III. To achieve equilibrium, the solution with biomass is vigorously agitated for about 16 hr.

IV. The biomass is then separated from the solution in the fourth phase (centrifugation, filtration).

V.   The solution is then analyzed for metal ions after being treated with biosorption (C_f).

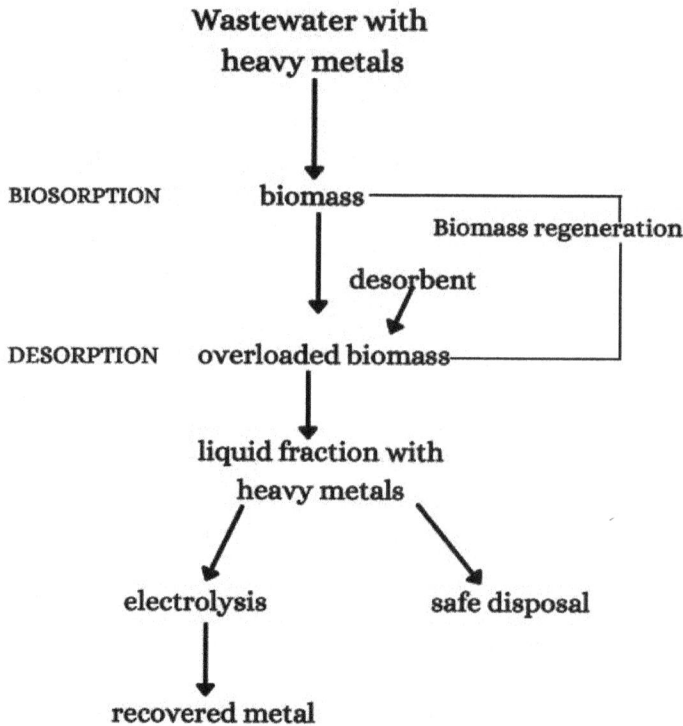

Figure 4.2. Removal and recovery of heavy metals from wastewater by microbial biomass.

The following formula (Eq. 4.1) is used to compute biosorption in mg metal per g biomass:

$$q = \frac{v(c_i - c_f)}{M}$$

Eq. 4.1

where:

q   – number of adsorbed metals (mgg^{-1})

V   – is the sample quantity of the solution (in millilitres).

C_i – metal ion concentration in solution at the start (mg L^{-1})

C_f – metal ion concentration in solution at the end (mg L^{-1})

M   – mass of biomass (g)

4.5 Bioaccumulation

Different elements from the surrounding environment are absorbed and retained by living organisms. Bioaccumulation is the process of harmful metals or organic compounds being bound inside a cell structure. The biological activity of the biomass is critical in the bioaccumulation process (Loppi et al. 2020). As a result of their metabolic functions, cells must be alive in order to adsorb pollutants. The procedure should be kept under constant control in order to get the intended results. During bioaccumulation, metal ions are absorbed by the entire cell. Metals enter the cells of living beings through the same pathways as nutrients do. Heavy metals are absorbed by unicellular organisms with the help of essential minerals like calcium and magnesium (Louati et al. 2020). Plants absorb nutrients from water or soil through their roots, whereas mammals absorb nutrients through their digestive or respiratory systems. There are two stages of bioaccumulation. Metal ions are bonded on the cell surface in the first stage. The metabolically passive phase of the process is identical to the biosorption mechanism. Metal ions are then transferred into the cell. It is possible only when the cells are metabolically active in the second half of this process (Maal Bared 2020).

The quantity of biomass grows if correct conditions for organism growth are maintained in the second stage. In comparison to biosorption, this allows (Nguyen et al. 2020) for the binding of larger amounts of metal ions (Maal Bared 2020). Precipitation can lead to metal accumulation in microbial biomass. It is critical, however, that precipitated metals were linked to the cell. *Citrobacter*, for example, generates phosphatase, an acidic enzyme. Heavy metals are precipitated in the cell wall in the form of barely soluble acidic phosphate when this enzyme is present. Denitrification caused by *Achromobactin dentifrices* contributes to environmental acidity and hence increases metal precipitation. Polysaccharides found in the cell walls of fungus and algae, such as cellulose, chitin and alginates, also help to capture metals (Naik and Dubey 2013).

4.6 Comparison of Biosorption and Bioaccumulation

Biosorption is a metabolically gradual process that does not involve any metabolic activity. As a result, the biomass used in the biosorption process is not alive. Bioaccumulation takes place in the presence of living cells and necessitates increased nutrition and energy inputs. There is also a need to avoid the detrimental effects of heavy metals on cells, which could stymie the process (Parsania et al. 2021). As a result, heavy metals with low concentrations that impede microbial growth should be determined prior to bioaccumulation. Taking all these factors into account, bioaccumulation as a heavy metals removal approach is more complicated and costly (Parsania et al. 2021). A comparison of these two methodologies may be seen in Table 4.1.

Table 4.1. Comparison of Biosorption and Bioaccumulation.

Characteristics	Biosorption	Bioaccumulation
Cost	Low in most instances. The industrial waste can be used to generate biomass. Most of the cost is spent on transportation and biosorbent production.	Usually, it is quite high. The method is performed in the presence of living organisms, which must be preserved.
pH	The sorption capacity of heavy metals is influenced by the pH of the liquid. The process, however can occur in a wide pH range.	A considerable shift in pH can have a major impact on living cells.
Selectivity	Poor modifications/biomass transformation, on the other hand, can increase it.	Compared to biosorption, this is a significant improvement.
Rate of removal	Many mechanisms occur at a rapid rate.	In comparison to biosorption, the pace is slower. It takes a long time for intracellular accumulation to occur.
Regeneration and reuse	In many cycles, biosorbents can be regenerated and reused.	Reuse is limited due to intercellular accumulation.
Recovery of metals	Heavy metal recovery is feasible with the right eluent.	Biomass cannot be used for other purposes even if that is possible.
Energy demand	Low	Cell growth necessitates the use of energy.

4.7 The Impact of Environmental factors on Biosorption and Bioaccumulation

Many environmental parameters influence the efficiency of heavy metal biosorption and bioaccumulation, including pH, temperature, contact time, concentration, biomass age and the inclusion of other ions in a solution.

4.7.1 pH

As biosorption and bioaccumulation are analogous to ion exchange in some ways, the pH of a solution has a big impact on heavy metal elimination. The amount of binding sites available on the surface of cells is affected by pH (Deng et al. 2011). The accessible binding sites in a cell bind to hydrogen cations in a solution when the pH is low. As a result, the number of accessible places is limited, and fewer metal cations can be adsorbed. However, when pH rises, so does the amount of active areas with a negative charge that attract cations (Devatha and Shivani 2020).

4.7.2 Temperature

Temperature affects the stability of metal ions in a solution as well as metal-cell complexes. Temperatures between 20 and 35 degrees celsius, on the other hand, have little effect on biosorption and bioaccumulation (Dey et al. 2020). Higher temperatures improve biomass sorption capability, but they can also damage the sorption material.

4.7.3 Contact Time

Contact time between biomass and a metal-containing solution affects bioremediation. Both are quick processes, with most metals being adsorbed right away (Fabre et al. 2020). Starting from the moment the biomass is exposed to the solution, equilibrium is reached within the first few minutes.

4.7.4 Concentration and Age of Biomass

The degree of metal removal from the solution is also significant when there is a large concentration of biomass in the solution. However, at high biomass concentrations, the ratio of bound metals to dry matter is relatively low. Higher biomass concentrations result in the creation of larger cell aggregates, which can disrupt the reactor's balance (El Sayed and El Sayed 2020). The characteristics of cell walls, which are critical for heavy metal adsorption, can be influenced by the age of the biomass. The link between biomass age and heavy metal adsorption is not understood well, according to several observations. Older cultures may have a broader ability to remove metals than younger cultures, depending on the organisms used during the sorption process and biomagnification, or vice versa. (Elahian et al. 2017).

4.7.5 Inclusion of Other Ions in the Solution

Wastewaters are contaminated with a number of pollutants, including multiple types of metals, which affect biosorption dynamics. Metal biosorption might be hampered by the presence of other dissolved substances in a solution (Du et al. 2016). This is due to competition for binding places on the surface of cells between ions of metals that have been eliminated and other ions.

4.8 Application of Biosorption and Bioaccumulation

➤ The elimination of metal cations from fluids is mainly accomplished through biosorption and bioaccumulation.

➤ Wastewaters from the metallurgical sector, rinsed waters from electroplating, metal polishing, printed circuit board manufacture, mining operations, leachates, surface and ground waters are among the effluents that can be treated by both biosorption and bioaccumulation.

➤ Biosorption and bioaccumulation can remove a wide range of sorbates, including Al, Cd, Cr, Co, Cu, Au, Fe, Pb, Mn, Hg, Mo, Ni, Ag, U, V and Zn.

➤ During the 1980s and 1990s, pilot installations and a few commercial-scale units were built in the United States and Canada. The applicability of biosorption as a basis for metal sequestering/ recovery procedures was confirmed in these pilot plants, particularly in the case of uranium. It was put to test as part of a biotechnologically based uranium production plan that included *in situ* bioleaching. These pilot plants let researchers discover the drawbacks of employing biosorption with inactive microbial biomass in an industrial setting, owing to the high expense of converting the biomass into a suitable biosorbent material (Ying et al. 2008).

➤ The detrimental influence of co-ions in the solution on the immobilized microbial biomass's uptake of the targeted metals, as well as the biological material's diminished resilience, make recycling and reuse of the biosorbent even more challenging.

➤ Biosorption, on the other hand, is a process with a few distinct properties. It has great effectiveness in sequestering dissolved metals from very dilute complicated solutions. As a result, biosorption is an excellent contender for the treatment of complicated wastewaters with high volume but low concentration. Additionally, it has recently been demonstrated that in biological reactors with metabolically active microbial cells, biosorption works in tandem with other metabolically induced mechanisms such as bioprecipitation and bio reduction. As a result, in any metal-bearing water-treatment method based on the interactions of microbial cells with soluble metal species, biosorption should always be considered a metal immobilization process.

➤ It is feasible to use biomass as a source of specialized binding molecules that can be used to separate valuable biomolecules from a mixture. This would allow for a single-step recovery while reducing the number of traditional separation steps.

➤ Another strategy is to use biomass as a carrier of highly accessible microelements in livestock diets, substituting microelements linked to the biomass via biosorption or bioaccumulation for microelements provided in the form of inorganic salts. The focus would be on bio binding rather than bio removal in this case.

4.9 Conclusions and Prospects

With reference to the presented analysis of biosorption and bioaccumulation potentials, the following conclusions can be formulated:

Heavy metals are among the pollutants that are the most difficult to eliminate as a complement to currently employed physical and chemical approaches. When using biotechnological methods to remove heavy metals from the environment, several physicochemical factors should be considered, including temperature, pH, biomass contact time with a solution containing metals, biomass concentration and age and toxicity when using living organisms. Physical and chemical changes, as well as biomass immobilization, can be used to improve metal removal efficiency. Stirred tank reactors, fixed-bed reactors, and fluidized-bed reactors are the most often used reactors. One of the most appealing aspects of using biosorbents is that they can be regenerated.

The removal of heavy metals from the environment is critical for preserving a healthy and safe ecosystem. While physicochemical methods are frequently employed for comparable reasons, biological processes, namely biosorption/bioaccumulation, appear to be a potential alternative option in terms of prices, technological requirements, metal recovery efficiency, energy requirements and environmental consequences. When biosorption and bioaccumulation were compared, the former was revealed to be superior owing to the usage of dead/inactive biomaterials, which limit metal uptake via the cell wall. Due to the variety of the microbial cell surface, selective adsorption remains difficulty. Furthermore, when using live cells for heavy metal removal, genetic engineering may be required to improve microbial strains' metal tolerances. More research is needed to get a better understanding of heavy metal removal by these two methods.

References

Abdul Jaffar, A. H., M. Tamilselvi, A. S. Akram, M. L. Kaleem Arshan and V. Sivakumar. 2015. Comparative study on bioremediation of heavy metals by solitary ascidian, *Phallusia nigra*, between Thoothukudi and Vizhinjam ports of India. Ecotoxicol. Environ. Saf. 121: 93–99.
Ahmed, D. A. E., S. F. Gheda and G. A. Ismail. 2021. Efficacy of two seaweeds dry mass in bioremediation of heavy metal polluted soil and growth of radish (*Raphanus sativus* L.) plant. Environ. Sci. Pollut. Res. 28: 12831–12846.
Al-Ansari, M. M., H. Benabdelkamel, R. H. AlMalki, A. M. Abdel Rahman, E. Alnahmi and A. Masood. 2021. Effective removal of heavy metals from industrial effluent wastewater by a multi metal and drug resistant *Pseudomonas aeruginosa* strain RA-14 using integrated sequencing batch reactor. Environ. Res. 199: 111240.
Arroyo-Herrera, I., B. Roman-Ponce, A. L. Resendiz-Martinez, S. P. Estrada-de Los, E. T. Wang and M. S. Vasquez-Murrieta. 2021. Heavy-metal resistance mechanisms developed by bacteria from *Lerma-Chapala* basin. Arch. Microbiol. 203: 1807–1823.
Bai, H., S. Wei, Z. Jiang, M. He, B. Ye and G. Liu. 2019. Pb (II) bioavailability to algae (*Chlorella pyrenoidosa*) in relation to its complexation with humic acids of different molecular weight. Ecotoxicol. Environ. Saf. 167: 1–9.
Bai, L., H. Xu, C. Wang, J. Deng and H. Jiang. 2016. Extracellular polymeric substances facilitate the biosorption of phenanthrene on cyanobacteria *Microcystis aeruginosa*. Chemosphere. 162: 172–180.
Baselga-Cervera, B., J. Romero-Lopez, C. Garcia-Balboa, E. Costas and V. Lopez-Rodas. 2018. Improvement of the uranium sequestration ability of a *Chlamydomonas* sp. (ChlSP Strain) isolated from extreme uranium mine tailings through selection for potential bioremediation Application. Front. Microbiol. 9: 523.
Bindschedler, S., T. Q. T. Vu Bouquet, D. Job, E. Joseph and P. Junier. 2017. Fungal biorecovery of gold from E-waste. Adv. Appl. Microbiol. 99: 53–81.
Boriova, K., S. Cernansky, P. Matus, M. Bujdos, A. Simonovicova and M. Urik. 2019. Removal of aluminium from aqueous solution by four wild-type strains of *Aspergillus niger*. Bioprocess. Biosyst. Eng. 42: 291–296.
Cachada, A., R. Pereira, E. F. da Silva and A. C. Duarte. 2014. The prediction of PAHs bioavailability in soils using chemical methods: state of the art and future challenges. Sci. Total Environ. 472: 463–480.
Chamekh, A., O. Kharbech, R. Driss-Limam, C. Fersi, M. Khouatmeya and R. Chouari. 2021. Evidences for antioxidant response and biosorption potential of *Bacillus* simplex strain 115 against lead. World J. Microbiol. Biotechnol. 37: 44.

Chen, S. H., Y. L. Cheow, S. L. Ng, S. L and A. S. Y. Ting. 2019. Mechanisms for metal removal established via electron microscopy and spectroscopy: a case study on metal tolerant fungi *Penicillium simplicissimum*. J. Hazard. Mater. 362: 394–402.

Chen, X., M. Zheng, G. Zhang, F. Li, H. Chen and Y. Leng. 2020. The nature of dissolved organic matter determines the biosorption capacity of Cu by algae. Chemosphere 252: 126465.

Chojnacka, K. 2010. Biosorption and bioaccumulation—the prospects for practical applications. Environ. Int. 36: 299–307.

Danouche, M., G. N. El and A. H. El. 2021. Phycoremediation mechanisms of heavy metals using living green microalgae: physicochemical and molecular approaches for enhancing selectivity and removal capacity. Heliyon 7: e07609.

Deng, Z., L. Cao, H. Huang, X. Jiang, W. Wang, Y. Shi et al. 2011. Characterization of Cd- and Pb-resistant fungal endophyte *Mucor* sp. CBRF59 isolated from rapes (*Brassica chinensis*) in a metal-contaminated soil. J. Hazard. Mater. 185: 717–724.

Devanesan, S. and M. S. AlSalhi. 2021. Effective removal of Cd(2+), Zn(2+) by immobilizing the non-absorbent active catalyst by packed bed column reactor for industrial wastewater treatment. Chemosphere 277: 130230.

Devatha, C. P. and S. Shivani. 2020. Novel application of maghemite nanoparticles coated bacteria for the removal of cadmium from aqueous solution. J. Environ. Manage. 258: 110038.

Dey, P., A. Malik, A. Mishra, D. K. Singh, B. M. von and N. Jehmlich. 2020. Mechanistic insight to mycoremediation potential of a metal resistant fungal strain for removal of hazardous metals from multimetal pesticide matrix. Environ. Pollut. 262: 114255.

Du, J., P. Sun, Z. Feng, X. Zhang and Y. Zhao. 2016. The biosorption capacity of biochar for 4-bromodiphengl ether: study of its kinetics, mechanism, and use as a carrier for immobilized bacteria. Environ. Sci. Pollut. Res. 23: 3770–3780.

El Sayed, M. T. and A. S. A. El-Sayed. 2020. Tolerance and mycoremediation of silver ions by *Fusarium solani*. Heliyon. 6: e03866.

Elahian, F., S. Reiisi, A. Shahidi and S. A. Mirzaei. 2017. High-throughput bioaccumulation, biotransformation, and production of silver and selenium nanoparticles using genetically engineered *Pichia pastoris*. Nanomed. 13: 853–861.

Fabre, E., M. Dias, M. Costa, B. Henriques, C. Vale, C. B. Lopes et al. 2020. Negligible effect of potentially toxic elements and rare earth elements on mercury removal from contaminated waters by green, brown and red living marine macroalgae. Sci. Total Environ. 72: 138133.

Fargasova, A., I. Ondrejkovicova, Z. Kramarova and Z. Faberova. 2010. Changes in physiological activity of algae *Desmodesmus quadricauda* after active bioaccumulation of newly prepared and characterized Fe(III) complexes with pyridine-3-carboxamide (pca) by living algal cells. Bioresour. Technol. 101: 6410–6416.

Flouty, R. and G. Estephane. 2012. Bioaccumulation and biosorption of copper and lead by a unicellular algae *Chlamydomonas reinhardtii* in single and binary metal systems: a comparative study. J. Environ. Manage. 111: 106–114.

Gajda-Meissner, Z., K. Matyja, D. M. Brown, M. G. J. Hartl and T. F. Fernandes. 2020. Importance of surface coating to accumulation dynamics and acute toxicity of copper nanomaterials and dissolved copper in *Daphnia magna*. Environ. Toxicol. Chem. 39: 287–299.

Garg, S. K., M. Tripathi and T. Srinath. 2012. Strategies for chromium bioremediation of tannery effluent. Rev. Environ. Contam. Toxicol. 217: 75–140.

Gerber, U., R. Hubner, A. Rossberg, E. Krawczyk-Barsch and M. L. Merroun. 2018. Metabolism-dependent bioaccumulation of uranium by *Rhodosporidium toruloides* isolated from the flooding water of a former uranium mine. PLoS. One. 13: e0201903.

Hansda, A., V. Kumar and A. Mishra. 2016. A comparative review towards potential of microbial cells for heavy metal removal with emphasis on biosorption and bioaccumulation. World J. Microbiol. Biotechnol. 32: 170.

Hlihor, R. M., M. Diaconu, F. Leon, S. Curteanu, T. Tavares and M. Gavrilescu. 2015. Experimental analysis and mathematical prediction of Cd(II) removal by biosorption using support vector machines and genetic algorithms. N. Biotechnol. 32: 358–368.

Ibuot, A., R. E. Webster, L. E. Williams and J. K. Pittman. 2020. Increased metal tolerance and bioaccumulation of zinc and cadmium in *Chlamydomonas reinhardtii* expressing a AtHMA4 C-terminal domain protein. Biotechnol. Bioeng. 117: 2996–3005.

Irshad, S., Z. Xie, S. Mehmood A. Nawaz, A. Ditta and Q. Mahmood. 2021. Insights into conventional and recent technologies for arsenic bioremediation: a systematic review. Environ. Sci. Pollut. Res. 28: 18870–18892.

Jin, C. S., R. J. Deng, B. Z. Ren, B. L. Hou and A. S. Hursthouse. 2020. Enhanced biosorption of Sb(III) onto living *Rhodotorula mucilaginosa* strain DJHN070401: Optimization and Mechanism. Curr. Microbiol. 77: 2071–2083.

Joshi, P. K., A. Swarup, S. Maheshwari, R. Kumar and N. Singh. 2011. Bioremediation of heavy metals in liquid media through fungi isolated from contaminated sources. Indian J. Microbiol. 51: 482–487.

Kadukova, J. 2016. Surface sorption and nanoparticle production as a silver detoxification mechanism of the freshwater alga *Parachlorella kessleri.* Bioresour. Technol. 216: 406–413.

Kumar, P. A., Y. S. Tseng, S. R. Rani, C. W. Chen, J. S. Chang and D. C. Di. 2021. Novel application of microalgae platform for biodesalination process: A review. Bioresour. Technol. 337: 125343.

Kumar, R., S. Singh and O. V. Singh. 2007. Bioremediation of radionuclides: emerging technologies. OMICS. 11: 295–304.

Leong, Y. K. and J. S. Chang. 2020. Bioremediation of heavy metals using microalgae: Recent advances and mechanisms. Bioresour. Technol. 303: 122886.

Loppi, S., A. Vannini, F. Monaci, D. Dagodzo, F. Blind, M. Erler et al. 2020. Can chitin and chitosan replace the lichen *Evernia prunastri* for environmental biomonitoring of Cu and Zn Air contamination? Biology. (Basel), 9.

Louati, I., J. Elloumi-Mseddi, W. Cheikhrouhou, B. Hadrich, M. Nasri, S. Aifa et al. 2020. Simultaneous cleanup of Reactive Black 5 and cadmium by a desert soil bacterium. Ecotoxicol. Environ. Saf. 190: 110103.

Maal-Bared, R. 2020. Operational impacts of heavy metals on activated sludge systems: the need for improved monitoring. Environ. Monit. Assess. 192: 560.

Naik, M. M. and S. K. Dubey. 2013. Lead resistant bacteria: lead resistance mechanisms, their applications in lead bioremediation and biomonitoring. Ecotoxicol. Environ. Saf. 98: 1–7.

Nguyen, T. Q., V. Sesin, A. Kisiala and R. J. N. Emery. 2020. The role of phytohormones in enhancing metal remediation capacity of algae. Bull. Environ. Contam. Toxicol. 105: 671–678.

Parsania, S., P. Mohammadi and M. R. Soudi. 2021. Biotransformation and removal of arsenic oxyanions by *Alishewanella agri* PMS5 in biofilm and planktonic states. Chemosphere 284: 131336.

Pushkar, B., P. Sevak, S. Parab and N. Nilkanth. 2021. Chromium pollution and its bioremediation mechanisms in bacteria: a review. J. Environ. Manage. 287: 112279.

Qiu, J., X. Song, S. Li, B. Zhu, Y. Chen, L. Zhang et al. 2021. Experimental and modeling studies of competitive Pb (II) and Cd (II) bioaccumulation by *Aspergillus niger.* Appl. Microbiol. Biotechnol. 105: 6477–6488.

Reddy, K., N. Renuka, S. Kumari and F. Bux. 2021. Algae-mediated processes for the treatment of antiretroviral drugs in wastewater: Prospects and challenges. Chemosphere 280: 130674.

Singh, S. and V. Kumar. 2020. Mercury detoxification by absorption, mercuric ion reductase, and exopolysaccharides: a comprehensive study. Environ. Sci. Pollut. Res. 27: 27181–27201.

Thesai, A. S., G. Nagarajan, S. Rajakumar, A. Pugazhendhi and P. M. Ayyasamy. 2021. Bioaccumulation of fluoride from aqueous system and genotoxicity study on *Allium cepa* using *Bacillus licheniformis*. J. Hazard. Mater. 407: 124367.

Yaashikaa, P. R., K. P. Senthil, S. Varjani and A. Saravanan. 2020. Rhizoremediation of Cu(II) ions from contaminated soil using plant growth promoting bacteria: an outlook on pyrolysis conditions on plant residues for methylene orange dye biosorption. Bioeng. 11: 175–187.

Ying T. H., D. Ishii, A. Mahara, S. Murakami, T. Yamaoka, K. Sudesh, R. Samian, M. Fujita, M. Maeda and T. Iwata. 2008. Scaffolds from electrospun polyhydroxyalkanoate copolymers: fabrication, characterization, bioabsorption and tissue response. Biomaterials. 29(10): 1307–17.

CHAPTER 5

PAHs in Terrestrial Environment and their Phytoremediation

Sandip Singh Bhatti,[1] Astha Bhatia,[2] Gulshan Bhagat,[2] Simran Singh,[2] Salwinder Singh Dhaliwal,[3] Vivek Sharma,[3] Vibha Verma,[3] Rui Yin[4] and Jaswinder Singh[5],*

5.1 Introduction

Polycyclic Aromatic Hydrocarbons (PAHs) are organic compounds that are present everywhere in one's surroundings. PAHs have high melting and boiling points which facilitates them to remain solid at room temperature. Other properties of PAHs include low vapor pressure and very low aqueous solubility which decrease with an increase in molecular weight. The increase in the molecular weight of PAHs is also related to their resistance towards oxidation and reduction processes. According to IUPAC (International Union of Pure and Applied Chemistry), phenanthrene and anthracene are the simplest PAHs (Rengarajan et al. 2015).

5.2 What are PAHs?

Polycyclic hydrocarbons (PAHs) are persistent and widely spread organic pollutants that are found in soil, sludge, water and air. Chemically these contain multiple aromatic (benzene) rings. They are both naturally occurring and man-made. These are formed by the partial combustion of carbon source materials like forest fires, volcanic eruptions (natural), fossil fuels, petroleum refining and other anthropogenic activities. Due to anthropogenic factors, PAHs are released into aquatic systems, deposited in surface water, soils, sediments and air (Honda and Suzuki 2020, Phan Thi et al. 2020).

[1] Department of Chemistry, Lovely Professional University, Phagwara, 144001-India.
  Email: singh.sandip87@gmail.com, sandip.28707@lpu.co.in
[2] Department of Botanical and Environmental Sciences, Guru Nanak Dev University, Amritsar, 143005-India.
[3] Department of Soil Science, Punjab Agricultural University, Ludhiana, Punjab, 141004, India.
[4] Institute of Ecology, College of Urban and Environmental Sciences, and Key Laboratory for Earth Surface Processes of the Ministry of Education, Peking University, 100871, Beijing, China.
[5] Department of Zoology, Khalsa College, Amritsar, 143005-India.
* Corresponding author: singhjassi75@yahoo.co.in

PAHs released from the industries into the marine ecosystems are the main source of contaminated marine ecosystems, whereas PAHs in the air are mainly from two sources which include stationary (e.g., industrial production including power plant, coking plant, boiler and waste incinerator and residential combustion like smoking, cooking, etc.) and mobile sources (e.g., exhausts from railways, motor vehicle, aircraft, etc.). PAHs in food (or the food cycle) is a major health concern (Premnath et al. 2021). PAHs in the soil through the precipitation is carried to the surface/groundwater and incorporated into crops, which after human consumption is accumulated in human and other organisms via the food chain (Wang et al. 2018).

In terrestrial environment, PAHs are covered by the SOM (Soil Organic Matter) and soil minerals due to their hydrophobic, lipophilic and semi volatile properties resulting in their decreased degradation by microorganisms and finally their accumulation into the soil in large quantities (Mazarji et al. 2021). Due to the carcinogenic toxicity of PAHs, there are strict regulations to limit their release into the natural sources (Chatzimichail et al. 2021).

Out of 100 detected PAHs (parent and alkylated derivatives) 16 are listed in Clean Water Act, 1972 (United States of America) which are presented in Table 5.1, whereas eight are listed by European Commission (2013) due to their potential risk to mankind and ecological well-being (Table 5.2) (Ofori et al. 2021).

Table 5.1. List of 16 PAHs under Clean Water Act, 1972 (USEPA 2014).

1. Chrysene	9. Napthalene
2. Acenaphthylene	10. Benzo[b]fluoranthene
3. Acenaphthene	11. Benzo[k]fluoranthene
4. Phenanthrene	12. Benzo[a]pyrene
5. Anthracene	13. Benzo[a]anthracene
6. Fluoranthene	14. Dibenz [a,h]anthracene
7. Fluorene	15. Benzo[g,h,i]perylene
8. Pyrene	16. Indo[1,2,3-cd] pyrene

Table 5.2. List of 8 PAHs under Water Framework Directive (European Commission 2013).

1. Napthalene	5. Anthracene
2. Benzo[b]fluoranthene	6. Fluoranthene
3. Benzo[k]fluoranthene	7. Indeno [1,2,3-cd]
4. Benzo[a]pyrene	8. Benzo[g,h,i]perylene

5.2.1 Sources of PAHs in Soil

PAHs are mainly produced when there is incomplete combustion of coal, wood, gas, oil and garbage. Petroleum product spillage and various domestic and industrial activities are known to produce PAHs (Singh and Haritash 2019). PAHs can be released in the environment and can be found dispersed in soil, air and water (Wang et al. 2018, Singh and Haritash 2019). Since PAHs are hydrophobic and lipophilic in nature, soil remains a vital sink for PAHs. There are studies which report that about 90% of PAHs can be stored in soil (Wang et al. 2018). When water contaminated with PAHs is used to irrigate soil, it also acts as a source of PAHs in soil (Tsibart and Gennadiev 2013). PAHs can also be incorporated in crops through soil (containing PAHs) which can result in accumulation of PAHs in human and other organisms through food chains (Wang et al. 2018). There is thus, a dire need to monitor the concentration of PAHs in soils. Sources of PAHs in soil can be mainly classified into two categories: (1) Natural sources; (2) Anthropogenic sources (Tsibart and Gennadiev 2013). Natural sources of PAH refer to various cosmic, geological and biological factors which favor the formation of PAHs (Tsibart and Gennadiev 2013). On the other hand, anthropogenic sources of PAHs include various human activities which contribute to the production of PAHs in soil. Production of cement, aluminium, asphalt, creosote and various petrochemical industries come under anthropogenic sources of PAHs in soil.

Based on the origin of production of PAHs, sources of PAHs can be categorized into three types, i.e., pyrogenic, petrogenic and biogenic. Pyrogenic PAHs are produced when there is incomplete combustion of fossil fuels at very high temperatures under anaerobic conditions. Petrogenic PAHs are mainly present in crude and refined petroleum. These PAHs enter the soil during storage, transport and leakage from oil refineries. Biogenic PAHs are produced by microorganisms, algae, phytoplanktons and plants (Abdel-Shafy and Mansour 2016).

5.2.2 Effects of PAHs in Terrestrial Ecosystem

PAHs polluted soil can cause various health implications in humans and in other organisms (Abdel-Shafy and Mansour 2016). Many PAHs are mutagenic and carcinogenic in nature. They act as immunotoxins to various living organisms including microorganisms. Contamination of soil by PAHs can affect the population and activity of microorganisms living in the soil (Singh and Haritash 2019). Overall microbial diversity in soil can also be affected as PAHs may have toxic effects towards the microorganisms. When soil is highly contaminated with PAHs some adverse effects can be seen such as tumors in invertebrate animals. High concentrations of PAHs in soil can also affect development, immunity and reproduction of animals inhabiting that area. PAHs can be absorbed by mammals through dermal contact with PAHs polluted soil (Tsibart and Gennadiev 2013). PAHs can be absorbed by plants via roots and then transported to other parts of the plants. However, phytotoxicity induced due to PAHs is of rare occurrence. PAHs can also enter food chain through accumulation in plants. Water run-off from PAHs contaminated soils into the water bodies can cause exposure of PAHs to seafood and fishes (Abdel-Shafy and Mansour 2016). In some studies, it is reported that long term exposure of skin to PAHs causes skin cancer in animals.

Thus high levels of PAHs can prove fatal for various terrestrial animals. PAHs can leach from soil and can contaminate underground water. Bathing with groundwater contaminated with PAHs can cause skin problems. Dermal contact with soil contaminated with PAHs can cause redness of skin, peeling or blistering. The health effects caused due to exposure of PAHs depend on the:

i. Quantity of PAHs entering the body

ii. Duration of exposure to PAHs

iii. Response of body to PAHs

Some tests are available to check the presence of PAHs in blood and urine. These tests cannot predict the potential side effects of PAHs that entered the body. Cancer is the far most alarming side effect of PAHs present in soil. Many PAHs present in soil are slightly mutagenic *in-vitro* but their derivatives can be potent mutagens. The various effects of PAHs in terrestrial ecosystems are given in Figure 5.1.

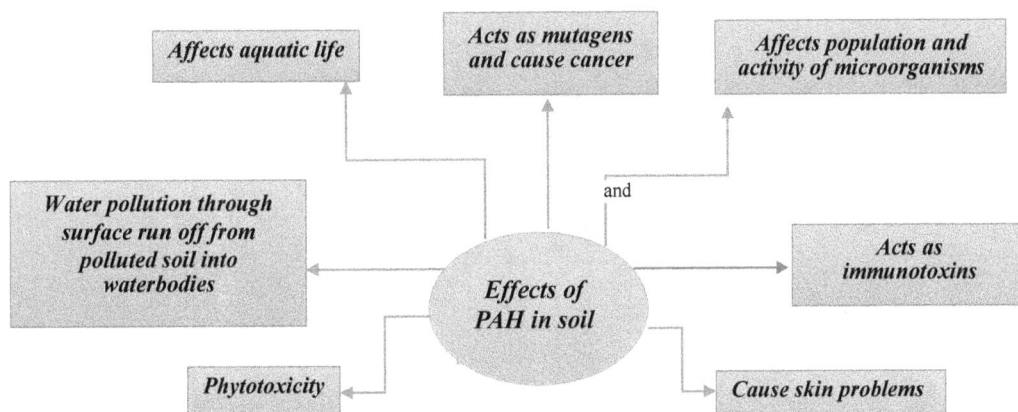

Figure 5.1. Overview of effects of PAHs in soil.

5.2.3 *PAHs in Terrestrial Ecosystems as Reported in the Literature*

Several studies have been conducted to determine the levels of PAHs in different types of soils of India and other countries that are presented in the Table 5.3. The natural concentration of PAHs present in the soil is in the range of 0.001–0.010 µg g^{-1} or by Dutch standards 0.02–0.05 µg g^{-1} DW (Wang et al. 2010). The concentration of PAHs increases due to anthropogenic inputs in soils around industrial belts, urban and traffic areas. The level of PAHs can be as high as 200000 µg kg^{-1} (DW) in the soil near oil-refineries and in traffic-affected soils it is less than 2000 µg kg^{-1} (DW) (Rengarajan et al. 2015). In some studies, the health risks of PAHs were also assessed. For such assessment, the USEPA (2005) health risk model is used. PAHs are very toxic, mutagenic and carcinogenic even at very low concentrations. The carcinogenic potencies of PAHs are measured by determining TEFs (Toxic Equivalency Factors).

5.2.4 *Overview of Remediation measures for PAHs in Soil*

PAHs have many environmental implications, as they are resistant to environmental degradation. The reasons behind their resistance are their hydrophobic nature, low water solubility, low sorption, low volatility, high oxidative resistance and low bioavailability. They are ubiquitous pollutants. PAHs can enter the human body by three major ways: inhalation, consumption and epidermal. They are toxic, mutagenic, carcinogenic, teratogenic and potent immunosuppressants and hence have adverse impacts on human health and other living organisms. PAHs contaminated-soil negatively affects the soil microbial community. Microbial activity and biomass get reduced due to the presence of PAHs in soil pores. It is investigated that the population of microorganisms of phyla *Actinobacteria Alphaproteobacteria Chlorfexi, Crenarchaeota* and *Deltaproteobacteria* gets reduced after the addition of pyrene in soil (Ren et al. 2015). Therefore, the accumulation of PAHs in the soil is of great concern. Keeping this in mind, many methods have been developed for the removal of PAHs from the soil. An overview of remediation methods for PAHs is depicted in Figure 5.2 and Table 5.4. Physical and chemical methods like solvent extraction, chemical oxidation, air sparging, adsorption, thermal desorption, photo-oxidation and electrokinetic remediation are available for remediation. But these methods are expensive, labor-intensive, inefficient and can form toxic intermediates (Gan et al. 2009). Thus bioremediation and phytoremediation are the two emerging potential approaches to mitigate the effects of PAHs. In bioremediation, living organisms (mainly microorganisms) and their products (mostly enzymes), are used to remove persistent pollutants like PAHs. The catabolic actions of enzymes convert complex PAHs (both high molecular weight and low molecular weight)

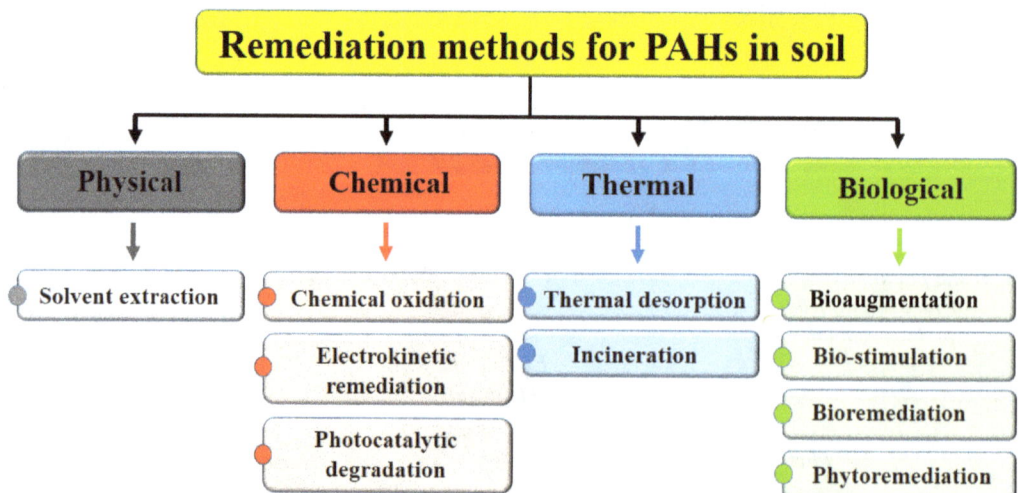

Figure 5.2. Overview of remediation methods for PAHs in soil.

Table 5.3. PAHs detected in soils in different geographical regions of the world.

Sr. No.	Types of Soil	Sites	Sampling time/study year	Extraction methods	No. PAHs	Range of PAHs	References
1	Air-port soil	Indira Gandhi International Airport, Delhi, India	November 2005–May 2006	Ultra-sonication method	12	2394–7529 (ng g⁻¹)	Ray et al. 2008
2	Traffic soil	Delhi, India	Winter season, 2006	Ultra-sonication Method	16	1062–9652 (µg kg⁻¹)	Agarwal 2009
3	Forest-fire region soil	South Korea	2001	Accelerated Solvent Extraction (ASE)	16	153–1,570 (ng g⁻¹)	Kim et al. 2011
4	Soil from brick manufacturing site (San Nicolás)	Mexico (North America)	-	Soxhlet extraction	13	7–1384 (ng g⁻¹)	Barrán-Berdón et al. 2012
5	Urban soil	Kurukshetra, India	June 2012	Ultra-sonication	16	19.1–2538.0 (µg kg⁻¹)	Kumar et al. 2013
6	Roadside soil	Punjab, Jalandhar, India	2011–2012 Summer Winter Autumn	Soxhlet extraction	16	18.17(µg g⁻¹) 4.04 (µg g⁻¹) 16.38(µg g⁻¹)	Kumar et al. 2014
7	Industrial soil	Ulsan city South Korea	July 2010	Soxhlet extraction	16	65-12,000 (ng g⁻¹)	Kwon and Choi 2014
8	Urban soil	Kogarah Bay, Sydney Australia	July 2012 to January 2013	Solvent extraction method	15	0.4–7.49 (mg kg⁻¹)	Nguyen et al. 2014
9	Sub-surface soil	Yellow River delta, China	May 2006–2008	Soxhlet extraction	16	27–753 (ng g⁻¹)	Yuan et al. 2014
10	Agricultural soil of coal production area	Xinzhou, China	June 2010	ASE	16	N.D.–782 (ng g⁻¹)	Zhao et al. 2014
11	Urban soil	Guwahati city, Brahmaputra valley, India	May 2011 (monsoon) November 2011 (Post-monsoon) February 2011 (Pre-monsoon)	Ultrasonic agitation		799.7–19 492 (ng g⁻¹) 1271.4–51299.7 (ng g⁻¹) 1639.7–29492.3 (ng g⁻¹)	Hussain and Hoque 2015

Table 5.3 contd. ....

...Table 5.3 contd.

Sr. No.	Types of Soil	Sites	Sampling time/study year	Extraction methods	No. PAHs	Range of PAHs	References
12	Eastern Himalayan soil	Guwahati, Tezpur, Dibrugarh, Itanagar, India	2014	Soxhlet Extraction	16	15.3 to 4762 (ng g^{-1})	Devi et al. 2016
13	Urban soil near industrial area	Lanzhou, China	July 2012	Soxhlet extraction	22	115 to 12,100 (µg kg^{-1})	Jiang et al. 2016
14	Urban traffic soil	Jharkhand, India	November 2014	Ultra-sonication method	13	1.019–10.856 (µg g^{-1})	Suman et al. 2016
15	Industrial soil	Dilovasi, Turkey	February 2015–Feb 2016	Ultrasonication	15	49–10512 (µg g^{-1})	Cetin et al. 2017a
16	Urban soil	Istanbul, Turkey	September–Dec 2014	Ultrasonic extraction	16	684 ± 609 (µg kg^{-1})	Cetin et al. 2017b
17	Around chemical plant Agricultural soil Roadside soil Park green space soil	Shanxi, China	September 2016	Ultrasonication	16	3.87–76.0 (mg kg^{-1}) 35.4–116.0 (mg kg^{-1}) 5.93–66.5 (mg kg^{-1})	Jiao et al. 2017
18	Exposed-lawn soil of urban parks	Guangzhou, China	May 2014	Ultrasonic extraction	16	76.44–890.85 (ng g^{-1})	Ke et al. 2017
19	Agricultural soil near industrial area Surface soil Sub-soil	Yangtze River delta region, China	2015	Ultrasonic extraction	16	189.5–1070.4 (ng g^{-1}) 103.8–743.7 (ng g^{-1})	Wang et al. 2017
20	Waste-water irrigated subsurface soil Ground-water irrigated subsurface soil	Tongliao, China	September 2012	Soxhlet extraction	16	103.28–479.32 (ng g^{-1}) 140.46–418.32 (ng g^{-1})	Zhang et al. 2017
21	Oil refinery sludge	Haldia, Barauni and Guwahati, India	–	Ultrasonication method	13	67.023–95.21 (µg g^{-1})	Tarafdar and Sinha 2018
22	Metropolitan soil	Naples Southern Italy	December 2014–Feb 2015	Accelerated Solvent Extraction (ASE)	20	94.9 (ng g^{-1})	Qu et al. 2019
23	Coal mines soil	Dhanbad, India	October–December 2016	Ultra-sonication method	13	8.256–12.562 (µg g^{-1})	Tarafdar and Sinha 2019

No.	Soil type	Location	Season/Period	Method	Number	Concentration	Reference
24	Urban soil	Tezpur, Brahmaputra valley India	2015 Pre-monsoon (March) Monsoon (July) Post-monsoon (December)	Ultra-sonication	16	242.68–2414.37 (ng g⁻¹) 134.25–773.79 (ng g⁻¹) 594.73–7961.22 (ng g⁻¹)	Deka et al. 2020
25	Urban soil near thermal power plants	Delhi, India	pre-winter (October 2011) post-winter (March 2012)	Ultra-sonication	7	956–2348 (µg g⁻¹) 1230–2714 (µg g⁻¹)	Gupta and Kumar 2020
26	Dumpsite surface soil	Chennai, India	-	Microwave assisted extraction system	16	0.62–3649 (ng g⁻¹)	Rajan et al. 2021
27	Urban soil	Industrial sites, Jamshedpur Bokaro, India	January–mid March 2020	Solvent extraction method	16	2223.1–11265.9 (ng g⁻¹) 729.6–5358.9 (ng g⁻¹)	Ambade et al. 2022
28	Solid waste dumpsite soil	Awka, Nigeria	Dry season Wet season	-	-	4.73 ± 5.69 (µg g⁻¹) 6.83 ± 10.58 (µg g⁻¹)	Aralu et al. 2022
29	Power plants soil	Bijie, Guizhou province of China		Soxhlet extraction	16	0.2–12.2 (mg kg⁻¹)	Liu et al. 2022
30	Coal based soil (road dust)	Chattisgarh sites: Raipur, Bhilai, Korba, India	May 2008–May 2015	Solvent extraction method	16	8689–87458 (ng g⁻¹)	Nayak et al. 2022
31	Roadside soil	Lucknow, India	-	-	15	478.94–8164.07 (ng g⁻¹)	Shukla et al. 2022

Table 5.4. Different remediation measures for PAH contaminated soils.

Sr. No.	Remediation Method	Details
1.	Solvent extraction method/Soil washing	In the solvent extraction method, as the name suggests, solvents are used to separate out compounds based on the solubility in the solvent. In this technique, PAHs are removed or washed from the soil with the help of solvents or surfactants. As PAHs are hydrophobic, they get readily dissolved in organic solvents. Individual or a mixture of different organic solvents, vegetable oils (sunflower oil, vegetable oil) and cyclodextrins can be used for PAHs extraction. The solvent extraction method is a two-step process, involving desorption and elution. The mixture of cyclohexane and ethanol, (3:1), mixture of ethyl acetate, acetone and water, (5:4:1), mixture of 1-pentanol (5%), water (10%) and ethanol (85%) can be used as solvents for PAHs extraction (Singh and Haritash 2019). However, these solvents can be toxic, so selection of solvents is important before proceeding. β-cyclodextrin (BCD), methyl β-cyclodextrin (MCD) and hydroxypropyl-β-cyclodextrin (HPCD) are three cyclodextrins which are used for PAHs flushing from soil. They are non-toxic and biodegradable. To enhance the efficiency of washing of PAHs surfactants can be used, known as surfactant-aided soil washing. Non-ionic surfactants such as Tween 40, Tween 80, T-Maz 80, Brij 30, CA 620 are used in PAHs washing (Gan et al. 2009).
2.	Thermal treatment	Heat is used in thermal treatment to destroy or volatilize the PAHs into gases which are then collected for further *ex-situ* treatment. In thermal desorption, the maximum temperature of 450°C is applied to increase the vapor pressure of PAHs to convert them into a gaseous form. In microwave frequency heating, microwave energy is converted into heat energy to remove PAHs via volatilization. Vitrification is another thermal technique, in which extreme temperatures of the range 1600–2000°C is used to melt the contaminants in soil. The temperature is provided through an electric current via molybdenum electrodes.
3.	Chemical treatment	Chemical treatment can be of different types viz; chemical oxidation, electrokinetic remediation method or photocatalytic degradation. In the chemical oxidation method, there is an involvement of electrons or redox reactions. Electrons are transferred from one chemical to another, converting more toxic chemicals to fewer toxic ones. It involves the use of different oxidants such as Ozone, Fenton's reagent, hydrogen peroxide, persulfate, potassium permanganate and peroxymonosulfate for PAHs remediation. In the electrokinetic method, direct low voltage electric current is applied to remove soil contaminants. Contaminants get accumulated towards electrodes, which are then collected for further treatment. The electrokinetic method can be combined with different surfactants and solvents to increase the solubility and desorption of contaminants. In the photocatalytic degradation method, photocatalysts are used for oxidizing reactions or photo reactions for the degradation of PAHs in soil. The function of photocatalysts is to enhance the rate of photodegradation. TiO_2, and H_2O_2 are photocatalysts that help in the remediation of PAHs (De Boer and Wagelmans 2016).
4.	Biological treatment	Different types of biological remediation methods are available and are being developed. Bioremediation, biostimulation, bioaugmentation, phytoremediation, landfarming, composting and bioventing are biological techniques for PAHs remediation. In bioremediation, living organisms mainly potential natural microorganisms that have the capability (enzymatic machinery) of degradation, are used to convert toxic organic pollutants to CO_2 and H_2O. *Acinetobacter, Aeromonas, Alcaligenes, Alcanivorax, Arthrobacter, Bacillus, Corynebacterium, Enterobacter, Microbulbifer, Micrococcus, Mycobacteria, Paenibacillus, Pseudomonas, Ralstonia Sphingomonas* and *Xanthomonas* are the reported bacterial genera that are capable of PAHs degradation. Bioremediation through fungi is known as mycoremediation. *Aspergillus, Bjerkandera adusta, Irpex lacteus, Phanerochaete chrysosporium, Pleurotus ostreatus* are the reported PAHs degrading fungi. In biostimulation, nutrients and oxygen (electron acceptor) are added to contaminated soil to stimulate the degradative activity of microorganisms. Nutrients can be carbon, organic biostimulants (phycocyanin) nitrogen (NH_4CL), phosphorous (NaH_2PO_4) (Singh and Haritash 2019). In bioventing, air or oxygen is supplied to contaminated soil through wells to trigger the growth of PAHs degrading microorganisms. In the landfarming technique, contaminated soil is excavated from the site and transported to the landfarming site, where by the degradative action of the microorganisms' PAHs are degraded slowly. In composting, mesophilic and thermophilic microbes, degrade contaminants at high temperature, i.e., 55–66°C. Compost beds (mushroom compost) are prepared for this. Phytoremediation is the process where plants and associated microbes with some soil amendments and agronomic techniques are used to accumulate, immobilize, adsorb, absorb, degrade or volatilize the harmful contaminants into less harmful ones. Phytoremediation in combination with other techniques, increases the remediation efficiency. The detailed account of phytoremediation method is discussed below.

into less toxic forms under both aerobic as well as anaerobic conditions. In aerobic degradation, CO_2 and water are formed as by-products, while in anaerobic conditions, methane is formed. Some of the bacterial species involved in PAHs remediation are *Pseudomonas, Rhodococcus, Bacillus* and *Mycobacterium* (Imam et al. 2022). Gamma-proteobacteria and actinobacteria are the two major bacterial lineages that degrade PAHs in PAH-contaminated soils (Chaudhary et al. 2015). Apart from bacteria, fungi are also employed for remediation known as mycoremediation due to their co-metabolic activity (Srivastava and Kumar 2019).

Phytoremediation is a broader term, it includes many techniques where plants are used to remove, detoxify or immobilize contaminants like PAHs in soil. It is even more efficient, eco-friendly and cost-effective than bioremediation. Phytoextraction, rhizoextraction, phytovolatilization, phyto stabilization, phyto stimulation, phyto transformation, phyto assimilation, phyto reduction, etc., are various green technologies of phytoremediation. Plants involved in the phytoremediation process are called "hyperaccumulators" or "phytoremediators". The process in which plants remove contaminants from soil by storing them in tissues is called phytoextraction/rhizoextraction/phytofilteration. In the phytovolatilization process, plants remove the contaminants through volatilization. In phyto stabilization, plants reduce the mobility of contaminants. Phyto stabilization followed by the addition of microbes degrade the immobile contaminants. The physical, chemical or biological remediation methods can be employed both *in-situ* and *ex-situ*. In *in-situ*, remediation is done at the original contaminated soil whereas in *ex-situ*, contaminated soil is transferred to another location for remediation. Bionano-remediation is also an emerging technique for PAHs degradation. The use of nanoparticles like Nano zero-valent iron and metal oxides, carbon-based and polymer-based materials with biotechnological tools could be a promising approach for PAHs remediation. It is observed that the addition of nanoparticles in soil positively affects the mobility of PAHs, decreases the phytotoxicity and microbial flora remain unaffected (Mazarji et al. 2021). Combinations of physical-biological, chemical-biological or all the three (physical-chemical-biological) known as integrated methods are also used as remediation techniques for PAHs (Patel et al. 2020). Although various remediation technologies are available for PAH contaminated soils, but out of them, the most suitable and sustainable technology is the phytoremediation (Dolatabadi et al. 2021).

5.3 Phytoremediation of PAHs from Soil

Phytoremediation is a process that involves the use of plant/parts of plants to remove, degrade or stabilize the pollutants from soil/water. By the process of phytoremediation (phytodegradation, rhizodegradation, phytoextraction, rhizofilteration, etc.) different PAHs can be removed (Bose et al. 2022). The term "Phytoremediation" refers to the technique that uses plants to clean up soil and water contaminated with harmful substance (Ashraf et al. 2019, Dhaliwal et al. 2020). Phytoremediation has gained the attention of the scientific community for environmental remediation as it provides an alternative to the conventional methods that are energy intensive, use expensive chemicals and instruments and alter the soil properties (Ali et al. 2013, Dal Corso et al. 2019). Thus, to achieve site-specific remedial goals, phytoremediation is considered as a cost-effective and sustainable approach that has a profound restoration effect without disturbing the native microflora (Wan et al. 2016, Nedjimi 2021). The technology uses green plants (trees, shrubs, grasses and aquatic plants) because of their transport capacity and accumulation of contaminants which enables them to remove, degrade or isolate toxic environmental contaminants (Yan et al. 2020). These contaminants include heavy metals, organic compounds such as polyaromatic hydrocarbons (PAHs) and radioactive substances in soil or water. Till date, various plants have been used for remediation of various contaminants (Table 5.5).

Table 5.5. Application of Phytoremediation for removing PAHs from soil.

Type of phytoremediation	Plant	Contaminant	References
Phytostabilization	*Robinia pseudoacacia* Nyirsegi	Phenanthrene	Wawra et al. 2018
Phytodegradation	*Erythrina crista*-galli L.	Petroleum-contaminated soil	de Farias et al. 2009
	Lupinus luteus	PAHs	Gutiérrez-Ginés et al. 2014
	Acorus calamus	Phenanthrene and pyrene	Jeelani et al. 2017
	Maize	Phenanthrene and pyrene	Houshani et al. 2021
	Vallisneria spiralis and *Hydrilla verticillate*	Phenanthrene and pyrene	He and Chi 2019
	Sorghum	Pyrene	Salehi et al 2020
Rhizodegradation	*Avicennia marina*	Pyrene	Jia et al. 2016
	Mangrove (*Kandelia candel* (L.) Druce)	Phenanthrene and pyrene	Lu et al. 2011
	Maize (*Zea Mays* L.) Inoculated with *Piriformospora Indica*	Petroleum-contaminated soil	Zamani et al. 2018
	Rhizophora mangle	Σ16 PAHs	Verâne et al. 2020
	Melia azadirachta with bacteria *Bacillus flexus* and *Paenibacillus* sp. S118	Benzo(a)pyrene	Kotoky and Pandey 2020
	Lolium multiflorum	Total petroleum hydrocarbon	Hussain et al. 2022

5.3.1 *Mechanism of Phytoremediation*

The mechanism of the phytoremediation process varies with the chemical properties of the contaminant as well as plant characteristics (Figure 5.3). Thus, different strategies under phytoremediation have been discussed below.

5.3.1.1 *Phytoextraction*

In phytoextraction, contaminants are absorbed by roots followed by their translocation and accumulation in their aboveground biomass (Sreelal and Jayanthi 2017).

Screening of suitable plant species is the key and most straightforward strategy for successful phytoextraction, i.e., the plant must be efficient in accumulating contaminants in the aerial parts. Besides hyperaccumulation, the plant to act as eminently suitable for phytoextraction must also possess traits like (1) rapid growth and production of large biomass; (2) vast root systems; (3) easy cultivation and harvesting management; (4) preferably be repulsive to herbivores to avoid the entrance in food chain (Seth 2012). However, natural hyperaccumulating plants lack these characteristics thus limiting the phytoextraction potential (Chaney et al. 2005). To overcome the problem, research has focused to modify or engineer large biomass producing non-hyperaccumulator plants to achieve the above-mentioned attributes. To date, numerous hyperaccumulator plants ranging from annual herbs to perennial shrubs and trees, have been used for phytoextraction. Phytoextraction is considered advantageous as it does not alter the landscape, preserves the ecosystem and is cost-effective, thus considered as the most commercially promising technique. However, several factors such as lower bioavailability and absorption of metal in the roots limit the metal's phytoextraction by plants. However, the technique has been so far used for heavy metals (Jacobs et al. 2017, Guo et al. 2020).

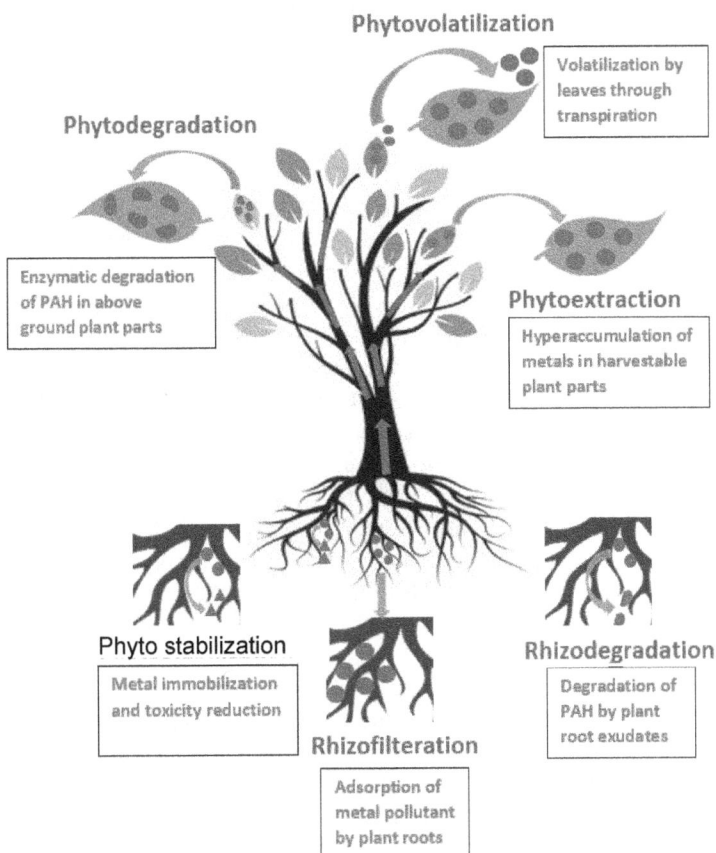

Figure 5.3. Different mechanisms for phytoremediation.

5.3.1.2 Phyto Stabilization

In phyto stabilization, the contaminants are immobilized in the root system through adsorption on roots, adsorption and accumulation by roots or precipitation in the rhizosphere. The process supresses the pollutants mobility, preventing their migration by leaching, erosion or dispersion along with soil, water or air and thus attenuate their bioavailability in the food chain (Radziemska et al. 2017, Zgorelec et al. 2020). Phyto stabilization does not physically eliminate the pollutants from the soil, but rather causes the deactivation and immobilization of their potential ions, thereby preventing further migration to the food chain. For an effective phyto stabilization process, selection of suitable plants that offer certain characteristics is necessary. Plants should grow rapidly under field conditions, possess high tolerance to the soil conditions and vast root system. Moreover, the plants must be easy to maintain under site-specific conditions and must have a relatively long life (Muthusaravanan et al. 2018). The efficiency of the phyto stabilization process is generally affected by various factors such as contaminant to be removed, soil cover, rhizosphere modification and root system of the plant. Vegetation cover interrupts the direct impact of wind and rain, increases the soil compaction thus reducing the velocity of runoff, and acts as an insulator against temperature variation. Rhizosphere induces changes such as acidification, microbial activity and release of organic acids alter the soil properties and thus the transformation, bioavailability and mobility of contaminants which affects the phyto stabilization of contaminated sites. The plant root anatomy is a critical factor for potential application of this technique. The higher root biomass producing plants with dense and deep roots are desirable for phyto stabilization as they can exploit larger volumes of contaminated soils. Fibrous roots provide greater surface area for contaminant absorption and plant–microbe interactions. Remediation potential of black locust (*Robinia pseudoacacia* Nyirsegi) on soil contaminated with polycyclic aromatic hydrocarbons was also studied. The concentration of phenanthrene was reduced to greater extent with the integrated application of black locust and soil amendments after 1 yr of plant growth. The uptake of trace elements Cu, As, Cd, Pb and Sb to leaves was low (Wawra et al. 2018).

5.3.1.3 Phytovolatilization

Phytovolatilization has been observed for numerous contaminants including volatile organic compounds and inorganic contaminants. Based on the possible mechanisms, the technique has been divided into two categories, i.e., direct phytovolatilization by stems or leaves and indirect phytovolatilization from the root zone. In direct phytovolatilization, the pollutants are absorbed by roots, transported through the xylem and excreted to the atmosphere from the aerial plant parts in the volatile form. However, in indirect phytovolatilization process, volatile organic contaminants flux increases from the subsurface due to plant roots activities, e.g., increasing soil permeability, chemical transport via hydraulic redistribution or water table fluctuations (Limmer and Burken 2016). This strategy may prove advantageous as the contaminant can be transformed from more a toxic form to a relatively lesser toxic substance, however, there might be the possibility that the resultant form is still potentially toxic, and thus resettle into the environment. The phytovolatilization can be applied for contaminants present in soil, sediment or water, particularly, for organic contaminants. Till date, this technique has been used for heavy metals (Shrestha et al. 2006, Sakakibara et al. 2010).

5.3.1.4 Phytodegradation

'Phytodegradation' is a subset technique of phytoremediation under which organic contaminants are absorbed by the roots followed by degradation due to catalytic action of enzymes which are involved in metabolism of the plant (Newman and Reynolds 2004, Sharma and Pandey 2014). The enzymes involved in the phytodegradation process are dehalogenase, peroxidase, nitroreductase, nitrilase and phosphatase (Chatterjee et al. 2013). Phytodegradation is primarily used for the remediation sites by contamination with organic pollutants like and PAHs (Muthusaravanan et al. 2018). To date, numerous plants have been used for phytodegradation of organic contaminants. *Lupinus luteus* in association with endophytic bacteria has shown immense phytodegradation potential in landfill soils of the Iberian Peninsula contaminated with PAHs (Gutiérrez-Ginés et al. 2014). He and Chi (2019) investigated the phytodegradation of phenanthrene and pyrene using aquatic plants, *Vallisneria spiralis* and *Hydrilla verticillata*, in PAH-polluted sediments. Among them, sediments planted with *V. spiralis* showed the highest dissipation of phenanthrene and pyrene (85.9 and 79.1%) as compared to sediments planted with *H. verticillata* and unplanted sediment. Phytodegradation of phenanthrene and pyrene using the maize plant has been confirmed using GC-MS analysis. The degradation rate of phenanthrene was found to be faster than that of pyrene and prominently occurred in the roots (Houshani et al. 2021).

5.3.1.5 Rhizofiltration

'Rhizofiltration' or 'phytofiltration' refers to the process that uses both terrestrial and aquatic plants in adsorbing contaminants in the surrounding root zone (rhizosphere), concentrating and precipitating them on or within the root (Rahman and Hasegawa 2011). The plants' roots are harvested after becoming saturated. The plant species which possess high tolerance towards metal toxicity and have a high surface area for absorption of metal are preferred (e.g., *Salix* spp., *Populus* spp., *Brassica* spp.). Terrestrial plants have been reported as a suitable candidate for rhizofilteration as they possess a more developed and fibrous root structure, thus provide a higher surface area for absorption of contaminants (Cule et al. 2016). This strategy can be applied for the remediation of contaminated sites loaded with heavy metals and radionuclides. The factors that limit the application of this technique are pH adjustment, hydroponic cultivation in a greenhouse, frequent harvests and proper disposal of the plants (Kristanti et al. 2020).

Research studies demonstrated that sunflowers have shown greater potential for phytoremediation of soil contaminated with radionuclides (caesium and strontium) in Chernobyl, Ukraine. The results indicated that Cs was accumulated in roots, whereas Sr uptake was recorded in the shoots (Prasad 2007). Another study reported that rhizofiltration using and *Phaseolus vulgaris* has shown remediation of uranium contaminated groundwater (Lee and Yang 2010). The aromatic medicinal plant, *Plectranthus amboinicus*, has been evaluated as a candidate for rhizofilteration of lead-containing wastewater. The results indicated that *P. amboinicus* showed potential tolerance towards Pb toxicity and accumulates Pb maximum in the roots, whereas translocation in the stem and leaf was limited (Ignatius et al. 2014). In hydroponic experiments, *Typha angustifolia* and *Acorus calamus* (aquatic) and *Pandanus amaryllifolius* (terrestrial) plants were evaluated for rhizofiltration of Cd and Zn. Among these, *T. angustifolia* showed maximum tolerance towards HM toxicity without loss in dry biomass production. With maximum HM accumulation in roots, *T. angustifolia* was found to be the most suitable candidate plant for phytoremediation in constructed wetlands and aquatic plant systems (Woraharn et al. 2021).

5.3.1.6 Rhizodegradation

Rhizodegradation is an emerging technique for remediation of contaminated soil that involves plant roots, plant-supplied nutrients and soil microorganisms. This process occurs through the plant-supplied substrates such as bacteria, fungi and yeasts which favors the growth of microbial communities in the rhizosphere to break down organic pollutants (Allamin et al. 2020, Rajkumari et al. 2021). The factors affecting the rhizodegradation efficiency of processes are the type of contaminant, the ability of microflora to degrade the contaminants and the bioavailability of pollutants (Noroozi et al. 2017, Patel and Patra 2017). Plant roots provide air to the soil and release exoenzymes and nutrients through root exudates. The advantages of rhizodegradation are complete mineralization of the contaminant, lesser translocation of the contaminant to other plant parts or in the atmosphere, low installation and maintenance costs (Dos Santos and Maranho 2018). However, there exist some disadvantages of the process as it is a slow process and effective only on the surface of contamination (20–25 cm of depth); the limited depth of the roots and requirement of fertilizers by plants.

Various studies have reported the effective remediation of soils via rhizodegradation contaminated including PAHs, pesticides and solvents containing benzene ring. Plants such as *Avicennia marina* (Jia et al. 2016) or *Lolium multiflorum* (Hussain et al. 2022) have shown immense potential towards the degradation of PAH following the rhizodegradation mechanism. The efficiency of *Rhizophora mangle* to extract PAHs in mangrove sediment contaminated with crude oil has also been studied. The soil planted with *Rhizophora mangle* showed higher removal of 16 PAHs as compared to natural attenuation (Verâne et al. 2020).

5.3.2 Parameters that affect the Phytoremediation Efficiency

To achieve maximum efficiency of the phytoremediation process, different aspects of certain parameters should be taken care of. These include the type of contaminant, selection of plant species, type of soil and other biotic and abiotic conditions (Mudgal et al. 2010).

5.3.2.1 Contaminant Characteristics and Concentration

To date, phytoremediation technology has been used for the removal of a wide range of soil contaminants including inorganic and organic contaminants (Henry 2000). For organic pollutants, the hydrophobicity affects the efficiency of its uptake by the plants to a greater extent. The remediation of soil contaminated with moderately hydrophobic pollutants has been successfully reported due to efficient uptake and translocation by the plant (Cunningham and Ow 1996). Inorganic pollutants generally include heavy metals and the efficiency of phytoremediation process for removal of one metal may vary as compared to a removal of mixture of metals (Dushenkov et al. 1995). Apart from the contaminant type, the concentration of the contaminant should also be estimated prior to the phytoremediation to prevent the toxicity effects on healthy plants. The non-aqueous phase pollutants create adverse effects on plant growth. Pollutants with low bioavailability are difficult to extract with phytoremediation process (EPA 2000).

5.3.2.2 Selection of Plant Species

It requires substantial information regarding the plant biological system before deciding its suitability for phytoremediation of a given area. The root system of the plants significantly affects the phytoremediation potential. Fibrous roots offer the advantage of greater surface area for adsorption and hence extraction of the pollutant. On the other hand, tap roots possess central large roots extending in the deeper soil layers, thus can extract the pollutant from the deeper layers (Schwab 1998). The desirable root depth for phytoremediation varies for different plants. Non-woody plant roots have shown efficient removal with feet depth, whereas tree roots less than 10–20 feet are found to be most effective (Gatliff 1994). The scrutiny of seed source and seed health is also imperative before commencing the phytoremediation process. Seeds from local regions are preferred as these adapt to the environmental conditions easily (EPA 2000).

5.3.2.3 Climatic Conditions and site Accessibility

Climatic conditions such as temperature, moisture, rainfall and sunlight influence seed germination and plant growth to a major extent. Thus, as a suitable candidate for phytoremediation, the adaptability of the plant under climatic variations must be considered (EPA 2000). Consideration of accessibility to the site is imperative before the commencement of the process as the consumption of these plants by livestock or general public may lead to severe health consequences. Thus, edible plants are not usually preferred for remediation (Ghavzan and Trivedy 2005).

Conclusively, phytoremediation is an economic, sustainable and environment friendly process for soil reclamation. Moreover, the process is catalyzed by natural solar driven pumps and their associated metabolic processes. It is also environmentally benign as it reduces water losses by reducing its evaporation, limits soil erosion and prevents run-off that occurs due to heavy rain or flooding.

5.4 Future Perspectives

Undoubtedly, the phytoremediation technology has proved beneficial over the conventional methods for soil reclamation. However, there exists a gap between the process and its practical application.

- The strategy has been applied in laboratory and greenhouse tests, but its application extensively at field studies should be tested.

- Another major problem of the process is the effective extraction, the contaminant is located shallow enough for the plant roots to reach it, which is not possible in case of each contaminant.

- Longer lifecycle duration of most plants to attain the maturity stage also limits the application of this process. Nowadays plants with a short lifecycle are being scrutinized to overcome this problem.

- Extremely high levels of contaminants pose adverse effects on the plants' growth and may even led to plant death. Thus, the process is effective only for lower concentrations of contaminant.

References

Abdel-Shafy, H. I. and M. S. Mansour. 2016. A review on polycyclic aromatic hydrocarbons: source, environmental impact, effect on human health and remediation. Egypt. J. Pet. 25(1): 107–123.

Agarwal, T. 2009. Concentration level, pattern and toxic potential of PAHs in traffic soil of Delhi, India. J. Hazard. Mater. 171(1-3): 894–900.

Ali, H., E. Khan and M. A. Sajad. 2013. Phytoremediation of heavy metals—concepts and applications. Chemosphere. 91(7): 869–881.

Allamin, I. A., M. I. E. Halmi, N. A. Yasid, S. A. Ahmad, S. R. S. Abdullah and Y. Shukor. 2020. Rhizodegradation of petroleum oily sludge-contaminated soil using *Cajanus cajan* increases the diversity of soil microbial community. Sci. Rep. 10(1): 1–11.

Ambade, B., S. S. Sethi and M. R. Chintalacheruvu. 2022. Distribution, risk assessment, and source apportionment of polycyclic aromatic hydrocarbons (PAHs) using positive matrix factorization (PMF) in urban soils of East India. Environ. Geochem. Health 1–15. https://doi.org/10.1007/s10653-022-01223-x.

Aralu, C. C., P. A. C. Okoye, K. G. Akpomie, H. O. Chukwuemeka-Okorie and H. O. Abugu. 2022. Polycyclic aromatic hydrocarbons in soil situated around solid waste dumpsite in Awka, Nigeria. Toxin Rev. 1–10. https://doi.org/10.1080/15569543.2021.2022700.

Ashraf, S., Q. Ali, Z. A. Zahir, S. Ashraf and H. N. Asghar. 2019. Phytoremediation: Environmentally sustainable way for reclamation of heavy metal polluted soils. Ecotoxicol. Environ. Saf. 174: 714–727.

Barrán-Berdón, A. L., V. Garcia Gonzalez, G. Pedraza Aboytes, I. Rodea-Palomares, A. Carrillo-Chavez, H. Gomez-Ruiz and B. Verduzco Cuellar. 2012. Polycyclic aromatic hydrocarbons in soils from a brick manufacturing location in central Mexico. Rev. Int. Contam. Ambient. 28(4): 277–288.

Bose, V. G., K. S. Shreenidhi and J. A. Malik. 2022. Phytoremediation of PAH-contaminated areas. pp. 141–156. *In*: Advances in Bioremediation and Phytoremediation for Sustainable Soil Management. Springer, Cham.

Cetin, B., S. Yurdakul, M. Keles, I. Celik, F. Ozturk and C. Dogan. 2017a. Atmospheric concentrations, distributions and air-soil exchange tendencies of PAHs and PCBs in a heavily industrialized area in Kocaeli, Turkey. Chemosphere. 183: 69–79.

Cetin, B., F. Ozturk, M. Keles and S. Yurdakul. 2017b. PAHs and PCBs in an Eastern mediterranean megacity, Istanbul: their spatial and temporal distributions, air-soil exchange and toxicological effects. Environ. Pollut. 220: 1322–1332.

Chaney, R. L., J. S. Angle, M. S. McIntosh, R. D. Reeves, Y. M. Li, E. P. Brewer, K. Y. Chen, R. J. Roseberg, H. Perner, E. C. Synkowski and C. L. Broadhurst. 2005. Using hyperaccumulator plants to phytoextract soil Ni and Cd. Z. Naturforsch C 60(3-4): 190–198.

Chatterjee, S., A. Mitra, S. Datta and V. Veer. 2013. Phytoremediation protocols: An overview. *In*: D. Gupta (eds.). Plant-Based Remediation Processes. Soil Biol. 35. https://doi.org/10.1007/978-3-642-35564-6_1.

Chatzimichail, S., F. Rahimi, A. Saifuddin, A. J. Surman, S. D. Taylor-Robinson and A. Salehi-Reyhani. 2021. Hand-portable HPLC with broadband spectral detection enables analysis of complex polycyclic aromatic hydrocarbon mixtures. Commun. Chem. 4(1): 1–14.

Chaudhary, P., H. Sahay, R. Sharma, A. K. Pandey, S. B. Singh, A. K. Saxena and L. Nain. 2015. Identification and analysis of polyaromatic hydrocarbons (PAHs)—biodegrading bacterial strains from refinery soil of India. Environ. Monit. Assess. 187(6): 1–9.

Cule, N., D. Vilotic, M. Nesic, M. Veselinovic, D. Drazic and S. Mitovic. 2016. Phytoremediation potential of *Canna indica* L. in water contaminated with lead. Fresenius Environ. Bull. 25(11): 3728–3733.

Cunningham, S. D. and D. W. Ow. 1996. Promises and prospects of phytoremediation. Plant Physiol. 110(3): 715–719.

Dal Corso, G., E. Fasani, A. Manara, G. Visioli and A. Furini. 2019. Heavy metal pollutions: state of the art and innovation in phytoremediation. Int. J. Mol. Sci. 20(14): 3412.

De Boer, J. and M. Wagelmans. 2016. Polycyclic aromatic hydrocarbons in soil–practical options for remediation. CLEAN–Soil Air Water. 44(6): 648–653.

Deka, J., N. Baul, P. Bharali, K. P. Sarma and R. R. Hoque. 2020. Soil PAHs against varied land use of a small city (Tezpur) of middle Brahmaputra Valley: Seasonality, sources, and long-range transport. Environ. Monit. Assess. 192: 357.

Devi, N. L., I. C. Yadav, Q. Shihua, Y. Dan, G. Zhang and P. Raha. 2016. Environmental carcinogenic polycyclic aromatic hydrocarbons in soil from Himalayas, India: Implications for spatial distribution, sources apportionment and risk assessment. Chemosphere. 144: 493–502.

Dhaliwal, S. S., J. Singh, P. K. Taneja and A. Mandal. 2020. Remediation techniques for removal of heavy metals from the soil contaminated through different sources: a review. Environ. Sci. Pollut. Res. 27(2): 1319–1333.

Dolatabadi, N., S. Mohammadi Alagoz, B. Asgari Lajayer and E. D. van Hullebusch. 2021. Phytoremediation of polycyclic aromatic hydrocarbons-contaminated soils. pp. 419–445. *In*: Climate Change and the Microbiome. Springer, Cham.

Dos Santos, J. J. and L. T. Maranho. 2018. Rhizospheric microorganisms as a solution for the recovery of soils contaminated by petroleum: a review. J. Environ. Manag. 210: 104–113.

Dushenkov, V., P. N. Kumar, H. Motto and I. Raskin. 1995. Rhizofiltration: the use of plants to remove heavy metals from aqueous streams. Environ. Sci. Technol. 29(5): 1239–1245.

EPA Environmental Protection Agency. 2000. Introduction to Phytoremediation. EPA-report EPA/600/R-99/107. http://clu-in.org/techpubs.html.

European Commission. 2013. No. SANCO/12571/2013 of 19 November 2013. Guidance document on analytical quality control and validation procedures for pesticide residues analysis in food and feed, Supersedes SANCO/12495/2011. Implemented by 01/01/2014.

de Farias, V., L. T. Maranho, E. C. de Vasconcelos, M. A. da Silva Carvalho Filho, L. G. Lacerda, J. A. M. Azevedo, A. Pandey and C. R. Soccol. 2009. Phytodegradation potential of *Erythrina crista-galli* L., Fabaceae, in petroleum-contaminated soil. Appl. Biochem. Biotechnol. 157(1): 10–22. https://doi.org/10.1007/s12010-009-8531-1.

Gan, S., E. V. Lau and H. K. Ng. 2009. Remediation of soils contaminated with polycyclic aromatic hydrocarbons (PAHs). J. Hazard. Mater. 172(2-3): 532–549.

Gatliff, E. G. 1994. Vegetative remediation process offers advantages over traditional pump-and-treat technologies. Remediat. J. 4(3): 343–352.

Ghavzan, N. J. and R. K. Trivedy. 2005. Environmental pollution control by using phytoremediation technology. Poll. Res. 24(4): 875–884.

Guo, Y., C. Qiu, S. Long, H. Wang and Y. Wang. 2020. Cadmium accumulation, translocation, and assessment of eighteen *Linum usitatissimum* L. cultivars growing in heavy metal contaminated soil. Int. J. Phytoremediat. 22(5): 490–496.

Gupta, H. and R. Kumar. 2020. Distribution of selected polycyclic aromatic hydrocarbons in urban soils of Delhi, India. Environ. Technol. Innov. 17: 100500.

Gutiérrez-Ginés, M. J., A. J. Hernández, M. I. Pérez-Leblic, J. Pastor and J. Vangronsveld. 2014. Phytoremediation of soils co-contaminated by organic compounds and heavy metals: bioassays with *Lupinus luteus* L. and associated endophytic bacteria. J. Environ. Manag. 143: 197–207.

He, Y. and J. Chi. 2019. Pilot-scale demonstration of phytoremediation of PAH-contaminated sediments by *Hydrilla verticillata* and *Vallisneria spiralis*. Environ. Technol. 40(5): 605–613.

Henry, J. R. 2000. An overview of the phytoremediation of lead and mercury. pp. 1–31. Washington, DC: US Environmental Protection Agency, Office of Solid Waste and Emergency Response, Technology Innovation Office.

Honda, M. and N. Suzuki. 2020. Toxicities of polycyclic aromatic hydrocarbons for aquatic animals. Int. J. Environ. Res. Public Health. 17(4): 1363.

Houshani, M., S. Y. Salehi-Lisar, R. Motafakkerazad and A. Movafeghi. 2021. Proposed pathways for phytodegradation of phenanthrene and pyrene in maize (*Zea Mays* L.) using GC-MS analysis. Research Square. DOI: https://doi.org/10.21203/rs.3.rs-1110084/v1.

Hussain, F., A. H. A. Khan, I. Hussain, A. Farooqi, Y. S. Muhammad, M. Iqbal, M. Arslan and S. Yousaf. 2022. Soil conditioners improve rhizodegradation of aged petroleum hydrocarbons and enhance the growth of *Lolium multiflorum*. Environ. Sci. Pollut. Res. 29(6): 9097–9109.

Hussain, K. and R. R. Hoque. 2015. Seasonal attributes of urban soil PAHs of the Brahmaputra Valley. Chemosphere. 119: 794–802.

Ignatius, A., V. Arunbabu, J. Neethu and E. V. Ramasamy. 2014. Rhizofiltration of lead using an aromatic medicinal plant *Plectranthus amboinicus* cultured in a hydroponic nutrient film technique (NFT) system. Environ. Sci. Pollut. Res. 21(22): 13007–13016.

Imam, A., S. K. Suman, P. K. Kanaujia and A. Ray. 2022. Biological machinery for polycyclic aromatic hydrocarbons degradation: a review. Bioresour. Technol. 343: 126121.

Jacobs, A., T. Drouet, T. Sterckeman and N. Noret. 2017. Phytoremediation of urban soils contaminated with trace metals using *Noccaea caerulescens*: comparing non-metallicolous populations to the metallicolous 'Ganges' in field trials. Environ. Sci. Pollut. Res. 24(9): 8176–8188.

Jeelani, N., W. Yang, L. Xu, Y. Qiao, S. An and X. Leng. 2017. Phytoremediation potential of *Acorus calamus* in soils co-contaminated with cadmium and polycyclic aromatic hydrocarbons. Sci. Rep. 7: 8028.

Jia, H., H. Wang, H. Lu, S. Jiang, M. Dai, J. Liu and C. Yan. 2016. Rhizodegradation potential and tolerance of *Avicennia marina* (Forsk.) Vierh in phenanthrene and pyrene contaminated sediments. Mar. Pollut. Bull. 110(1): 112–118.

Jiang, Y., U. J. Yves, H. Sun, X. Hu, H. Zhan and Y. Wu. 2016. Distribution, compositional pattern and sources of polycyclic aromatic hydrocarbons in urban soils of an industrial city, Lanzhou, China. Ecotoxicol. Environ. Saf. 126: 154–162.

Jiao, H., Q. Wang, N. Zhao, B. Jin, X. Zhuang and Z. Bai. 2017. Distributions and sources of polycyclic aromatic hydrocarbons (PAHs) in soils around a chemical plant in Shanxi, China. Int. J. Environ. Res. Public Health 14(10): 1198.

Ke, C. L., Y. G. Gu and Q. Liu. 2017. Polycyclic aromatic hydrocarbons (PAHs) in exposed-lawn soils from 28 urban parks in the megacity Guangzhou: occurrence, sources, and human health implications. Arch. Environ. Contam. Toxicol. 72(4): 496–504.

Kim, E. J., S. D. Choi and Y. S. Chang. 2011. Levels and patterns of polycyclic aromatic hydrocarbons (PAHs) in soils after forest fires in South Korea. Environ. Sci. Pollut. Res. 18(9): 1508–1517.

Kotoky, R. and P. Pandey. 2020. Rhizosphere mediated biodegradation of benzo(A)pyrene by surfactin producing soil bacilli applied through *Melia azadirachta* rhizosphere. Int. J. Phytoremediat. 22(4): 363–372.

Kristanti, R. A. W. J. Ngu, A. Yuniarto and T. Hadibarata. 2021. Rhizofiltration for removal of inorganic and organic pollutants in groundwater: a review. Biointerafce Res. Appl. Chem. 4: 12326–12347.

Kumar, A. V., N. C. Kothiyal, S. Kumari, R. Mehra, A. Parkash, R. R. Sinha, S. K. Tayagi and R. Gaba. 2014. Determination of some carcinogenic PAHs with toxic equivalency factor along roadside soil within a fast developing northern city of India. J. Earth Syst. Sci. 123(3): 479–489.

Kumar, B., V. K. Verma, S. Kumar and C. S. Sharma. 2013. Probabilistic health risk assessment of polycyclic aromatic hydrocarbons and polychlorinated biphenyls in urban soils from a tropical city of India. J. Environ. Sci. Health, Part A 48(10): 1253–1263.

Kwon, H. O. and S. D. Choi. 2014. Polycyclic aromatic hydrocarbons (PAHs) in soils from a multi-industrial city, South Korea. Sci. Total Environ. 470: 1494–1501.

Lee, M. and M. Yang. 2010. Rhizofiltration using sunflower (*Helianthus annuus* L.) and bean (*Phaseolus vulgaris* L. var. vulgaris) to remediate uranium contaminated groundwater. J. Hazard. Mater. 173(1-3): 589–596.

Limmer, M. and J. Burken. 2016. Phytovolatilization of organic contaminants. Environ. Sci. Technol. 50(13): 6632–6643.

Liu, J., S. Zhang, J. Jia, M. Lou, X. Li, S. Zhao, W. Chen, B. Xiao and Y. Yu. 2022. Distribution and source apportionment of polycyclic aromatic hydrocarbons in soils at different distances and depths around three power plants in Bijie, Guizhou Province. Polycycl. Aromat. Compd. https://doi.org/10.1080/10406638.2022.2039232.

Lu, H., Y. Zhang, B. Liu, J. Liu, J. Ye and C. Yan. 2011. Rhizodegradation gradients of phenanthrene and pyrene in sediment of mangrove (*Kandelia candel* (L.) Druce). J. Hazard. Mater. 196: 263–269.

Mazarji, M., T. Minkina, S. Sushkova, S. Mandzhieva, G. N. Bidhendi, A. Barakhov and A. Bhatnagar. 2021. Effect of nanomaterials on remediation of polycyclic aromatic hydrocarbons-contaminated soils: a review. J. Environ. Manag. 284: 112023.

Mudgal, V., N. Madaan and A. Mudgal. 2010. Heavy metals in plants: phytoremediation: plants used to remediate heavy metal pollution. Agric. Biol. J. North Am. 1(1): 40–46.

Muthusaravanan, S., N. Sivarajasekar, J. S. Vivek, T. Paramasivan, M. Naushad, J. Prakashmaran, V. Gayathri and O. K. Al-Duaij. 2018. Phytoremediation of heavy metals: mechanisms, methods and enhancements. Environ. Chem. Lett. 16(4): 1339–1359.

Nayak, Y., S. Chakradhari, K. S. Patel, R. K. Patel, S. Yurdakul, H. Saathoff and P. Martín-Ramos. 2022. Distribution, variations, fate and sources of polycyclic aromatic hydrocarbons and carbon in particulate matter, road dust, and sediments in central India. Polycycl. Aromat. Compd. https://doi.org/10.1080/10406638.2022.2026991.

Nedjimi, B. 2021. Phytoremediation: a sustainable environmental technology for heavy metals decontamination. SN Appl. Sci. 3: 286. https://doi.org/10.1007/s42452-021-04301-4.

Newman, L. A. and C. M. Reynolds. 2004. Phytodegradation of organic compounds. Curr. Opin. Biotechnol. 15(3): 225–230.

Nguyen, T. C., P. Loganathan, T. V. Nguyen, S. Vigneswaran, J. Kandasamy, D. Slee, G. Stevenson and R. Naidu. 2014. Polycyclic aromatic hydrocarbons in road-deposited sediments, water sediments, and soils in Sydney, Australia: comparisons of concentration distribution, sources and potential toxicity. Ecotoxicol. Environ. Saf. 104: 339–348.

Noroozi, M., M. A. Amozegar, R. Rahimi, S. A. Shahzadeh Fazeli and G. Bakhshi Khaniki. 2017. The isolation and preliminary characterization of native cyanobacterial and microalgal strains from lagoons contaminated with petroleum oil in Khark Island. Biological Journal of Microorganism 5(20): 33–41.

Ofori, S. A., S. J. Cobbina, A. Z. Imoro, D. A. Doke and T. Gaiser. 2021. Polycyclic Aromatic Hydrocarbon (PAH) pollution and its associated human health risks in the Niger delta region of Nigeria: A Systematic Review. Environ. Process. 8(2): 455–482.

Patel, A. and D. D. Patra. 2017. A sustainable approach to clean contaminated land using terrestrial grasses. pp. 305–331. *In*: Phytoremediation Potential of Bioenergy Plants. Springer, Singapore.

Patel, A. B., S. Shaikh, K. R. Jain, C. Desai and D. Madamwar. 2020. Polycyclic aromatic hydrocarbons: sources, toxicity, and remediation approaches. Front. Microbiol. 11: 562813.

Phan Thi, L. A., N. T. Ngoc, N. T. Quynh, N. V. Thanh, T. T. Kim, D. H. Anh and P. H. Viet. 2020. Polycyclic aromatic hydrocarbons (PAHs) in dry tea leaves and tea infusions in Vietnam: contamination levels and dietary risk assessment. Environ. Geochem. Health. 42(9): 2853–2863.

Prasad, M. N. V. 2007. Sunflower (*Helinathus annuus* L.)—a potential crop for environmental industry/girasol (*Helianthus annuus* L.)-cultivo potencial para la industria ecológica/le tournesol (*Helianthus annuus* L.) culture potentielle dans l'industrie écologique. Helia. 30(46): 167–174.

Premnath, N., K. Mohanrasu, R. G. R. Rao, G. H. Dinesh, G. S. Prakash, V. Ananthi, K. Ponnuchamy, G. Muthusamy and A. Arun. 2021. A crucial review on polycyclic aromatic Hydrocarbons-Environmental occurrence and strategies for microbial degradation. Chemosphere. 280: 130608.

Qu, C., S. Albanese, A. Lima, D. Hope, P. Pond, A. Fortelli, N. Romano, P. Cerino, A. Pizzolante and B. De Vivo. 2019. The occurrence of OCPs, PCBs, and PAHs in the soil, air, and bulk deposition of the Naples metropolitan area, southern Italy: implications for sources and environmental processes. Environ. Int. 124: 89–97.

Radziemska, M., M. D. Vaverková and A. Baryła. 2017. Phytostabilization—management strategy for stabilizing trace elements in contaminated soils. Int. J. Environ. Res. Public Health. 14(9): 958.

Rahman, M. A. and H. Hasegawa. 2011. Aquatic arsenic: phytoremediation using floating macrophytes. Chemosphere. 83(5): 633–646.

Rajan, S., K. R. Rex, M. Pasupuleti, J. Muñoz-Arnanz, B. Jiménez and P. Chakraborty. 2021. Soil concentrations, compositional profiles, sources and bioavailability of polychlorinated dibenzo dioxins/furans, polychlorinated biphenyls and polycyclic aromatic hydrocarbons in open municipal dumpsites of Chennai city, India. Waste Manag. 131: 331–340.

Rajkumari, J., Y. Choudhury, K. Bhattacharjee and P. Pandey. 2021. Rhizodegradation of pyrene by a non-pathogenic *Klebsiella pneumoniae* isolate applied with *Tagetes erecta* L. and changes in the rhizobacterial community. Front. Microbiol. 12: 593023.

Ray, S., P. S. Khillare, T. Agarwal and V. Shridhar. 2008. Assessment of PAHs in soil around the international airport in Delhi, India. J. Hazard. Mater. 156(1-3): 9–16.

Ren, G., W. Ren, Y. Teng and Z. Li. 2015. Evident bacterial community changes but only slight degradation when polluted with pyrene in a red soil. Front. Microbiol. 6: 22.

Rengarajan, T., P. Rajendran, N. Nandakumar, B. Lokeshkumar, P. Rajendran and I. Nishigaki. 2015. Exposure to polycyclic aromatic hydrocarbons with special focus on cancer. Asian Pac. J. Trop. Biomed. 5(3): 182–189.

Sakakibara, M., A. Watanabe, M. Inoue, S. Sano and T. Kaise. 2010. Phytoextraction and phytovolatilization of arsenic from As-contaminated soils by *Pteris vittata*. Vol. 121, p. 26. *In*: Proceedings of the Annual International Conference on Soils, Sediments, Water And Energy.

Salehi, N., A. Azhdarpoor and M. Shirdarreh. 2020. The effect of different levels of leachate on phytoremediation of pyrene-contaminated soil and simultaneous extraction of lead and cadmium. Chemosphere. 246: 125845.

Schwab, A. P. 1998. Phytoremediation of soils contaminated with PAHs and other petroleum compounds. *In*: Beneficial effects of vegetation in contaminated soils workshop, Kansas State University, Manhattan, KS.

Seth, C. S. 2012. A review on mechanisms of plant tolerance and role of transgenic plants in environmental clean-up. Bot. Rev. 78(1): 32–62.

Sharma, P. and S. Pandey. 2014. Status of phytoremediation in world scenario. Int. J. Environ. Bioremediat. Biodegrad. 2(4): 178–191.

Shrestha, B., S. Lipe, K. A. Johnson, T. Q. Zhang, W. Retzlaff and Z. Q. Lin. 2006. Soil hydraulic manipulation and organic amendment for the enhancement of selenium volatilization in a soil–pickleweed system. Plant and Soil. 288(1): 189–196.

Shukla, S., R. Khan, P. Bhattacharya, S. Devanesan and M. S. AlSalhi. 2022. Concentration, source apportionment and potential carcinogenic risks of polycyclic aromatic hydrocarbons (PAHs) in roadside soils. Chemosphere. 292: 133413.

Singh, S. K. and A. K. Haritash. 2019. Polycyclic aromatic hydrocarbons: soil pollution and remediation. Int. J. Sci. Environ. Technol. 16(10): 6489–6512.

Sreelal, G. and R. Jayanthi. 2017. Review on phytoremediation technology for removal of soil contaminant. Indian J. Sci. Res. 14(1): 127–130.

Srivastava, S. and M. Kumar. 2019. Biodegradation of polycyclic aromatic hydrocarbons (PAHs): a sustainable approach. pp. 111–139. *In*: Sustainable Green Technologies for Environmental Management. Springer, Singapore.

Suman, S., A. Sinha and A. Tarafdar. 2016. Polycyclic aromatic hydrocarbons (PAHs) concentration levels, pattern, source identification and soil toxicity assessment in urban traffic soil of Dhanbad, India. Sci. Total Environ. 545: 353–360.

Tarafdar, A. and A. Sinha. 2018. Public health risk assessment with bioaccessibility considerations for soil PAHs at oil refinery vicinity areas in India. Sci. Total Environ. 616: 1477–1484.

Tarafdar, A. and A. Sinha. 2019. Health risk assessment and source study of PAHs from roadside soil dust of a heavy mining area in India. Arch. Environ. Occup. Health. 74(5): 252–262.

Tsibart, A. S. and A. N. Gennadiev. 2013. Polycyclic aromatic hydrocarbons in soils: sources, behavior, and indication significance (a review). Eurasian Soil Sci. 46(7): 728–741.

USEPA. 2005. Guidelines for carcinogen risk assessment. EPA/630/P-03/001F2005: risk assessment forum. US Environmental Protection Agency, Washington, DC.

USEPA. 2014. Priority pollutant list. https://www.epa.gov/sites/production/files/2015-09/documents/priority-pollutant-list-epa.pdf. Accessed on April 08, 2022.

Verâne, J., N. C. Dos Santos, V. L. da Silva, M. de Almeida, O. M. de Oliveira and Í. T. Moreira. 2020. Phytoremediation of polycyclic aromatic hydrocarbons (PAHs) in mangrove sediments using Rhizophora mangle. Mar. Pollut. Bull. 160: 111687.

Wan, X., M. Lei and T. Chen. 2016. Cost-benefit calculation of phytoremediation technology for heavy-metal-contaminated soil. Sci. Total Environ. 563: 796–802.

Wang, D., J. Ma, H. Li and X. Zhang. 2018. Concentration and potential ecological risk of PAHs in different layers of soil in the petroleum-contaminated areas of the Loess Plateau, China. Int. J. Environ. Res. Public Health. 15(8): 1785.

Wang, J., X. Zhang, W. Ling, R. Liu, J. Liu, F. Kang and Y. Gao. 2017. Contamination and health risk assessment of PAHs in soils and crops in industrial areas of the Yangtze River Delta region, China. Chemosphere. 168: 976–987.

Wang, R., G. Liu, C. L. Chou, J. Liu and J. Zhang. 2010. Environmental assessment of PAHs in soils around the Anhui Coal District, China. Arch. Environ. Contam. Toxicol. 59(1): 62–70.

Wawra, A., W. Friesl-Hanl, M. Puschenreiter, G. Soja, T. Reichenauer, C. Roithner and A. Watzinger. 2018. Degradation of polycyclic aromatic hydrocarbons in a mixed contaminated soil supported by phytostabilisation, organic and inorganic soil additives. Sci. Total Environ. 628: 1287–1295.

Woraharn, S., W. Meeinkuirt, T. Phusantisampan and P. Chayapan. 2021. Rhizofiltration of cadmium and zinc in hydroponic systems. Water Air Soil Pollut. 232(5): 1–17.

Yan, A., Y. Wang, S. N. Tan, M. L. Mohd Yusof, S. Ghosh and Z. Chen. 2020. Phytoremediation: a promising approach for revegetation of heavy metal-polluted land. Front. Plant Sci. 11: 359.

Yuan, H., T. Li, X. Ding, G. Zhao and S. Ye. 2014. Distribution, sources and potential toxicological significance of polycyclic aromatic hydrocarbons (PAHs) in surface soils of the Yellow River Delta, China. Mar. Pollut. Bull. 83(1): 258–264.

Zamani, J., M. A. Hajabbasi, M. R. Mosaddeghi, M. Soleimani, M. Shirvani and R. Schulin. 2018. Experimentation on degradation of petroleum in contaminated soils in the root zone of maize (*Zea Mays* L.) inoculated with *Piriformospora indica*. Soil Sediment Contam. 27: 13–30.

Zgorelec, Z., N. Bilandzija, K. Knez, M. Galic and S. Zuzul. 2020. Cadmium and mercury phytostabilization from soil using *Miscanthus giganteus*. Sci. Rep. 10(1): 1–10.

Zhang, S., H. Yao, Y. Lu, X. Yu, J. Wang, S. Sun, M. Liu, D. Li, Y.F. Li and D. Zhang. 2017. Uptake and translocation of polycyclic aromatic hydrocarbons (PAHs) and heavy metals by maize from soil irrigated with wastewater. Sci. Rep. 7(1): 1–11.

Zhao, L., H. Hou, Y. Shangguan, B. Cheng, Y. Xu, R. Zhao, Y. Zhang, X. Hua, X. Huo and X. Zhao. 2014. Occurrence, sources, and potential human health risks of polycyclic aromatic hydrocarbons in agricultural soils of the coal production area surrounding Xinzhou, China. Ecotoxicol. Environ. Saf. 108: 120–128.

CHAPTER 6

Fungal Strategies for the Remediation of Polycyclic Aromatic Hydrocarbons

Nitu Gupta,[1] Sandipan Banerjee,[2] Apurba Koley,[3]
Aman Basu,[4] Nayanmoni Gogoi,[1] Raza Rafiqul Hoque,[1]
Narayan Chandra Mandal[2] and
Srinivasan Balachandran[3],*

6.1 Introduction

A progressive increase in industrialization over several decades has caused the release of various potentially hazardous contaminants such as xenobiotic compounds and heavy metals, that severely affect the environment, i.e., soil, air and water. In these conditions, nowadays, environmental pollutants are becoming a global concern for our ecosphere. These contaminants exhibit versatile harmful effects on the human body and ecosystem. The significant human health hazards associated with Polycyclic Aromatic Hydrocarbons (PAHs) include carcinogenic, mutagenic, teratogenic and immune suppressants. It can rapidly assimilate in the gastrointestinal tract of mammals because of its lipid-dissolvable nature (Abdel-Shafy and Mansour 2016). Such toxic substances can be bioaccumulated in the food chain due to their non-polar, hydrophobic and lipophilic nature (Masih et al. 2012). Several findings reveal that PAHs induce tumors in humans. This carcinogenic characteristic of PAHs is co-relatable with their molecular structure. Structurally, carcinogenic PAHs have a bay or K locale, showing a high affinity towards mammalian DNA. As a result, PAHs-linked DNA adducts transformed the typical cell into a tumorigenic cell (Weis et al. 1998). Hence, more attention should be paid to the remediation and restoration approaches for cleaning up the contaminated environment from these pollutants. In this context, PAHs are considered one of the prime concerns for the ecological malaise. PAHs are characterized as hydrophobic organic compounds that are chemically constituted by two or more benzenoid rings. They are omnipresent,

[1] Department of Environmental Science, Tezpur University, Napaam, Tezpur, Assam, 784028, India.
[2] Mycology and Plant Pathology Laboratory, Department of Botany, Visva-Bharati, Santiniketan 731235, West Bengal, India.
[3] Department of Environmental Studies, Visva-Bharati, Santiniketan-731235, West Bengal, India.
[4] Department of Biology, York University, Canada.
* Corresponding author: s.balachandran@visva-bharati.ac.in

and their concentration is elevated in the vicinity of the industries associated with petroleum and gas production. On the other hand, anthropogenic combustion is the key contributor to PAHs pollution (Banerjee and Mandal 2020, Ghosal et al. 2016).

PAHs compounds constitute two or more aromatic rings structurally arranged linearly, clusterly or angularly. Generally, PAH compounds are comprised of carbon-hydrogen atoms; additionally, nitrogen, oxygen and sulfur atoms can also be involved in forming heterocyclic aromatic compounds. Broadly PAHs can be categorized into two groups: Low-Molecular-Weight PAHs (LMW-PAHs) having less than four benzenoid rings whereas High-Molecular-Weight PAHs (HMW-PAHs) with more than four benzenoid rings. The stability of the PAHs primarily depends on the arrangement of the aromatic ring; the linear arrangement of aromatic rings (LMW-PAHs) represents the instability and exhibits fewer recalcitrance characteristics, whereas the angular arrangement of aromatic rings (HMW-PAHs) is highly stable and exhibits more recalcitrance features of PAHs compounds (Blumer 1976). In addition to this, according to United States Environmental Protection Agency (USEPA), 16 PAHs are classified as priority environmental pollutants and seven PAHs are categorized as potent human carcinogens, known as carcinogenic PAHs (USEPA 2002).

PAHs are omnipresent environmental contaminants, generated from the partial burning of fossil fuels. Air mass movement and transboundary deposition play a significant role in the distribution of PAH in the environment. Through long-range transport, rural areas are also getting affected, which are usually situated far from the actual origin of the PAHs. Soil and street dust are the primary sinks for the deposition of atmospheric PAHs. Furthermore, this opens the gateway for PAHs to enter the aquatic ecosystem. PAHs are preferentially fragmented and assembled in the particle state of sediments in aquatic environments due to their hydrophobic nature. In this way, PAHs are present throughout the multi-compartment structure of the ecosystem, paving the path for various exposure routes to these carcinogens (Hussain et al. 2018). Broadly, PAHs can evolve from two types of sources, i.e., naturogenic or anthropogenic sources. The naturogenic sources include forest fires, volcanic eruptions, petroleum spills, bacterial and algal synthesis and decomposition of litter fall (Abdel-Shafy and Mansour 2016). Various sources of PAHs and their exposure routes to human are reported by many researchers. The major contributors of PAHs in the ecosphere are anthropogenic sources such as the combustion of wood gas, fuels, crude oil and industrial wastes (Hussain et al. 2018). PAHs produced from anthropogenic sources are mainly categorized into pyrogenic, petrogenic and biological. Pyrogenic mediated PAHs are produced from partial burning of organic matter subjected to high-temperature ranges from 350 to 1200°C under anaerobic conditions. Examples of pyrogenic processes include the partial combustion of motors fuels in automobiles and coal is thermal distilled into coal tar and coke, thermal cracking of oil deposits asphalt creation. Therefore, pyrogenic PAHs released in the open air are accumulated more in urban regions. The PAHs generated from the petrogenic process are similar to pyrogenic except regarding petroleum processing (Nayak et al. 2022).

In this case, bioremediation is nature's own green machinery that acts consistently as a good cleanup, cost-effective, energy-efficient and eco-sustainable alternative equipment in comparison to the physicochemical techniques viz, soil replacement, soil washing and flushing, chemical reduction and oxidation, incineration and thermal desorption, vitrification, encapsulation, immobilization, electrokinetic remediation, nonthermal plasma technology, etc. (Mandree et al. 2021, Kuppusamy et al. 2017). The main objective of the bioremediation approach is based on the mineralization of these hazardous compounds into non-toxic compounds and can be achieved by employing bioremediating representatives like plants (phytoremediation), earthworms (vermiremediation), as well as microorganisms, i.e., bacteria, yeast, fungi and algae. Such microbial agents are well equipped with their enzymatic systems to bio transform the PAHs compounds into either carbon dioxide (CO_2) or partially degraded non-toxic metabolites, i.e., byproduct where no CO_2 is liberated (Cerniglia 1993). The microbial PAHs metabolism is generally achieved by a variety of mechanisms based on the enzymatic depository systems.

In regard to microbial PAHs bioremediating representatives, bacteria are extensively documented as organisms being used for bioremediation practices due to their biochemical as well as genetic flexibility (Banerjee et al. 2017). However recently, mycoremediation has been acknowledged more (Conejo-Saucedo et al. 2019, Treu and Falandysz 2017). Fungi are the most suitable agent for PAHs biotransformation and are preferred above bacteria, algae or plants for various reasons. The association of the mycelia meshwork of fungi is capable of penetrating soils and accessing the soil void space. In comparison to unicellular bacteria, the fungal mycelium act as a unit, proliferates like a net or mesh-like structure or formulates rhizomorphs around hazardous materials and can propagate in an indeterminate way until the resources are accessible (Banerjee and Mandal 2020). In addition, the key equipment of fungi is that they habitually produce a greater number of bioactive molecules, especially the extracellular enzymes which are accountable for the degradation of macromolecules like PAHs (Mougin et al. 2009, Ritz and Young 2004, Osono et al. 2003). Further, ligninolytic fungi gained great attention due to their enzymatic appliances, which are not only associated with lignin degradation but also employed during the mycoremediation practices in PAHs contaminated environment (Gadd and Gadd 2001). Apart from these enzymatic approaches, this chapter is also focused on the diverse fungal strategies such as the production of biosurfactants, biochar immobilized mycoremediation and the contribution of rhizospheric and endophytic fungi to remediate the PAHs.

6.2 Bibliographic Analysis

In this bibliographic study, a total of 1641 articles have been extracted from the Web of Science with the combinations of keywords such as PAHs, biodegradation, mycoremediation, white-rot fungi, enzyme, peroxidase, biochar, biosurfactant and so on, which are relevant to construct this chapter. All the keywords have been analyzed with the help of VOS viewer software. Here the co-occurrence keywords have been chosen up to 17, and a fractionalization network of linkage map was constructed displaying the frequently searched keywords in those articles, which is illustrated in Figure 6.1. A total of five clusters were found in this keyword co-occurrence analysis. Each cluster and co-occurrence keywords are represented by specific size and color, and the straight lines connecting them display the anecdote links. From this network analysis of the keywords, it was discovered that these articles primarily addressed the following five clusters or aspects: (1) different PAHs and their mycoremediation strategies, i.e., biosorption, biosurfactant, biochar, fungal consortium, etc. (red cluster) (2) fungal enzymes responsible for PAHs degradation like laccase, lignin peroxidase, manganese peroxidase, etc. (green cluster) (3) fungal members with PAHs degrading ligninolytic enzymatic activity (yellow cluster) (4) fungal metabolism in the biotransformation of the PAHs (sky blue cluster) (5) mycoremediation with enzymatic and surfactant based approaches (purple cluster). This keywords-oriented text mining study explored the fungal members and their strategies involved in the remediation of PAHs. At the same time, it is also demonstrated that the rhizospheric as well as arbuscular mycorrhizal fungi are the key appliances found in support of the mycoremediation processes. Therefore, the connecting bond between these diverse fungal members and their strategies assisted in the construction of this current study.

6.3 Fungal Strategies mediated PAHs Remediation

Due to the high population growth and excess usage of petroleum resources, the number of PAHs released from street transportation, industrial, commercial and residential origin are increasing exponentially day by day. Additionally, in urban areas spontaneous contamination is occurring from the fossil fuel industries, motor vehicles, electricity generation plants, etc. (Figure 6.2) (Ball and Truskewycz 2013). Further, the structural rigidity and human health impacts of PAHs (Table 6.1) gained the attention of several environmental specialists. Therefore, remediating the PAHs contaminated areas is indeed needed to restore the ecological health to a regular condition as this hazardous pollution is detrimental to ecological damages. In this context, bioremediation appears

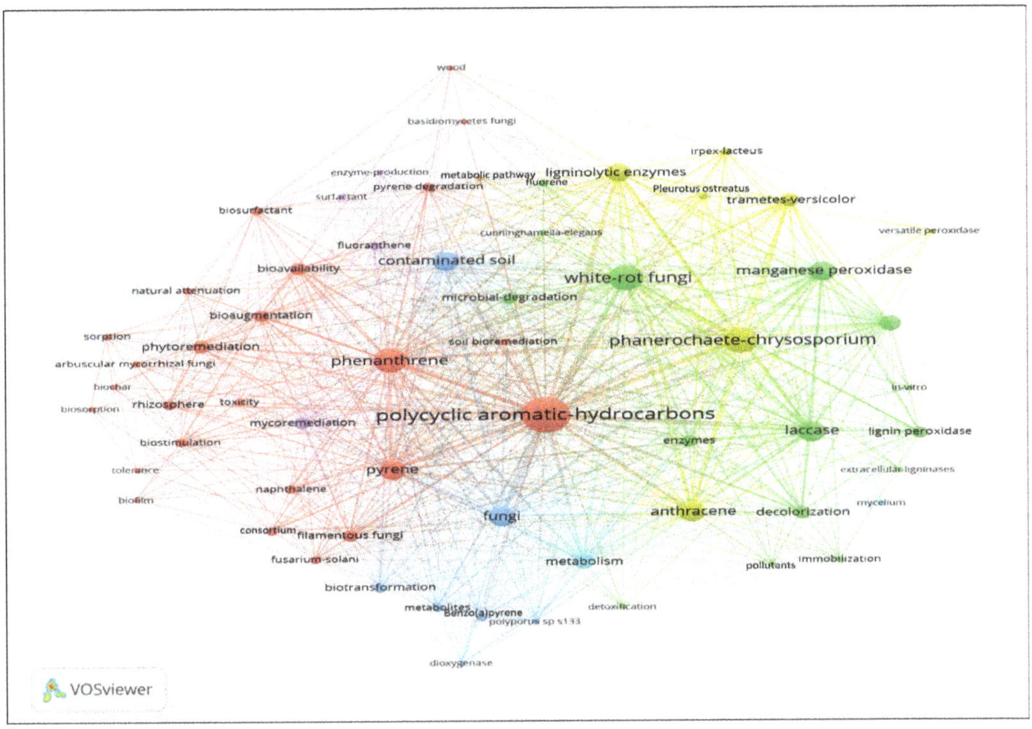

Figure 6.1. Keywords co-occurrence network analysis using VOS viewer software.

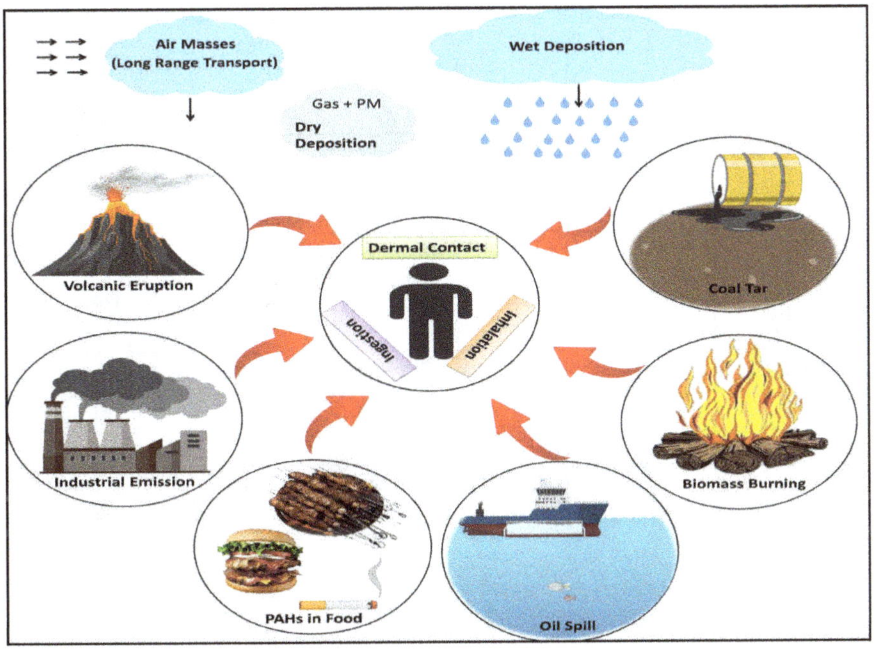

Figure 6.2. Sources and routes of PAHs exposure to humans.

as nature's own reclamation or biorestoration practice where such pollutants are either degraded or transformed into less toxic composites by various living organisms. Usually microorganisms play an indispensable role in altering the pollutants into CO_2, water, and cell biomass (Langenbach 2013). Examples of such bioremediation processes are usually known by different formulations

Table 6.1. Structure of PAHs, their molecular weight, solubility and health risk (Adopted from Pathak et al. 2022, Hussain et al. 2018).

PAHs	No. of Rings	Molecular weight (g/mol)	Solubility (mg/L)	Health risk	Structure
Naphthalene	2	128.17	3.93	Liver and kidney failure, anemia and skin inflammation	
Acenaphthene	3	154.21	1.93	Carcinogenicity, infant asthma and infection in liver and kidney	
Acenaphthylene	3	152.1	3.93	Animal carcinogenicity	
Anthracene	3	178.23	0.076	Diarrhea, nausea, eye infection and skin irritation	
Phenanthrene	3	178.23	1.2	Carcinogenicity and skin irritation	
Pyrene	4	202.25	0.077	Changes in morphology of blood cell	
Fluorene	3	166.22	1.68–1.98	Eye and skin infection	
Fluoranthene	4	202.25	0.2–2.6	Acute animal toxicity	
Benzo[a]anthracene	4	228.3	0.01	Animal carcinogenicity and skin irritation	
Chrysene	4	228.29	0.0028	Carcinogenicity, emesis, diarrhea, eye and skin infection, heart problems	
Benzo[b]fluoranthene	5	252.3	0.0012	Carcinogenicity, jaundice, liver and kidney failure	
Benz[k]fluoranthene	5	252.3	0.00076	Carcinogenicity, liver and kidney failure	
Benzo[a]pyrene	5	252.3	0.0023	Animal carcinogenicity and eye infection	

Table 6.1 contd. ...

...Table 6.1 contd.

PAHs	No. of Rings	Molecular weight (g/mol)	Solubility (mg/L)	Health risk	Structure
Dibenz[*a,h*] anthracene	5	278.4	0.0005	Carcinogenicity and toxicity to aquatic life	
Benzo[*ghi*] perylene	6	276.3	0.00026	Toxicity to aquatic life, lung and skin	
Indeno[1,2,3-*cd*] pyrene	6	276.3	0.062	Carcinogenicity, nausea, emesis and diarrhea	

viz, biosorption used dead or alive microbial organic matter, bioaccumulation assisted by microbes, biostimulation is ameliorating on-site microbial community, phytoremediation associated with plants, bioaugmentation is the unnatural incorporation of microbial communities and rhizoremediation is the interaction between plants and microbes. Among the bioremediation practices, microbes (algae, fungi, bacteria) mediated remediation is the most effective in PAHs mineralization and numerous documentations have reported more than 100 genera and 200 species of microorganisms. Where, fungi are the most predominant group with 103 genera among the other microbial representatives, i.e., bacteria (79 genera), algae (19 genera) and cyanobacteria (9 genera) documented from a different environment and responsible for effective PAHs degradation (Premnath et al. 2021). Fungi and bacteria utilize versatile strategies for remediating PAHs in contaminated soil. The first stage of the bacterial PAHs degradation mechanism includes oxidation as well as hydroxylation of PAHs. The detoxification operation is carried out as the initial stage in fungal PAHs mineralization. Individually, fungal members like *Aspergillus* sp., *Fusarium oxysporum* and *Trichocladium canadense* can also mineralize LMW-PAHs. However, in contrast to the LMW-PAHs, HMW-PAHs, i.e., pyrene (PYR) and phenanthrene (PHE) can be efficiently mineralized by the *Penicillium* sp., *A. terreus* and *Verticillium* sp. (Biswas et al. 2015). Moreover, as a consortium of bacteria and fungi, rather than individuals, proficiently enhance the rate of PAH degradation (Winquist et al. 2014, Li and Li 2008). On the other hand, fungal consortia of *A. flavus, A. fumigatus, A. nomius, Trichoderma asperellum* and *Rhizomucor variabilis* can also escalate the remediation of Benzo(a) pyrene [B(a)p], PYR and PHE (Tripathi et al. 2017). Universally fungal agents are equipped with several remediating strategies like enzymatic activities, the production of biosurfactants and the utilization of rhizospheric and endophytic fungal communities.

6.3.1 Enzymatic Activity

The notable evidence of fungal bioremediation is the synthesis of a vast number of biochemically active biomolecules, primarily extracellular enzymes, i.e., Laccase, Cytochrome P450 monooxygenase (CYP450), Lignin Peroxidase (LiP), Dioxygenase, Manganese peroxidase (MnP), Versatile Peroxidase (VP), and Dye-decolorizing peroxidases, etc., by fungi. Primarily, there are two fungal phenomenal bioremediation-oriented mechanisms, i.e., lignin catabolism-dependent mechanism (ligninolytic fungi) and CYP450 system-dependent system (non-ligninolytic fungi) (Srivastava and Kumar 2019). Ligninolytic fungi are proficient in synthesizing extracellular enzymes viz, LiP and MnP which are utilized in PAHs mineralization, apart from intracellular enzymes. Ligninolytic enzymes, break down such contaminants and produce free radicals. These free radicals then oxidize various aromatic molecules, resulting in different quinones (Kadri et al. 2017). Non-ligninolytic fungi are usually inefficient in extracellular enzyme production and intracellular enzymes such as CYP450 play a significant role in PAHs degradation (Dutta and Laha 2022). The predominant fungal phyla that are highly considered as a potential bioagent for PAHs remediation belong to Ascomycota (non-ligninolytic fungi), followed by Basidiomycota (ligninolytic fungi) and the least by Zygomycota that is categorized under the non-ligninolytic members (Banerjee and Mandal 2020, Passarini et al. 2011). In this, diversified fungal agents which have the potential to eliminate the PAHs from contaminated soil and their mode of action have been documented in (Table 6.2) and from this analysis a similar pattern of dominant fungal phyla was observed, where Ascomycota appeared as the most dominant member (64%) followed by Basidiomycota (33%) and Zygomycota (3%), respectively (Figure 6.3).

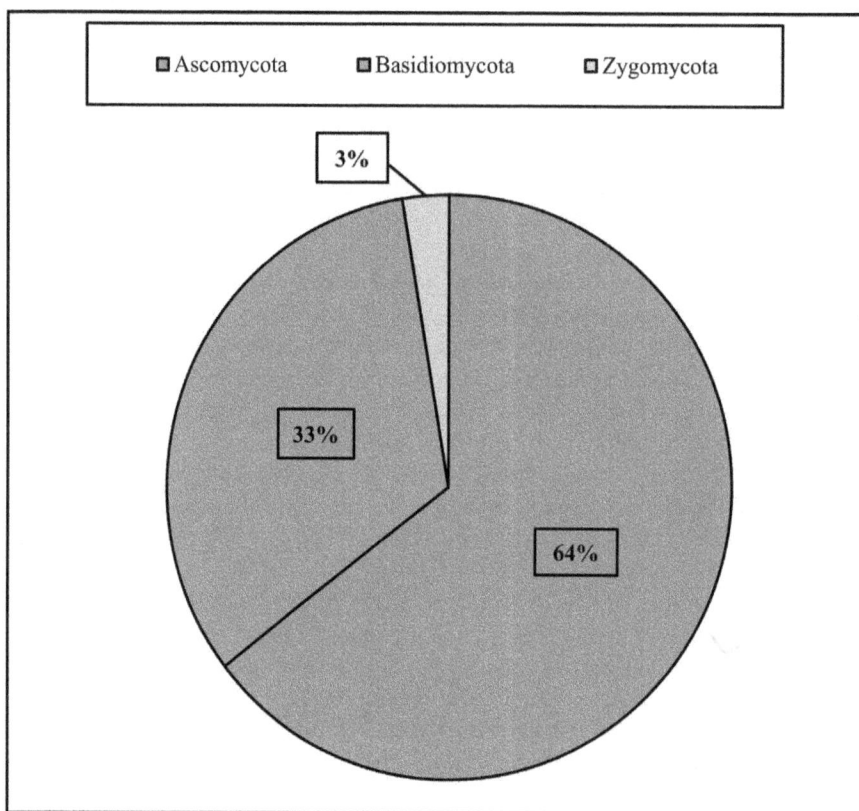

Figure 6.3. Diversity of fungal representatives involved in PAHs biodegradation.

6.3.1.1 PAHs Degradation by Ligninolytic Fungi

In the last couple of decades, ligninolytic fungi have been thoroughly investigated for their PAHs remediating potentialities. Phenomenally they produce extracellular enzymes such as laccase and peroxidases which play the most crucial role in PAHs mineralization (Kadri et al. 2017). In this aspect, most extracellular ligninolytic enzymes-producing fungi belong to Basidiomycota (Winquist et al. 2014). White-Rot Fungi (WRF) also play a crucial role in PAHs degradation, e.g., *Bjerkandera adusta, Pleurotus ostreatus, Phanerochaete chrysosporium, Trametes versicolor, Irpex lacteus* (Bezalel et al. 1996, Vyas et al. 1994, Field et al. 1992). Some investigations show that

Table 6.2. Fungal mechanisms for enzymatic bioremediation of PAHs.

Fungi	Phylum	Substrate	Mechanism	References
Lasiodiplodia theobromae	Ascomycota	Benzo(a)pyrene	Enzymatic cleavage	Cao et al. 2020
Phanerochaete chrysosporium	Basidiomycota	Phenanthrene, fluoranthene, and benzo(a)pyrene	Enzymatic cleavage	Guo et al. 2019
Tolypocladium sp. *Xylaria* sp.	Ascomycota Ascomycota	Pyrene Benzo(a)pyrene	Cytochrome P450 monooxygenase	Vasconcelos et al. 2019
Penicillium canescens Cladosporium cladosporioides Fusarium solani Talaromyces helices	Ascomycota Ascomycota Ascomycota Ascomycota	Benzo(a)pyrene	Ligninolytic enzymes	Fayeulle et al. 2019
Aspergillus sp.	Ascomycota	Naphthalene, phenanthrene, and pyrene	Intra and extracellular enzymes	Al-Hawash et al. 2019
Ganoderma lucidum	Basidiomycota	Phenanthrene and pyrene	Ligninolytic enzymes	Agrawal et al. 2018
Penicillium oxalicum	Ascomycota	Anthracene, phenanthrene, dibenzothiophene and dibenzofuran	Oxygenase enzyme	Aranda et al. 2017
Phanerochaete chrysosporium	Basidiomycota	Pyrene	Aerobic fungal enzymatic system	Aydin and Karacay 2017
Trichoderma longibrachiatum	Ascomycota	Phenanthrene	Anerobic biodegradation	
Trametes polyzona	Ascomycota	Phenanthrene, fluorene, and pyrene	Ligninolytic enzymes	Teerapatsakul et al. 2016
Pleurotus ostreatus	Basidiomycota	Fluorene and fluoranthene	Ligninolytic enzymes	Pozdnyakova et al. 2016
Penicillium oxalicum, Nigrospora oryzae Aspergillus oryzae A. aculeatus	Ascomycota Ascomycota Ascomycota Ascomycota	Naphthalene and phenanthrene	Endophytism	Kannangara et al. 2016
Phanerochaete chrysosporium Irpex lacteus Pleurotus ostreatus	Basidiomycota Basidiomycota Basidiomycota	Anthracene	Ring cleavage	Jove et al. 2016
Pleurotus sajor-caju	Basidiomycota	Pyrene and chrysene	Ligninolytic enzymes	Saiu et al. 2016
Mycoaciella bispora	Basidiomycota	Anthracene	Ligninolytic enzymes	Lee et al. 2015
Punctularia strigosozonata	Basidiomycota	Phenanthrene	Cytochrome P450 monooxygenase	Young et al. 2015
Pycnoporus sanguineus	Basidiomycota	Anthracene and pyrene	Cytochrome P450 monooxygenase	Zhang et al. 2015
Fusarium sp.	Ascomycota	Chrysene	Ligninolytic enzymes	Hidayat and Tachibana 2015

Table 6.2 contd. ...

...Table 6.2 contd.

Fungi	Phylum	Substrate	Mechanism	References
Peniophora incarnate	Basidiomycota	Phenanthrene, fluoranthene, and pyrene	Laccase and manganese peroxidase activity	Lee et al. 2014
Phlebia brevispora	Basidiomycota	Anthracene	Enzymatic catalysis	
Pycnoporus sanguineus	Basidiomycota	Anthracene and pyrene	Laccase activity	Li et al. 2014
Tinea versicolor	Basidiomycota	phenanthrene and pyrene	Enzymatic cleavage	Rosales et al. 2013
Anthracophyllum discolor	Basidiomycota	mixture of benzopyrene, phenanthrene, and pyrene	Manganese peroxide activity and lignin peroxide activity	Acevedo et al. 2011
Mucor racemosus	Zygomycota	Benzo(a)pyrene	Non-ligninolytic enzymatic cleavage	Passarini et al. 2011
Ceratobasidum stevensii	Basidiomycota	Phenanthrene	Ligninolytic enzymatic activity	Dai et al. 2010
Aspergillus flavus *Paecilomyces farinosus*	Ascomycota Ascomycota	Benzo(a)pyrene	Cytochrome P450 monooxygenase	Romero et al. 2010
Bjerkandera adusta	Basidiomycota	Anthracene and other PAHs	Versatile peroxidase	Eibes et al. 2010
Aspergillus sp. *Fusarium oxysporum* *Trichocladium canadense* *Trichoderma* sp. *Verticillium* sp.	Ascomycota Ascomycota Ascomycota Ascomycota Ascomycota	Naphthalene, phenanthrene, chrysene, perylene, naphthol, pyrene and decacyclene	Ligninolytic activity	Silva et al. 2009
Daedalea elegans *Fomitopsis palustris*	Basidiomycota Basidiomycota	Naphthalene and anthracene	Ligninolytic activity	Arun et al. 2008
Fusarium solani	Ascomycota	Benzopyrene	Cytochrome P450 monooxygenase	Koukkou and Drainas 2008
Irpex lacteus	Basidiomycota	Benzoanthracene and benzopyrene	Ligninolytic activity	Cajthaml et al. 2008
Pleurotus ostreatus	Basidiomycota	Phenanthrene	Ring-opening reaction by cytochrome P450 monooxygenase	Cajthaml et al. 2006
Phanerochaete chrysosporium	Basidiomycota	Anthracene	Ring-fission reaction	
Phanerochaete chrysosporium *Pleurotus pulmonarius*	Basidiomycota Basidiomycota	Naphthalene	Manganese peroxide activity	D'Annibale et al. 2005
Penicillium ochrochloron	Ascomycota	Pyrene	Ring cleavage	Saraswathy and Hallberg 2005
Aspergillus terreus	Ascomycota	Phenanthrene, anthracene, pyrene, and benzo[a]pyrene	Cytochrome P450 monooxygenase	Capotorti et al. 2005
Penicillium canescens *P. janczewskii* *P. montanense* *P. simplicissimum* *P. restrictum*	Ascomycota Ascomycota Ascomycota Ascomycota Ascomycota	Fluorene	Ring cleavage	Garon et al. 2004

Table 6.2 contd. ...

...*Table 6.2 contd.*

Fungi	Phylum	Substrate	Mechanism	References
Cladosporium sphaerospermum	Ascomycota	Naphthalene, acenaphthylene, acenaphthene, fluorene, phenanthrene, anthracene, pyrene, benz(a) anthracene, chrysene, benzo(b) fluoranthene, benzo(k)fluoranthene, benzo(a) pyrene, dibenz(a,h) anthracene, indeno(1,2,3-c,d) pyrene and benzo(ghi)perylene	Laccase activity	Potin et al. 2004
Cunninghamella elegans	Zygomycota	Phenanthrene	Cytochrome P450 monooxygenase	Lisowska and Dlugonski 2003
Trichoderma harzianum *Penicillium simplicissimum* *P. janthinellum* *P. funiculosum* *P. terrestre*	Ascomycota Ascomycota Ascomycota Ascomycota Ascomycota	Pyrene	Cytochrome P450 monooxygenase	Saraswathy and Hallberg 2002
Penicillium chrysogenum *P. italicum*	Ascomycota Ascomycota	Fluorene	Enzymatic cleavage	Garon et al. 2002
Penicillium janthinellum	Ascomycota	Benz[a]pyrene	Enzymatic cleavage	Boonchan et al. 2000
Penicillium chrysogenum *P. aurantiogriseum* *P. crustosum* *P. decumbens* *P. griseofulvum* *P. janczewskii* *P. janthinellum* *P. roqueforti* *P. rugulosum* *P. simplicissimum* *P. velutinum*	Ascomycota Ascomycota Ascomycota Ascomycota Ascomycota Ascomycota Ascomycota Ascomycota Ascomycota Ascomycota Ascomycota	Pyrene	Cytochrome P450 monooxygenase	Ravelet et al. 2000
Fusarium solani	Ascomycota	Benzo[a]pyrene	Cytochrome P450 monooxygenase	Rafin et al. 2000

white-rot fungus can degrade LMW-PAHs as well as HMW-PAHs with removal efficiencies of up to 58–73 and 21–26%, respectively (Kariyawasam et al. 2021, Leonardi et al. 2007). Utilization of extracellular enzymes laccases and peroxidases accelerate the B(a)P metabolized into B(a) P-1,6, -3,6-quinones or B(a)P-6,12-quinones intermediates observed in fungal species such as *P. chrysosporium, Cunninghamella elegans, A. ochraceus, T. versicolor* and *P. cinnabarinus* illustrated in Figure 6.4 (Majcherczyk et al. 1998, Datta and Samanta 1988, Haemmerli et al. 1986, Cerniglia and Gibson 1980). Another report of WRF *P. ostreatus* in immobilized commercial pellets that can degrade a significant number of PAHs such as 69.1% of Benzo(a)anthracene [B(a)A], 29.7% of chrysene (CHY), 39.7% of Benzo(b)fluoranthene [B(b)F], 32.8% of Benzo (k)fluoranthene [B(k)F], 85.2% of B(a)P and 80% of the total PAHs (Covino et al. 2010). Furthermore, Bhattacharya et al. (2017) discovered that the biodegradation of B(a)P was enhanced when it was subjected to

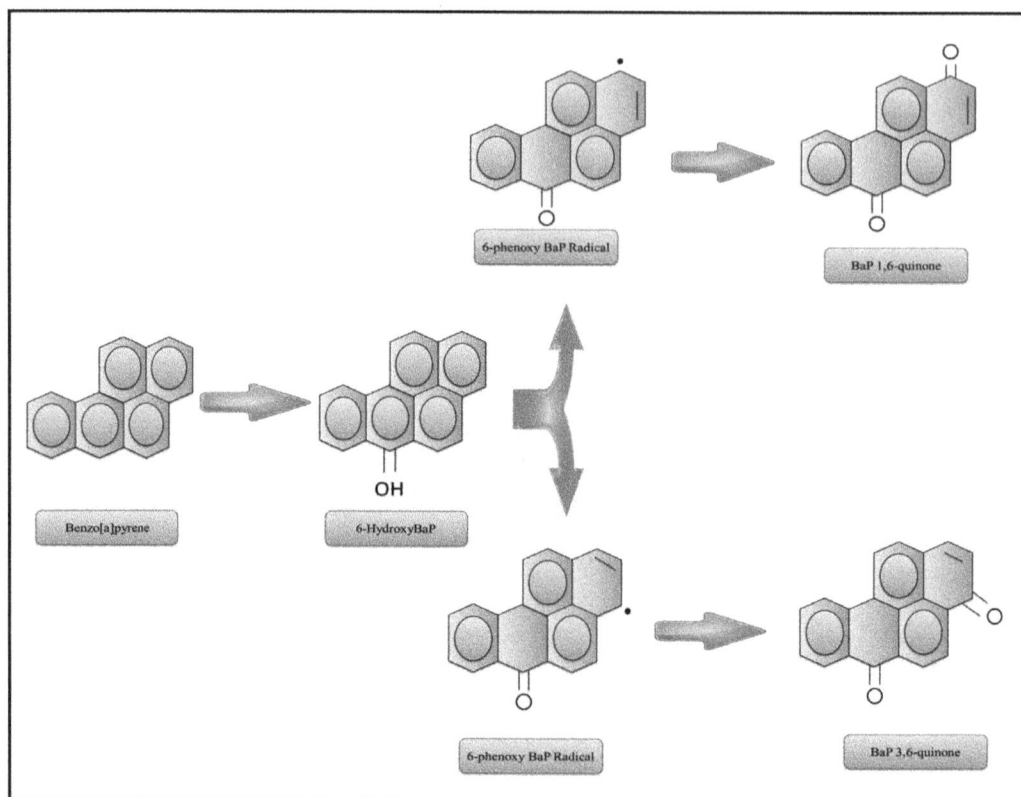

Figure 6.4. B(a)P mineralization by *P. chrysosporium, C. elegans, A. ochraceus, T. versicolor* and *P. cinnabarinus* (Majcherczyk et al. 1998, Datta and Samanta 1988, Haemmerli et al. 1986, Cerniglia and Gibson 1980).

consortia of *P. ostreatus* with *Penicillium chrysogenum* by 86.15% removal rate and *P. ostreatus* with *Pseudomonas aeruginosa* by 75.1% removal rate, whereas *P. ostreatus* PO-3 only remediated 64.3%.

On the other hand, the ligninolytic fungi or the WRF are known for their magnificent PAHs-degrading abilities with their potential intra and extracellular enzymes. In WRF mediated PAHs remediation process, PAHs oxidation begins with the formation of dihydrodiol by a multi-component dioxygenase enzyme system which further undergoes ortho or meta ring cleavage pathways and produces metabolites such as catechol and protocatechuate (Sipilä et al. 2010). Furthermore, some fungi undergo more extensive PAH decomposition, leading to the breakdown of the benzenoid rings and the release of CO_2. Enzymes involved in WRF-mediated PAHs breakdown mainly belong to the LiP, laccase, MnP and VP (Gupta and Pathak 2020). The mechanisms involved in PAHs degradation by both ligninolytic and non-ligninolytic fungi have been depicted in Figure 6.5.

6.3.1.1.1 Laccase

Extracellular enzymes, especially laccase, are crucial features that contribute to PAHs cleanup. Laccases are benzenediol: oxygen oxidoreductases and are metalloproteins in nature, which belong to the polyphenol oxidases and consist of four copper atoms of altered types in their catalytic site (Baldrian 2006). Laccases are one of the key ligninolytic biomolecules that oxidize a broad range of aromatic compounds holding phenolic moieties, benzenothiols, aromatic amines and hydroxylindols (Dhagat and Jujjavarapu 2022). Such enzymes utilize the molecular oxygen as an electron acceptor and organic or inorganic metal complexes as substrates (Zimmerman et al. 2008). Additionally, laccases use oxygen as an oxidizing agent and reduce it into water molecules (Tavares et al. 2006). Laccases are produced by diversified living systems like plants, fungi, bacteria and insects (Show et al. 2022). Among microbes, laccases of bacterial origin are reported from some potent

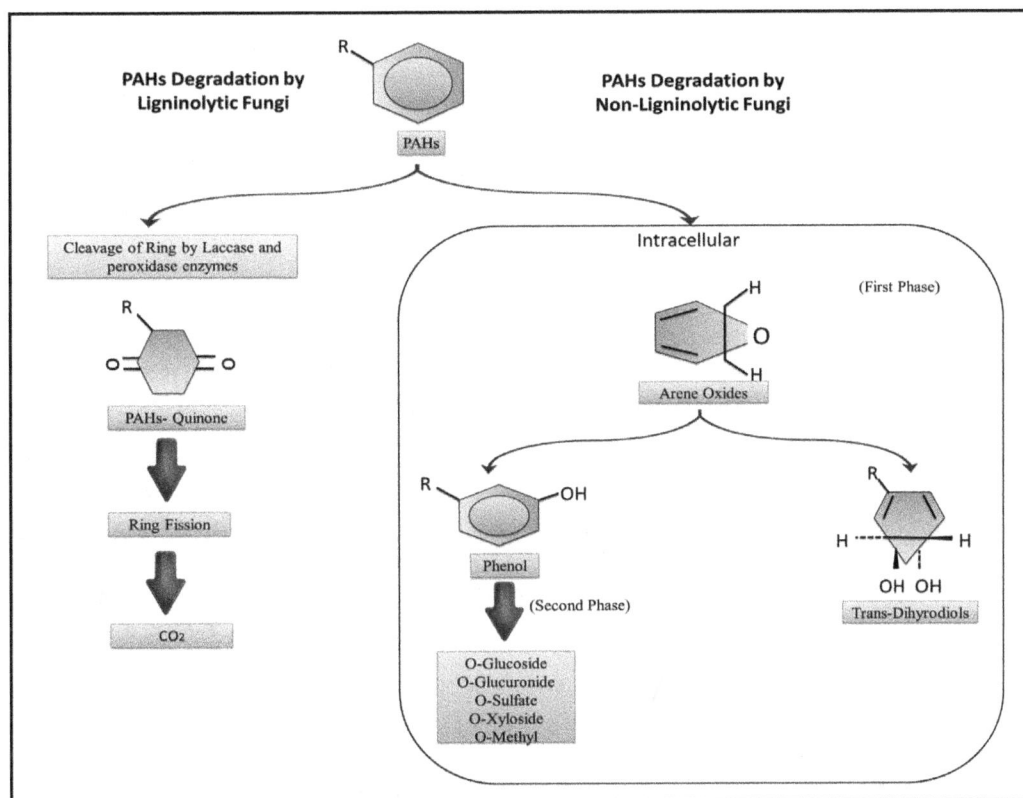

Figure 6.5. Ligninolytic and non-ligninolytic fungal mechanism responsible for PAHs degradation.

members viz, *Bacillus subtilis*, *Azospirillum lipoferum*, *Sinorhizobium meliloti* and *Streptomyces lavendulae*, whereas in the fungal system, intracellular and extracellular laccases are documented for a different type of PAHs remediation (Banerjee and Mandal 2020, Pawlik et al. 2016). Several reports have demonstrated that fungi utilize PAHs via the production of an extracellular enzyme like laccases resulting in PAHs degradation with CO_2 as a byproduct. A significant amount of PAHs removal capability by laccase-producing WRF *Anthracophyllum discolor* was documented with 62% of PHE, 73% of anthracene (ANTH), 54% of fluoranthene (FLU), 60% of PYR and 75% of B(a)P degrading potentiality (Acevedo et al. 2011). In a recent study, a WRF strain *Trametes versicolor* exhibited excellent remediational attributes like 81.0% CHY and 91.0% B(a)P degradation with 37.8 Ug^{-1} laccase producing potentiality (Vipotnik et al. 2022). In addition, another report of a WRF *Pycnoporus sanguineus* with 2516.7 UL^{-1} of laccase production ability displayed 90.1% B(a)A and 45.6% PHE degrading capacity (Li et al. 2018). Further, laccase produced by *T. versicolor* was also found to be able to remediate contaminants such as acenaphthene (ACE), B(b)F and PYR (Noman et al. 2019). In this regard, Punnapayak et al. (2009) reported that laccase-producing WRF *Ganoderma lucidum* can also efficiently degrade ACE and acenaphthylene.

6.3.1.1.2 Peroxidases

Peroxidases are the glycosylated extracellular enzyme that requires hydrogen peroxide to catalyze lignin and organopollutant compounds like PAHs (Thurston 1994). The peroxidases are classified into two categories based on their substrate interaction—MnP which is the most effective reducing substrate and LiP which catalyzes both non-aromatic and aromatic compounds (Ten Have and Teunissen 2001). Hydrogen peroxide (H_2O_2) is essential for the activation of the MnP and LiP are synthesized by fungal producers during their metabolic processes. In such metabolic processes, H_2O_2 acts as an oxidizing agent that oxidizes the PAHs and transforms them into a substrate

(Deshmukh et al. 2016). VPs also fall in this peroxidase isozymes family, capable to operate on a number of substrates. Several scientific reports have been documented regarding the PAHs degradation by fungal peroxidases.

6.3.1.1.2.1 Manganese peroxidase

Manganese peroxidases are the heme peroxidase, principally oxidize the Mn^{2+} ions that persist within the xenobiotic compounds such as PAHs, into Mn^{3+} which is an extra reactive form and fixed by fungal chelator, i.e., oxalic acids. Though, Mn^{3+} performs as a low molecular weight molecule that functions as a soluble redox mediator, which disintegrates both the phenolic and nonphenolic contaminants and generates free radicals that have the propensity to break reluctantly. Structurally, MnPs are comprised of two α-helices with haem in between them, two Ca^{2+} ions, and five disulfide bridges (Sutherland 1992). MnP has pronounced industrial applications viz, beverage, biofuel, pulp and paper, textile and food industries in addition to PAHs degradation (Karigar and Rao 2011, Chowdhary et al. 2019). An example of such fungal MnP-mediated PAHs degradation was documented from *I. lacteus*, where the experimented PAHs were ANTH and PYR (Kadri et al. 2017). Another example of WRF is *G. lucidum*, which can produce 47,444 UL^{-1} and 50,977 of MnP for the degradation of 99.65 and 99.58% of 20 mg L^{-1} PHE and PYR, respectively (Agrawal et al. 2018). In addition to this, 7.21 U g^{-1} MnP production was also observed from a WRF *Agrocybe aegerita* for the mineralization of 50 mM fluorene (FLR), PYR, CHY and B(a)P (Vipotnik et al. 2021). Further, *Trametes* sp. is a WRF that can degrade a significant amount of FLR, FLU, PYR, PHE and ANTH mixture with 69.7 UL^{-1} of MnP production (Zhang et al. 2016).

6.3.1.1.2.2 Lignin peroxidase

Lignin peroxidases of fungal origin can oxidize most PAHs and requires manganese and hydrogen peroxide to function. Generally, LiP can oxidize a number of phenolic and nonphenolic substances. Such LiP has two locations for Ca^{2+} binding and glycosylation and in addition to that have four disulfide bridges. All these integral conformations sustain the enzyme's 3-D structure. Eight minor and major-helices, structured into two domains, make up the globular configuration of LiPs, which contains the active center. This active center constitutes the haem-chelater: ferric ion. Due to its higher redox potential, LiPs can oxidize compounds that other peroxidases are unable to oxidize. Several steps are involved in each catalytic cycle of LiPs. Firstly, H_2O_2 produces a radical cation as an intermediate by oxidizing the native enzyme known as ferryloxo porphyrin. Subsequently, the first step succeeds by a pair of single electron reduction stages by an electron donor compounds, i.e., veratryl alcohol to generate an intermediate complex as well as a radical cation. Now, oxidation of another veratryl alcohol molecule occurs through such a transient composite and thereafter the transient composite converts to its native state to begin a new catalytic cycle of LiP (Sigoillot et al. 2012, Wong 2009, Choinowski et al. 1999). Additionally, LiP has a non-specific type of mechanism toward the substrates. One such example is ANTH bioremediation by utilizing LiP of *P. chrysosporium* with metabolites like 9,10-anthraquinone and phthalic, illustrated in Figure 6.6 (Pozdnyakova 2012). In another instance, *G. lucidum*, a member of WRF can produce 3613 UL^{-1} and 3283 UL^{-1} of LiP, which is responsible for 99.65 and 99.58% mineralization of 20 mg L^{-1} PHE and PYR, respectively (Agrawal et al. 2018). Further, 85.9% remediation potentiality with 2419 UL^{-1} LiP production was also observed from *Fusarium* sp. in response to the 5 mg L^{-1} mixture of HMW-PAHs [FLU, PYR, B(a)A, CHY, B(b)F, B(k)F, B(a)P, dibenzo(a,h)anthracene, benzo(g,hi) perylene and indeno(1,2,3-cd)pyrene] (Zhang et al. 2020). Interestingly, 2.24 ± 0.09 Ug^{-1} and 2.08 ± 0.04 Ug^{-1} of LiP production from *A. aegerita* was documented with 85.62 and 85.% FLR, 83.64 and 80.42% PYR, 79.24 and 60.32% CHY, 80 and 85.74% B(a)P degradation when treated with kiwi peels and peanut shells on solid state fermentation condition, respectively (Vipotnik et al. 2021). In a recent study, Omoni et al. (2022) demonstrated that 2.00 Ug^{-1} and 1.10 Ug^{-1} of LiP production were achieved by the WRF *P. chrysosporium* and *P. ostreatus* for utmost PHE degradation.

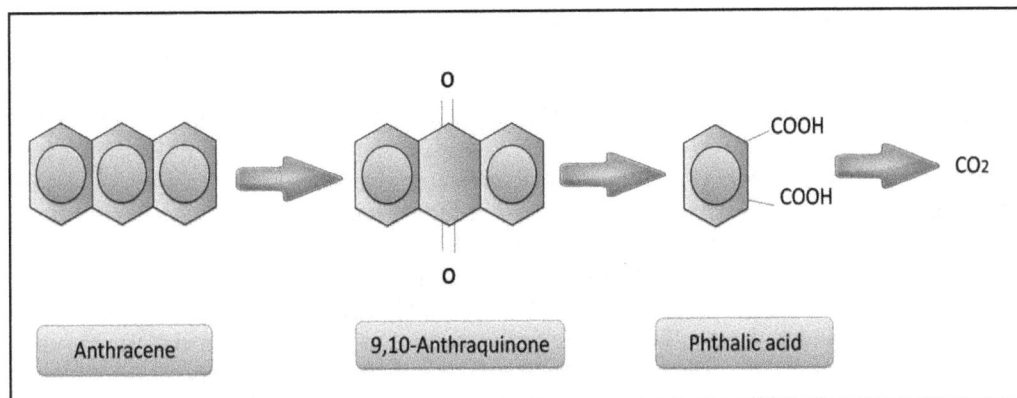

Figure 6.6. Bioremediation of anthracene by *P. chrysosporium* (Pozdnyakova 2012).

6.3.1.1.2.3 Versatile peroxidase

Versatile Peroxidases (VP) belong to the peroxidase family and are glycoproteins in nature which consist of 11–12 α-helices, a pair of Ca^{2+} binding sites, four disulfide bonds, a haem pocket and an Mn^{2+} binding site (Perez-Boada et al. 2005). It can be used in bioremediation of both aromatic and aliphatic contaminants. It involves the oxidative reaction of Mn^{2+}, aromatic compounds and methoxybenzenes like LiP and MnP. VP is more efficient than other peroxidases as it exhibits various substrate specificities and can function even in manganese-deficient conditions. This enables VP to be a potent biocatalyst and opens new doors in biotechnological techniques for the bioremediation of organic pollutants (Ruiz-Duenas et al. 2007). Examples of WRF which produced VP enzyme are *B. adusta* and *P. eryngii* (Naghdi et al. 2018). Additionally, a significant amount of PAHs degradation of ANTH, PYR and B(a)P was documented from the *B. adusta* by its VP enzymatic appliances (Wang et al. 2003). Another example of VP-mediated PHE degradation was also reported from a member of WRF *T. versicolor* (Collins et al. 1996). Further, Bogan et al. (1996) observed that the utmost FLR degradation was achieved by a WRF *P. chrysosporium*. CHY remediation was also documented from the *P. ostreatus* with its VP activity (Nikiforova et al. 2010). *P. ostreatus* was also reported for its PYR degrading capability through VP producing ability (Pozdnyakova et al. 2010). Multiple PAHs remediating potentiality was also achieved by the VP producing WRF *Nematoloma frowardii* (Sack et al. 1997).

6.3.1.2 PAHs Degradation by Non-ligninolytic Fungi

Ligninolytic fungi primarily grow on woody substances, which partially restrict their growth in soil. Hence their potentiality for PAHs biotransformation in the soil condition is quite ambiguous (Steffen et al. 2002). Therefore, non-ligninolytic fungi with intracellular enzymes, especially CYP450, can metabolize PAHs along with only substrate. Detoxification of organic pollutants by non-ligninolytic fungi is a predominant characteristic of their ability to pass through the cell wall where cell membrane-mediated enzymes such as cytochrome P450 monooxygenase and hydrolase operate the mineralization process (Marco-Urrea et al. 2015). The enzymatic mechanism involves the catalysis and hydrophobic (water-insoluble) PAHs transformed into less-toxic partially hydrophilic (water-soluble) intermediates via the epoxide ring opening mechanism, which cleaves the aromatic structure of PAHs and leads to the formation of arene oxide. This can be further transformed into phenol and trans-dihydrodiols, which are catalyzed by the hydrolase enzyme (Sutherland 1992).

Chrysosporium pannorum, C. elegans and *A. niger* are among such examples, involved in PAH degradation with such a mechanism. The biotransformation of PAHs by non-ligninolytic fungi depicts a typical sequential pattern having two significant phases. The first phase causes the production of dihydroxy-quinone and dihydrodiol-derivatives, which further conjugates in the second phase with O-glucoside, O-sulfate, O-glucuronide, O-methyl and O-xyloside. These metabolites are water-soluble and non-hazardous when compared with their parental PAHs compounds, hence lowering health-risk. *C. elegans* exhibit a sulfur addition process in which naphthalene (NAP) and ANTH are converted into 1-naphthyl and 1-anthryl sulfate metabolites, respectively (Lisowska and Długoński 2003). On the other hand, enhancement in PAHs degradation by accelerating the PAHs solubility can be achieved by conjugation process. *P. chrysosporium* shows glucose as PAHs conjugates. According to Sutherland (1992), 7,12-dimethylbenzanthracene can be biotransformed by *C. elegans* producing trans-3,4-dihydrodiol metabolites. It can be further metabolized into phenol carried out by *Mucor* sp., *P. chrysogenum. S. racemosum* biosystem via conjugating with glucuronic acid. ANTH biodegradation can be accomplished by various fungi such as *Fusarium* sp., *Aspergillus* sp., *Penicillium* sp., *Trichoderma* sp., *Ulocladium chartarum* and *Absidia cylindrospora* (Giraud et al. 2001). In addition to this, the non-ligninolytic fungi involved in pherenthrene metabolism usually belong to *Alternaria* sp., *F. culmorum, P. janczewskii, C. elegans, Cladosporium herbarum* and *T. hamatum*, etc. (Schmidt et al. 2010).

6.3.1.2.1 Cytochrome P450

Cytochrome P450 monooxygenase is a heme thiolate intra-cellular biocatalyst produced by various species of fungi and bacteria. Usually, CYP450 accomplishes as terminal monooxygenases which catalyze a wide range of biochemical reactions such as carbon hydroxylations, heteroatom-oxygenations, dealkylations, epoxidations, reductions and dehalogenations (Bernhardt 2006). Phenomenal eukaryotic Cytochrome P450 systems consist of P450 monooxygenase and P450 oxidoreductase and both these are typically membrane-associated (Zhuo and Fan 2021). WRF has employed P450 monooxygenase systems for the oxidative decontamination and eradication of several xenobiotic compounds, especially various organic pollutants like PAHs. It helps in converting water-insoluble compounds into water-soluble compounds and cleaves the organic pollutants by incorporating molecular oxygen. Most of the non-ligninolytic fungi can secrete this class of enzyme.

Examples of CYP450 producing fungi are *P. chrysosporium, Cochliobolus lunatus* and *A. niger* which are evidenced by the bioremediation of PAHs. A significant amount (up to 89%) of B(a)P degradation by the WRF *P. chrysosporium* has also been documented (Bhattacharya et al. 2013). PHE degradation via cytochrome P450 system of *P. sanguineus* was also observed by Li et al. (2018). Multiple PAH degrading capability and their transformation from PYR to 1-hydroxypyrene; PHE to 3-phenanthrol, 4-phenanthrol, 9-phenanthrol; and B(a)P to 3-hydroxy B(a)P by the activity of cytochrome P450 from *P. chrysosporium* has also been reported (Syed et al. 2010). Further, Syed et al. (2011 and 2013) reported HMW-PAH degradation by the similar enzymatic potentiality of *P. chrysosporium*. In addition to that, the conversion of PHE to phenanthrene trans-9,10-dihydrodiol was shown by the *P. chrysosporium* with its same enzymatic activities (Ning et al. 2010). Ostrem Loss et al. (2019) investigated that PHE was bioremediated by the non-ligninolytic fungi, *P. chrysosporium*, by utilizing CYP450 as the key enzyme. Despite the fungal origin, researchers explored the enhanced CYP450 production by the development of recombinant strain and subsequently gene expression in white-rot fungus in response to HMW-PAHs; for example, the expression of genes CYP450-pc-1 and CYP450-pc-2 in *P. chrysosporium* (Lin et al. 2022). In another study, the recombinant strain CYP63A2 of *P. chrysosporium* revealed ameliorated oxidation capacity towards HMW-PAHs (Syed et al. 2013).

6.4 Production of Biosurfactants

Recent bioremediation studies emphasize on the use of biosurfactants having a microbial origin that have attributed to the mineralization of PAHs in the environment (Guo and Wen 2021, Mekwichai et al. 2020, Karlapudi et al. 2018). Greater biodegradable potentiality, lower toxicity, foam formation and eco-sustainable nature of biosurfactants promote their application in the remediation of PAHs over chemical surfactants (Pi et al. 2017). Microbial biosurfactants are surface-active compounds made up of amphiphilic domains consisting of both a hydrophilic (polar) head as well as hydrophobic (non-polar) tail, which bind with the complementary polarity compounds by lowering the surface tension and interface pressure of solutions (John et al. 2021, Kour et al. 2021). Synthesis of biosurfactants by various microbes enhances the solubility and causes ease to mineralize non-polar compounds such as PAHs via mobilization, micelle formation and solubilization, that accelerate their bio-accessibility to the microbial community in the environment (Imam et al. 2022, Pourfadakari et al. 2021, Maia et al. 2019, Dell'Anno et al. 2018). Based on the molecular structure, biosurfactants are categorized into two types (1) Low-molecular-weight biosurfactants and (2) High-molecular weight biosurfactants. LMW-Biosurfactants are primarily composed of lipopeptides, phospholipids, lipoproteins, fatty acids, glycolipids and neutral lipids, which enhance the surface area of organic pollutants and accelerate their bioavailability in the microbial cleanup process (Henkel and Hausmann 2019, Yuliani et al. 2018). HMW-Biosurfactants include polysaccharides, lipopolysaccharides, proteins and lipoproteins which in turn can readily interact with various surfaces and act as an emulsifier of pollutants like PAHs (Fenibo et al. 2019, McClements and Gumus 2016).

Phellinus sp. and *Polyporus sulphureus* belong to the Basidiomycota phylum and are capable of mineralizing HWM-PAHs by producing a high amount of biosurfactant. Biosurfactants secreted by *Phellinus* sp. and *P. sulphureus* were achieved to remediate the oil spillage by the production of biosurfactants with dry weight 3.4 g/L and 2.6 g/L where PAHs degradation potentiality were evidenced as the emulsion index were 122 and 106%, respectively (Arun and Eyini 2011). Wang et al. (2008) suggested that the biosurfactant rhamnolipid enhanced the functionality of extracellular enzymes such as LiP and MnP synthesized by *P. chrysosporium* and laccase secreted by *P. simplicissimum*. Additionally, *P. ostreatus* can degrade both LMW-PAHs as well as HMW-PAHs facilitated with an emulsifier agent which also escalates the action of VP (Nikiforova et al. 2009).

6.5 Biochar Immobilized Fungal Administration

Biochar is a pyrolyzed carbonaceous substance that can be produced from biomass (woody waste) through thermal treatment under anaerobic conditions. Biochar is environmentally friendly, of low cost and a sustainable administrator, its features are primarily used for organic and inorganic pollutant remediation. Biochar that has been totally carbonized and generated at temperatures exceeding 500°C acts as a superior adsorbent for organic contaminants because of its vast surface area, microporosity, hydrophobicity, C-N ratio and high pH value (Jatav et al. 2021). Features that make biochar a promising strategy for the bioremediation of organic hydrocarbons like PAHs are (i). Biochar can readily change its surface volume and quantity of oxygen molecules having functional groups, (ii). Affinity towards organic hydrocarbon through the π-π bond, hydrogen bond and covalent bond, (iii). Reduction in the bioavailability and solubility of organic pollutants such as PAHs (cause immobilization of PAHs), and (iv). Microbial augmentation and phytoremediation assisted with biochar enhancing its pollutant remediation efficiency by many folds. García-Delgado et al. (2015) investigated that biochar immobilized fungal agent with *P. ostreatus* exhibited 55, 40, 31 and 89% biodegradation of HMW-PAHs like B(a)P, B(a)A, CHRY and DBA, respectively. Biochar immobilized with *P. chrysosporium* hyphae was able to accelerate the rate of abatement of PAHs like PHE with more than 90% removal efficiency (Zhang et al. 2022).

6.6 Utilization of Rhizospheric and Endophytic Fungi

Numerous documents have been recorded describing the plant uptake and bioaccumulation of organic hydrocarbons (PAHs) from the soil (Anderson et al. 1993). Rhizospheric and endophytic microbes play a critical part in the removal of PAHs from the soil, exhibiting an eco-friendly, low-cost and sustainable approach. The rhizospheric microorganisms reside in the vicinity of plant roots and get nourished by the root exudates, whereas endophytic microorganisms are the residents of plant internal tissue. Arbuscular mycorrhizal fungi symbiotically associated with most herbaceous plants are omnipresent. Arbuscular mycorrhizae perform several roles in plant growth, such as enhancing nutrient and minerals uptake, drought resistance and immunization from plant diseases (Pramanik et al. 2021, Hildebrandt et al. 2007). Arbuscular mycorrhizae such as *Glomus mosseae* and *G. etunicatum* associated with ryegrass roots efficiently degrade PAHs such as FLR and PHE (Gao et al. 2010). Chulalaksananukul et al. (2006) reported that *Fusarium* sp., isolated from the leaves of *Pterocarpus macrocarpus*, capable of degrading B(a)P by 70% of 100 ppm concentration. Kannangara et al. (2016) identified different endophytic fungi such as *P. oxalicum, Nigrospora oryzae, A. oryzae* and *A. aculeatus* which are very efficient at degrading NAP as well as PHE above than 85% of removal efficiency.

6.7 Conclusion

Nowadays, bioremediation approaches for environmental pollutants like PAHs of contaminated soil have received significant importance in the scientific community as they are not easily eliminated by natural processes and can reside for decades in the environment. In accordance with this, PAHs are associated with various health issues such as cancer, kidney failure, lung malfunction, etc. In this context, among the different PAHs remediation approaches, fungal bioremediation approaches are the most promising techniques for the future as it promotes sustainability and cost-effectivity. Fungal representatives like Basidiomycota and Ascomycota are the predominant communities, involved in biodegradation of different types of LMW and HMW PAHs compounds. Accelerated biotransformation of PAHs via mycoremediation strategies such as enzymatic metabolism, biosurfactant production, biochar immobilized fungal administration and utilization of rhizospheric and endophytic fungi can be harnessed as nature's own medicine to combat with such polluting ecological malaises. Utilization of these fungal strategies can accelerate the bioremediation of PAHs and can open a "novel biotechnological green-window" for eco-sustainable mother earth.

Acknowledgement

Sandipan Banerjee and Nitu Gupta thank the Department of Biotechnology, Govt. of India, for granting DBT Twinning Project and Research Fellowship [No. BT/PR25738/NER/95/1329/2017 dated December 24, 2018]. Apurba Koley is thankful to the BBSRC, United Kingdom, for granting funding from the BEFWAM project: Bioenergy, Fertilizer and Clean water from Invasive Aquatic macrophytes [Grant Ref: BB/S011439/1] for financial support and research fellowship.

References

Abdel-Shafy, H. I. and M. S. Mansour. 2016. A review on polycyclic aromatic hydrocarbons: source, environmental impact, effect on human health and remediation. Egypt. J. Pet. 25(1): 107–123.

Acevedo, F., L. Pizzul, M. del Pilar Castillo, R. Cuevas and M. C. Diez. 2011. Degradation of polycyclic aromatic hydrocarbons by the Chilean white-rot fungus *Anthracophyllum discolor*. J. Hazard. Mater. 185(1): 212–219.

Agrawal, N., P. Verma and S. K. Shahi. 2018. Degradation of polycyclic aromatic hydrocarbons (phenanthrene and pyrene) by the ligninolytic fungi *Ganoderma lucidum* isolated from the hardwood stump. Bioresour. Bioprocess. 5(1): 1–9.

Anderson, T. A., E. A. Guthrie and B. T. Walton. 1993. Bioremediation in the rhizosphere. Environ. Sci. Technol. 27(13): 2630–2636.

Aranda, E., P. Godoy, R. Reina, M. Badia-Fabregat, M. Rosell, E. Marco-Urrea and I. García-Romera. 2017. Isolation of Ascomycota fungi with capability to transform PAHs: Insights into the biodegradation mechanisms of *Penicillium oxalicum*. Int. Biodeterior. Biodegrad. 122: 141–150.

Arun, A., P. P. Raja, R. Arthi, M. Ananthi, K. S. Kumar and M. Eyini. 2008. Polycyclic aromatic hydrocarbons (PAHs) biodegradation by basidiomycetes fungi, *Pseudomonas* isolate, and their cocultures: comparative *in vivo* and *in silico* approach. Appl. Biochem. Biotechnol. 151(2): 132–142.

Arun, A. and M. Eyini. 2011. Comparative studies on lignin and polycyclic aromatic hydrocarbons degradation by basidiomycetes fungi. Bioresour. Technol. 102(17): 8063–8070.

Aydin, S., H. A. Karaçay, A. Shahi, S. Gökçe, B. Ince and O. Ince. 2017. Aerobic and anaerobic fungal metabolism and Omics insights for increasing polycyclic aromatic hydrocarbons biodegradation. Fungal Biol. Rev. 31(2): 61–72.

Baldrian, P. 2006. Fungal laccases–occurrence and properties. FEMS Microbiol. Rev. 30(2): 215–242.

Ball, A. and A. Truskewycz. 2013. Polyaromatic hydrocarbon exposure: an ecological impact ambiguity. Environ. Sci. Pollut. Res. 20(7): 4311–4326.

Banerjee, S., T. K. Maiti and R. N. Roy. 2017. Protease production by thermo-alkaliphilic novel gut isolate *Kitasatospora cheerisanensis* GAP 12.4 from *Gryllotalpa africana*. Biocatal. Biotransformation. 35(3): 168–176.

Banerjee, S. and N. C. Mandal. 2020. Fungal Bioagents in the remediation of degraded soils. In Microbial Services in Restoration Ecology. pp. 191–205. Elsevier.

Bernhardt, R. 2006. Cytochromes P450 as versatile biocatalysts. J. Biotechnol. 124(1): 128–145.

Bezalel, L., Y. Hadar, P. P. Fu, J. P. Freeman and C. E. Cerniglia. 1996. Initial oxidation products in the metabolism of pyrene, anthracene, fluorene, and dibenzothiophene by the white rot fungus *Pleurotus ostreatus*. Appl. Environ. Microbiol. 62: 2554–2559.

Bhattacharya, S. S., K. Syed, J. Shann and J. S. Yadav. 2013. A novel P450-initiated biphasic process for sustainable biodegradation of benzo [a] pyrene in soil under nutrient-sufficient conditions by the white rot fungus *Phanerochaete chrysosporium*. J. Hazard. Mater. 261: 675–683.

Bhattacharya, S., A. Das, M. Palaniswamy and J. Angayarkanni. 2017. Degradation of benzo [a] pyrene by *Pleurotus ostreatus* PO-3 in the presence of defined fungal and bacterial co-cultures. J. Basic Microbiol. 57(2): 95–103.

Biswas, B., B. Sarkar, R. Rusmin and R. Naidu. 2015. Bioremediation of PAHs and VOCs: advances in clay mineral–microbial interaction. Environ. Int. 85: 168–181.

Blumer, M. 1976. Polycyclic aromatic compounds in nature. Sci. Am. 234: 34–45.

Bogan, B. W., R. T. Lamar and K. E. Hammel. 1996. Fluorene oxidation *in vivo* by *Phanerochaete chrysosporium* and *in vitro* during manganese peroxidase-dependent lipid peroxidation. Appl. Environ. Microbiol. 62(5): 1788–1792.

Boonchan, S., M. L. Britz and G. A. Stanley. 2000. Degradation and mineralization of high-molecular-weight polycyclic aromatic hydrocarbons by defined fungal-bacterial cocultures. Appl. Environ. Microbiol. 66(3): 1007–1019.

Cajthaml, T., P. Erbanová, V. Šašek and M. Moeder. 2006. Breakdown products on metabolic pathway of degradation of benz [a] anthracene by a ligninolytic fungus. Chemosphere. 64(4): 560–564.

Cajthaml, T., P. Erbanová, A. Kollmann, Č. Novotný, V. Šašek and C. Mougin. 2008. Degradation of PAHs by ligninolytic enzymes of *Irpex lacteus*. Folia Microbiol. 53(4): 289–294.

Cao, H., C. Wang, H. Liu, W. Jia and H. Sun. 2020. Enzyme activities during benzo(a) pyrene degradation by the fungus *Lasiodiplodia theobromae* isolated from a polluted soil. Sci. Rep. 10: 865. https://doi.org/10.1038/s41598-020-57692-6.

Capotorti, G., P. Cesti, A. Lombardi and G. Guglielmetti. 2005. Formation of sulfate conjugates metabolites in the degradation of phenanthrene, anthracene, pyrene and benzo [a] pyrene by the ascomycete *Aspergillus terreus*. Polycycl. Aromat. Comp. 25(3): 197–213.

Cerniglia, C. E. and D. T. Gibson. 1980. Oxidation of benzo [a] pyrene by the filamentous fungus *Cunninghamella elegans*. J. Biol. Chem. 254(23): 12174–12180.

Cerniglia, C. E. 1993. Biodegradation of polycyclic aromatic hydrocarbons. Curr. Opin. Biotechnol. 4(3): 331–338.

Chowdhary, P., G. Shukla, G. Raj, L. F. R. Ferreira and R. N. Bharagava. 2019. Microbial manganese peroxidase: a ligninolytic enzyme and its ample opportunities in research. SN Appl. Sci. 1(1): 1–12.

Choinowski, T., W. Blodig, K. H. Winterhalter and K. Piontek. 1999. The crystal structure of lignin peroxidase at 1.70 Å resolution reveals a hydroxy group on the Cβ of tryptophan 171: a novel radical site formed during the redox cycle. J. Mol. Biol. 286(3): 809–827.

Chulalaksananukul, S., G. M. Gadd, P. Sangvanich, P. Sihanonth, J. Piapukiew and A. S. Vangnai. 2006. Biodegradation of benzo (a) pyrene by a newly isolated *Fusarium* sp. FEMS Microbiol. Lett. 262(1): 99–106.

Collins, P. J. and A. D. Dobson. 1996. Oxidation of fluorene and phenanthrene by Mn (II) dependent peroxidase activity in whole cultures of *Trametes (Coriolus) versicolor*. Biotechnol. Lett. 18(7): 801–804.

Conejo-Saucedo, U., D. R. Olicón-Hernández, T. Robledo-Mahón, H. P. Stein, C. Calvo and E. Aranda. 2019. Bioremediation of polycyclic aromatic hydrocarbons (PAHs) contaminated soil through fungal communities. In Recent Advancement in White Biotechnology Through Fungi. pp. 217–236. Springer, Cham.

Covino, S., K. Svobodová, M. Čvančarová, A. D'Annibale, M. Petruccioli, F. Federici et al. 2010. Inoculum carrier and contaminant bioavailability affect fungal degradation performances of PAH-contaminated solid matrices from a wood preservation plant. Chemosphere. 79(8): 855–864.

D'Annibale, A., M. Ricci, V. Leonardi, D. Quaratino, E. Mincione and M. Petruccioli. 2005. Degradation of aromatic hydrocarbons by white-rot fungi in a historically contaminated soil. Biotechnol. Bioeng. 90: 723–731.

Dai, C. C., L. S. Tian, Y. T. Zhao, Y. Chen and H. Xie. 2010. Degradation of phenanthrene by the endophytic fungus *Ceratobasidum stevensii* found in *Bischofia polycarpa*. Int. Biodeterior. 21(2): 245–255.

Datta, D. and T. B. Samanta. 1988. Effect of inducers on metabolism of benzo (a) pyrene *in vivo* and *in vitro*: analysis by high pressure liquid chromatography. Biochem. Biophys. Res. Commun. 155(1): 493–502.

Dell'Anno, F., C. Sansone, A. Ianora and A. Dell'Anno. 2018. Biosurfactant-induced remediation of contaminated marine sediments: current knowledge and future perspectives. Mar. Environ. Res. 137: 196–205.

Deshmukh, R., A. A. Khardenavis and H. J. Purohit. 2016. Diverse metabolic capacities of fungi for bioremediation. Indian J. Microbiol. 56(3): 247–264.

Dhagat, S. and S. E. Jujjavarapu. 2022. Utility of lignin-modifying enzymes: a green technology for organic compound mycodegradation. J. Chem. Technol. Biotechnol. 97(2): 343–358.

Dutta, S. and S. Laha. 2022. An approach toward the biodegradation of PAHs by microbial consortia. In Development in Wastewater Treatment Research and Processes. pp. 383–406. Elsevier.

Eibes, G., C. McCann, A. Pedezert, M. T. Moreira, G. Feijoo and J. M. Lema. 2010. Study of mass transfer and biocatalyst stability for the enzymatic degradation of anthracene in a two-phase partitioning bioreactor. Biochem. Eng. J. 51(1-2): 79–85.

Fayeulle, A., E. Veignie, R. Schroll, J. C. Munch and C. Rafin. 2019. PAH biodegradation by telluric saprotrophic fungi isolated from aged PAH-contaminated soils in mineral medium and historically contaminated soil microcosms. J. Soils Sediments. 19(7): 3056–3067.

Fenibo, E. O., G. N. Ijoma, R. Selvarajan and C. B. Chikere. 2019. Microbial surfactants: The next generation multifunctional biomolecules for applications in the petroleum industry and its associated environmental remediation. Microorganisms. 7(11): 581.

Field, J. A., E. deJong, G. F. Costa and J. A. M. deBont. 1992. Biodegradation of polycyclic aromatic hydrocarbons by new iso-lates of white-rot fungi. Appl. Environ. Microbiol. 58: 2219–2226.

Gadd, G. M. and G. M. Gadd (Eds.). 2001. Fungi in bioremediation (No. 23). Cambridge University Press.

Gao, Y., Z. Cheng, W. Ling and J. Huang. 2010. *Arbuscular mycorrhizal* fungal hyphae contribute to the uptake of polycyclic aromatic hydrocarbons by plant roots. Bioresour Technol. 101(18): 6895–6901.

García-Delgado, C., I. Alfaro-Barta and E. Eymar. 2015. Combination of biochar amendment and mycoremediation for polycyclic aromatic hydrocarbons immobilization and biodegradation in creosote-contaminated soil. J. Hazard. Mater. 285: 259–266.

Garon, D., S. Krivobok, D. Wouessidjewe and F. Seigle-Murandi. 2002. Influence of surfactants on solubilization and fungal degradation of fluorene. Chemosphere. 47(3): 303–309.

Garon, D., L. Sage and F. Seigle-Murandi. 2004. Effects of fungal bioaugmentation and cyclodextrin amendment on fluorene degradation in soil slurry. Biodegradation. 15(1): 1–8.

Ghosal, D., S. Ghosh, T. K. Dutta and Y. Ahn. 2016. Current state of knowledge in microbial degradation of polycyclic aromatic hydrocarbons (PAHs): a review. Front Microbiol. 7: 1–27.

Giraud, F., P. Guiraud, M. Kadri, G. Blake and R. Steiman. 2001. Biodegradation of anthracene and fluoranthene by fungi isolated from an experimental constructed wetland for wastewater treatment. Water Res. 35(17): 4126–4136.

Guo, J., X. Liu, X. Zhang, J. Wu, C. Chai, D. Ma et al. 2019. Immobilized lignin peroxidase on Fe3O4@ SiO2@ polydopamine nanoparticles for degradation of organic pollutants. Int. J. Biol. Macromol. 138: 433–440.

Guo, J. and X. Wen. 2021. Performance and kinetics of benzo (a) pyrene biodegradation in contaminated water and soil and improvement of soil properties by biosurfactant amendment. Ecotoxicol. Environ. Saf. 207: 111292.

Gupta, S. and B. Pathak. 2020. Mycoremediation of polycyclic aromatic hydrocarbons. In Abatement of Environmental Pollutants. pp. 127–149. Elsevier.

Haemmerli, S. D., M. S. Leisola, D. Sanglard and A. Fiechter. 1986. Oxidation of benzo (a) pyrene by extracellular ligninases of *Phanerochaete chrysosporium*. Veratryl alcohol and stability of ligninase. J. Biol. Chem. 261(15): 6900–6903.

Henkel, M. and R. Hausmann. 2019. Diversity and classification of microbial surfactants. Biobased surfactants. pp. 41–63. AOCS Press.

Hidayat, A. and S. Tachibana. 2015. Simple screening for potential chrysene degrading fungi. KnE Life Sci., 364–370.

Hildebrandt, U., M. Regvar and H. Bothe. 2007. *Arbuscular mycorrhiza* and heavy metal tolerance. Phytochemistry. 68: 139–146.

Hussain, K., R. R. Hoque, S. Balachandran, S. Medhi, M. G. Idris, M. Rahman and F. L. Hussain. 2018. Monitoring and risk analysis of PAHs in the environment. Handbook of Environmental Materials Management, 1–35.

Imam, A., S. K. Suman, P. K. Kanaujia and A. Ray. 2022. Biological machinery for polycyclic aromatic hydrocarbons degradation: A review. Bioresour. Technol. 343: 126121.

Jatav, H. S., V. D. Rajput, T. Minkina, S. K. Singh, S. Chejara, A. Gorovtsov et al. 2021. Sustainable approach and safe use of biochar and its possible consequences. Sustainability. 13: 10362.

John, W. C., I. O. Ogbonna, G. M. Gberikon and C. C. Iheukwumere. 2021. Evaluation of biosurfactant production potential of *Lysinibacillus fusiformis* MK559526 isolated from automobile-mechanic-workshop soil. Braz. J. Microbiol. 52: 663–674.

Jove, P., M. À. Olivella, S. Camarero, J. Caixach, C. Planas, L. Cano and F. X. De Las Heras. 2016. Fungal biodegradation of anthracene-polluted cork: a comparative study. J. Environ. Sci. Health. Part A, 51(1): 70–77.

Kadri, T., T. Rouissi, S. K. Brar, M. Cledon, S. Sarma and M. Verma. 2017. Biodegradation of polycyclic aromatic hydrocarbons (PAHs) by fungal enzymes: a review. J. Environ. Sci. 51: 52–74.

Kannangara, S., P. Ambadeniya, L. Undugoda and K. Abeywickrama. 2016. Polyaromatic hydrocarbon degradation of moss endophytic fungi isolated from *Macromitrium* sp. in Sri Lanka. J. Agric. Sci. Technol. 6(03): 171–182.

Karigar, C. S. and S. S. Rao. 2011. Role of microbial enzymes in the bioremediation of pollutants: a review. Enzyme Res.

Kariyawasam, T., G. S. Doran, J. A. Howitt and P. D. Prenzler. 2021. Polycyclic aromatic hydrocarbon contamination in soils and sediments: sustainable approaches for extraction and remediation. Chemosphere. 132981.

Karlapudi, A. P., T. C. Venkateswarulu, J. Tammineedi, L. Kanumuri and B. K. Ravuru. 2018. Role of biosurfactants in bioremediation of oil pollutiona review. Petroleum, 241–249.

Koukkou, A. I. and C. Drainas. 2008. Addressing PAH biodegradation in Greece: biochemical and molecular approaches. IUBMB life. 60(5): 275–280.

Kour, D., T. Kaur, R. Devi, A. Yadav, M. Singh, D. Joshi and A. K. Saxena. 2021. Beneficial microbiomes for bioremediation of diverse contaminated environments for environmental sustainability: present status and future challenges. Environ. Sci. Pollut. Res. 28(20): 24917–24939.

Kuppusamy, S., P. Thavamani, K. Venkateswarlu, Y. B. Lee, R. Naidu and M. Megharaj. 2017. Remediation approaches for polycyclic aromatic hydrocarbons (PAHs) contaminated soils: Technological constraints, emerging trends and future directions. Chemosphere. 168: 944–968.

Langenbach, T. 2013. Persistence and bioaccumulation of persistent organic pollutants (POPs). pp. 305–329. *In*: Patil, Y. B. and P. Rao [ed.]. Applied bioremediation—active and passive approaches. DOI: 10.5772/56418.

Lee, H., Y. Jang, Y. S. Choi, M. J. Kim, J. Lee, H. Lee et al. 2014. Biotechnological procedures to select white rot fungi for the degradation of PAHs. J. Microbiol. Methods. 97: 56–62.

Lee, H., S. Y. Yun, S. Jang, G. H. Kim and J. J. Kim. 2015. Bioremediation of polycyclic aromatic hydrocarbons in creosote-contaminated soil by *Peniophora incarnata* KUC8836. Bioremediat. J. 19(1): 1–8.

Leonardi, V., V. Sasek, M. Petruccioli, A. D'Annibale, P. Erbanová and T. Cajthaml. 2007. Bioavailability modification and fungal biodegradation of PAHs in aged industrial soils. Int. Biodeterior. Biodegradation. 60(3): 165–170.

Li, X. and P. Li. 2008. Biodegradation of aged polycyclic aromatic hydrocarbons (PAHs) by microbial consortia in soil and slurry phases. J. Hazard. Mater. 150(1): 21–26.

Li, X., Y. Wang, S. Wu, L. Qiu, L. Gu, J. Li et al. 2014. Peculiarities of metabolism of anthracene and pyrene by laccase-producing fungus *Pycnoporus sanguineus* H 1. Biotechnol. Appl. Biochem. 61(5): 549–554.

Li, X., Y. Pan, S. Hu, Y. Cheng, Y. Wang, K. Wu and S. Yang. 2018. Diversity of phenanthrene and benz [a] anthracene metabolic pathways in white rot fungus *Pycnoporus sanguineus* 14. Int. Biodeterior. Biodegradation. 134: 25–30.

Lin, S., J. Wei, B. Yang, M. Zhang and R. Zhuo. 2022. Bioremediation of organic pollutants by white rot fungal cytochrome P450: The role and mechanism of CYP450 in biodegradation. Chemosphere. 134776.

Lisowska, K. and J. Długoński. 2003. Concurrent corticosteroid and phenanthrene transformation by filamentous fungus *Cunninghamella elegans*. J. Steroid Biochem. Mol. Biol. 85(1): 63–69.

Maia, M., A. Capao and L. Procópio, 2019. Biosurfactant produced by oil-degrading *Pseudomonas putida* AM-b1 strain with potential for microbial enhanced oil recovery. Bioremediat J. 23(4): 302–310.

Majcherczyk, A., C. Johannes and A. Hüttermann. 1998. Oxidation of polycyclic aromatic hydrocarbons (PAH) by laccase of *Trametes versicolor*. Enzyme Microb. Technol. 22(5): 335–341.

Mandree, P., W. Masika, J. Naicker, G. Moonsamy, S. Ramchuran and R. Lalloo. 2021. Bioremediation of polycyclic aromatic hydrocarbons from industry contaminated soil using indigenous *Bacillus* spp. Processes. 9(9): 1606.

Marco-Urrea, E., I. García-Romera and E. Aranda. 2015. Potential of non-ligninolytic fungi in bioremediation of chlorinated and polycyclic aromatic hydrocarbons. N. Biotechnol. 32(6): 620–628.

Masih, J., R. Singhvi, K. Kumar, V. K. Jain and A. Taneja. 2012. Seasonal variation and sources of polycyclic aromatic hydrocarbons (PAHs) in indoor and outdoor air in a semi-arid tract of northern India. Aerosol Air Qual. Res. 12(4): 515–525.

McClements, D. J. and C. E. Gumus. 2016. Natural emulsifiers—Biosurfactants, phospholipids, biopolymers, and colloidal particles: molecular and physicochemical basis of functional performance. Adv. Colloid Interface Sci. 234: 3–26.

Mekwichai, P., C. Tongcumpou, S. Kittipongvises and N. Tuntiwiwattanapun. 2020. Simultaneous biosurfactant-assisted remediation and corn cultivation on cadmium contaminated soil. Ecotoxicol. Environ. Saf. 192: 110298.

Mougin, C., H. Boukcim and C. Jolivalt. 2009. Soil bioremediation strategies based on the use of fungal enzymes. In Advances in Applied Bioremediation. pp. 123–149. Springer, Berlin, Heidelberg.

Naghdi, M., M. Taheran, S. K. Brar, A. Kermanshahi-Pour, M. Verma and R. Y. Surampalli. 2018. Removal of pharmaceutical compounds in water and wastewater using fungal oxidoreductase enzymes. Environ. Pollut. 234: 190–213.

Nayak, Y., S. Chakradhari, K. S. Patel, R. K. Patel, S. Yurdakul, H. Saathoff and P. Martín-Ramos. 2022. Distribution, variations, fate and sources of polycyclic aromatic hydrocarbons and carbon in particulate matter, road dust, and sediments in Central India. Polycycl. Aromat. Compd. 1–23.

Nikiforova, S. V., N. N. Pozdnyakova and O. V. Turkovskaya. 2009. Emulsifying agent production during PAHs degradation by the white rot fungus *Pleurotus ostreatus* D1. Curr. Microbiol. 58(6): 554–558.

Nikiforova, S. V., N. N. Pozdnyakova, O. E. Makarov, M. P. Chernyshova and O. V. Turkovskaya. 2010. Chrysene bioconversion by the white rot fungus *Pleurotus ostreatus* D1. Microbiol. 79(4): 456–460.

Ning, D., H. Wang, C. Ding and H. Lu. 2010. Novel evidence of cytochrome P450-catalyzed oxidation of phenanthrene in *Phanerochaete chrysosporium* under ligninolytic conditions. Biodegradation. 21(6): 889–901.

Noman, E., A. Al-Gheethi, R. M. S. R. Mohamed and B. A. Talip. 2019. Myco-remediation of xenobiotic organic compounds for a sustainable environment: a critical review. Top. Curr. Chem. 377(3): 1–41.

Omoni, V. T., C. N. Ibeto, A. J. Lag-Brotons, P. O. Bankole and K. T. Semple. 2022. Impact of lignocellulosic waste-immobilised white-rot fungi on enhancing the development of 14C-phenanthrene catabolism in soil. Sci. Total Environ. 811: 152243.

Osono, T., Y. Ono and H. Takeda. 2003. Fungal ingrowth on forest floor and decomposing needle litter of *Chamaecyparis obtusa* in relation to resource availability and moisture condition. Soil Biol. Biochem. 35(11): 1423–1431.

Ostrem Loss, E. M., M. K. Lee, M. Y. Wu, J. Martien, W. Chen, D. Amador-Noguez and J. H. Yu. 2019. Cytochrome P450 monooxygenase-mediated metabolic utilization of benzo [a] pyrene by *Aspergillus* species. mBio 10: 00558–19.

Passarini, M. R., L. D. Sette and M. V. Rodrigues. 2011. Improved extraction method to evaluate the degradation of selected PAHs by marine fungi grown in fermentative medium. J. Braz. Chem. Soc. 22: 564–570.

Pathak, S., A. K. Sakhiya, A. Anand, K. K. Pant and P. Kaushal. 2022. A state-of-the-art review of various adsorption media employed for the removal of toxic Polycyclic aromatic hydrocarbons (PAHs): an approach towards a cleaner environment. J. Water Process. Eng. 47: 102674.

Pawlik, A., M. Wójcik, K. Rułka, K. Motyl-Gorzel, M. Osińska-Jaroszuk, J. Wielbo and G. Janusz. 2016. Purification and characterization of laccase from *Sinorhizobium meliloti* and analysis of the lacc gene. Int. J. Biol. Macromol. 92: 138–147.

Perez-Boada, M., F. J. Ruiz-Duenas, R. Pogni, R. Basosi, T. Choinowski, M. J. Martínez and A. T. Martínez. 2005. Versatile peroxidase oxidation of high redox potential aromatic compounds: site-directed mutagenesis, spectroscopic and crystallographic investigation of three long-range electron transfer pathways. J. Mol. Biol. 354(2): 385–402.

Pi, Y., B. Chen, M. Bao, F. Fan, Q. Cai, L. Ze and B. Zhang. 2017. Microbial degradation of four crude oil by biosurfactant producing strain *Rhodococcus* sp. Bioresour Technol. 232: 263–269.

Potin, O., E. Veignie and C. Rafin 2004. Biodegradation of polycyclic aromatic hydrocarbons (PAHs) by *Cladosporium sphaerospermum* isolated from an aged PAH contaminated soil. FEMS Microbiol. Ecol. 51(1): 71–78.

Pourfadakari, S., M. Ahmadi, N. Jaafarzadeh, A. Takdastan, S. Ghafari and S. Jorfi. 2021. Remediation of PAHs contaminated soil using a sequence of soil washing with biosurfactant produced by *Pseudomonas aeruginosa* strain PF2 and electrokinetic oxidation of desorbed solution, effect of electrode modification with Fe3O4 nanoparticles. J. Hazard. Mater. 379: 120839.

Pozdnyakova, N. N., S. V. Nikiforova, O. E. Makarov, M. P. Chernyshova, K. E. Pankin and O. V. Turkovskaya. 2010. Influence of cultivation conditions on pyrene degradation by the fungus *Pleurotus ostreatus* D1. World J. Microbiol. Biotechnol. 26(2): 205–211.

Pozdnyakova, N. N. 2012. Involvement of the ligninolytic system of white-rot and litterdecomposing fungi in the degradation of polycyclic aromatic hydrocarbons. Biotechnol. Res. Int. 1–20.

Pozdnyakova, N. N., M. P. Chernyshova, V. S. Grinev, E. O. Landesman, O. V. Koroleva and O. V. Turkovskaya. 2016. Degradation of fluorene and fluoranthene by the basidiomycete *Pleurotus ostreatus*. Appl. Biochem. Microbiol. 52(6): 621–628.

Pramanik, K., S. Banerjee, D. Mukherjee, K. K. Saha, T. K. Maiti and N. C. Mandal. 2021. Beneficial role of plant growth-promoting rhizobacteria in bioremediation of heavy metal (loid)-contaminated agricultural fields. In Microbes: The Foundation Stone of the Biosphere. pp. 441–495. Springer, Cham.

Premnath, N., K. Mohanrasu, R. G. R. Rao, G. H. Dinesh, G. S. Prakash, V. Ananthi and A. Arun. 2021. A crucial review on polycyclic aromatic Hydrocarbons-Environmental occurrence and strategies for microbial degradation. Chemosphere. 280: 130608.

Punnapayak, H., S. Prasongsuk, K. Messner, K. Danmek and P. Lotrakul. 2009. Polycyclic aromatic hydrocarbons (PAHs) degradation by laccase from a tropical white rot fungus *Ganoderma lucidum*. Afr. J. Biotechnol. 8(21).

Rafin, C., O. Potin, E. Veignie, L. H. Sahraoui and M. Sancholle. 2000. Degradation of benzo [a] pyrene as sole carbon source by a non-white rot fungus, *Fusarium solani*. Polycycl. Aromat. Comp. 21(1/4): 311–330.

Ravelet, C., S. Krivobok, L. Sage and R. Steiman. 2000. Biodegradation of pyrene by sediment fungi. Chemosphere. 40(5): 557–563.

Ritz, K. and I. M. Young. 2004. Interactions between soil structure and fungi. Mycologist. 18(2): 52–59.

Romero, M. C., M. I. Urrutia, H. E. Reinoso and M. M. Kiernan. 2010. Benzo [a] pyrene degradation by soil filamentous fungi. J. Yeast Fungal Res. 1(2): 025–029.

Rosales, E., M. Pazos and M. Ángeles Sanromán. 2013. Feasibility of solid-state fermentation using spent fungi-substrate in the biodegradation of PAHs. CLEAN–Soil, Air, Water. 41(6): 610–615.

Ruiz-Duenas, F. J., M. Morales, M. Pérez-Boada, T. Choinowski, M. J. Martínez, K. Piontek and Á. T. Martínez. 2007. Manganese oxidation site in *Pleurotus eryngii* versatile peroxidase: a site-directed mutagenesis, kinetic, and crystallographic study. Biochemistry. 46(1): 66–77.

Sack, U., M. Hofrichter and W. Fritsche. 1997. Degradation of polycyclic aromatic hydrocarbons by manganese peroxidase of *Nematoloma frowardii*. FEMS Microbiol. Lett. 152(2): 227–234.

Saiu, G., S. Tronci, M. Grosso, E. Cadoni and N. Curreli. 2016. Biodegradation of polycyclic aromatic hydrocarbons by *pleurotus sajor-caju*. Chem. Eng. Trans. 49: 487–492.

Saraswathy, A. and R. Hallberg. 2002. Degradation of pyrene by indigenous fungi from a former gasworks site. FEMS Microbiol. Lett. 210(2): 227–232.

Saraswathy, A. and R. Hallberg. 2005. Mycelial pellet formation *by Penicillium ochrochloron* species due to exposure to pyrene. Microbiol. Res. 160(4): 375–383.

Schmidt, S. N., J. H. Christensen and A. R. Johnsen. 2010. Fungal PAH-metabolites resist mineralization by soil microorganisms. Environ. Sci. Technol. 44(5): 1677–1682.

Show, B. K., S. Banerjee, A. Banerjee, R. GhoshThakur, A. K. Hazra, N. C. Mandal and S. Chaudhury. 2022. Insect gut bacteria: a promising tool for enhanced biogas production. Rev. Environ. Sci. Biotechnol., 1–25.

Sigoillot, J. C., J. G. Berrin, M. Bey, L. Lesage-Meessen, A. Levasseur, A. Lomascolo and E. Uzan-Boukhris. 2012. Fungal strategies for lignin degradation. Adv. Bot. Res., 263–308.

Silva, I. S., M. Grossman and L. R. Durrant. 2009. Degradation of polycyclic aromatic hydrocarbons (2–7 rings) under microaerobic and very-low-oxygen conditions by soil fungi. Int. Biodeterior. Biodegrad. 63(2): 224–229.

Sipilä, T. P., P. Väisänen, L. Paulin and K. Yrjälä. 2010. *Sphingobium* sp. HV3 degrades both herbicides and polyaromatic hydrocarbons using ortho-and meta-pathways with differential expression shown by RT-PCR. Biodegradation. 21(5): 771–784.

Srivastava, S. and M. Kumar. 2019. Biodegradation of polycyclic aromatic hydrocarbons (PAHs): a sustainable approach. In Sustainable Green Technologies for Environmental Management. pp. 111–139. Springer, Singapore.

Steffen, K. T., M. Hofrichter and A. Hatakka. 2002. Purification and characterization of manganese peroxidases from the litter-decomposing basidiomycetes *Agrocybe praecox* and *Stropharia coronilla*. Enzyme Microb. Technol. 30(4): 550–555.

Sutherland, J. B. 1992. Detoxification of polycyclic aromatic hydrocarbons by fungi. J. Ind. Microbiol. Biotech. 9(1): 53–61.

Syed, K., H. Doddapaneni, V. Subramanian, Y. W. Lam and J. S. Yadav. 2010. Genome-to-function characterization of novel fungal P450 monooxygenases oxidizing polycyclic aromatic hydrocarbons (PAHs). Biochem. Biophys. Res. Commun. 399(4): 492–497.

Syed, K., A. Porollo, Y. W. Lam and J. S. Yadav. 2011. A fungal P450 (CYP5136A3) capable of oxidizing polycyclic aromatic hydrocarbons and endocrine disrupting alkylphenols: role of Trp129 and Leu324. PloS one. 6(12): 28286.

Syed, K., A. Porollo, Y. W. Lam, P. E. Grimmett and J. S. Yadav. 2013. CYP63A2, a catalytically versatile fungal P450 monooxygenase capable of oxidizing higher-molecular-weight polycyclic aromatic hydrocarbons, alkylphenols, and alkanes. Appl. Environ. Microbiol. 79(8): 2692–2702.

Tavares, A. P. M., M. A. Z. Coelho, M. S. M. Agapito, J. A. P. Coutinho and A. Xavier. 2006. Optimization and modeling of laccase production by *Trametes versicolor* in a bioreactor using statistical experimental design. Appl. Biochem. Biotechnol. 134: 233–248.

Teerapatsakul, C., C. Pothiratana, L. Chitradon and S. Thachepan. 2016. Biodegradation of polycyclic aromatic hydrocarbons by a thermotolerant white rot fungus Trametes polyzona RYNF13. J. Gen. Appl. Microbiol. 62(6): 303–312.

Ten Have, R. and P. J. M. Teunissen. 2001. Oxidative mechanisms 1952 involved in lignin degradation by white-rot fungi. Chem. Rev. 1953 101: 3397–3414.

Thurston, C. F. 1994. The structure and function of fungal laccases. Microbiology. 140(1): 19–26.

Treu, R. and J. Falandysz. 2017. Mycoremediation of hydrocarbons with basidiomycetes—a review. J. Environ. Sci. Health B. 52(3): 148–155.

Tripathi, V., S. A. Edrisi, B. Chen, V. K. Gupta, R. Vilu, N. Gathergood and P. C. Abhilash. 2017. Biotechnological advances for restoring degraded land for sustainable development. Trends Biotechnol. 35(9): 847–859.

USEPA. 2002. Polycyclic Oganic Matter, US Environmental Protection Agency. (http://www.epa.gov/ttn/atw/hlthef/polycycl.html). Accessed 06 May 2009.

Vasconcelos, M. R. S., G. A. L. Vieira, I. V. R. Otero, R. C. Bonugli-Santos, M. V. N. Rodrigues, V. L. G. Rehder et al. 2019. Pyrene degradation by marine-derived ascomycete: process optimization, toxicity, and metabolic analyses. Environ. Sci. Pollut. Res. 26: 12412–12424. https://doi.org/ 10.1007/s11356-019-04518-2.

Vipotnik, Z., M. Michelin and T. Tavares. 2022. Biodegradation of chrysene and benzo [a] pyrene and removal of metals from naturally contaminated soil by isolated *Trametes versicolor* strain and laccase produced thereof. Environ. Technol. Innov. 28: 102737.

Vipotnik, Z., M. Michelin and T. Tavares. 2021. Ligninolytic enzymes production during polycyclic aromatic hydrocarbons degradation: effect of soil pH, soil amendments and fungal co-cultivation. Biodegradation. 32(2): 193–215.

Vyas, B. R. M., S. Bakowski, V. Sasek and M. Matucha. 1994. Degradation of anthracene by selected white-rot fungi. FEMS Microbiol. Ecol. 14: 65–70.

Wang, R. Y., J. X. Liu, H. L. Huang, Z. Yu, X. M. Xu and G. M. Zeng. 2008. Effect of rhamnolipid on the enzyme production of two species of lignin-degrading fungi. J. Hunan Univ. Nat. Sci. 35(10): 70–74.

Wang, Y., R. Vazquez-Duhalt and M. A. Pickard. 2003. Manganese–lignin peroxidase hybrid from *Bjerkandera adusta* oxidizes polycyclic aromatic hydrocarbons more actively in the absence of manganese. Can. J. Microbiol. 49(11): 675–682.

Wong, D. W. 2009. Structure and action mechanism of ligninolytic enzymes. Appl. Biochem. Biotechnol. 157(2): 174–209.

Weis, L. M., A. M. Rummel, S. J. Masten, J. E. Trosko and B. L. Upham. 1998. Bay or baylike regions of polycyclic aromatic hydrocarbons were potent inhibitors of Gap junctional intercellular communication. Environ. Health Perspect. 106(1): 17–22.

Winquist, E., K. Björklöf, E. Schultz, M. Räsänen, K. Salonen, F. Anasonye et al. 2014. Bioremediation of PAH-contaminated soil with fungi–from laboratory to field scale. Int. Biodeterior. Biodegradation. 86: 238–247.

Young, D., J. Rice, R. Martin, E. Lindquist, A. Lipzen, I. Grigoriev and D. Hibbett. 2015. Degradation of bunker C fuel oil by white-rot fungi in sawdust cultures suggests potential applications in bioremediation. PloS one. 10(6): 0130381.

Yuliani, H., M. S. Perdani, I. Savitri, M. Manurung, M. Sahlan, A. Wijanarko and H. Hermansyah. 2018. Antimicrobial activity of biosurfactant derived from *Bacillus subtilis* C19. Energy Procedia. 153: 274–278.

Zhang, H., S. Zhang, F. He, X. Qin, X. Zhang and Y. Yang. 2016. Characterization of a manganese peroxidase from white-rot fungus *Trametes* sp. 48424 with strong ability of degrading different types of dyes and polycyclic aromatic hydrocarbons. J. Hazard. Mater. 320: 265–277.

Zhang, S., Y. Ning, X. Zhang, Y. Zhao, X. Yang, K. Wu et al. 2015. Contrasting characteristics of anthracene and pyrene degradation by wood rot fungus *Pycnoporus sanguineus* H1. Int. Biodeterior. Biodegrad. 105: 228–232.

Zhang, X., X. Wang, C. Li, L. Zhang, G. Ning, W. Shi and Z. Yang. 2020. Ligninolytic enzyme involved in removal of high molecular weight polycyclic aromatic hydrocarbons by *Fusarium* strain ZH-H2. Environ. Sci. Pollut. Res. 27(34): 42969–42978.

Zhang, Y., X. Xiao, X. Zhu and B. Chen. 2022. Self-assembled fungus-biochar composite pellets (FBPs) for enhanced co-sorption-biodegradation towards phenanthrene. Chemosphere. 286: 131887.

Zhuo, R. and F. Fan. 2021. A comprehensive insight into the application of white rot fungi and their lignocellulolytic enzymes in the removal of organic pollutants. Sci. Total Environ. 778: 146132.

Zimmerman, A. R., D. H. Kang, M. Y. Ahn, S. Hyun and M. K. Banks. 2008. Influence of a soil enzyme on iron-cyanide complex speciation and mineral adsorption. Chemosphere. 70(6): 1044–1051.

CHAPTER 7

Microbe-Assisted Bioremediation of Pesticides from Contaminated Habitats
Current Status and Prospects

Karen Reddy,[1] Shisy Jose,[1] Tufail Fayaz,[2]
Nirmal Renuka,[2,]* Sachitra Kumar Ratha,[3]
Sheena Kumari[1] and Faizal Bux[1]

7.1 Introduction

The rapid growth of the global pesticide market, driven by widespread use of pesticides in agriculture and non-agriculture sectors, has led to the introduction of numerous pesticide residues into the environment (Mali et al. 2022). Pesticides are recalcitrant and non-biodegradable, thus when applied to farmlands, gardens and other vegetation, they often remain toxic for years (Gonçalves and Delabona 2022). They are also carcinogenic in nature and are banned in many countries due to the risk created by their presence in the environment (Singh et al. 2020). In addition to polluting soil and crops, they also pose a threat to ground water and other aquatic environments (Castelo-Grande et al. 2010, Lehmann et al. 2018). Most pesticides reach destinations other than their intended target, even though each is designed to eliminate a specific pest (Huang et al. 2018, Mali et al. 2022).

Several technologies have been developed and applied to contaminated sites to eliminate the adverse environmental effects of pesticides. These include, physical treatments (adsorption and percolator filters) and chemical treatments (advanced oxidation) (Satish et al. 2017). Even though these methods seem promising, they are not cost effective and have several disadvantages. As a result, a strategic plan for reducing agrochemical use and implementing sustainable farming practices is essential (Gupta et al. 2016, Sun et al. 2020, Avila et al. 2021).

One such avenue that has been explored is bioremediation and degradation combined with microbes. It is a cost effective and environmentally friendly technology of soil and water reclamation

[1] Institute for Water and Wastewater Technology, Durban University of Technology, Durban, South Africa-4001.
[2] Algal Biotechnology Laboratory, Department of Botany, Central University of Punjab, Bathinda, India-151401.
[3] Phycology Laboratory, CSIR-National Botanical Research Institute, Lucknow, India-226001.
* Corresponding author: renuka.bot@gmail.com

(Subashchandrabose et al. 2013). The two processes use living organisms to metabolize xenobiotics, recalcitrant and toxic materials found in soil, water and sediments (Avila et al. 2021). There has been extensive research on the biological removal of pesticides by microorganisms, including bacteria (*Arthrobacter, Bacillus, Corynebacterium, Flavobacterium, Pseudomonas* and *Rhodococcus*), algae/cyanobacteria (*Chlorella, Scenedesmus, Synechococcus, Phormidium, Nostoc* sp., *Oscillatoria, Anabaena* sp., and *Aulosira*) and fungi (*Penicillium, Aspergillus, Fusarium* and *Trichoderma*) (Nicolopoulou-Stamati et al. 2016). Several species display a strong pesticide degradation activity and are highly adaptable (Subashchandrabose et al. 2013, Huang et al. 2018, Tarla et al. 2020). With advanced metabolic mechanisms, microorganisms can use or transform pesticide molecules into non-toxic or less hazardous forms of metabolites (Huang et al. 2018). This chapter aims to describe the current status of pesticide pollution, and the types and mechanisms involved in microbe-assisted bioremediation of pesticides.

7.2 Types of Pesticides

The term "pesticide" is used to describe a variety of insecticides, herbicides, fungicides, rodenticides and garden chemicals. They may be grouped according to the type of pest to be controlled and by their mode of action, penetration, chemical composition and toxicity (Table 7.1) (Castelo-Grande et al. 2010). Pesticides are classified according to their chemical nature as organochlorine pesticides, organophosphorus pesticides, carbamates and other pesticides such as pyrethroids, triazine, and neonicotinoids (Nicolopoulou-Stamati et al. 2016). The use of pesticides creates significant risks to the environment and non-target organisms, such as beneficial soil microorganisms, plants, fish and birds. The main classes of pesticides are considered as follows:

7.2.1 Organochlorine Pesticides

Organochlorine pesticides (OCPs) are chlorinated hydrocarbons that have been extensively used since the 1940s for commercial agriculture, mosquito, termite and tsetse fly control (Nicolopoulou-Stamati et al. 2016). The widely known organochlorine pesticides are dichlorodiphenyltrichloroethane, dieldrin, endosulfan, heptachlor, dicofol and methoxychlor (Nicolopoulou-Stamati et al. 2016). These chemical substances act upon the pest's nervous system by disrupting the enzyme that regulates neurotransmitters (Huang et al. 2018). After application, they remain in the environment for up to three decades and accumulate in water, food and sediments (Kafilzadeh 2015). Many studies suggest the toxic potential of OCPs to human health including reproductive failure, birth defects, endocrine disorders, lipid metabolism, immune system dysfunction and cancer (Lehmann et al. 2018, Tarla et al. 2020, Bose et al. 2021, Gonçalves and Delabona 2022, Mali et al. 2022). Consequently, in many countries, several OCPs are no longer permitted due to health and environmental concerns.

7.2.2 Organophosphorus Pesticides

Organophosphorus pesticides (OPPs) are known as one of the major groups of pesticides widely used in agriculture to control pests from exhibiting undesirable behavior. These include esters of phosphoric acids as an alternative to organochlorine in pesticides (Mali et al. 2022). The most widely used OPPs are glyphosate, chlorpyrifos, monocrotophos, acephate, bromophos, ectophos, leptophos, quinalphos, malathion, parathion, dimethoate, tribufos, merphos and trichlorofon (Foong et al. 2020). Studies have reported that these chemicals persist in soil for up to 10 to 360 d (Singh and Walker 2006). Thus, posing increased environmental risk to flora and fauna in terrestrial and aquatic ecosystems. Many ecosystems have been contaminated by OPPs, and they present threats such as cardiovascular disease, dementia, neurological problems, autoimmune disorders, negative effects on the reproductive system and other adverse effects on human health (Subashchandrabose et al. 2013, Gonçalves and Delabona 2022).

Table 7.1. An overview of commonly used pesticides.

Types of pesticides	Chemical classification and mechanism of action	Names of common pesticides	References
Insecticides	Organochlorines, organophosphorus, pyrethroids, carbamates, neonicotinoids • Acts upon insects' nervous system	Acephate, aldrin, azinphos-methyl, bromophos, carbaryl, carbofuran, carbosulfan, chlorpyrifos, chlordane coumaphos, DDT, deltamethrin, diazinon, dimethoate.	Singh and Walker 2006, Huang et al. 2008, Nicolopoulou-Stamati et al. 2016, Mali et al. 2022
Herbicides	Triazines, urea • Inhibits photosynthesis	Acetanilides, asulam, baraban, bentazon, butylate, chlorbromuron, chloropropham, chlorbrupham, cycloate, diallate, diuron, EPTC, glyphosate, karbutilate, linuron	Castelo-Grande et al. 2010, Alamgir Zaman Chowdhury et al. 2013, Bhandari, 2017, Huang et al. 2018
Fungicides	Organochlorines, organomercurials • Respiration and sterol biosynthesis inhibitors	Azadirachtin, copper sulfate, ferbam, mancozeb, maneb, pentachlorophenol, thiram	Zhang et al. 2011, Bhandari, 2017
Algicides	Dicarboxylic acid, copper sulfate, copper chelates • Targets specific physiological process	Bacillamide, copper sulfate, copper chelates, endothal, phorate nostocarboline	Castelo-Grande et al. 2010
Bactericides	Concentrated alcohol, ether, aldehydes, phenolic compounds • Inhibit bacterial protein synthesis pathway	Broponol, blue copper, chlorothalonil, copper hydrochloride, copper oxychloride, dithane, dithiocarbamates, formaldehyde, mancozeb, metalaxyl, methyl phosphorus, polytrin, ridomil, triazoles, thiocarbamates	Singh and Walker 2006, Castelo-Grande et al. 2010, Huang et al. 2018, Tarla et al. 2020
Nematicides	Carbamates • Damages specific and vital processes in the nematode tissues	Adicarb, carbofuran, chloropicrin, furadan, paladin, oxamyl, telone II, temik, vapam, vydate	Castelo-Grande et al. 2010, Hajihassani et al. 2018
Rodenticides	Organobromines • Interferes with the activation of blood clotting factors in the liver	Brodifacoum, bromethalin, carbon monoxide, chlorophacinone, cholecalciferol, diphacinone, hydrogen cyanide, methyl bromide, sulfur dioxide	Castelo-Grande et al. 2010, DeClementi and Sobczak 2012

DDT – Dichlorodiphenyltrichloroethane, EPTC – Ethyl dipropylthiocarbamate

7.2.3 Carbamates

Carbamate compounds are commonly used as insecticides (esters of carbamic acid). While, the derivatives of carbamic acid, thiocarbamic acid and dithiocarbamic acid are used as herbicides (Gupta 2014). Regularly known carbamate pesticides are aldicarb, bendiocarb, carbofuran, carbosulfan, carbaryl, methomyl, oxamyl, propoxur and ziram (Gupta 2014, Huang et al. 2018). In soil and water, carbamates tend to hydrolyze easily, thus resulting in low levels of persistence in the environment. The direct application of carbamates in soil can cause a significant reduction in the microflora, worms and some of the compounds are toxic to mammals and birds (Basheer et al. 2009). Additionally, the excessive use of carbamates can disrupt enzymes that regulate acetylcholine, which is a neurotransmitter in the nervous system. They cause disorders in reproductive and cellular metabolic functions in animals and humans (Huang et al. 2018, Gonçalves and Delabona 2022).

7.2.4 Pyrethroids, Triazine, and Neonicotinoids

Pyrethroids are synthetic insecticides that are widely used in agriculture and the public health sector. These are considered to be among the safer compounds currently available (Kolaczinski and Curtis 2004). Pyrethroids are found in many commercial products used to control insects, including household insecticides, pet sprays and shampoos (Nicolopoulou-Stamati et al. 2016). These include permethrin, resmethrin, sumithrin, cypermethrin, chlorfenvinphos, deltamethrin, fenvalerate, flumethrin and ivermectin (Huang et al. 2018). These compounds are neurotoxic, and extremely toxic to fish and other insects but show less toxicity towards mammals and land birds (Kolaczinski and Curtis 2004, Ray and Fry 2006). Triazines are widely used in agriculture as herbicides for controlling weeds. Commonly used triazines are atrazine, simazine and ametryn. Among these, atrazine is widely used as an herbicide, however, it has been banned from production to preserve the environment (Nicolopoulou-Stamati et al. 2016). These herbicides are endocrine-disrupting and carcinogenic, having long-term residual effects and creating health and environmental risks (Zhang et al. 2021). Neonicotinoids are now the world's most widely used insecticides (Hoshi et al. 2014). It is water-soluble, easily absorbed by plants and transported throughout their tissues for protection (Goulson 2013). Regularly known neonicotinoids are imidacloprid, thiacloprid and guadipyr. Studies have concluded that these insecticides do not present a high risk to non-targeted organisms; however, they can have possible adverse effects on the endocrine and reproductive systems of animals (Bal et al. 2012).

7.3 Current Status of Pesticide Pollution

In the past decade, the global production of pesticides, or so-called plant protection products, has been growing at a significant rate. In 1990, the production was 2.3 million tons, but in 2018, it reached 4.1 million tons and this growth is expected to continue during the next decade (Degrendele et al. 2022). While pesticides protect food production and cater to global food demand, they are also ubiquitous environmental pollutants that negatively impact water quality, biodiversity and human health (Landrigan et al. 2018). As a result of their toxicity and carcinogenicity, pesticides have always been deemed potentially harmful to living systems and have been a matter of debate and concern. Over the last few years, a high concentration of pesticides has been detected in environments such as agricultural, forest regions and surrounding water, across the globe. Recent data reports by Tang et al. (2021), documented the world geography of environmental risk of pollution by pesticides in 168 countries. Worldwide, more than 70% of agricultural land is polluted by pesticides. The risk of pollution was determined by pesticide residues exceeding the no-effect concentration in the environment, and it was considered high risk if the residues exceeded this by three orders of magnitude. In Figure 7.1, the land area subjected to low quantity and high variability of water supply and high risk of pollution by pesticide mixtures is demonstrated for the top 30 high- and low-income countries. About 34% of high-risk areas are in high-biodiversity regions, 5% are in water-scarce areas, and 19% are in low- and lower-middle-income countries. Among the regions of concern are South Africa, China, Australia and India due to high pesticide pollution risks, high biodiversity and water scarcity. Therefore, in order to transition to a low-pesticide future, global mandates for pesticide management and responsible food consumption should be implemented to reduce pesticide application at a global scale.

7.4 Microbe Assisted Bioremediation and their Prospects

Bioremediation of pesticides can be achieved by using various biological systems, as microorganisms (bacteria, fungi, cyanobacteria and algae) are major ways involved in the breakdown of pesticides. Microbe-assisted bioremediation is considered eco-friendly and a green option for converting pesticides into non-toxic or less toxic metabolites. It is a low-cost technology approach that is generally accepted by the public and can be carried out most often on site (Azubuike et al. 2016).

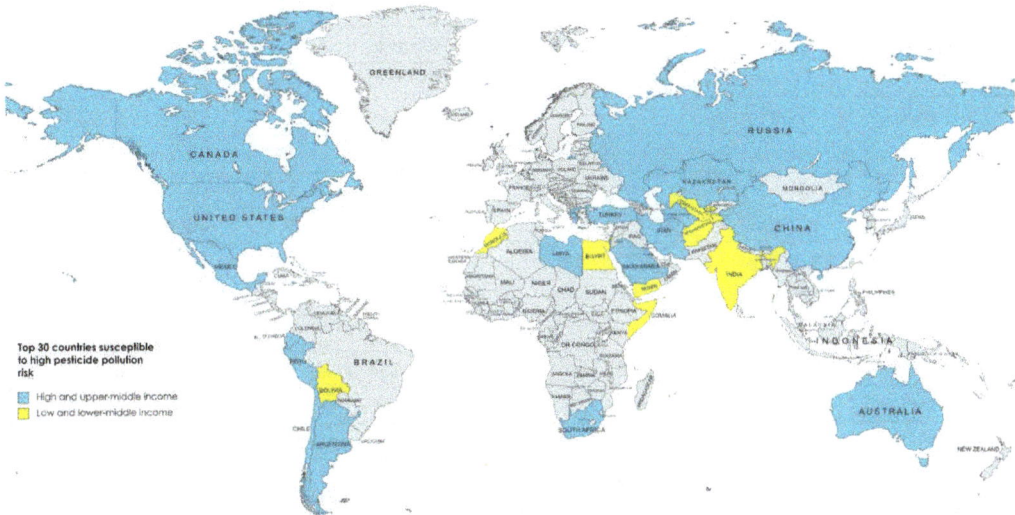

Figure 7.1. The top 30 countries susceptible to high pesticide pollution risk (Tang et al. 2021).

A bioremediation method can be classified into three subcategories: *in-situ*, *ex-situ* solid or *ex-situ* slurry. *In-situ* techniques treat soils and associated groundwater on-site without excavation, while *ex-situ* techniques require excavation prior to treatment (Shukla et al. 2010). Three fundamental principles guide the choice of an appropriate bioremediation technology, namely the ability of the pollutant to be transformed biologically, the accessibility of the contaminant to microorganisms and the possibility of optimizing biological activity (Dua et al. 2002). Biological activity is achieved as microorganisms use the contaminants as nutrients or energy sources. By introducing microorganisms with desired catalytic capabilities or by supplementing nutrients (nitrogen and phosphorus), electron acceptors (oxygen) and substrates (methane, phenol, and toluene), the activity of the microbe can be further stimulated (Shukla et al. 2010). However, the appropriate degrading enzymes and several environmental factors also contribute to the remediation of pesticides. The possible roles and applications of microbes for the remediation of pesticides are further summarized.

7.4.1 Bacteria-assisted Degradation

In nature, pesticides continue for many years and travel along various food chains (Kalyabina et al. 2021). The fate of pesticides in the ecosystem is determined by their physico-chemical properties and inherent biodegradability as a result of their structure (Bernardino et al. 2012). Many pesticides have insoluble properties, that can cause residues to remain in the environment and then negatively affect the ecosystem. In order to minimize the impact of pesticides in the environment, several methods have been developed for pesticide remediation. The traditional methods for pesticide remediation (physical, chemical and physico-chemical degradation) can, however, produce secondary pollutants that cause additional harm to the environment. Furthermore, the increased cost involved in using these techniques makes them unfeasible for pesticide removal from contaminated sites (Rajmohan et al. 2020). Thus, bioremediation of pesticides by using beneficial living organisms has been explored as an alternative (Deng and Wang 2016). Bioremediation of pesticide-polluted sites or samples via microorganisms has become a desired method for decontamination. It is an efficient, cost-effective technique for remediating pesticides without causing any harm to the environment, as they metabolize pesticides for nutrients and release CO_2 and H_2O during the process (Huang et al. 2018). Several microorganisms, including bacteria, have been isolated and used to degrade toxic pesticides from contaminated sites (Sarker et al. 2021). Many potential bacterial genera, such as *Pseudomonas*, *Bacillus*, *Acinetobacter*, *Klebsiella* and *Burkholderia* are effective in pesticide

Table 7.2. The remediation of pesticides by common bacterial strains from the environment.

Bacterial strain	Pesticide	Remediation (%)	Retention time	References
Acinetobacter	Diazinon	88.27	20d	Amani et al. 2018
Pseudomonas	Chlorpyrifos	65	6d	Ajaz et al. 2012
	Endosulfan	70–80	5d	Zaffar et al. 2018
	DDT	67.55	7d	Powthong et al. 2016
	Atrazine	99.9	2d	Cai et al. 2003
Bacillus	Fipronil	73	42d	Mandal et al. 2013
	Mesotrione	99	5h	Sun et al. 2020
	DDT	67.55	7d	Powthong et al. 2016
	Imidacloprid	25.36–45.48	25d	Sabourmoghaddam et al. 2015
Klebsiella	Imidacloprid	78	7d	Phugare et al. 2013
Burkholderia	Dieldrin	39	7d	Matsumoto et al. 2008
	Endrin	74	14d	Matsumoto et al. 2008

DDT – Dichlorodiphenyltrichloroethane; d – days; hr – hours

bioremediation of diverse chemical classes (Singh et al. 2020). Pesticides are remediated by bacteria based on species specificity and several abiotic factors, such as temperature, pH, nutrient content, moisture and humidity (Huang et al. 2018). Among pesticides, bacteria are the main degraders of organochlorines. These are synthetic organic compounds containing at least one covalently bonded chlorine atom and are insecticides primarily composed of carbon, hydrogen and chlorine. Some of the most known organochlorine pesticides include, dieldrin, aldrin, lindane, endosulfan, dichlorodiphenyltrichloroethane (DDT) and hexachlorocyclohexane (HCH). Pesticides containing organophosphates (imidacloprid, diazinon and chlorpyrifos) are also among the chemical groups investigated for degradation by bacteria (Jayaraj et al. 2016).

A few earlier reports (Table 7.2) had suggested the potential of *Pseudomonas* sp. for the remediation of insecticides and herbicides. Endosulfan (insecticide) was bioremediated by *Pseudomonas fluorescens* during a 5 d laboratory study. The results of the study showed that up to 80% of endosulfan could be remediated (Zaffar et al. 2018). In another study, approximately 99.9% of atrazine (herbicide) was remediated after 2 d incubation with *Pseudomonas* sp. There has also been research into the use of bacterial mixtures for pesticide remediation (Cai et al. 2003). When a mixture of *Pseudomonas stutzeri*, *Pseudomonas aeruginosa* and *Bacillus firmus* was used in a week-long laboratory study, approximately 68% of the DDT (insecticide) was remediated. Imidacloprid, another insecticide, was remediated by *Klebsiella pneumonia* (78%) and *Bacillus subtilis* (25.36–45.48%) during 7-d and 25-d experiments, respectively (Phugare et al. 2013, Sabourmoghaddam et al. 2015). Additionally, *Burkholderia* and *Acinetobacter* were found to be capable of remediating pesticides during tests conducted over a week. Almost all the potential bacterial strains were isolated from soil samples, indicating the cost-effectiveness of using indigenous bacteria for pesticide remediation.

Bacteria can interact, both chemically and physically, with substances, leading to structural changes or complete degradation of the target molecule (Ortiz-Hernández et al. 2013). Bacteria can transform or degrade pesticides into less toxic or non-toxic forms (McGuinness and Dowling 2009). This is commonly known as a detoxification mechanism (Figure 7.2), where bacteria produce intracellular and extracellular enzymes (Singh et al. 2020). The activity of enzymes depends on the metabolic potential of the bacteria to detoxify or transform the pollutants, which depends on both accessibility and bioavailability (Ramakrishnan et al. 2011). There could be three phases to the metabolism of pesticides. Phase one involves the transformation of the initial properties of the parent compound through oxidation, reduction or hydrolysis. This transformation produces a more water-soluble and usually less toxic product than the parent. In the second phase, a pesticide or pesticide

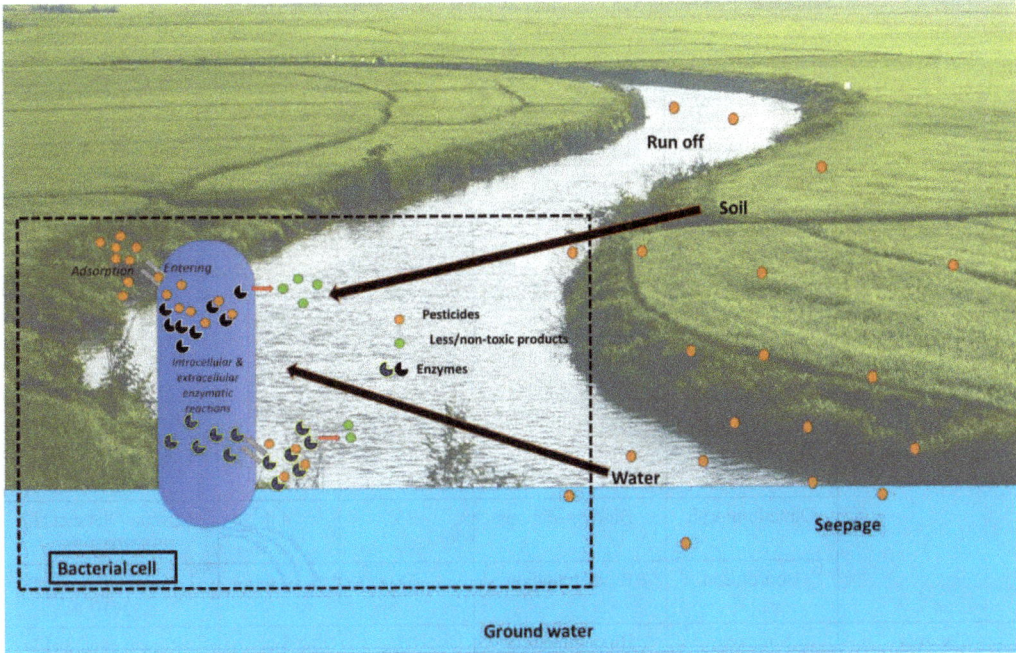

Figure 7.2. Mechanisms of pesticide degradation in bacterial system.

metabolite is conjugated to a sugar or amino acid, that increases its water solubility and reduces its toxicity. During the third phase, phase two metabolites are converted into secondary conjugates, which are also non-toxic (Ramakrishnan et al. 2011). The overall metabolism of pesticides by enzymes proves to be favorable, as enzymes are more resistant to environmental conditions, which makes it easier for them to remove pesticides more efficiently from affected areas (Huang et al. 2008). Therefore, it is imperative to identify species of bacteria that can thrive in extreme conditions and completely degrade pesticides. A combination of mechanistic approaches and microbial genetic advancement can enhance sustainable pesticide bioremediation by identifying the most effective bacterial strains or bacterial consortia.

7.4.2 Phyco-assisted Degradation

The persistence of pesticides creates a grave threat to the environment, and their removal is crucial in maintaining stability in the ecosystem (Bodin 2014). For many years, biological approaches have been used in the removal of inorganic pollutants such as pesticides. There is a growing interest in cyanobacteria and micro- and macroalgae as potent organisms for removing pesticides. It has been shown that these organisms easily adapt to environmental changes and can grow efficiently under various stress conditions (Mata et al. 2010). A wide range of photosynthetic algae exist, ranging from single cells to multicellular organisms (Singh et al. 2020). They can grow in the presence and absence of light and utilize organic carbon as an energy source (Subashchandrabose et al. 2013). As part of a sustainable future, these photosynthetic microorganisms can be used for the remediation of pesticides at contaminated sites. Research on phycoremediation technology, i.e., the utilization of microalgae/cyanobacteria and their consortia as bioremediating agents, has proved to have promising potential. Several authors have reported that common algal strains can remove pesticides from contaminated environments, mainly by biosorption, bioaccumulation and biodegradation (Table 7.3) (Verasoundarapandian et al. 2022).

Pesticide biosorption occurs in both living and dead photosynthetic organisms, as this mechanism is not dependent on energy for the removal of pesticides in contaminated environments.

Table 7.3. Remediation of pesticides by common algal/cyanobacterial strain.

Algal/ cyanobacterial Strain	Pesticide	Mode of action	Remediation (%)	Retention time	References
Nostoc	Malathion	Biodegradation	91	52 d	Ibrahim et al. 2014
Coleofasciculus	Chlorpyrifos	Biodegradation	90	21 d	Vijayan et al. 2020
Fischerella	Methyl parathion	Bioadsorption	~ 80	5 d	Tiwari et al. 2017
Chlamydomonas	Trichlorfon	Biodegradation	100	10 d	Wan et al. 2020
	Fluroxypyr	Bioaccumulation/ biodegradation	57	5 d	Zhang et al. 2011
	Prometryne	Bioaccumulation/ biodegradation	30-40	4 d	Jin et al. 2012
	Atrazine	Bioaccumulation	14-36		Kabra et al. 2014
Scenedesmus	Pyrimethanil	Biodegradation	10	4 d	Dosnon-Olette et al. 2010
	Dimethomorph	Biodegradation	24	4 d	Dosnon-Olette et al. 2010
	Isoproturon	Biodegradation	58	4 d	Dosnon-Olette et al. 2010
Chlorella	Simazine	Biodegradation	97	5 d	Hussein et al. 2017
	Pendimethalin	Biosorption	88	5 d	Hussein et al. 2017

d – days

In the biosorption process, the liquid and solid phases contain the dissolved or suspended pesticides to be absorbed. It could be defined as the attachment of potentially toxic pesticides to the surface of the photosynthetic strain. Pesticides are biosorbed in a passive and metabolically independent process that occurs faster than bioaccumulation (Verasoundarapandian et al. 2022). Strains such as *Chlorella* and *Fischerella* have been reported to remediate > 80% of pesticides through biosorption (Hussein et al. 2017, Tiwari et al. 2017). While bioaccumulation studies on fluroxypyr, prometryne and atazarine, have shown that members of *Chlorophyta* are able to accumulate pesticides during 4–5-d intervals but at a lower rate (< 60%) (Jin et al. 2012, Kabra et al. 2014, Zhang et al. 2021). The bioaccumulation ability of organisms such as algae is determined by their lipid content, which is influenced by their growth conditions and cell distribution (Sakurai et al. 2016). In addition to bioaccumulating pesticides, these photosynthetic organisms are also able to transform and degrade pesticides into non-toxic or less toxic compounds. Pesticide degradation is influenced by microorganism types, optimal environmental conditions and the metabolic activity of various enzymes (hydrolase, phosphatase, phosphodiesterase, oxygenase, esterase, transferase and oxidoreductases) (Verasoundarapandian et al. 2022). The probable biosorption, bioaccumulation and biodegradation processes for the removal of pesticides in photosynthetic microorganisms are shown in Figure 7.3.

Phycoremediation technology offers additional benefits such as minimized greenhouse gas emissions from the environment and biomass reuse (Renuka et al. 2018, Reddy et al. 2021, Renuka et al. 2021). Therefore, phycoremediation of pesticides can be a promising integrated and sustainable approach for eco-friendly and efficient removal of pesticides from contaminated areas with additional environmental benefits (Singh et al. 2020).

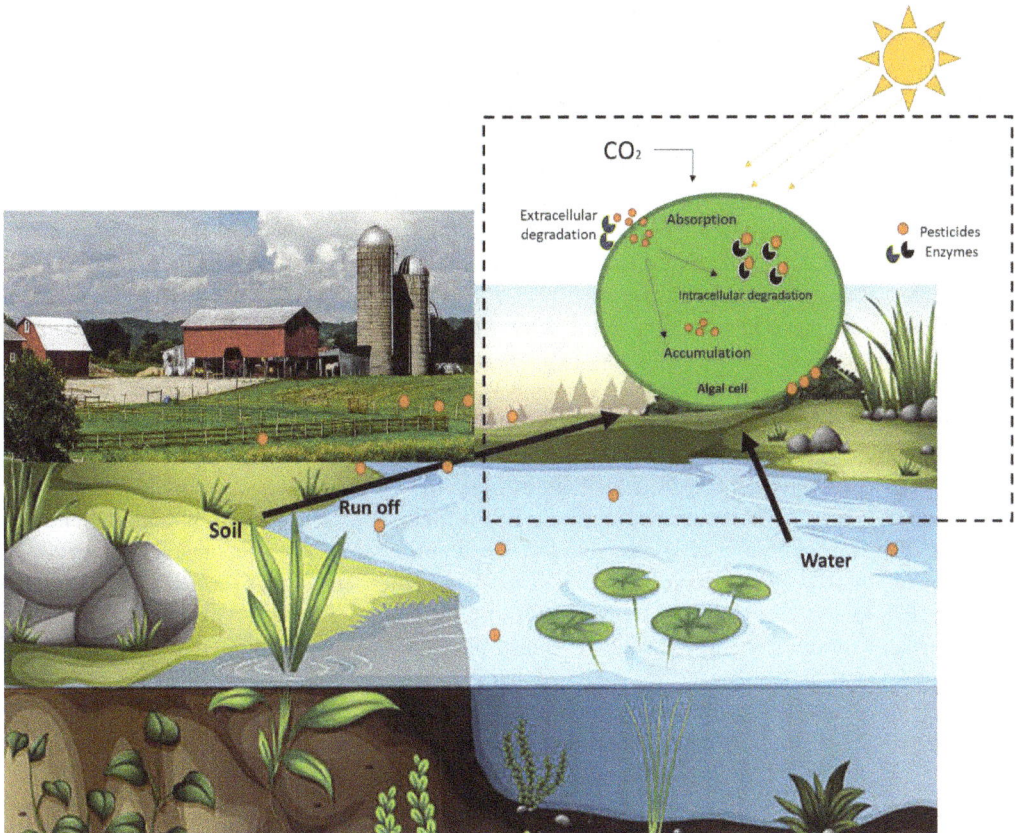

Figure 7.3. Mechanisms of pesticide remediation in algae.

7.4.3 Myco-assisted Degradation

Fungi are ubiquitous eukaryotic heterotrophic living organisms present in various habitats (Sanchez and Demain 2017, Singh et al. 2020). Several fungi can remove harmful chemicals from the environment, including pesticides, industrial wastes, dyes and personal care products (Deshmukh et al. 2016). Fungal mycelium facilitates deep penetration into soil, allowing many pollutants to be transported or translocated to cells (Sagar and Singh 2011). Therefore, fungi can access the contaminated area in the deeper layers and degrade compounds that are inaccessible to bacteria (Harms et al. 2011). Since fungal mycelium has low specificity of catabolic enzymes and can utilize non-specific organic compounds, it is more suitable for degrading pesticides (Harms et al. 2011, Chen et al. 2012).

Several studies have been conducted to isolate and characterize fungal strains that can remediate pesticides (Shahi et al. 2016, Saez et al. 2018). Different fungal species are reported to remediate pesticides using different processes and mechanisms (Figure 7.3). Bioremediation of many pesticides such as oxytetracycline, cypermethrin, carbofuran, hexachlorocyclohexane, chlorophenol and petroleum-based pesticides has been reported to be achieved using fungi such as *Trametes versicolor* (Chen et al. 2014, Mir-Tutusaus et al. 2014), *Pleurotus ostreatus* (Sadiq et al. 2015) and *Ganoderma lucidum* (Kaur et al. 2016). *Trametes versicolor* is also known for its ability to degrade carbamate pesticides (methomyl, methocarb and aldicarb) when grown in a defined liquid medium (Rodríguez-Rodríguez et al. 2017). *Phlebia* species such as *Lenzites betulinus* and

Irprex lacteus demonstrated degradation rates greater than 70%, when grown in the presence of chlorpyrifos at a temperature of 30°C and pH 7 (Wang et al. 2020). In another study, Fang et al. (2008), reported the removal of chlorpyrifos by the fungal strain *verticillium* sp. with high removal efficiency. Kulshrestha and Kumari (2011), explored the potential removal of chlorpyrifos in both mixed and pure fungi. They reported that mixed cultures could remove up to 300 mg L^{-1} chlorpyrifos. Another widely used insecticide of interest, allethrin, has a direct impact on human health. Bhatt et al. (2020) isolated *Fusarium proliferatum* CF2 from an allethrin contaminated agriculture site. The isolate could completely degrade the allethrin in 6 d. Endosulfan is degraded by a variety of fungal strains, including *Paecilomyces variotii*, *Paecilomyces lilacinus* (Hernández-Ramos et al. 2019), *Aspergillus sydoni* (Goswami et al. 2009), *Pleurotus eryngi* (Wang et al. 2018), *Aspergillus niger*, *Penicillium chrysogenum* and *Aspergillus flavus* (Ahmad 2020). Endosulfan degradation was also reported using *Trametes hirsute* (Kamei et al. 2011). Two pathways were reported, i.e., hydrolysis and oxidation of endosulfan that convert it to endosulfan diol and endosulfan sulfate, respectively. In an aqueous medium and soil, the bacterial and fungal consortia could remove endosulfan effectively (Abraham and Silambarasan 2014).

Several studies have reported the efficacy of fungi to degrade different pesticides (Table 7.4). *Phanerochaete chrysosporium* grown in 10 mg L^{-1} of thiamethoxam resulted in degradation rates of 49 and 98% during 15 and 25 d, respectively (Chen et al. 2021). An estimated 71% of triclosan was degraded by *Aspergillus versicolor* (Ertit Taştan and Dönmez 2015). *Trichoderma harzianum* CBMAI 167 isolated from a marine environment was grown in 50 mg L^{-1} of pentachlorophenol (PCP), which resulted in complete degradation to pentachloroanisole (PCA) after incubation for 7 d (Vacondio et al. 2015). It was also observed that PCA was degraded by *Trichoderma harzianum* CBMAI 167. When *Bjerkandera adusta* and *Anthracophyllum discolor* species were grown in PCP contaminated soil (100, 250 and 350 mg kg^{-1}), both fungi degraded PCP, however, the highest

Table 7.4. Pesticide removal efficiency of various fungal strains from the environment.

Fungal Strain	Pesticide	Pesticide concentration (mg L^{-1})	Remediation (%)	Reaction time	References
Phanerochaete chrysosporium	Thiamethoxam	10	49	10 d	Chen et al. 2021
		10	98	25 d	
Fusarium oxysporum Strain JASA1	Malathion	0.4	100	8 d	Peter et al. 2015
Aspergilus fumigatus and Byssochlamys spectabilis consortia	Chlorpyrifos	10	98.4	30 d	Kumar et al. 2021
Fusarium proliferatum Strain CF2	Allethrin	50	100	6 d	Bhatt et al. 2020
Aspergillus vesicolor	Trichlosan	7.5	71.91	5 d	Ertit Taştan and Dönmez 2015
Aspergillus niger	Endosulfan	35	100	5 d	Bhalerao and Puranik 2007
Aspergillus niger 13 MK640786	Diazinon	25	82	7 d	Hamad 2020
Ganoderma lucidum GL2	Lindane,	0.004–0.040	75.5	28 d	Kaur et al. 2016
Trichoderma harzianum CBMAI 167	Pentachlorophenol	50	100	7 d	Vacondio et al. 2015

d – days

removal (95%) of PCP was attained by *Anthracophyllum discolor* after an incubation of 28 d (Rubilar et al. 2007).

Abdel-Fattah Mostafa et al. (2022), isolated six fungal strains from contaminated soil and determined their removal efficiency for diazinon. The collected soil samples were found to contain 0.106 mg kg^{-1} of the pesticide diazinon. Among different fungal strains, *B. antenatal* was more effective in removing diazinon (83.88%) followed by *Trichoderma viride* (80.26%), *A. niger* (78.22%), *Rhizopus stolonifer* (77.36%) and *Fusarium graminearum* (75.43%) in 10 d. Strains such as *Aspergillus sydowii* CBMAI 935 and *Trichoderma* sp. were able to reduce 63 and 70% of chlorpyrifos after 30 d of incubation, respectively (Alvarenga et al. 2015). *Ganoderma lucidum* GL-2 strain, showed the potential for the degradation of lindane (organochlorine pesticide) when grown on rice bran substrate (Kaur et al. 2016). Biodegradation of lindane was higher (59.4%) using *F. solani* when compared to *F. poae* (56.7%) (Sagar and Singh 2011). It has been reported that *Phlebia tremellosa*, *Phlebia brevispora* and *Phlebia acanthocystis* could eliminate roughly 71, 74 and 90% of heptachlor, respectively, in a period of 14 d (Xiao et al. 2011). Alvarenga et al. (2014) studied the efficiency of *Aspergillus sydowii* CBMAI 934, and *Penicillium decaturense* CBMAI 1234 for methyl parathion removal. *A. sydowii* CBMAI 934 was able to remove 100% of methyl parathion in 20 d, whereas *P. decaturense* CBMAI 1234 degraded 100% methyl parathion in 30 d (Alvarenga et al. 2014).

Different fungal species have been reported to effectively remediate pesticides from various contaminated environments. However, several factors, such as species type, temperature, incubation time and growth medium, influence the rate and efficiency of remediation of agrochemicals. In addition, accelerated co-metabolism can be achieved by implementing the use of microbial consortia for the effective bioremediation of many pesticides. Therefore, mixed consortia or synergistic effects of the rhizosphere encourage and require further research for improving the efficiency of bioremediation of pesticides.

7.5 Future Aspects and Conclusion

The overall evaluation of microbial bioremediation suggests a constructive outlook for microbe-assisted pesticide remediation. Nevertheless, more scientific investigations into the technical aspects of microbial bioremediation systems are needed. The major restriction in microbial pesticide bioremediation is the long response time, which is usually due to the growth of microbes not being supported by environmental conditions. The accessibility and bioavailability of pesticides is another major challenge that limits the efficiency of bioremediation. Researchers are continuously exploring advanced bioengineering techniques to develop bioremediation tools that could be applied to explore the molecular basis of bioremediation and improve its efficiency. In depth studies using molecular approaches, including metatranscriptomics, metagenomics, metaproteomics and metabolomics, are required to reveal the interactions within communities and between different species in the microbe-assisted bioremediation systems. Genetic engineering is a crucial systemic technology to modify the metabolic pathways of microbes for improved remediation. In order to improve the practical feasibility, some novel concepts for degrading pesticides using integrated processes and genetic modifications to improve microbial based bioremediation technologies are also recommended.

Acknowledgment

The authors hereby acknowledge the Durban University of Technology, South Africa (UID: 84166), Central University of Punjab, Bathinda, India (CUPB/Acad./2022/1194), and CSIR-National Botanical Research Institute, Lucknow, India for providing facilities.

References

Abdel-Fattah Mostafa, A., M. T. Yassin, T. M. Dawoud, F. O. Al-Otibi and S. R. M. Sayed. 2022. Mycodegradation of diazinon pesticide utilizing fungal strains isolated from polluted soil. Environ. Res. 212: 113421.

Abraham, J. and S. Silambarasan. 2014. Biomineralization and formulation of endosulfan degrading bacterial and fungal consortiums. Pestic. Biochem. Physiol. 116: 24–31.

Ahmad, K. S. 2020. Remedial potential of bacterial and fungal strains (*Bacillus subtilis, Aspergillus niger, Aspergillus flavus* and *Penicillium chrysogenum*) against organochlorine insecticide Endosulfan. Folia Microbiol. (Praha). 65(5): 801–810.

Ajaz, M., S. A. Rasool, S. K. Sherwani and T. A. Ali. 2012. High profile chlorpyrifos degrading pseudomonas putida mas-1 from indigenous soil: gas chromatographic analysis and molecular characterization. Int. J. Basic Appl. Med. Sci. (IJBMSP). 2(2).

Alamgir Zaman Chowdhury, M., A. N. M. Fakhruddin, M. Nazrul Islam, M. Moniruzzaman, S. H. Gan and M. Khorshed Alam. 2013. Detection of the residues of nineteen pesticides in fresh vegetable samples using gas chromatography–mass spectrometry. Food Control. 34(2): 457–465.

Alvarenga, N., W. G. Birolli, M. H. Seleghim and A. L. Porto 2014. Biodegradation of methyl parathion by whole cells of marine-derived fungi *Aspergillus sydowii* and *Penicillium decaturense*. Chemosphere. 117: 47–52.

Alvarenga, N., W. G. Birolli, M. Nitschke, M. O. de O Rezende, M. Seleghim and A. Porto. 2015. Biodegradation of chlorpyrifos by whole cells of marine-derived fungi *Aspergillus sydowii* and *Trichoderma* sp. J. Microb. Biochem. Technol. 7: 133–139.

Amani, F., A. A. Safari Sinegani, F. Ebrahimi and S. Nazarian. 2018. Biodegradation of chlorpyrifos and diazinon organophosphates by two bacteria isolated from contaminated agricultural soils. Biol. J. Microorganism. 7(28): 27–39.

Avila, R., A. Peris, E. Eljarrat, T. Vicent and P. Blánquez. 2021. Biodegradation of hydrophobic pesticides by microalgae: transformation products and impact on algae biochemical methane potential. Sci. Total Environ. 754: 142114.

Azubuike, C. C., C. B Chikere and G. C. Okpokwasili. 2016. Bioremediation techniques–classification based on site of application: principles, advantages, limitations and prospects. World J. Microbiol. Biotechnol. 32(11): 180.

Bal, R.., M. Naziroğlu, G. Türk, Ö. Yilmaz, T. Kuloğlu, E. Etem and G. Baydas. 2012. Insecticide imidacloprid induces morphological and DNA damage through oxidative toxicity on the reproductive organs of developing male rats. Cell Biochem. Funct. 30(6): 492–499.

Basheer, C., A. A. Alnedhary, B. S. M. Rao and H. K. Lee. 2009. Determination of carbamate pesticides using micro-solid-phase extraction combined with high-performance liquid chromatography. J. Chromatogr. A. 1216(2): 211–216.

Bernardino, J., A. M. Bernardette, M. Ramrez-Sandoval and D. Domnguez-Oje. 2012. Biodegradation and bioremediation of organic pesticides. pp. 253–272. *In*: R. P. Soundararajan [Ed.]. Pesticides: Recent Trends in Pesticide Residue Assay. London: IntechOpen.

Bhalerao, T. S. and P. R. Puranik. 2007. Biodegradation of organochlorine pesticide, endosulfan, by a fungal soil isolate, *Aspergillus niger*. Int. Biodeterior. Biodegradation. 59(4): 315–321.

Bhandari, G. 2017. Mycoremediation: an eco-friendly approach for degradation of pesticides. pp. 119–131. *In:* R. Prasad [Ed.]. Mycoremediation and Environmental Sustainability. (1). Cham: Springer International Publishing.

Bhatt, P., W. Zhang, Z. Lin, S. Pang, Y. Huang and S. Chen. 2020. Biodegradation of allethrin by a novel fungus *Fusarium proliferatum* strain CF2, Isolated from Contaminated Soils. Microorganisms. 8(4): 593.

Bodin, H. 2014. Phycoremediation of pesticides using microalgae. MSc thesis, Swedish University of Agricultural Sciences, Sweden.

Bose, S., P. S. Kumar and D.-V. N. Vo. 2021. A review on the microbial degradation of chlorpyrifos and its metabolite TCP. Chemosphere. 283: 131447.

Cai, B., Y. Han, B. Liu, Y. Ren and S. Jiang. 2003. Isolation and characterization of an atrazine-degrading bacterium from industrial wastewater in China. Lett. Appl. Microbiol. 36(5): 272–276.

Castelo-Grande, T., P. A. Augusto, P. Monteiro, A. M. Estevez and D. Barbosa. 2010. Remediation of soils contaminated with pesticides: a review. Int. J. Environ. Anal. Chem. 90(3-6): 438–467.

Chen, A., G. Zeng, G. Chen, C. Zhang, M. Yan, C. Shang, X. Hu, L. Lu, M. Chen, Z. Guo and Y. Zuo. 2014. Hydrogen sulfide alleviates 2,4-dichlorophenol toxicity and promotes its degradation in *Phanerochaete chrysosporium*. Chemosphere. 109: 208–212.

Chen, A., W. Li, X. Zhang, C. Shang, S. Luo, R. Cao and D. Jin. 2021. Biodegradation and detoxification of neonicotinoid insecticide thiamethoxam by white-rot fungus *Phanerochaete chrysosporium*. J. Hazard Mater. 417: 126017.

Chen, S., C. Liu, C. Peng, H. Liu, M. Hu and G. Zhong. 2012. Biodegradation of chlorpyrifos and its hydrolysis product 3,5,6-trichloro-2-pyridinol by a new fungal strain *Cladosporium cladosporioides* Hu-01. PLoS One, 7(10): e47205.

DeClementi, C. and B. R. Sobczak. 2012. Common rodenticide toxicoses in small animals. Vet Clin. North Am. Small Anim. Pract. 42(2): 349–360.

Degrendele, C., J. Klánová, R. Prokeš, P. Přibylová, P. Šenk, M. Šudoma, M. Röösli, M. A. Dalvie and S. Fuhrimann. 2022. Current use pesticides in soil and air from two agricultural sites in South Africa: implications for environmental fate and human exposure. Sci. Total Environ. 807(Pt 1): 150455.

Deng, Y.-J. and Wang, S. Y. 2016. Synergistic growth in bacteria depends on substrate complexity. J. Microbiol. 54(1): 23–30.

Deshmukh, R., A. A. Khardenavis and H. J. Purohit. 2016. Diverse metabolic capacities of fungi for bioremediation. Indian J. Microbiol. 56(3): 247–264.

Dosnon-Olette, R., P. Trotel-Aziz, M. Couderchet and P. Eullaffroy. 2010. Fungicides and herbicide removal in Scenedesmus cell suspensions. Chemosphere. 79(2): 117–123.

Dua, M., A. Singh, N. Sethunathan and A. Johri. 2002. Biotechnology and bioremediation: successes and limitations. Appl. Microbiol. Biotechnol. 59(2): 143–152.

Ertit Taştan, B. and G. Dönmez. 2015. Biodegradation of pesticide triclosan by A. versicolor in simulated wastewater and semi-synthetic media. Pestic Biochem Physiol. 118: 33–37.

Fang, H., Y. Q. Xiang, Y. J. Hao, X. Q. Chu, X. D. Pan, J. Q. Yu and Y. L. Yu. 2008. Fungal degradation of chlorpyrifos by *Verticillium* sp. DSP in pure cultures and its use in bioremediation of contaminated soil and pakchoi. Int. Biodeterior. Biodegradation. 61(4): 294–303.

Foong, S. Y., N. L. Ma, S. S. Lam, W. Peng, F. Low, B. H. Lee, A. K. Alstrup and C. Sonne, 2020. A recent global review of hazardous chlorpyrifos pesticide in fruit and vegetables: prevalence, remediation and actions needed. J. Hazard. Mater. 400: 123006.

Gonçalves, C. R. and P. D. S. Delabona. 2022. Strategies for bioremediation of pesticides: challenges and perspectives of the Brazilian scenario for global application – A review. Environ. Adv. 8: 100220.

Goswami, S., K. Vig and D. K. Singh. 2009. Biodegradation of α and β endosulfan by *Aspergillus sydoni*. Chemosphere. 75(7): 883–888.

Goulson, D. 2013. An overview of the environmental risks posed by neonicotinoid insecticides. J. Appl. Ecol. 50(4): 977–987.

Gupta, M., S. Mathur, T. K. Sharma, M. Rana, A. Gairola, N. K. Navani and R. Pathania. 2016. A study on metabolic prowess of *Pseudomonas* sp. RPT 52 to degrade imidacloprid, endosulfan and coragen. J. Hazard Mater. 301: 250–258.

Gupta, R. C. 2014. Carbamate pesticides. pp. 661–664. *In:* P. Wexler [Ed.]. Encyclopedia of Toxicology (Third Edition). Oxford. Academic Press.

Hajihassani, A., K. S. Lawrence and G. B. Jagdale. 2018. Plant parasitic nematodes in Georgia and Alabama Plant parasitic nematodes in sustainable agriculture of North America. pp. 357–391. Springer.

Hamad, M. T. M. H. 2020. Biodegradation of diazinon by fungal strain *Apergillus niger* MK640786 using response surface methodology. Environ. Technol. Innov. 18: 100691.

Harms, H., D. Schlosser and L. Y. Wick. 2011. Untapped potential: exploiting fungi in bioremediation of hazardous chemicals. Nat. Rev. Microbiol. 9(3): 177–192.

Hernández-Ramos, A. C., S. Hernández and I. Ortíz. 2019. Study on endosulfan-degrading capability of *Paecilomyces variotii*, *Paecilomyces lilacinus* and *Sphingobacterium* sp. in liquid cultures. Bioremed. J. 23(4): 251–258.

Hoshi, N., T. Hirano, T. Omotehara, J. Tokumoto, Y. Umemura, Y. Mantani, T. Tanida, K. Warita, Y. Tabuchi, T. Yokoyama and H. Kitagawa. 2014. Insight into the mechanism of reproductive dysfunction caused by neonicotinoid pesticides. Biol. Pharm. Bull. 37(9): 1439–1443.

Huang, D.-L., G.-M. Zeng, C.-L. Feng, S. Hu, X.-Y. Jiang, L. Tang, F. F. Su, Y. Zhang, W. Zeng and H. L. Liu. 2008. Degradation of lead-contaminated lignocellulosic waste by *Phanerochaete chrysosporium* and the reduction of lead toxicity. Environ. Sci. Technol. 42(13): 4946–4951.

Huang, Y., Xiao, L., Li, F., Xiao, M., Lin, D., Long, X. and Wu, Z. 2018. Microbial degradation of pesticide residues and an emphasis on the degradation of cypermethrin and 3-phenoxy benzoic acid: a review. Molecules. 23(9): 2313.

Hussein, M., A. Abdullah, N. Badr El-Din and E. Mishaqa. 2017. Biosorption potential of the microchlorophyte *Chlorella vulgaris* for some pesticides. J. Fertil. Pestic. 8(01): 1000177.

Ibrahim, W. M., M. A. Karam, R. M. El-Shahat and A. A. Adway. 2014. Biodegradation and utilization of organophosphorus pesticide malathion by cyanobacteria. Biomed Res. Int. 392682.

Jayaraj, R., P. Megha and P. Sreedev. 2016. Organochlorine pesticides, their toxic effects on living organisms and their fate in the environment. Interdiscip. Toxicol. 9(3-4): 90–100.

Jin, Z. P., K. Luo, S. Zhang, Q. Zheng and Yang, H. 2012. Bioaccumulation and catabolism of prometryne in green algae. Chemosphere. 87(3): 278–284.

Kabra, A. N., M.-K. Ji, J. Choi, J. R. Kim, S. P. Govindwar and B.-H. Jeon. 2014. Toxicity of atrazine and its bioaccumulation and biodegradation in a green microalga, *Chlamydomonas mexicana*. Environ. Sci. Pollut. Res. 21(21): 12270–12278.

Kafilzadeh, F. 2015. Assessment of organochlorine pesticide residues in water, sediments and fish from Lake Tashk, Iran. Achiev. Life Sci. 9(2): 107–111.

Kalyabina, V. P., E. N. Esimbekova, K. V. Kopylova and V. A. Kratasyuk. 2021. Pesticides: formulants, distribution pathways and effects on human health—a review. Toxicol. Rep. 8: 1179–1192.

Kamei, I., K. Takagi and R. Kondo. 2011. Degradation of endosulfan and endosulfan sulfate by white-rot fungus *Trametes hirsuta*. J. Wood Sci. 57(4): 317–322.

Kaur, H., S. Kapoor and G. Kaur. 2016. Application of ligninolytic potentials of a white-rot fungus *Ganoderma lucidum* for degradation of lindane. Environ. Monit. Assess. 188(10): 588.

Kolaczinski, J. H. and C. F. Curtis 2004. Chronic illness as a result of low-level exposure to synthetic pyrethroid insecticides: a review of the debate. Food Chem. Toxicol. 42(5): 697–706.

Kulshrestha, G. and A. Kumari. 2011. Fungal degradation of chlorpyrifos by *Acremonium* sp. strain (GFRC-1) isolated from a laboratory-enriched red agricultural soil. Biol. Fertil. Soils. 47(2): 219–225.

Kumar, A., A. Sharma, P. Chaudhary and S. Gangola. 2021. Chlorpyrifos degradation using binary fungal strains isolated from industrial waste soil. Biologia. 76(10): 3071–3080.

Landrigan, P. J., R. Fuller, N. J. Acosta, O. Adeyi, R. Arnold, A. B. Baldé, R. Bertollini, S. Bose-O'Reilly, J. I. Boufford, P. N. Breysse and T. Chiles. 2018. The Lancet Commission on pollution and health. The Lancet. 391(10119): 462–512.

Lehmann, E., M. Fargues, J.-J. Nfon Dibié, Y. Konaté and L. F. de Alencastro. 2018. Assessment of water resource contamination by pesticides in vegetable-producing areas in Burkina Faso. Environ. Sci. Pollut. Res. 25(4): 3681–3694.

Mali, H., C. Shah, B. H. Raghunandan, A. S. Prajapati, D. H. Patel, U. Trivedi and R. B. Subramanian. 2022. Organophosphate pesticides an emerging environmental contaminant: pollution, toxicity, bioremediation progress, and remaining challenges. J. Environ. Sci. 127: 234–250.

Mandal, K., B. Singh, M. Jariyal and V. K. Gupta. 2013. Microbial degradation of fipronil by *Bacillus thuringiensis*. Ecotoxicol. Environ. Saf. 93: 87–92.

Mata, T. M., A. A. Martins and N. S. Caetano. 2010. Microalgae for biodiesel production and other applications: a review. Renew. Sustain. Energy Rev. 14(1): 217–232.

Matsumoto, E., Y. Kawanaka, S. J. Yun and H. Oyaizu. 2008. Isolation of dieldrin- and endrin-degrading bacteria using 1,2-epoxycyclohexane as a structural analog of both compounds. Appl. Microbiol. Biotechnol. 80(6): 1095–1103.

McGuinness, M. and D. Dowling. 2009. Plant-associated bacterial degradation of toxic organic compounds in soil. Int. J. Environ. Res. Public Health. 6(8): 2226–2247.

Mir-Tutusaus, J. A., M. Masís-Mora, C. Corcellas, E. Eljarrat, D. Barceló, M. Sarrà, G. Caminal, T. Vicent and C. E. Rodríguez-Rodríguez. 2014. Degradation of selected agrochemicals by the white rot fungus *Trametes versicolor*. Sci. Total Environ. 500: 235–242.

Nicolopoulou-Stamati, P., S. Maipas, C. Kotampasi, P. Stamatis and L. Hens. 2016. Chemical pesticides and human health: the urgent need for a new concept in agriculture. Public Health Front. 4: 148.

Nile, A. S., Kwon, Y. D. and Nile, S. H. 2019. Horticultural oils: possible alternatives to chemical pesticides and insecticides. Environ. Sci. Pollut. Res. 26(21): 21127–21139.

Ortiz-Hernández, M. L., E. Sánchez-Salinas, E. Dantán-González and M. L. Castrejón-Godínez. 2013. Pesticide biodegradation: mechanisms, genetics and strategies to enhance the process. Biodegradation-life of Science. 10: 251–287.

Peter, L., A. Gajendiran, D. Mani, S. Nagaraj and J. Abraham. 2015. Mineralization of malathion by *Fusarium oxysporum* strain JASA1 isolated from sugarcane fields. Environ. Prog. Sustain. Energy. 34(1): 112–116.

Phugare, S. S., D. C. Kalyani, Y. B. Gaikwad and J. P. Jadhav. 2013. Microbial degradation of imidacloprid and toxicological analysis of its biodegradation metabolites in silkworm (*Bombyx mori*). Chem. Eng. J. 230: 27–35.

Powthong, P., B. Jantrapanukorn and P. Suntornthiticharoen. 2016. Isolation, identification and analysis of DDT-degrading bacteria for agriculture area improvements. J. Food Agric. Environ. 14(1): 131–136.

Rajmohan, K. S., R. Chandrasekaran and S. Varjani 2020. A Review on occurrence of pesticides in environment and current technologies for their remediation and management. Indian J. Microbiol. 60(2): 125–138.

Ramakrishnan, B., M. Megharaj, K. Venkateswarlu, N. Sethunathan and R. Naidu 2011. Mixtures of environmental pollutants: effects on microorganisms and their activities in soils. Rev. Environ. Contam. Toxicol. 211: 63–120.

Ray, D. E. and J. R. Fry. 2006. A reassessment of the neurotoxicity of pyrethroid insecticides. Pharmacol. Ther. 111(1): 174–193.

Reddy, K., N. Renuka, S. Kumari and F. Bux. 2021. Algae-mediated processes for the treatment of antiretroviral drugs in wastewater: Prospects and challenges. Chemosphere. 280: 130674.

Renuka, N., A. Gulde, R. Prasanna, P. Singh and F. Bux. 2018. Microalgae as multi-functional options in modern agriculture: current trends, prospects and challenges. Biotechnol. Adv. 36(4): 1255–1273.

Renuka, N., S. K. Ratha, F. Kader, I. Rawat and F. Bux. 2021. Insights into the potential impact of algae-mediated wastewater beneficiation for the circular bioeconomy: a global perspective. J. Environ. Manage. 297: 113257.

Rodríguez-Rodríguez, C. E., K. Madrigal-León, M. Masís-Mora, M. Pérez-Villanueva and J. S. Chin-Pampillo. 2017. Removal of carbamates and detoxification potential in a biomixture: fungal bioaugmentation versus traditional use. Ecotoxicol. Environ. Saf. 135: 252–258.

Rubilar, O., G. Feijoo, C. Diez, T. A. Lu-Chau, M. T. Moreira and J. M. Lema. 2007. Biodegradation of Pentachlorophenol in Soil Slurry Cultures by *Bjerkandera adusta* and *Anthracophyllum discolor*. Ind. Eng. Chem. Res. 46(21): 6744–6751.

Sabourmoghaddam, N., M. P. Zakaria and D. Omar. 2015. Evidence for the microbial degradation of imidacloprid in soils of Cameron Highlands. J. Saudi Soc. Agric. Sci. 14(2): 182–188.

Sadiq, S., M. Haq, I. Ahmad, K. Ahad, A. Rashid and N. Rafiq. 2015. Bioremediation potential of white rot fungi, *Pleurotus* spp. against organochlorines. J. Bioremed. Biodeg. 6(308): 2.

Saez, J. M., A. L. Bigliardo, E. E. Raimondo, G. E. Briceño, M. A. Polti and C. S. Benimeli. 2018. Lindane dissipation in a biomixture: effect of soil properties and bioaugmentation. Ecotoxicol. Environ. Saf. 156: 97–105.

Sagar, V. and D. Singh. 2011. Biodegradation of lindane pesticide by non white-rots soil fungus *Fusarium* sp. World J. Microbiol. Biotechnol. 27(8): 1747–1754.

Sakurai, T., M. Aoki, X. Ju, T. Ueda, Y. Nakamura, S. Fujiwara, T. Umemura, M. Tsuzuki and A. Minoda. 2016. Profiling of lipid and glycogen accumulations under different growth conditions in the sulfothermophilic red alga *Galdieria sulphuraria*. Bioresour. Technol. 200: 861–866.

Sanchez, S. and A. L. Demain. 2017. Bioactive products from fungi. pp. 59–87. *In*: M. Puri [Ed.]. Food Bioactives. Springer, Cham.

Sarker, A., R. Nandi, J.-E. Kim and T. Islam. 2021. Remediation of chemical pesticides from contaminated sites through potential microorganisms and their functional enzymes: prospects and challenges. Environ. Technol. Innov. 23: 101777.

Satish, G. P., D. M. Ashokrao and S. K. Arun. 2017. Microbial degradation of pesticide: a review. Afr. J. Microbiol. Res. 11(24): 992–1012.

Shahi, A., S. Aydin, B. Ince and O. Ince. 2016. The effects of white-rot fungi *Trametes versicolor* and *Bjerkandera adusta* on microbial community structure and functional genes during the bioaugmentation process following biostimulation practice of petroleum contaminated soil. Int. Biodeterior. Biodegrad. 114: 67–74.

Shukla, K. P., N. K. Singh and S. Sharma 2010. Bioremediation: developments, current practices and perspectives. Genet. Eng. Biotechnol. J. 3: 1–20.

Singh, B. K. and A. Walker. 2006. Microbial degradation of organophosphorus compounds. FEMS Microbiol. Rev. 30(3): 428–471.

Singh, D. V., R. Ali, M. Kulsum and R. A. Bhat. 2020. Ecofriendly approaches for remediation of pesticides in contaminated environs. pp. 173–194. *In*: R. Bhat, K. Hakeem and N. Saud Al-Saud [Eds.]. Bioremediation and Biotechnology, Vol 3. Springer, Cham.

Subashchandrabose, S. R., B. Ramakrishnan, M. Megharaj, K. Venkateswarlu and R. Naidu 2013. Mixotrophic cyanobacteria and microalgae as distinctive biological agents for organic pollutant degradation. Environ. Int. 51: 59–72.

Sun, Y., M. Kumar, L. Wang, J. Gupta and D. C. W. Tsang. 2020. 13—Biotechnology for soil decontamination: opportunity, challenges, and prospects for pesticide biodegradation. pp. 261–283. *In*: F. Pacheco-Torgal, V. Ivanov and D. C. W. Tsang (Eds.). Bio-Based Materials and Biotechnologies for Eco-Efficient Construction: Woodhead Publishing.

Tang, F. H. M., M. Lenzen, A. McBratney and F. Maggi. 2021. Risk of pesticide pollution at the global scale. Nat. Geosci. 14(4): 206–210.

Tarla, D. N., L. E. Erickson, G. M. Hettiarachchi, S. I. Amadi, M. Galkaduwa, L. C. Davis, A. Nurzhanova and V. Pidlisnyuk. 2020. Phytoremediation and bioremediation of pesticide-contaminated soil. Appl. Sci. 10(4): 1217.

Tiwari, B., S. Chakraborty, A. K. Srivastava and A. K. Mishra. 2017. Biodegradation and rapid removal of methyl parathion by the paddy field cyanobacterium *Fischerella* sp. Algal Res. 25: 285–296.

Vacondio, B., W. G. Birolli, I. M. Ferreira, M. H. R. Seleghim, S. Gonçalves, S. P. Vasconcellos and A. L. Porto. 2015. Biodegradation of pentachlorophenol by marine-derived fungus *Trichoderma harzianum* CBMAI 1677 isolated from ascidian *Didemnun ligulum*. Biocatal. Agric. Biotechnol. 4(2): 266–275.

Verasoundarapandian, G., Z. S. Lim, S. B. Radziff, S. H. Taufik, N. A. Puasa, N. A. Shaharuddin, F. Merican, C. Y. Wong, J. Lalung and S. A. Ahmad. 2022. Remediation of pesticides by microalgae as feasible approach in agriculture: bibliometric strategies. Agronomy, 12(1): 117.

Vijayan, N. P., S. H. Ali, H. Madathilkovilakathu and S. Abdulhameed. 2020. Chlorpyrifos-degrading cyanobacterium–*Coleofasciculus chthonoplastes* isolated from paddy field. Int. J. Environ. Stud. 77(2): 307–317.

Wan, L., Y. Wu, H. Ding and W. Zhang. 2020. Toxicity, biodegradation, and metabolic fate of organophosphorus pesticide trichlorfon on the freshwater algae *Chlamydomonas reinhardtii*. J. Agric. Food Chem. 68(6): 1645–1653.

Wang, X., L. Song, Z. Li, Z. Ni, J. Bao and H. Zhang. 2020. The remediation of chlorpyrifos-contaminated soil by immobilized white-rot fungi. J. Serbian Chem. Soc. 85(7): 857–868.

Wang, Y., B. Zhang, N. Chen, C. Wang, S. Feng and H. Xu. 2018. Combined bioremediation of soil co-contaminated with cadmium and endosulfan by *Pleurotus eryngii* and *Coprinus comatus*. J. Soils Sediments. 18(6): 2136–2147.

Xiao, P., T. Mori, I. Kamei and R. Kondo. 2011. Metabolism of organochlorine pesticide heptachlor and its metabolite heptachlor epoxide by white rot fungi, belonging to genus *Phlebia*. FEMS Microbiol. Lett. 314(2): 140–146.

Zaffar, H., R. Ahmad, A. Pervez and T. A. Naqvi. 2018. A newly isolated *Pseudomonas* sp. can degrade endosulfan via hydrolytic pathway. Pestic. Biochem. Physiol. 152: 69–75.

Zhang, B., C. Li, Y. Zhang, M. Yuan, J. Wang, J. Zhu, J. Ji and Y. Ma. 2021. Improved photocatalyst: elimination of triazine herbicides by novel phosphorus and boron co-doping graphite carbon nitride. Sci. Total Environ. 757: 143810.

Zhang, S., C. B. Qiu, Y. Zhou, Z. P. Jin and H. Yang. 2011. Bioaccumulation and degradation of pesticide fluroxypyr are associated with toxic tolerance in green alga *Chlamydomonas reinhardtii*. Ecotoxicology. 20(2): 337–347.

CHAPTER 8

Pseudomonas putida
An Environment Friendly Bacterium

Sneha S. Das and *Gunderao H. Kathwate**

8.1 Introduction

The world's population is rapidly increasing. The current global population is estimated to be 7.97 billion people as of 2020. This has obvious consequences for the world in which we live. To meet the needs of a growing population, soil, water and air pollution are unavoidable. The use of living organisms to degrade pollutants in order to prevent pollution, restore natural surroundings and prevent further pollution is known as bioremediation. This process employs microorganisms such as bacteria, fungi, algae and others. Pollutants including pesticides are metabolized and used as carbon or nitrogen sources for survival. Microorganisms can also be used to synthesize a number of metabolites, including anabolites and catabolites. Recombinant DNA techniques can be used to modify wild-type organisms.

Microorganisms can also be used to synthesize a range of metabolites, including pesticide and other xenobiotic compounds. Traditional methods and recombinant DNA techniques are being used to improve the productivity of wild-type organisms. This can be accomplished through the selection of mutants with desired phenotypes in controlled environmental conditions. Microbial techniques have attempted to compete with chemical product synthesis. Rising prices for petroleum, a non-renewable resource, necessitates the use of microbial production. First, advances in metabolic engineering techniques have resulted in engineered microorganisms that produce an excess of metabolites and intermediates. Another significant advancement is "heterologous expression of genes", which involves introducing genes from other organisms into microbial cells, that then express these foreign genes.

Second, the compounds produced by microorganisms are highly specific in terms of structure, position and stereochemistry. As a result, the cost of the downstream process is reduced. Third, unlike chemical synthesis, microbial processes do not use toxic heavy metals as catalysts, and the operating temperatures and pressures are lower. Microorganisms can produce valuable chemicals as well as degrade toxic pollutants. Sophisticated recombinant DNA techniques are used for targeted genetic manipulations, changing existing pathways or heterologous expression of genes in a new external pathway (Santos and Stephanopoulos 2008). However, the relationship between phenotype and genotype in organisms is extremely complex. To achieve the desired phenotype, a combination of genome editing tools are used.

Department of Biotechnology, Savitribai Phule Pune University, Pune 411007.
* Corresponding author: santoshkathwate@gmail.com

The purpose of genetic improvement would be to develop non-growing but metabolically active sedentary cells that would reroute the metabolic fluxes away from growth and towards product formation. However, when it comes to products such as recombinant proteins that are intimately related to the process of growth, it could be challenging to identify the genes that need to be knocked-out or knocked-in in order to achieve the phenotype that is desired. In order to get around the challenge, inverse metabolic engineering is the best option. Inverse metabolic engineering is the process of elucidating a metabolic engineering technique through the identification, construction or calculation of a desired phenotype, the identification of the genetic or specific environmental elements conferring that the phenotype, and the intentional genetic or environmental alteration of another strain or organism to give that phenotype. Inverse Metabolic Engineering (IME) is divided into three steps:

i.   Choosing or creating strains with the desired phenotype.

ii. Investigating the impact of genetic and environmental factors on the phenotype.

iii. Transferring the phenotype to a different organism.

An ideal balance between phenotypic expression and cell viability is essential since genes are being overexpressed. The expression of several enzymes should be balanced (Koffas et al. 2003, Pitera et al. 2007). All the enzymes involved in a pathway must work together. An imbalance expression leads to accumulation of intermediate metabolites, potentially impairing cell viability.

This chapter will briefly discuss y how genetic modifications can be made in microbial cells to achieve the desired phenotypic expression. Some prior knowledge about the pathways and the metabolites is important prerequisite.

The applications of *Pseudomonas* spp. described in this chapter are divided into two parts. To begin with, it entails the use of engineered microbes to degrade pollutants in the environment. Pesticides of the class organophosphate and others are discussed here, including Benzene/toluene/p-xylene (BTX), phenylurea, methyl parathion and cadmium; 1,2,3-trichloropropane; phenol; chlorpyrifos; carbofuran; c-hexachlorocyclohexane (c-HCH) or lindane; S-triazines; polycyclic aromatic hydrocarbons (PAHs); Diuron; Naphthol. Second, it includes ethylene, polyhydroxyalkanoate, 2-methylcitric acid, isoprenoid, long-chain polyunsaturated fatty acids, n-butanol, para-hydroxyl benzoic acid, and enzyme production using pollutants as a substrate (Figure 8.1). Many conventional chemical synthesis methods rely on depleting petroleum as a prerequisite. As a result, microorganisms are being used in the clean and green synthesis of a wide range of petroleum products.

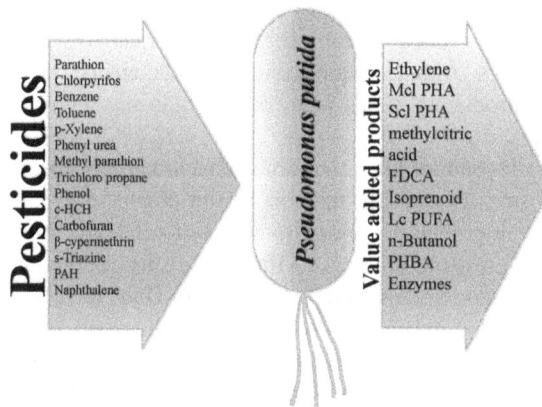

Figure 8.1. Utilization of *Pseudomonas putida* for the degradation of pesticides and production of value-added products, PAH-Polycyclic aromatic hydrocarbons, c-HCH-c Hexachlorocyclohexane, mcl-PHA, medium chain length polyhydroxyalkanoates, short chain length polyhydroxyalkanoates, FDCA-2,5 furandicarboxyalic acid, Lc-PUFA-long chain polyunsaturated fatty acid, PHBA-para-Hydroxy benzoic acid.

Microorganisms found in nature may have some restriction on degradation of substrate and synthesis of a product. Such microorganism especially bacteria can be improved by making a change in the genes using different tools available today. Furthermore, utilization of constitutive promoter libraries can modulate gene expression. They have a wide range of expression levels and can control multiple genes at the same time (Alper et al. 2005b, Jensen and Hammer 1998). This kind of system also eliminates the need for inducible systems. It is extremely useful in large-scale product production. Secondary structures of mRNA, RNase cleavage sites and RBS sequestering sequences are the post-transcription modulators and control the expression of multiple genes under the control of single promotor. Tunable Intergenic Regions (TIGR) placed between coding regions aid in post-transcriptional manipulation of phenotype expression (Pfleger et al. 2006).

8.2 Techniques for strain Improvement

Microorganisms with/without desired characters can be improved by using various methods modulating sequence in the genome by adding or removing the nucleotides. Such changes can be made using evolutionary engineering and genetic engineering. The concept of evolutionary engineering is based on the idea that organisms naturally evolve and undergo mutations in order to do so (Sauer 2001). As the frequency of mutation is very low in evolutionary engineering, mutagenic agents, physical or chemical can cause changes in an organism's DNA. This includes the use of nitrosoguanidine, ultraviolet light, etc., to introduce point mutations and increase the population's variation (Adrio and Demain 2006). Site directed mutation is carried out due to prior knowledge of functioning of the gene. Required mutation can be implemented using different methods revolving around thermocycler mediated polymerizing reactions. Using recombinant DNA technology, random or site-specific genetic variations can be introduced into the microorganism. This can be accomplished through transposon mutagenesis. Transposition is the process of moving a chromosomal segment from one location to another. For example, the incorporation of an antibiotic marker sequence into the organism's genome via transposon (Alper et al. 2005a). Tn5 transposon is used for random chromosomal integration, and *Tn7* transposon is used for targeted chromosomal integration, homologous recombination (RecET-mediated recombination can be used for greater efficiency), and *Cre/loxO* integration (Aoyama et al. 2005). *Pseudomonas putida* KT2440 rDNA is a favourable locus for biosynthetic gene cluster expression (Bojanovič et al. 2017). Antisense RNA inhibition is another approach (antisense cDNA is used to downregulate the gene expression) to control the expression of a particular gene (Bojanovič et al. 2017). In some cases, instead of down regulating the gene expression, overexpression is necessary. Gene overexpression can be accomplished by inserting the expression vector plasmid into the organism. In plasmid vectors, single or multiple biosynthesis genes or operons are introduced (Aoyama et al. 2005).

Site specific recombination can generate a scar after deletion mutation in the form of FRT (Flp/Flp recombination target) or lox sites. Thus, the recognition sites recombine in the chromosome generating deletion or inversion mutations. Furthermore, the selection method fails to select resistance cassette deleted mutants. A counter-selective marker is a method of genetic modification that does not use markers. This technique can be used to positively select double recombinants for plasmid excision. Counter-selective marker systems employ sucrose, fusaric acid and streptomycin as substrates for mutant selection (Reyrat et al. 1998, Graf and Altenbuchner 2011). A novel counter-selective marker system is the *upp* gene, which codes for UPRTase from *Bacillus subtilis*, *Enterococcus faecalis* and other bacteria. UPRTase converts 5-flurouracil to 5-fluro-UMP, which is then converted to 5-fluro-dUMP. The latter is harmful to cells. This is particularly useful for the deletion or insertion of large DNA fragments. Meganucleases, Zinc Finger Nucleases and Transcription activator-like effector nucleases are homing endonucleases that recognize and cleave longer DNA sequences (12 to 40 base pairs). ZFNs and TALENs are fusion proteins made up of a DNA binding domain engineered by scientists and a nonspecific nuclease (Sharma and Shukla

2022). TALENs and ZFNs can recognize one and three base pairs with modular regulatory proteins, respectively. CRISPR (Clustered Regularly Interspaced Short Palindromic Repeats)-Cas is an adaptive immune system that protects bacteria from foreign nucleic acids. CRISPR-Cas systems are a type of adaptive immune defence mechanism found in prokaryotes. This mechanism can be used to create specific double-stranded DNA breaks. NHEJ or HDR oversee repairing these sites (Sharma and Shukla 2022). CRISPR has several benefits, including high precision, convenience of design, fast turnaround time and is inexpensive. The disadvantage of CRISPR is that it may cause cell death in bacteria that lack the NHEJ repair system. To combat this, recombinase proteins such as SSr and -Red are used (Caldwell and Bell 2019). For example, 2,5--furandicarboxylic acid (FDCA) is a renewable alternative to petroleum-based terephthalic acid and is used to make polyurethanes, polyamides and polyesters (Pham et al. 2020). As a result, FDCA is a critical component of industrial production. A number of enzymes convert 5-hydroxymethyl furfural (HMF) to FDCA (Caldwell and Bell 2019, Pham et al. 2020 Zhang et al. 2002, Biot-Pelletier and Martin 2014). Recombinant *P. putida* S12 expressing HMF/furfural oxireductase can produce FDCA where CRISPR-Cas was used to induce double-stranded breaks. Double-stranded breaks were then used to recombine HMF/furfural oxidoreductase through Red-mediated recombineering. The recombinant strain was extremely efficient, converting approximately 88–85% of HMF to FDCA and frequent one-step stable gene integration was possible into all chromosomes of polyploid *P. putida* S12. As a result, the most stable strain of *P. putida* S12 was produced for the synthesis of FDCA. Although the stable integration of genes and efficient degradation was observed in this recombinant strain, higher concentration HMF (100 mM) was affecting the conversion. Initial increased density of biomass rescues the inhibition with increased conversion 100 mM and 150 mM HMF to 96 to 75% of FDCA (Hsu et al. 2020).

Genome shuffling is another method usually used for strain improvement at the industrial level. It is a combination of DNA shuffling and directed evolution used to improve strains (Zhang et al. 2002). DNA shuffling has successfully achieved directed molecular evolution of several genes and pathways (Biot-Pelletier and Martin 2014). For genome shuffling, the protoplast fusion method is employed where the fusion of two cells with different genetic traits achieves the desired modified phenotype. Even intergenic hybridization between *Aspergillus niger* and *Penicillium digitatum* results in increased verbenol production (Rao et al. 2003). The consequence of protoplast fusion is a library of strains with accumulated mutations. These strains can now be screened for phenotypes of interest. As a result, it is a very useful tool, and more hybrid strains can be obtained than through protoplast fusion as more parents are involved and saves a lot of assays and time (Zhang et al. 2002).

Pseudomonas are the gram-negative saprophytic in most of the soil dwelling bacteria. One of the most important features of *Pseudomonas* sps., is degradation of almost all type of carbon skeletons. The Recombinant DNA Advisory Committee has approved *Pseudomonas putida* as one of the popular bacteria for genetic engineering and certified it as a biosafe host for heterologous gene expression (Nelson et al. 2002). They exhibit rapid and robust growth. *P. putida* is resistant to very high concentrations of toxic especially xenobiotic compounds due to solvent resistance mechanisms mediated by the interaction of cellular factors (Ramos et al. 2002). Some examples include fine-tuning lipid fluidity to adjust membrane functions, activating stress response systems and inducing efflux pumps. *P. putida* is also metabolically extremely adaptable. It has a ED/EMP cycle, a distinct metabolic pathway that employs enzymes from the Entner-Doudoroff (ED), pentose phosphate and Embden-Meyerhof-Parnas (EMP) pathways (Nikel et al. 2015). Due to the availability of the ED/EMP cycle, it is now possible to build completely different biochemistries using new methodologies. In this chapter, the use of genetically modified *Pseudomonas* sp. especially *P. putida* for compound synthesis and degradation to aid in pollution control will be described.

8.3 Degradation of Pesticides

Advances in biodegradation of toxic pesticides using microorganisms are being made with prior knowledge of microbial metabolism and genetic techniques. Liquid cultures are used to test microbes for the optimization of temperature, pH, organic nutrients, substrate bioavailability and concentration that play an important role in bioremediation in soil. But at the same time, this is not enough to overcome the problem of pesticide pollution. Using genetic techniques, genes from microbes with desired phenotypes should be transferred to industrially useful strains. As a result, critical enzymes for bioremediation can be economically and successfully produced. Non-native microbial bioremediators are used in bioaugmentation. It can degrade hydrocarbons, pesticides, explosives and other substances (Monga et al. 2021). The number of soil microorganisms, their enzymatic capabilities and the pace of pesticide degradation in response to changes in soil conditions play a major role in the elimination of pesticides from the environment (for example, pH). Pesticide degrading genes are usually found on plasmids. Carbofuran, atrazine, 2,4D and parathion degrading genes have been identified (Gong et al. 2016, Neumann et al. 2004, Dejonghe et al. 2000, Rani and Lalithakumari 1994). In addition, a deeper understanding of microbial genomes and proteomics can facilitate the manipulation of microorganisms for the purpose of more effective bioremediation of contaminants (Kondakova and Cronan 2019, Mandalakis et al. 2013).

8.3.1 Organophosphates Pesticides

Organophosphate Pesticides (OP) are insecticides and herbicides, and have contaminated groundwater. They are strong cholinesterase inhibitors and thus toxic to humans (Serdar et al. 1989, DeFrank 1991). OPs are neurotoxins that inhibit acetylcholine esterase, causing the CNS function to be disrupted, paralysis of the muscle and death (Carvalho et al. 2003). As a result, they are more toxic to vertebrates (Malhat and Nasr 2011). To detoxify OP, chemical treatment, photodecomposition, incineration and other methods can be used, but they are not environmentally friendly as they produce organophosphate triesterases, which could be toxic. On the other hand, microbial detoxification is cheap and environment friendly. Bacteria such as *Flavobacterium* sp. (Yoshida and Sathunathan 1973), *Arthobacter* sp. (Mallick et al. 1999) and *P. diminuta* are the prominent Ops degraders (Serdar et al. 1989). Similarly, the soil microorganism *P. putida* was also found to mineralize OP, parathion (DeFrank 1991). Parathion is a well-known example of OP, which is degraded into diethyl thiophosphate (DETP) and para-nitrophenol (PNP). DETP is non-toxic, and it can be metabolized by microbial cultures (Shelton and Somich 1988). This reaction is catalyzed by organophosphate (encoded by the opd gene). As PNP is carcinogenic, it should be metabolized. *Bacillus, Arthrobacter* (Bhushan et al. 2000), and *Pseudomonas* perform this function (Bang 1997). *Pseudomonas* sp. ENV2030 has operons that encode PNP transformation to beta-ketoadipate via hydroquinone (Zylstra et al. 2000). To reduce the concentration of parathion, *opd* and PNP degrading operons are introduced into *P. putida* KT2442. The organophosphorus hydrolase (OPH) enzyme, which is encoded by the *oph* gene, the *mpd* gene and/or the *opd* gene in different strains, aids in the detoxification of OPs (Bigley and Raushel 2013). Chlorpyrifos (O, O-diethyl O-(3,5,6-trichloro-2-pyridyl phosphorothioate) is a pesticide used in both the home and agriculture. The *oph* gene oversees the chlorpyrifos degradation. Chemical mutations by ethylmethane sulphonate (EMS) have been used to improve the yield of bacterial enzymes in order to improve bioremediation by bacterial strains. AchM15 and AchMS1 mutants degraded 63 and 82.03% chlorpyrifos, respectively, compared to 35% in the wild type (El-sayed et al. 2019).

8.3.2 Benzene, Toluene and p-xylene (BTX)

BTX indicates the presence of benzene, toluene and p-xylene. These water-soluble petroleum components have poisoned the most of the bodies of potable water. A single microbe can mineralize all three chemicals in a combination of the *tod* and *tol* metabolic pathways (Worsey and Williams 1975, Zylstra et al. 1988).

As a result of the inability of catechol 2,3-dioxygenase in the tod pathway to attack 3,6-dimethylpyrocatechol produced from p-xylene and xylene oxygenase in the athway to employ benzene as a substrate, it is possible that complete biodegradation of the BTX combination through these two pathways will not be possible (Worsey and Williams 1973, Zylstra et al. 1988, Gibson et al. 1970, Gibson et al. 1974). Consequently, an existing route must be restructured. In order to manipulate the genes for two pathways in *P. putida*, a hybrid metabolic pathway was constructed around the crucial metabolic step. In *P. putida*, an attempt was made to mineralize BTX by a hybrid route. To construct the hybrid pathway, a bridging step between the tod and tol pathways was identified. The tol pathway destroys cis-glycol chemicals, which are metabolic intermediates immediately before the catechol molecules in the tod pathway. As the only source of carbon, BCG was 90% degraded (Lee et al. 1994). *P. putida* TB101 was generated by introducing the TOL plasmid pWW0 into *P. putida* F39/D, a *P. putida* FI variant incapable of converting cis-glycol chemicals to catechols. Toluate-cis-glycol dehydrogenase in *P. putida* TB101 diverted the metabolic flux of BTX into the tod pathway at the level of cis-glycol molecules, resulting in the simultaneous mineralization of BTX mixture without accumulation of any metabolic intermediates (Lee et al. 1994). Toluate-cis-glycol dehydrogenase, which is encoded on the TOL plasmid of *P. putida* mt-2, was considered to be responsible for the degradation of these cis-glycol compounds due to the comparable chemical structure of its original substrates, benzoate-cis-glycol (BACG) and TACG. BCG was utilized as a substrate during the toluate-cis-glycol dehydrogenase experiment. The observation shows that the reaction product produced by BCG's toluate-cis-glycol dehydrogenase may be destroyed by catechol 2,3-dioxygenase, the enzyme that follows toluate-cis-glycol dehydrogenase in the tol pathway. In *P. putida* TBlOl, a hybrid pathway was developed for the complete mineralization of benzene, toluene and p-xylene. Using *P. putida* TBlOl, simultaneous biodegradation of BTX combinations was accomplished with maximum specific degradation rates of 0.27, 0.86 and 2.89 mgmg^{-1} biomass/h for benzene, toluene, and p-xylene, respectively (Lee et al. 1994).

8.3.3 Phenylurea

Phenylureas like diuron linuron isoproturon and chlorotoluron were once used to control weeds, but they have now contaminated drinking water. Diuron is a phenylurea herbicide that is used to control weeds in a number of crops (Moretto et al. 2019). It has been found in soil and water (Giacomazzi and Cochet 2004). Diuron has been linked to kidney disease, haematopoiesis and hemolytic anaemia (Ihlaseh-Catalano et al. 2014). In high salinity conditions, the CASB3 strain of *Stenotrophomonas rhizophila* was used to degrade diuron. Fifty mg L^{-1} diuron was completely degraded in 48 to 120 hr (Silambarasan et al. 2020). *Arthrobacter globiformis* D47 can 'partially' degrade herbicide diuron via urea carbonyl group hydrolysis (Cullington and Walker 1999, Turnbull et al. 2001). The concentration of 3,4-dichloroaniline (DCA) rises as the diuron degrades. A very small amount of DCA can be converted to 3,3' 4,4' tetrachloroazobenzene and the remaining bulk of the DCA continue bound to the organic matter with slow half-life for several years. *Pseudomonas putida* strain was enriched for the demineralization of DCA. The pathway investigation study suggested biodegradion through 3,4 dichloromuconate, 3 chlorobutenolide, 3 chlorolevulinic acid, 3 chloromelyelacetate 3chloro 4 ketadipate to succinate (You and Bartha 1982).

8.3.4 Methyl Parathion

Microbes are also decomposers that improve soil fertility by degrading pesticides and immobilizing heavy metals. Toxic pollutants include organophosphate pesticides such as Methyl Parathion (MP) and heavy metals like cadmium. MP-degrading enzymes are found in microbes such as *Pseudomonas* sp. A3 (Ramanathan and Lalithakumari 1999), *Plesiomonas* sp. M6 (Zhongli et al. 2001), and *Pseudomonas* sp. SMSP-1 (Shen et al. 2010). They convert Methylparathion to paranitrophenol

(PNP) and dimethyl phosphorothioate but are unable to degrade PNP further. These bacteria were used to clone *ophc2*, an organophosphorus hydrolase encoding gene. Cd can be immobilized by bacteria such as *P. aeruginosa* KUCd1 and *Pseudomonas* sp. strain RB (Zhang et al. 2016). A soil-dwelling bacterium was isolated having the potential of methyl parathion degradation ability. The bacteria were responsible for producing the enzyme organophosphorus acid anhydrase, which was responsible for the hydrolysis of methyl parathion into p-nitrophenol. Hydroquinone and 1, 2, 4-benzenetriol were produced eventually as a by-product of further degradation of p-nitrophenol. At the end Maleyl acetate was produced by cleaving the last ring component, 1, 2, 4-benzenetriol, which was catalyzed by the enzyme benzenetriol oxygenase (Rani and Lalithakumari 1994). *Pseudomonas* sp., isolated from a mixed population of methyl parathion and parathion-degrading bacteria, can hydrolyze the MP to p-nitrophenol and demands glucose and other carbon sources for growth (Chaudhry et al. 1988). *Flavobacterium* sp. isolate from the same mixed population could convert pnitrophenol to nitrite. It can take up to 48 hr for both bacteria to mineralize methyl parathion as their only carbon source. Surface expression of metal-binding proteins in host strains is an effective way to increase heavy metal immobilization. To maximize bioremediation, an EGFP-expressing X3 bacterium was metabolically engineered to have MP degrading genes as well as a Cd-immobilization phenotype (Zhang et al. 2016).

8.3.5 1,2,3-trichloropropane (TCP)

TCP is a solvent, a precursor of soil fumigants, a catalyst in the synthesis of dichloropropene or polyphsulfone liquid polymers (WHO 2003), and a toxic industrial waste classified as a persistent pollutant (Dvorak et al. 2014). *P. putida* K2440 was designed to degrade 1,2,3-trichloropropane (Gong et al. 2017). The microbe was engineered with haloalkane dehalogenase (Kulakova et al. 1997), haloalcohol dehalogenase (van Hylckama et al. 2001) from *Rhodococcus rhodochrous* NCIMB 13064 and epoxide hydrolase (Rink et al. 1997) from *Agrobacterium radiobacter* AD1. This engineered strain was capable of successfully converting up to 10 mg L^{-1} of TCP (Dvorak et al. 2014). The strain was further improved by deleting the *glpR* gene, which increased carbon flux in this pathway, and improved the degradation with expression of *Vitreoscilla haemoglobin*, in aerobic conditions. By deleting flagella-related genes, the energy charge and reducing power were improved (Stark et al. 2015, Kim et al. 2005). The most common biological detoxification mechanism is oxidative dehalogenation of pollutants, which is limited to oxic environments. As a result, organohalide degradation is difficult in anoxic conditions (Janssen et al. 2001). *P. putida* was engineered to contain several genes that participate in respiration by facultative anaerobes and aerotolerant bacteria. These genes include the haloalkane dehalogenase activity from *Pseudomonas pavonaceae* 170, the acetate kinase (*ackA*) gene of the facultative anaerobe *Escherichia coli*, and the pyruvate decarboxylase (*pdc*) and alcohol dehydrogenase II (*adhB*) of *Zymomonas mobilis*. Under anoxic circumstances, the recombinant strain proved successful in degrading pollutants such as 1,3-DCP (Nikel and de Lorenzo 2013). Another method involved the expression of *Vitrescilla haemoglobin* (VHb) in *P. putida* KTUe. This made it possible for the microbe to proliferate more quickly in environments with a scarcity of oxygen. The capacity of the strain to bind oxygen is improved by the presence of VHb.

8.3.6 Phenol

Phenol is widely used and is a unfavourably industrial waste produced by petroleum refineries (NPCS Board of Consultants and Engineers 2000). As it quickly penetrates to the skin and can cause irritation to the eyes, respiratory tract and can be fatal in chronic exposure. The use is unavoidable leaving an option of maintaining the permissive limit. The maximum allowed level is 2ppm and can be maintained below or neutralized to zero level using *P. putida*. *P. putida* FNCC-0071 cells immobilized in alginate beads were used to degrade phenol (Hannaford and Kuek 1999). About 600 ppm of phenol was degraded within 48 hr without affecting viability. Recently, a constant electric current was applied for the enhancement of phenol degradation. Basically *P. putida* and other microorganisms degrade via meta and para cleavage pathways influenced by the concentration, temperature and biogenic factors. Most of the redox biochemical reaction are facilitated by applying constant voltage across the bioreactor along with microbial community (Bandhyopadhyay et al. 1999).

8.3.7 Chlorpyrifos and Carbofuran

Carbofuran is a chemical that is used to kill insects and corn rootworm (Tomasek and Karns 1989). As a nitrogen source, *Achromobacter* species WM111 can hydrolyze N-methylcarbamate (Karns et al. 1986). The *mcd* gene encodes N-methylcarbamate hydrolase, which aids in degradation. This gene has been cloned and is being expressed in soil bacteria (Tomasek and Karns, 1989). Chlorpyrifos and carbofuran were also co-degraded in *P. putida* KT2440 by scarless counter-selectable marker heterologous expression of CP/carbofuran hydrolase genes (Gong et al. 2016). Compared to the CP/carbofuran, the digestion products' toxic effects were significantly diminished. Interestingly, CP/carbofuran were used as the only carbon source by the mutant for the growth and degradation rate was also found high compared to uninoculated soil.

8.3.8 Beta-cypermethrin

Beta-cypermethrin (β-CY) is a significant pyrethroid pesticide (Xiao et al. 2015). It is a toxin that affects the reproductive, immune, nervous and genetic systems. (Jin et al. 2011, McKinlay et al. 2008, Schettgen et al. 2002). The pesticide can be co-metabolized and degraded by *Aspergillus niger* YAT, *Bacillus licheniformis* B-1, *Brevibacterium aureum* and *Brevibacillus parabrevis* BCP-09 (Tran et al. 2013, Chen et al. 2013, Deng et al. 2015, Tang et al. 2018). However, soil degradation was limited. As a result, Loh and Wang (1997) discovered that *P. putida* ATCC 49451 can degrade 4-chlorophenol. The process was accelerated in the presence of sodium glutamate. But this process was hampered by glucose (Zhao et al. 2019a,b).

Lindane or c-hexachlorocyclohexane, c-HCH is a highly toxic pesticide that cause *Anabaena* to lose protective chaperons, inhibit photosynthesis and fix nitrogen (Bueno et al. 2004, Singh 1973, Nawab et al. 2003). In *Sphingomonas paucimobilis* B90, the linA2 gene encodes c-HCH dehydrochlorinase, which degrades the pesticide. Similarly, the pesticide is degraded by the linA gene in *P. paucimobilis* UT26 (Adhya et al. 1996, Nagata et al. 1997). In the presence of an inducible promoter PpsbA1, these genes were knocked in and overexpressed in *Anabaena*. In 12 hr, this strain could degrade 10 ppm of lindane isomers (Chaurasia et al. 2013).

8.3.9 s-Triazines

s-Triazines are a class of herbicides used to control weeds, such as atrazine. *Pseudomonas* sp., strain A uses ozone to oxidize atrazine, and the ozonated products are used by soil microbes (Karns et al. 1987, Kearney et al. 1988). S-triazine is also used as the only nitrogen source by *Pseudomonas* (A, D and F) and *Klebsiella pneumoniae* strain 90 and 99 (Cook and Hutter 1981). Aminohydrolases, which catalyze ring deamination, and amidohydrolases catalyze this conversion. These genes were discovered through transposon mutagenesis (Eaton and Karns 1991). Recently pure culture of *Pseudomonas* sp., ADP was identified which can mineralize the complete ring of atrazine to CO_2 and hydroxyatrazine and polar metabolites.

8.3.10 Polycyclic Aromatic Hydrocarbons (PAHs)

PAHs (Polycyclic Aromatic Hydrocarbons) are both mutagenic and carcinogenic (Jones et al. 1989). The combustion of organic matter, as well as the processing and use of fossil fuels, are the sources of these pollutants. *Pseudomonas* sp., strain IOCa11 demonstrated broad substrate specificity for PAHs including dibenzothiophene, benzo (α) pyrene, naphthalene and phenanthrene from oil-contaminated soil (Kumar et al. 2012). After nitosoguanidine treatment, phenotypically favourable strains were chosen for protoplast fusion (Dai et al. 2005, Singleton et al. 2009, Dai and Copley 2004, Gong et al. 2009). As a result, the SF-IOC11-16A strain created by genome shuffling was efficient in degradation of PAHs (Kumar et al. 2012).

8.3.11 Naphthalene

Naphthalene degradation is associated with *Pseudomonas putida* PpG7 (Dunn and Gunsalus 1973), *P. paucimobilis* Q1 (Kuhm et al. 1991) and *P. putida* ND69 (Song et al. 2018). In 48 hr, the latter can degrade up to 98% of 2mg L^{-1} naphthalene. Jia Yan et al. (2008) observed the degradation of naphthalene in *Pseudomonas* N7 in 2008; the degradation rate reached 95.66%. The naphthalene degrading gene in *P. putida* ND6 is found on pND 6-1, a 102 kb plasmid. pND 6-1 has a G + C content of 57%, which includes the oriV, the region related to plasmid replication and stable inheritance, and the naphthalene degradation gene region. For naphthalene degradation, there are 23 coding domains (Li et al. 2004). The genes involved are *nahG*, which codes for salicylic acid hydroxylase, and *nahR*, that codes for NahR, a transcriptional regulatory protein of the LysR family. The gene *nahY* encodes a naphthalene chemotactic protein, while the gene *nahV* encodes a salicylaldehyde dehydrogenase (Song et al. 2018).

8.3.12 Toluene

The strain *P. putida* DOT-T1E is rifampin resistant. It can withstand supersaturating toluene concentrations (Ramos-Gonza'lez et al. 2003). Toluene is used as a carbon and energy source (Ramos et al. 1995). The mechanism is the *tod*-pathway, which converts toluene into 3-methylcathecol, is then used in the Krebs cycle (Mosqueda et al. 1999). *P. mendocina* KR1 metabolizes toluene to p-cresol via the T4MO pathway using the *tmo* gene product (Whited and Gibson 1991). *Pcu* genes further oxidize this to 4-hydroxybenzoate. 4-HBA is a precursor to paraben and methylparaben, which are used in the production of liquid glass and antimicrobial agents (Huang et al. 1992, Soni et al. 2002). Toluene is also converted into 4-HBA by *P. putida* DOT-T1E. To prevent toluene and 4-HBA from being misrouted, the tod and beta-ketoadipate pathways were turned off to maximize product formation (Ramos-Gonza'lez et al. 2003).

8.4 Biosynthesis of Value-added Products

P. Putida is not only proven to demineralize the xenobiotic compounds like pesticides, but also been explored to synthesize value-added products. These products are synthesized mostly using waste material that may not be degraded by common bacteria. Some products are synthesized as intrinsic property and others after introducing heterologous genes from various microorganisms. Use of *P. putida* for synthesis of all the compounds discussed below by *P. putida* is one of the best eco-friendly options.

8.4.1 Production of Ethylene

Ethylene is a vital petroleum-derived raw material in the chemical industry (Kniel et al. 1980). However, as petroleum reserves deplete, alternatives should be found. *Pseudomonas syringae* and *Penicillium digitatum*, for example, produce ethylene from 2-oxoglutarate. There are two methods for producing ethylene. The first uses 2-oxo-4-methylthiobutriyric acid (KMBA) as a precursor, while the second uses 2-ketoglurate (Chagué et al. 2002). The ethylene-forming enzyme (EFE) is responsible for ethylene production. This gene was cloned from endogenous plasmids of *P. syringae* pv. glycinea. Through double crossover recombination, *efe* gene was integrated into the *P. putida* K2440 16S rDNA sites of *P. putida* KT2440 (Watanabe et al. 1998). It contains seven identical copies of 16S rDNA, so *efe* gene can be integrated multiple times without causing any harm; and the ethylene production can be increased even further. *P. putida* was given pMEFE1, a plasmid with a high copy number and a broad host range that carried *efe* from *P. syringae*. The recombinant's maximum ethylene production rate was 26 times greater than that of the parental *P. syringae*. The oxygen and carbon sources could be improved to increase yield even further (Wang et al. 2010).

8.4.2 Medium Chain Length PHA

A bioadaptive functional polymer is medium-chain-length polyhydroxyalkanoate (mcl-PHA). It is found in fluorescent *Pseudomonas* species that share the same rRNA homology as *P. putida*, *P. oleovorans* and *P. aeruginosa* (Kim et al. 2006). Mcl biosynthesis is controlled by the phaC1 and phaC2 genes (Rehm and Steinbüchel 1999, Steinbüchel and Hein 2001). The *phaC1* gene was overexpressed, resulting in a 2.86-fold increase in mcl biosynthesis. It produced larger, higher molecular weight and less crystalline mcl-PHA granules, which aided in the purification process. Overexpression of *phaC2* had no effect.

The engineered *P. putida* KCTC1639 overexpressing the *phaC1* gene encoding PHA synthase I was grown in a pH-stat fed-batch system, resulting in increased mcl-PHA concentration and content of 8.91 g L^{-1} and 70.5%, respectively (Kim et al. 2006). Under nitrogen-limited conditions, *P. menodocina* NK-01 produces mcl-PHA and alginate oligosaccharide from glucose. This strain's production was improved by using a counter-selective maker (Wang et al. 2015). Ferulic acid is a phenolic lignocellulose component. In *P. putida*, ferulic acid is converted into mcl-PHA via a non-beta-oxidative pathway. This conversion rate is extremely low. To overcome the limitations of conventional CRISPR-Cas9n—Red genome editing, a type II CRISPR-Cas9n—Red genome editing strategy was used (in bacteria with no or less NHEJ system, the cell viability goes down, mutation efficiencies of different targets are highly variable). The expression of nine genes was altered in order to improve the ferulic acid to PHA conversion (Volke et al. 2020).

8.4.3 Short Chain Length PHA

Thermoplastic polyhydroxyalkanoate (PHA) with a short chain length (scl) can be utilized as an alternative to plastics derived from petroleum (Gahlawat and Soni 2017). It has application in the field of medical as well as packaging (Kamravamanesh et al. 2018). PHAs that have shorter chain lengths are made up of 3-hydroxybutylate, also known as 3HB (Lemoigne 1926). The melting point of the polymer can be lowered by adding poly(3-hydroxybutyrate-co-3-hydroxyvalate) and 4-hydroxyvalate to scl-PHA. Additionally, the polymer's toughness can be increased, making it a very valuable material (Gahlawat and Soni 2017, Yang et al. 2012, Schmack et al. 1998). Levulinic acid, which is used in the production of 3 Hydroxyvalate and 4 Hydroxyvalate monomers, is not made from petroleum-based precursors; rather, it is produced from sugars generated from biomass, specifically glucose and xylose (Koller et al. 2017, Novackova et al. 2019). In *P. putida*, the breakdown of LA is controlled by a seven-gene operon called *lva* (lvaABCDEFG). *P. putida* strain SP01 was developed by removing the native *PhaC1*, *PhaC2* and *PhaZ* genes from *P. putida* strain EM42. This was done with the intention of inhibiting the synthesis of mcl-PHA and halting the degradation of scl-PHA. It was not necessary to apply an external inducer because the IvaA promoter is already stimulated by LA. The PHA synthase (PhaEC) of *Thiococcus pfennigii*, which has a broad substrate specificity, was only expressed in the LA catabolic pathway of *P. putida* in order to create scl-PHA from two intermediates (4HV-CoA and 3HV-CoA). This route is responsible for the degradation of fatty acids (Cha et al. 2020). *P. putida* has been genetically modified to produce a greater amount of PHA. The *gcd* gene, which encodes dehydrogenase, has been removed in order to divert flux towards products that are more desirable (Tiso et al. 2016).

8.4.4 2-methylcitric Acid and others

2-methylcitric acid (2-MC) has the potential to inhibit fast cancer cells. It can also be used in synthetic materials or pharmaceutical products as an emulsifier or polymer plasticizer (Ewering et al. 2006).

As the chemical synthesis of 2-MC requires hazardous solvents such as benzene, microorganisms are used. 2-MC was produced using *Ralstonia eutropha* H16 and *P. putida* KT2440 (Brämer and Steinbüchel 2001). The manipulated strains, *R. eutropha* DeltaacnM (Re) OmegaKmprpC (Pp) and *P. putida* DeltaacnM (Pp) OmegaKmprpC (Re), were created by inserting the 2-methylcitrate synthase gene, causing the 2-methyl-cis-aconitate hydratase to be disrupted (*acnM*). Due to the disruption of the *acnM*, there was an excessive generation of 2MC, which led to its build up. The maximum concentrations attained by the stains after 144 hr of cultivation were 7.2 g L^{-1}, which is equivalent to 26.5 mM, and 19.2 g L^{-1}, that is equivalent to 70.5 mM, respectively (Ewering et al. 2006). Guaiacol is one of the products of depolymerization of kraft lignin along with catechol benzoate and toluene. In recent years, more attention was diverted to convert guaiacol to a value added product like muconic acid. *P. putida* KT2440 is engineered for two step the conversion of guaiacol to muconic acid. Deletion of CatBC gene thereby blocking the catabolism to muconic and insertion of cytochrome P450 and ferredoxin reductase gene from *R. rhodochrous* enabling the conversion of guaiacol to catechol (Almqvist et al. 2021). Multiple natural enzymes in *P. putida* KT2440 are capable of utilizing vanillin as a substrate. Vanillin dehydrogenase and various aldehyde reductases are enzymes involved in the breakdown of vanillin into vanillyl alcohol and vanillic acid, respectively (Simon et al. 2014). GN442PP 2426, which was previously modified to manufacture vanillin from ferulic acid (Graf and Altenbuchner 2014, García-Hidalgo et al. 2020), may be a more ideal host strain than KT2440 for the generation of VA since vanillic acid can then be further assimilated via protocatechuate. This is because ferulic acid is converted into vanillin by genetically engineering KT2440. *P. putida* EM42, a genome-reduced variety of *P. putida* KT2440 with superior physiological features, was recently modified for growth on cellobiose (Dvořák). Cellobiose and glucose can be used together in the same metabolic pathway owing to a mutant (PP_1444) that lacks the periplasmic glucose dehydrogenase Gcd, but unfortunately the Δ*gcd* strain suffered from a significant growth defect. The growth defect was compensated by introduction of heterologous glucose (Glf from *Zymomonas mobilis*) and cellobiose (LacY from *Escherichia coli*) transporters with surprised production of pyruvate (Bujdoš et al. 2023)

8.4.5 Isoprenoid

It is a profitable molecule that has implications in the pharmaceutical, as well as the food and beverage industries (Arendt et al. 2016). Bacteria such as *P. putida* can tolerate larger amounts of isoprenoids (Mi et al. 2014). As a result, they can be utilized to satisfy an increasing demand. *P. putida* is utilized in the process of biotransformation of isoprenoids in order to get oxidation products of the plant monoterpene, limonene (Loeschcke and Thies 2015) or *de novo* biosynthesis of the monoterpene geranic acid (Mi et al. 2014) or the carotenoids zeaxanthin and -carotene. *P. putida* utilizes the methylerythritol 4-phosphate (MEP) pathway, whereas other bacteria utilize the unrelated mevalonate (MVA) process to produce acetyl-CoA. *P. putida* KT2440 was genetically modified to manufacture modest quantities of lycopene via the MEP route under the control of IPTG-induced stress regulated promoters. These promoters allowed to produce measurable levels of lycopene. The amount of lycopene that was produced by this strain increased by a factor of 50 (Hernandez-Arranz et al. 2019).

8.4.6 Long-chain Polysaturated Fatty Acids

In the treatment of cardiovascular disease, obesity and diabetes, long-chain polyunsaturated fatty acids like eicosapentaenoic acid (EPA) and docosahexaenoic acid (DHA), for example, are adopted (Lorente-Cebrian et al. 2013). Both EPA and DHA were first extracted from fish and fish oil, but both of those resources are becoming increasingly scarce (Lenihan-Geels et al. 2013). In marine species, polyketide synthase (PKS)-like enzymes and pfa biosynthetic gene clusters are responsible for the synthesis of long-chain polyunsaturated fatty acids (LC-PUFA) from acyl-CoA (Kaulmann and Hertweck, 2002, Napier, 2002, Wallis et al. 2002). *pfa* gene clusters are present in the proteobacteria *Shewanella pneumatophori* SCRC-2738, *Photobacterium profundum* SS9 (both of which are manufacturers of EPA) (Allen and Bartlett 2002, Metz et al. 2001) and *Moritella marina* MP-1 (which is a generator of DHA), amongst others. Both the *Sorangium cellulosum* species and the *Aetherobacter* sp. were found to contain two *pfa* gene clusters, which are responsible for the production of linoleic acid (Garcia et al. 2011, Gemperlein et al. 2014). These organisms are sluggish, tough to control, and challenging to manipulate genetically. As a result, *P. putida* is utilized as a stable host for the expression of *pfa* gene clusters, which ultimately leads to the creation of high quantities of LC-PUFA (Gemperlein et al. 2014, Gemperlein et al. 2016). In order to increase the amount of polyunsaturated fatty acids that are produced by marine species, transposon mutagenesis was utilized (Amiri-Jami et al. 2006).

8.4.7 n-Butanol

Currently, n-butanol is produced from petroleum compounds. It is used as a precursor in the production of polymers and paints (Cuenca et al. 2016). Microorganisms can also produce butanol from municipal solid waste and lignocellulose materials (Ezeji et al. 2007). Butanol can be produced at concentrations of 5mg L^{-1} in *P. putida* S12 by the heterologous expression of the *Clostridium acetobutylicum* pathway (Nielsen et al. 2009). The gene *glcB* encodes an enzyme in the glyoxylate shunt. The *glcB* mini-Tn5 mutant is mutated to prevent butanol assimilation.

8.4.8 Para-hydroxyl Benzoic Acid (PHBA)

It is a petroleum-derived chemical that is used in the chemical industry to produce liquid crystal polymers, as well as paraben preservatives in food, pharmaceutical and cosmetic products (Ibeh 2011). High temperatures and pressures are difficult to achieve (Yoshida and Nagasawa 2007). Research has been done on the production of PHBA in modified strains of *Saccharomyces cerevisiae* and *Escherichia coli* (Barker and Frost 2001), *Klebsiella pneumoniae* (Müller et al. 1995). The shikimate pathway produces aromatic amino acids (tryptophan, 2phenylalanine, tyrosine), quinones, secondary metabolites and other compounds (Karpf and Trussardi 2009, Koma et al. 2012, Gosset 2009). By adding extra versions of the shikimate pathway genes *aroA* (5-enolpyruvylshikimate 3-phosphate synthase), *aroL* (shikimate kinase), *aroB* (3-dehydroquinate synthase) and *aroF*, strain D2704 was able to obtain a titer of 12 g L^{-1} and a C-mole yield of 13% (feedback-insensitive DAHP synthase). As *P. putida* is resistant to PHBA, it is used in industrial production (Krömer et al. 2013, Ebert et al. 2011, Poblete-Castro et al. 2012). It only uses the ED pathway to produce more NADPH in order to prevent oxidative stress via the 3-dehydrogenase function (Ng et al. 2015, Chavarría et al. 2013). Furthermore, random nitrosoguanidine mutagenesis was used to improve the results. Pathways that degrade PHBA and those that compete with it have been removed. *ubiC* was not the only gene that was overexpressed; a feedback-resistant DAHP synthase that was encoded by aroGD146N was also overexpressed (Ebert et al. 2011).

8.4.9 Enzyme Production

Biofuels (bioethanol, biodiesel and bio hydrocarbon) are non-renewable fossil fuel alternatives (Koçar and Civaş 2013). However, the cost and competition with conventional petroleum limit the use of biofuels. Bio hydrocarbons produced by bacteria, fungi, algae and other microorganisms are concerned with various fatty-acid-derived hydrocarbons and their microbial sources (Liu and Li 2020). Cytochrome P450cam, for example, was isolated from *P. putida* (Ng et al. 2015). *CamC* is the gene that encodes cytochrome P450cam. This protein is overexpressed in *Escherichia coli* for large-scale production and structural studies (Gunsalus and Wagner 1978, Unger et al. 1986). Dimerization is the primary issue with the wild type P450cam. As a result, DTT was used to dissolve the disulphide bonds. To prevent dimerization, the cystine residue was later replaced with alanine via site-directed mutagenesis (Nickerson and Wong 1997). Strain *P. putida* LUA15.1 isolated from the rice rhizospheric soil and mutated for improved production of laccase enzyme using UV and chemical mutation. The mutant strain E4 was a hyper producer of laccase with 34.12 U.L^{-1} with 8.09 percent yield (Verma et al. 2016). Another strain of *P. putida* 922 was optimized to produce lipase with 24 U ml yield of enzyme in 48 hr (Khan et al. 2014). Along with chemical fertilizers *P. putida* have improved the production of soil enzyme and thereby improving crop yield (Nosheen et al. 2018).

8.5 Conclusion

The growing population will only continue to grow adding a large amount of waste that may not be degraded by natural degraders. As a result, the environment is in desperate need of remediation. Remediation can be accomplished in a number of ways. Using biological organisms or products to do the same is the most cost effective and least harmful to the environment. This is referred to as bioremediation. This chapter describes how *P. putida* can be used as a bioremediation tool. It is a very robust soil organism that is relatively easy to genetically modify. *P. putida* is rod shape, gram-negative motile bacteria, ubiquitous in the environment, especially in pollutant-contaminated water and soil. It thrives on almost all the carbon skeletons including plastic (Vague et al. 2019). This organism is naturally equipped with various kinds of enzymes making them highly important bacteria from the environmental point of view. Unlike *E. coli*, strains of *P. putida* are already adapted to the soil environment. They can grow and degrade pollutants in the soil or groundwater more easily. *P. putida* KT2440 has been bred to grow in anoxic conditions. As NADH produced during metabolism must be re-oxidized, oxygen is required. This strain has also been genetically modified to degrade pesticides (methyl parathion, chlorpyrifos, fenpropathrin, cypermethrin, carbofuran and carbaryl) (Gong et al. 2018). *Pseudomonas* strains that are resistant to extreme temperatures, pH and pressure can be created through further modification. As a result, the scope of bioremediation can be expanded to include harsher conditions. *Pseudomonas* has a wide range of applications due to its unique metabolism. It has demonstrated applications in both bioremediation and the production of compounds that will aid in the reduction of reliance on petroleum products.

References

Adhya, T. K., S. K. Apte, K. Raghu, N. Sethunathan and N. B. Murthy. 1996. Novel polypeptides induced by the insecticide lindane (gamma hexachlorocyclohexane) are required for its biodegradation by a *Sphingomonas paucimobilis* strain. Biochem. Biophys. Res. Commun. 221(3): 755–61.

Adrio, J. L. and A. L. Demain. 2006. Genetic improvement of processes yielding microbial products. FEMS Microbiol. Rev. 30(2): 187–214.

Allen, E. E. and D. H. Bartlett. 2002. Structure and regulation of the omega-3 polyunsaturated fatty acid synthase genes from the deep-sea bacterium *Photobacterium profundum* strain SS9The GenBank accession numbers for the sequences reported in this paper are AF409100 and AF467805. Microbiol. 148(6): 1903–1913.

Almqvist, H., H. Veras, K. Li, J. Garcia Hidalgo, C. Hulteberg, M. Gorwa-Grauslund, N. Skorupa Parachin and M. Carlquist. 2021. Muconic acid production using engineered *Pseudomonas putida* KT2440 and a guaiacol-rich fraction derived from kraft lignin. ACS Sustain. Chem. Eng. 9(24): 8097–8106.

Alper, H., C. Fischer, E. Nevoigt and G. Stephanopoulos. 2005a. Tuning genetic control through promoter engineering. Proc. Natl. Acad. Sci. USA. 102(36): 12678–12683.

Alper, H., Y. S. Jin, J. F. Moxley and G. Stephanopoulos. 2005b. Identifying gene targets for the metabolic engineering of lycopene biosynthesis in *Escherichia coli*. Metab. Eng. 7(3): 155–164.

Alshammari, A., V. N. Kalevaru, A. Bagabas and A. Martin. 2016. Production of ethylene and its commercial importance in the global market. pp. 82–115. *In*: H. Al-Megren (Ed.). Petrochemical Catalyst Materials Processes And Emerging Technologies. IGI Global.

Amiri-Jami, M., H. Wang, Y. Kakuda and M. W. Griffiths. 2006. Enhancement of polyunsaturated fatty acid production by Tn5 transposon in *Shewanella baltica*. Biotechnol. Lett. 28(15): 1187–1192.

Aoyama, M., K. Agari, G. H. Sun-Wada, M. Futai and Y. Wada. 2005. Simple and straightforward construction of a mouse gene targeting vector using *in vitro* transposition reactions. Nucleic Acids Res. 33(5): e52–e52.

Arendt, P., J. Pollier, N. Callewaert and A. Goossens. 2016. Synthetic biology for production of natural and new-to-nature terpenoids in photosynthetic organisms. Plant J. 87(1): 16–37.

Bandhyopadhyay, K., D. Das and B. R. Maiti. 1999. Solid matrix characterization of immobilized *Pseudomonas putida* MTCC 1194 used for phenol degradation. Appl. Microbiol. Biotechnol. 51(6): 891–895.

Bang, S. W. 1997. Molecular analysis of p-nitrophenol degradation by *Pseudomonas* sp. strain ENV2030. Ph.D. Thesis, Department of Environmental Sciences, Rutgers University, New Brunswick, New Jersey.

Barker, J. L. and J. W. Frost. 2001. Microbial synthesis of p-hydroxybenzoic acid from glucose. Biotechnol. Bioeng. 76(4): 376–390.

Bhushan, B., A. Chauhan, S. K. Samanta and R. K. Jain. 2000. Kinetics of biodegradation of p-nitrophenol by different bacteria. Biochem. Biophys. Res. Commun. 274(3): 626–630.

Bigley, A. N. and F. M. Raushel. 2013. Catalytic mechanisms for phosphotriesterases. Biochim. Biophys. Acta Proteins Proteom. 1834(1): 443–453.

Biot-Pelletier, D. and V. J. Martin. 2014. Evolutionary engineering by genome shuffling. Appl. Microbiol. Biotechnol. 98(9): 3877–3887.

Bojanovič, K., I. D'Arrigo and K. S. Long. 2017. Global transcriptional responses to osmotic. Appl. Environ. Microbiol. 83(7): e03236–16.

Brämer, C. O. and A. Steinbüchel. 2001. The methylcitric acid pathway in *Ralstonia eutropha*: new genes identified involved in propionate metabolism The GenBank accession numbers for the nucleotide sequences of the prp gene cluster are AF325554 and AF331923. Microbiol. 147(8): 2203–2214.

Bueno, M., M. F. Fillat, R. J. Strasser, R. Maldonado-Rodriguez, N. Marina, H. Smienk, C. Gómez-Moreno and F. Barja. 2004. Effects of lindane on the photosynthetic apparatus of the *Cyanobacterium anabaena*. Environ. Sci. Pollut. Res. Int. 11(2): 98–106.

Bujdoš, D., B. Popelářová, D. C. Volke, P. I. Nikel, N. Sonnenschein and P. Dvořák. 2023. Engineering of *Pseudomonas putida* for accelerated co-utilization of glucose and cellobiose yields aerobic overproduction of pyruvate explained by an upgraded metabolic model. Metabolic Eng. 75: 29–46.

Caldwell, B. J. and C. E. Bell. 2019. Structure and mechanism of the Red recombination system of bacteriophage λ. Prog. Biophys. Mol. Biol. 147: 33–46.

Carvalho, F. D., I. Machado, M. S. Martínez, A. Soares and L. Guilhermino. 2003. Use of atropine-treated *Daphnia magna* survival for detection of environmental contamination by acetylcholinesterase inhibitors. Ecotoxicol Environ Saf. 54(1): 43–46.

Cha, D., H. S. Ha and S. K. Lee. 2020. Metabolic engineering of *Pseudomonas putida* for the production of various types of short-chain-length polyhydroxyalkanoates from levulinic acid. Bioresour. Technol. 309: 123332.

Chagué, V., Y. Elad, R. Barakat, P. Tudzynski and A. Sharon. 2002. Ethylene biosynthesis in *Botrytis cinerea*. FEMS Microbiol. Ecol. 40(2): 143–149.

Chaudhry, G. R., A. N. Ali and W. B. Wheeler. 1988. Isolation of a methyl parathion-degrading *Pseudomonas* sp. that possesses DNA homologous to the opd gene from a *Flavobacterium* sp. Appl. Environ. Microbiol. 54(2): 288–93.

Chaurasia, A. K., T. K. Adhya and S. K. Apte. 2013. Engineering bacteria for bioremediation of persistent organochlorine pesticide lindane (γ-hexachlorocyclohexane). Bioresour. Technol. 149: 439–445.

Chavarría, M., P. I. Nikel, D. Pérez-Pantoja and V. de Lorenzo. 2013. The Entner–doudoroff pathway empowers *Pseudomonas putida* KT 2440 with a high tolerance to oxidative stress. Environ. Microbiol. 15(6): 1772–1785.

Chen, S., Y. H. Dong, C. Chang, Y. Deng, X. F. Zhang, G. Zhong, H. Song, M. Hu and L. H. Zhang. 2013. Characterization of a novel cyfluthrin-degrading bacterial strain *Brevibacterium aureum* and its biochemical degradation pathway. Bioresour. Technol. 132: 16–23.

Cook, A. M. and R. Huetter. 1981. s-Triazines as nitrogen sources for bacteria. J. Agric. Food Chem. 29(6): 1135–1143.

Cuenca, M. D. S., C. Molina-Santiago, M. R. Gómez-García and J. L. Ramos. 2016. A *Pseudomonas putida* double mutant deficient in butanol assimilation: a promising step for engineering a biological biofuel production platform. FEMS Microbiol. Lett. 363(5): fnw018.

Cullington, J. E. and A. Walker. 1999. Rapid biodegradation of diuron and other phenylurea herbicides by a soil bacterium. Soil Biol. Biochem. 31(5): 677–686.

Dai, M. and S. D. Copley. 2004. Genome shuffling improves degradation of the anthropogenic pesticide pentachlorophenol by *Sphingobium chlorophenolicum* ATCC 39723. Appl. Environ. Microbiol. 70(4): 2391–2397.

DeFrank, J. J. 1991. Organophosphorus cholinesterase inhibitors: detoxification by microbial enzymes. Applications of Enzyme Biotechnology. Springer, Boston, MA.

Dejonghe, W., J. Goris, S. El-Fantroussi, M. Höfte, P. DeVos, W. Verstraete and E. M. Top. 2000. Effect of dissemination of 2,4-dichlorophenoxyacetic acid (2,4-D) degradation plasmids on 2,4-D degradation and on bacterial community structure in two different soil horizons. Appl. Environ. Microbiol. 66(8): 3297–3304.

Deng, W., D. Lin, K. Yao, H. Yuan, Z. Wang, J. Li, L. Zou, X. Han, K. Zhou, L. He and X. Hu. 2015. Characterization of a novel β-cypermethrin-degrading *Aspergillus niger* YAT strain and the biochemical degradation pathway of β-cypermethrin. Appl. Microbiol. Biotechnol. 99(19): 8187–8198.

Dunn, N. W. and I. C. Gunsalus. 1973. Transmissible plasmid coding early enzymes of naphthalene oxidation in *Pseudomonas putida*. J. Bacteriol. 114(3): 974–979.

Dvořák, P. and V. de Lorenzo. 2018. Refactoring the upper sugar metabolism of *Pseudomonas putida* for co-utilization of cellobiose, xylose, and glucose. Metab. Eng. 48: 94–108.

Dvorak, P., S. Bidmanova, J. Damborsky and Z. Prokop. 2014. Immobilized synthetic pathway for biodegradation of toxic recalcitrant pollutant 1,2,3-Trichloropropane. Environ. Sci. Technol. 48(12): 6859–6866.

Eaton, R. W. and J. S. Karns. 1991. Cloning and comparison of the DNA encoding ammelide aminohydrolase and cyanuric acid amidohydrolase from three s-triazine-degrading bacterial strains. J. Bacteriol. 173(3): 1363–1366.

Ebert, B. E., F. Kurth, M. Grund, L. M. Blank and A. Schmid. 2011. Response of *Pseudomonas putida* KT2440 to increased NADH and ATP demand. Appl. Environ. Microbiol. 77(18): 6597–6605.

El-sayed, G. M., N. A. Abosereih, S. A. Ibrahim, A. El-Razik, B. Ashraf, M. A. Hammad and F. M. Hafez. 2019. Cloning of the *Organophosphorus Hydrolase* (oph) gene and enhancement of Chlorpyrifos degradation in the *Achromobacter xylosoxidans* Strain GH9OP via Mutation Induction. Jordan J. Biol. Sci. 12(3).

Ewering, C., F. Heuser, J. K. Benölken, C. O. Brämer and A. Steinbüchel. 2006. Metabolic engineering of strains of *Ralstonia eutropha* and *Pseudomonas putida* for biotechnological production of 2-methylcitric acid. Metab. Eng. 8(6): 587–602.

Ezeji, T., N. Qureshi and H. P. Blaschek. 2007. Butanol production from agricultural residues: impact of degradation products on *Clostridium beijerinckii* growth and butanol fermentation. Biotechnol. Bioeng. 97(6): 1460–1469.

Gahlawat, G. and S. K. Soni. 2017. Valorization of waste glycerol for the production of poly (3-hydroxybutyrate) and poly (3-hydroxybutyrate-co-3-hydroxyvalerate) copolymer by *Cupriavidus necator* and extraction in a sustainable manner. Bioresour. Technol. 243: 492–501.

Garcia, R., D. Pistorius, M. Stadler and R. Müller. 2011. Fatty acid-related phylogeny of myxobacteria as an approach to discover polyunsaturated omega-3/6 fatty acids. J. Bacteriol. 193(8): 1930–1942.

García-Hidalgo, J., D. P. Brink, K. Ravi, C. J. Paul, G. Lidénb and M. F. Gorwa-Grauslund. 2020. Vanillin production in *Pseudomonas*: whole-genome sequencing of *Pseudomonas* sp. strain 9.1 and reannotation of *Pseudomonas putida* CalA as a vanillin reductase. Appl. Environ. Microbiol. 86: e02442–19.

Gemperlein, K., G. Zipf, H. S. Bernauer, R. Müller and S. C. Wenzel. 2016. Metabolic engineering of *Pseudomonas putida* for production of docosahexaenoic acid based on a myxobacterial PUFA synthase. Metab. Eng. 33: 98–108.

Gemperlein, K., S. Rachid, R. O. Garcia, S. C. Wenzel and R. Müller. 2014. Polyunsaturated fatty acid biosynthesis in myxobacteria: different PUFA synthases and their product diversity. Chem. Sci. 5(5): 1733–1741.

Giacomazzi, S. and N. Cochet. 2004. Environmental impact of diuron transformation: a review. Chemosphere. 56(11): 1021–1032.

Gibson, D. T., G. E. Cardini, F. C. Maseles and R. E. Kallio. 1970. Oxidative degradation of aromatic hydrocarbons by microorganisms. Biochem. 9(7): 1631–1635.

Gibson, D. T., V. Mahadevan and J. F. Davey. 1974. Bacterial metabolism of para-and meta-xylene: oxidation of the aromatic ring. J. Bacteriol. 119(3): 930–936.

Gong, J., H. Zheng, Z. Wu, T. Chen and X. Zhao. 2009. Genome shuffling: progress and applications for phenotype improvement. Biotechnol. Adv. 27(6): 996–1005.

Gong, T., R. Liu, Y. Che, X. Xu, F. Zhao, H. Yu, C. Song, Y. Liu and C. Yang. 2016. Engineering *Pseudomonas putida* KT 2440 for simultaneous degradation of carbofuran and chlorpyrifos. Microb. Biotechnol. 9(6): 792–800.

Gong, T., X. Xu, Y. Che, R. Liu, W. Gao, F. Zhao, H. Yu, J. Liang, P. Xu, C. Song and C. Yang. 2017. Combinatorial metabolic engineering of *Pseudomonas putida* KT2440 for efficient mineralization of 1,2,3-trichloropropane. Sci. Rep. 7(1): 44896.

Gong, T., X. Xu, Y. Dang, A. Kong, Y. Wu, P. Liang, S. Wang, H. Yu, P. Xu and C. Yang. 2018. An engineered *Pseudomonas putida* can simultaneously degrade organophosphates. Sci. Total Environ. 628: 1258–1265.

Gosset, G. 2009. Production of aromatic compounds in bacteria. Curr. Opin. Biotechnol. 20(6): 651–658.

Graf, N. and J. Altenbuchner. 2011. Development of a method for markerless gene deletion in *Pseudomonas putida*. Appl. Environ. Microbiol. 77(15): 5549–5552.

Graf, N. and J. Altenbuchner. 2014. Genetic engineering of *Pseudomonas putida* KT2440 for rapid and high-yield production of vanillin from ferulic acid. Appl. Microbiol. Biotechnol. 98: 137–149.

Gunsalus, I. C. and G. C. Wagner. 1978. Bacterial P-450cam methylene monooxygenase components: cytochrome m, *putida*redoxin, and *putida*redoxin reductase. *In*: J. A. Tainer (Ed.). Methods in Enzymology. Academic Press.

Hannaford, A. M. and C. Kuek. 1999. Aerobic batch degradation of phenol using immobilized *Pseudomonas putida*. J. Ind. Microbiol. Biotechnol. 22(2): 121–126.

Hernandez-Arranz, S., J. Perez-Gil, D. Marshall-Sabey and M. Rodriguez-Concepcion. 2019. Engineering *Pseudomonas putida* for isoprenoid production by manipulating endogenous and shunt pathways supplying precursors. Microb. Cell Fact. 18(1): 41640.

Hsu, C. T. and Y. C. Kuo, Y. C. Liu and S. L. Tsai. 2020. Green conversion of 5-hydroxymethylfurfural to furan-2,5-dicarboxylic acid by heterogeneous expression of 5-hydroxymethylfurfural oxidase in *Pseudomonas putida* S12. Microb. Biotechnol. 13(4): 1094–1102.

Huang, K., Y. G. Lin and H. H. Winter. 1992. p-Hydroxybenzoate ethylene terephthalate copolyester: structure of high-melting crystals formed during partially molten state annealing. Polymer. 33(21): 4533–4537.

Ibeh, C. C. 2011. Thermoplastic Materials: Properties, Manufacturing Methods, and Applications. CRC Press.

Ihlaseh-Catalano, S. M., K. A. Bailey, A. P. F. Cardoso, H. Ren, R. C. Fry and J. L. V. deCamargo and D. C. Wolf. 2014. Dose and temporal effects on gene expression profiles of urothelial cells from rats exposed to diuron. Toxicology. 325: 21–30.

Janssen, D. B., J. E. Oppentocht and G. J. Poelarends. 2001. Microbial dehalogenation. Curr. Opin. Biotechnol. 12(3): 254–258.

Jensen, P. R. and K. Hammer. 1998. Artificial promoters for metabolic optimization. Biotechnol. Bioeng. 58(2-3): 191–195.

Jia Yan, H. Yin, J. S. Ye, H. Peng, B. Y. He, H. M. Qin, N. Zhang and J. Qiang. 2008. Characteristics and pathway of naphthalene degradation by *Pseudomonas* sp. N7. Huan Jing Ke Xue. 29(3): 756–762.

Jin, Y., S. Zheng and Z. Fu. 2011. Embryonic exposure to cypermethrin induces apoptosis and immunotoxicity in zebrafish (*Danio rerio*). Fish Shellfsh Immunol. 30(4-5):1049–1054.

Jones, K. C., J. A. Stratford, K. S. Waterhouse, E. T. Furlong, W. Giger, R. A. Hites, C. Schaffner and A. E. Johnston. 1989. Increases in the polynuclear aromatic hydrocarbon content of an agricultural soil over the last century. Environ. Sci. Technol. 23(1): 95–101.

Kamravamanesh, D., T. Kovacs, S. Pflügl, I. Druzhinina, P. Kroll, M. Lackner and C. Herwig. 2018. Increased poly-β-hydroxybutyrate production from carbon dioxide in randomly mutated cells of cyanobacterial strain *Synechocystis* sp. PCC 6714: mutant generation and characterization. Bioresour. Technol. 266: 34–44.

Karns, J. S., M. T. Muldoon, W. W. Mulbry, M. K. Derbyshire and P. C. Kearney. 1987. Use of microorganisms and microbial systems in the degradation of pesticides. pp. 156–170. *In*: H. M. LeBaron, R. O. Mumma, R. C. Honeycutt, J. H. Duesing, J. F. Phillips and M. J. Haas [Eds.]. Biotechnology in Agricultural Chemistry. American Chemical Society.

Karns, J. S., W. W. Mulbry, J. O. Nelson and P. C. Kearney. 1986. Metabolism of carbofuran by a pure bacterial culture. Pestic. Biochem. Physiol. 25(2): 211–217.

Karpf, M. and R. Trussardi. 2009. Efficient Access to oseltamivir phosphate (Tamiflu) via the O-Trimesylate of Shikimic Acid Ethyl Ester. Angew Chem. Int. Ed. Engl. 48(31): 5760–5762.

Kaulmann, U. and C. Hertweck. 2002. Biosynthesis of polyunsaturated fatty acids by polyketide synthases. Angew Chem. Int. Ed. Engl. 41(11): 1866–1869.

Kearney, P. C., M. T. Muldoon, C. J. Somich, J. M. Ruth and D. J. Voaden. 1988. Biodegradation of ozonated atrazine as a wastewater disposal system. J. Agric. Food Chem. 36(6): 1301–1306.

Khan, F. H. N., A. U. Rehman and Z. Hussain. 2014. Production and partial characterization of lipase from *Pseudomonas putida*. Ferment. Technol. 2: 112–119.

Kim, T. K., Y. M. Jung, M. T. Vo, S. Shioya and Y. H. Lee. 2006. Metabolic engineering and characterization of phaC1 and phaC2 genes from *Pseudomonas putida* KCTC1639 for overproduction of medium-chain-length polyhydroxyalkanoate. Biotechnol. Prog. 22(6): 1541–1546.

Kim, Y., D. A. Webster and B. C. Stark. 2005. Improvement of bioremediation by *Pseudomonas* and Burkholderia by mutants of the *Vitreoscilla hemoglobin* gene (vgb) integrated into their chromosomes. J. Ind. Microbiol. Biotechnol. 32(4): 148–154.

Kniel, L., O. Winter and K. Stork 1980. Ethylene: Keystone to the Petrochemical Industry (ChemicalIndustries). Marcel Dekker.

Koçar, G. and N. Civaş. 2013. An overview of biofuels from energy crops: current status and future prospects. Renew. Sustain. Energy Rev. 28: 900–916.

Koffas, M. A., G. Y. Jung and G. Stephanopoulos. 2003. Engineering metabolism and product formation in *Corynebacterium glutamicum* by coordinated gene overexpression. Metab. Eng. 5(1): 32–41.

Koller, M., P. Hesse, H. Fasl, F. Stelzer and G. Braunegg. 2017. Study on the effect of levulinic acid on whey-based biosynthesis of poly (3-hydroxybutyrate-co-3-hydroxyvalerate) by *Hydrogenophaga pseudoflava*. Appl. Food Biotechnol. 4(2): 65–78.

Koma, D., H. Yamanaka, K. Moriyoshi, T. Ohmoto and K. Sakai. 2012. Production of aromatic compounds by metabolically engineered *Escherichia coli* with an expanded shikimate pathway. Appl. Environ. Microbiol. 78(17): 6203–6216.

Kondakova, T. and J. E. Cronan. 2019. Transcriptional regulation of fatty acid cis–trans isomerization in the solvent-tolerant soil bacterium. Environ. Microbiol. 21(5): 1659–1676.

Krömer, J. O., D. Nunez-Bernal, N. J. Averesch, J. Hampe, J. Varela and C. Varela. 2013. Production of aromatics in *Saccharomyces cerevisiae*—a feasibility study. J. Biotechnol. 163(2): 184–193.

Kuhm, A. E., A. Stolz and H. J. Knackmuss. 1991. Metabolism of naphthalene by the biphenyl-degrading bacterium *Pseudomonas paucimobilis* Q1. Biodegrad. 2(2): 115–120.

Kulakova, A. N., M. J. Larkin and L. A. Kulakov.1997. The plasmid-located haloalkane dehalogenase gene from *Rhodococcus rhodochrous* NCIMB 13064. Microbiol. 143(1): 109–115.

Kumar, M., M. P. Singh and D. K. Tuli. 2012. Genome Shuffling of *Pseudomonas* sp. Ioca11 for improving degradation of polycyclic aromatic hydrocarbons. Adv. Microbiol. 2: 26–30.

Lee, J. Y., J. R. Roh and H. S. Kim. 1994. Metabolic engineering of *Pseudomonas putida* for the simultaneous biodegradation of benzene. Biotechnol. Bioeng. 43(11): 1146–1152.

Lemoigne, M. 1926. Products of dehydration and of polymerization of β-hydroxybutyric acid. Bull. Soc. Chem. Biol. 8: 770–782.

Lenihan-Geels, G., K. S. Bishop and L. R. Ferguson. 2013. Alternative sources of omega-3 fats: can we find a sustainable substitute for fish? Nutrients. 5(4): 1301–1315.

Li, W., J. Shi, X. Wang, Y. Han, W. Tong, L. Ma, B. Liu and B. Cai. 2004. Complete nucleotide sequence and organization of the naphthalene catabolic plasmid pND6-1 from *Pseudomonas* sp. Gene. 336(2): 231–240.

Liu, K. and S. Li. 2020. Biosynthesis of fatty acid-derived hydrocarbons: perspectives on enzymology and enzyme engineering. Curr. Opin. Biotechnol. 62: 41821.

Loeschcke, A. and S. Thies. 2015. *Pseudomonas putida*—a versatile host for the production of natural products. Appl. Microbiol. Biotechnol. 99(15): 6197–6214.

Loh, K. C. and S. J. Wang. 1997. Enhancement of biodegradation of phenol and a nongrowth substrate 4-chlorophenol by medium augmentation with conventional carbon sources. Biodegrad. 8(5): 329–338.

Lorente-Cebrian, S., A. G. V. Costa, S. Navas-Carretero, M. Zabala, J. A. Martinez and M. J. Moreno-Aliaga. 2013. Role of omega-3 fatty acids in obesity, metabolic syndrome, and cardiovascular diseases: a review of the evidence. J. Physiol. Biochem. 69(3): 633–651.

Malhat, F. and I. Nasr. 2011. Organophosphorus pesticides residues in fish samples from the River Nile tributaries in Egypt. Bull. Environ. Contam. Toxicol. 87(6): 689–692.

Mallick, K., K. Bharati, A. Banerji, N. A. Shakil and N. Sethunathan. 1999. Bacterial degradation of chlorpyrifos in pure cultures and in soil. Bull. Environ. Contam. Toxicol. 62(1): 48–54.

Mandalakis, M., N. Panikov, S. Dai, S. Ray and B. L. Karger. 2013. Comparative proteomic analysis reveals mechanistic insights into *Pseudomonas putida* F1 growth on benzoate and citrate. AMB Express. 3(1): 41275.

McKinlay, R., J. A. Plant, J. N. B. Bell and N. Voulvoulis. 2008. Endocrine disrupting pesticides: implications for risk assessment. Environ. Int. 34(2): 168–183.

Metz, J. G., P. Roessler, D. Facciotti, C. Levering, F. Dittrich, M. Lassner, R. Valentine, K. Lardizabal, F. Domergue, A. Yamada and K. Yazawa. 2001. Production of polyunsaturated fatty acids by polyketide synthases in both prokaryotes and eukaryotes. Science 293(5528): 290–293.

Mi, J., D. Becher, P. Lubuta, S. Dany, K. Tusch, H. Schewe, M. Buchhaupt and J. Schrader. 2014. *De novo* production of the monoterpenoid geranic acid by metabolically engineered *Pseudomonas putida*. Microb. Cell Factories. 13(1): 44866.

Monga, D., P. Kaur and B. Singh. 2021. Microbe mediated remediation of dyes, explosive waste and polyaromatic hydrocarbons, pesticides and pharmaceuticals. Curr. Res. Microb. Sci. 3: 100092.

Moretto, J. A. S., J. P. R. Furlan, A. F. T. Fernandes, A. Bauermeister, N. P. Lopes and E. G. Stehling. 2019. Alternative biodegradation pathway of the herbicide diuron. Int. Biodeterior. Biodegrad. 143: 104716.

Mosqueda, G., M. I. Ramos-González and J. L. Ramos. 1999. Toluene metabolism by the solvent-tolerant *Pseudomonas putida* DOT-T1 strain, and its role in solvent impermeabilization. Gene 232(1): 69–76.

Müller, R., A. Wagener, K. Schmidt and E. Leistner. 1995. Microbial production of specifically ring-13C-labelled 4-hydroxybenzoic acid. Appl. Microbiol. Biotechnol. 43(6): 985–988.

Nagata, Y. K. Miyauchi, J. Damborsky, K. Manova, A. Ansorgová and M. Takagi. 1997. Purification and characterization of a haloalkane dehalogenase of a new substrate class from a gamma-hexachlorocyclohexane-degrading bacterium. Appl. Environ. Microbiol. 63(9): 3707-3710.

Napier, J. A. 2002. Plumbing the depths of PUFA biosynthesis: a novel polyketide synthase-like pathway from marine organisms. Trends Plant Sci. 7(2): 51–54.

Nawab, A., A. Aleem and A. Malik. 2003. Determination of organochlorine pesicides in agricultural soil with special reference to γ-HCH degradation by *Pseudomonas* strains. Bioresour. Technol. 88(1): 41–46.

Nelson, K. E., C. Weinel, I. T. Paulsen, R. J. Dodson, H. Hilbert, V. A. P. Martinsdos Santos, D. E. Fouts, S. R. Gill, M. Pop, M. Holmes and L. Brinkac. 2002. Complete genome sequence and comparative analysis of the metabolically versatile *Pseudomonas putida* KT2440. Environ. Microbiol. 4(12): 799–808.

Neumann, G., R. Teras, L. Monson, M. Kivisaar, F. Schauer and H. J. Heipieper. 2004. Simultaneous degradation of atrazine and phenol by *Pseudomonas* sp. strain ADP: effects of toxicity and adaptation Appl. Environ. Microbiol. 70(4): 1907–1912.

Ng, C. Y., I. Farasat, C. D. Maranas and H. M. Salis. 2015. Rational design of a synthetic Entner–Doudoroff pathway for improved and controllable NADPH regeneration. Metab. Eng. 29: 86–96.

Nickerson, D. P. and L. L. Wong. 1997. The dimerization of *Pseudomonas putida* cytochrome P450cam: practical consequences and engineering of a monomeric enzyme. Protein Eng. 10(12): 1357–1361.

Nielsen, D. R., E. Leonard, S. H. Yoon, H. C. Tseng, C. Yuan and K. L. J. Prather. 2009. Engineering alternative butanol production platforms in heterologous bacteria. Metab. Eng. 11(4-5): 262–273.

Nikel, P. I. and V. de Lorenzo. 2013. Engineering an anaerobic metabolic regime in *Pseudomonas putida* KT2440 for the anoxic biodegradation of 1, 3-dichloroprop-1-ene. Metab. Eng. 15: 98–112.

Nikel, P. I., M. Chavarría, T. Fuhrer, U. Sauer and V. DeLorenzo. 2015. *Pseudomonas putida* KT2440 Strain metabolizes glucose through a cycle formed by enzymes of the Entner-Doudoroff, Embden-Meyerhof-Parnas, and pentose phosphate pathways. J. Biol. Chem. 290(43): 25920–25932.

Nosheen, A., H. Yasmin, R. Naz, A. Bano, R. Keyani and I. Hussain. 2018. *Pseudomonas putida* improved soil enzyme activity and growth of kasumbha under low input of mineral fertilizers. Soil Sci. Plant Nutr. 64(4): 520–525.

Novackova, I., D. Kucera, J. Porizka, I. Pernicova, P. Sedlacek, M. Koller, A. Kovalcik and S. Obruca. 2019. Adaptation of *Cupriavidus necator* to levulinic acid for enhanced production of P (3HB-co-3HV) copolyesters. Biochem. Eng. J. 151: 107350.

NPCS Board of Consultants and Engineers. 2000. Water and Air Effluents Treatment. pp. 317. India: Asia pacific Press business press.

Pfleger, B. F., D. J. Pitera, C. D. Smolke and J. D. Keasling. 2006. Combinatorial engineering of intergenic regions in operons tunes expression of multiple genes. Nat. Biotechnol. 24(8): 1027–1032.

Pham, N. N., C. Y. Chen, H. Li, M. T. T. Nguyen, P. K. P. Nguyen, S. L. Tsai, J. Y. Chou, T. C. Ramliand and Y. C. Hu. 2020. Engineering Stable *Pseudomonas putida* S12 by CRISPR for 2,5-Furandicarboxylic Acid (FDCA) Production. ACS Synth. Biol. 9(5): 1138–1149.

Pitera, D. J., C. J. Paddon, J. D. Newman and J. D. Keasling. 2007. Balancing a heterologous mevalonate pathway for improved isoprenoid production in *Escherichia coli*. Metab. Eng. 9(2): 193–207.

Poblete-Castro, I., J. Becker, K. Dohnt, V. M. Dos-Santos and C. Wittmann. 2012. Industrial biotechnology of *Pseudomonas putida* and related species. Appl. Microbiol. Biotechnol. 93(6): 2279–2290.

Ramanathan, M. P. and D. Lalithakumari. 1999. Complete mineralization of methylparathion by *Pseudomonas* sp. A3. Appl. Biochem. Biotechnol. 80(1): 1–12.

Ramos, J. L., E. Duque, M. J. Huertas and A. L. I. HaïDour. 1995. Isolation and expansion of the catabolic potential of a *Pseudomonas putida* strain able to grow in the presence of high concentrations of aromatic hydrocarbons. J. Bacteriol. 177(14): 3911–3916.

Ramos, J. L., E. Duque, M. T. Gallegos, P. Godoy, M. I. Ramos-Gonzalez, A. Rojas, W. Teránand A. Segura. 2002. Mechanisms of solvent tolerance in gram-negative bacteria. Annual Rev. Microbiol. 56(1): 743–768.

Ramos-González, M. I., A. Ben-Bassat, M. J. Campos and J. L. Ramos. 2003. Genetic engineering of a highly solvent-tolerant *Pseudomonas putida* strain for biotransformation of toluene to p-hydroxybenzoate. Appl. Environ. Microbiol. 69(9): 5120–5127.

Rani, N. L. and D. Lalithakumari. 1994. Degradation of methyl parathion by *Pseudomonas putida*. Can. J. Microbiol. 40(12): 1000–1006.

Rao, S. C., R. Rao and R. Agrawal. 2003. Enhanced production of verbenol, a highly valued food flavourant, by an intergeneric fusant strain of *Aspergillus niger* and *Penicillium digitatum*. Biotechnol. Appl. Biochem. 37(2): 145–147.

Rehm, B. H. and A. Steinbüchel. 1999. Biochemical and genetic analysis of PHA synthases and other proteins required for PHA synthesis. Int. J. Biol. Macromol. 25(1-3): 43525.

Reyrat, J. M., V. Pelicic, B. Gicquel and R. Rappuoli. 1998. Counter selectable markers: untapped tools for bacterial genetics and pathogenesis. Infect Immun. 66(9): 4011–4017.

Rink, R., M. Fennema, M. Smids, U. Dehmel and D. B. Janssen. 1997. Primary structure and catalytic mechanism of the epoxide hydrolase from *Agrobacterium radiobacterAD1*. J. Biol. Chem. 272(23): 14650–14657.

Santos, C. N. S. and G. Stephanopoulos. 2008. Combinatorial engineering of microbes for optimizing cellular phenotype. Curr. Opin. Chem. Biol. 12(2): 168–176.

Sauer, U. 2001. Evolutionary engineering of industrially important microbial phenotypes. Metab. Eng. 73: 129–169.

Schettgen, T., U. Heudorf, H. Drexler and J. Angerer. 2002. Pyrethroid exposure of the general population-is this due to diet. Toxicol. Lett. 134(1-3): 141–145.

Schmack, G., V. Gorenflo and A. Steinbüchel. 1998. Biotechnological production and characterization of polyesters containing 4-hydroxyvaleric acid and medium-chain-length hydroxyalkanoic acids. Macromolecules. 31(3): 644–649.

Serdar, C. M., D. C. Murdock and M. F. Rohde. 1989. Parathion hydrolase gene from *Pseudomonas* diminuta MG: Subcloning, complete nucleotide sequence, and expression of the mature portion of the enzyme in *Escherichia coli*. Nat. Biotechnol. 7(11): 1151–1155.

Sharma, B. and P. Shukla. 2022. Futuristic avenues of metabolic engineering techniques in bioremediation. Biotechnol. Appl. Biochem. 69(1): 51–60.

Shelton, D. R. and C. J. Somich. 1988. Isolation and characterization of coumaphos-metabolizing bacteria from cattle dip. Appl. Environ. Microbiol. 54(10): 2566–2571.

Shen, Y. J., P. Lu, H. Mei, H. J. Yu, Q. Hong and S. P. Li. 2010. Isolation of a methyl parathion-degrading strain *Stenotrophomonas* sp. SMSP-1 and cloning of the ophc2 gene. Biodegrad. 21(5): 785–792.

Silambarasan, S., P. Logeswari, A. Ruiz, P. Cornejo and V. R. Kannan. 2020. Influence of plant beneficial *Stenotrophomonas rhizophila* strain CASB3 on the degradation of diuron-contaminated saline soil and improvement of *Lactuca sativa* growth. Environ. Sci. Pollut. Res. Int. 27(28): 35195–35207.

Simon, O., I. Klaiber, A. Huber and J. Pfannstiel. 2014. Comprehensive proteome analysis of the response of *Pseudomonas putida* KT2440 to the flavor compound vanillin. J. Proteomics. 109: 212–227.

Singh, P. K. 1973. Effect of pesticides on blue-green algae. Arch. Mikrobiol. 89(4): 317–320.

Singleton, D. R., L. Guzmán-Ramirez and M. D. Aitken. 2009. Characterization of a polycyclic aromatic hydrocarbon degradation gene cluster in a phenanthrene-degrading Acidovorax strain. Appl. Environ. Microbiol. 75(9): 2613–2620.

Song, F., Y. Shi, S. Jia, Z. Tan and H. Zhao. 2018. Advances of naphthalene degradation in *Pseudomonas putida* ND6. Front. Bioeng. Biotechnol. 944(1): 20074.

Soni, M. G., S. L. Taylor, N. A. Greenberg and G. A. Burdock. 2002. Evaluation of the health aspects of methyl paraben: a review of the published literature. Food Chem. Toxicol. 40(10): 1335–1373.

Stark, B. C., K. R. Pagilla and K. L. Dikshit. 2015. Recent applications of *Vitreoscilla hemoglobin* technology in bioproduct synthesis and bioremediation. Appl. Microbiol. Biotechnol. 99(4): 1627–1636.

Steinbüchel, A. and S. Hein. 2001. Biochemical and molecular basis of microbial synthesis of polyhydroxyalkanoates in microorganisms. Adv. Biochem. Eng. Biotechnol. 71: 81–123.

Tang, J., B. Liu, T. T. Chen, K. Yao, L. Zeng, C. Y. Zeng and Q. Zhang. 2018. Screening of a beta-cypermethrin-degrading bacterial strain *Brevibacillus parabrevis* BCP-09 and its biochemical degradation pathway. Biodegrad. 29(6): 525–541.

Tiso, T., P. Sabelhaus, B. Behrens, A. Wittgens, F. Rosenau, H. Hayen and L. M. Blank. 2016. Creating metabolic demand as an engineering strategy in *Pseudomonas putida*–Rhamnolipid synthesis as an example. Adv. Energy Sci. Environ. Eng. 3: 234–244.

Tomasek, P. H. and J. S. Karns. 1989. Cloning of a carbofuran hydrolase gene from *Achromobacter* sp. strain WM111 and its expression in gram-negative bacteria. J. Bacteriol. 171(7): 4038–4044.

Tran, N. H., T. Urase, H. H. Ngo, J. Hu and S. L. Ong. 2013. Insight into metabolic and cometabolic activities of autotrophic and heterotrophic microorganisms in the biodegradation of emerging trace organic contaminants. Bioresour. Technol. 146: 721–731.

Turnbull, G. A., J. E. Cullington, A. Walker and J. Morgan. 2001. Identification and characterisation of a diuron-degrading bacterium. Biol. Fertil. Soils. 33: 472–476.

Unger, B. P., I. C. Gunsalus and S. G. Sligar. 1986. Nucleotide sequence of the *Pseudomonas putida* cytochrome P-450cam gene and its expression in *Escherichia coli*. J. Biol. Chem. 261(3): 1158–1163.

Vague, M., G. Chan, C. Roberts, N. A. Swartz and J. L. Mellies. 2019. *Pseudomonas* isolates degrade and form biofilms on polyethylene terephthalate (PET) plastic. bioRxiv. p: 647321.

van Hylckama Vlieg, J. E., L. Tang, J. H. LutjeSpelberg, T. Smilda, G. J. Poelarends, T. Bosma, A. E. van Merode, M. W. Fraaije and D. B. Janssen. 2001. Halohydrin dehalogenases are structurally and mechanistically related to short-chain dehydrogenases/reductases. J. Bacteriol. 183(17): 5058–5066.

Verma, A., K. Dhiman and P. Shirkot. 2016. Hyper-production of laccase by *Pseudomonas putida* LUA15. 1 through mutagenesis. J. Microbiol. Exp. 3(1): 00080.

Volke, D. C., L. Friis, N. T. Wirth, J. Turlin and P. I. Nikel. 2020. Synthetic control of plasmid replication enables target-and self-curing of vectors and expedites genome engineering of *Pseudomonas putida*. Metab. Eng. Commun. 10: e00126.

Wallis, J. G., J. L. Watts and J. Browse. 2002. Polyunsaturated fatty acid synthesis: what will they think of next? Trends Biochem. Sci. 27(9): 467–473.

Wang, J. P., L. X. Wu, F. Xu, J. Lv, H. J. Jin and S. F. Chen. 2010. Metabolic engineering for ethylene production by inserting the ethylene-forming enzyme gene (efe) at the 16S rDNA sites of *Pseudomonas putida* KT2440. Bioresour. Technol. 101(16): 6404–6409.

Wang, Y., C. Zhang, T. Gong, Z. Zuo, F. Zhao, X. Fan, C. Yang and C. Song. 2015. An upp-based markerless gene replacement method for genome reduction and metabolic pathway engineering in *Pseudomonas mendocina* NK-01 and *Pseudomonas putida* KT2440. J. Microbiol. Methods. 113: 27–33.

Watanabe, K., K. Nagahama and M. Sato. 1998. A conjugative plasmid carrying the efe gene for the ethylene-forming enzyme isolated from *Pseudomonas* syringae pv. Glycinea. Phytopathol. 88(11): 1205–1209.

Whited, G. M. and D. T. Gibson. 1991. Toluene-4-monooxygenase, a three-component enzyme system that catalyzes the oxidation of toluene to p-cresol in *Pseudomonas mendocina* KR1. J. Bacteriol. 173(9): 3010–3016.

WHO. 2003. Concise International Chemical Assessment Document 56:1,2,3-Trichloropropane. Geneva.

Worsey, M. J. and P. A. Williams. 1975. Metabolism of toluene and xylenes by *Pseudomonas* (*putida* (*arvilla*) mt-2: evidence for a new function of the TOL plasmid. J. Bacteriol. 124(1): 7–13.

Xiao, Y., S. Chen, Y. Gao, W. Hu, M. Hu and G. Zhong. 2015. Isolation of a novel beta-cypermethrin degrading strain *Bacillus subtilis* BSF01 and its biodegradation pathway. Appl. Microbiol. Biotechnol. 99(6): 2849–2859.

Yang, Y. H., C. J. Brigham, E. Song, J. M. Jeon, C. K. Rha and A. J. Sinskey. 2012. Biosynthesis of poly (3-hydroxybutyrate-co-3-hydroxyvalerate) containing a predominant amount of 3-hydroxyvalerate by engineered *Escherichia coli* expressing propionate-C o A transferase. 113(4): 815–823.

Yoshida, T. and N. Sethunathan. 1973. A Flat~ obacterium that degrades diazinon and parathion. Can. J. Microbiol. 19: 873–875.

Yoshida, T. and T. Nagasawa. 2007. Biological Kolbe-Schmitt carboxylation: possible use of enzymes for the direct carboxylation of organic substrates. *In*: T. Matsuda [Ed.]. Future Directions in Biocatalysis. Elsevier Science.

You, I. S. and R. Bartha. 1982. Metabolism of 3, 4-dichloroaniline by *Pseudomonas putida*. J. Agric. Food Chem. 30(2): 274–277.

Yu, S., M. R. Plan, G. Winter and J. O. Krömer. 2016. Metabolic engineering of *Pseudomonas putida* KT2440 for the production of para-Hydroxy Benzoic Acid. Front. Bioeng. Biotechnol. 4: 90.

Zhang, R., X. Xu, W. Chen and Q. Huang. 2016. Genetically engineered *Pseudomonas putida* X3 strain and its potential ability to bioremediate soil microcosms contaminated with methyl parathion and cadmium. Appl. Microbiol. Biotechnol. 100(4): 1987–1997.

Zhang, Y. X., K. Perry, V. A. Vinci, K. Powell, W. P. Stemmer and S. B. delCardayré. 2002. Genome shuffling leads to rapid phenotypic improvement in bacteria. Nature. 415(6872): 644–646.

Zhao, J., D. Jia, J. Du, Y. Chi and K. Yao. 2019a. Substrate regulation on co-metabolic degradation of β-cypermethrin by *Bacillus licheniformis* B-1. AMB Express. 9(1): 44866.

Zhao, J., X. Chen, H. X. Wei, J. Lv, C. Chen, X. Y. Liu, Q. Wen and L. M. Jia. 2019b. Nutrient uptake and utilization in Prince Rupprecht's larch (Larix principis-rupprechtii Mayr.) seedlings exposed to a combination of light-emitting diode spectra and exponential fertilization. Soil Sci. Plant Nutr. 65(4): 358–368.

Zhongli, C., L. Shunpeng and F. Guoping. 2001. Isolation of methyl parathion-degrading strain M6 and cloning of the methyl parathion hydrolase gene. Appl. Environ. Microbiol. 67(10): 4922–5.

Zylstra, G. J., W. R. McCombie, D. T. Gibson and B. A. Finette. 1988. Toluene degradation by *Pseudomonas putida* F1: genetic organization of the tod operon. Appl. Environ. Microbiol. 54(6): 1498–1503.

Zylstra, G. J., S. W. Bang, L. M. Newman and L. L. Perry. 2000. Microbial degradation of mononitrophenols and mononitrobenzoates. pp. 145–160 *In*: J. C. Spain, J. B. Hughes and H. J. Knackmuss [Eds.]. Biodegradation of Nitroaromatic Compounds and Explosives. CRC Press, FL.

CHAPTER 9

General Aspects/Case Studies on Sources and Bioremediation Mechanisms of Metal(loid)s

Manoj Kumar, Sushma K. Varma, Renju,
Neeraj Kumar Singh and *Rajesh Singh**

9.1 Introduction

With the growing world population, the consequential demand for industrial establishments to suit human requirements has caused a rise in pollution in air, land and aquatic ecosystems due to an overuse of accessible resources (Tarekegn et al. 2020). Environmental pollution by inorganic pollutants like heavy metal(loid)s is the main problem (Kumar and Singh 2017). Apart from Arsenic (As), Boron (B), and Selenium (Se), 'heavy metal(loid)s' refers to elements (metalloids and metals) having atomic densities larger than 6 g cm^{-3} in general. In this category, there are both biologically necessary (viz. Zn, Co, Mn, Cu and Cr) and non-essential components (viz. Hg, Cd and Pb). The necessary components with nutritional value for plants and animals. As these are required only in small quantities, are referred to as micronutrients. In the chemical industry, because they are phytotoxins and zootoxins, non-essential metal(loid)s are referred to as "toxic elements. Higher amounts of these non-essential metal(loids) are harmful to both plants and animals (Adriano 2001).

Heavy metal(loid)s contamination through natural and anthropogenic activities may deteriorate the environment and cause adverse effects on humans and animals (Douay et al. 2008, Shen et al. 2017, Liu et al. 2021, Adlane et al. 2020). Heavy metal(loid)s have become increasingly controversial due to their possible harmful health and environmental consequences, as well as their effects on international trade in a number of locations throughout the world. The increase of Cd in grazing animals' (mainly kidneys and liver) renders it unfit for human consumption; however, it also poses a threat to the export of offal products. The Cd bioaccumulation in potatoes, wheat and rice crops has significant consequences for the marketing of these crops on a local and international basis (Roberts et al. 1994, Mclaughlin et al. 1996, Kirkham, 2006, Makino et al. 2006, Mavropoulos et al. 2002, Pérez and Anderson 2009). Therefore, there is a worldwide alarm in ensuring that the heavy metal(loid)s content of food fulfills regulatory criteria and compares well to those of other countries. An organic amendment to soil can enhance the process of bioremediation, which is a

School of Environment and Sustainable Development, Central University of Gujarat, Gandhinagar, 382030, Gujarat, India.
* Corresponding author: rajeshsnain@gmail.com

natural phenomenon that uses fungi, bacteria and higher plants. It reduces the toxicity of soil by changing the soil environment through bioaugmentation and biostimulation processes. The organic amendments, higher plants and microorganisms include biological agents in the bioremediation of contaminated soil with metal(loids). These organisms can render metal(loid)s harmless by lowering their bioavailability and using chemical contaminants as a source of energy (Alexander 2000, Zhuang et al. 2007).

9.2 Sources of Environmental Contamination by Heavy Metals(loid)s

The quantity of metals contained in the parent material from which the soil evolved and chemical contributions from human activities make up the heavy metal(loids) total concentration in soil (contamination sources). Heavy metals(loids) are found in the atmosphere in raindrops, aerosol particles (< 30 mm diameter), agriculture fertilizers, agrichemicals, organic manure, livestock sludge and compost. Metal(loids) are present in the soil as parent material, anthropogenic material, ashes, mine wastes and demolition rubble. Metal(oids) leach into the soil profile by water and winds (Alloway 2013). These metal(loids) in higher concentrations cause a loss in agriculture yield (Alloway 2013).

9.2.1 Sulfidic Mine

The Earth's crust contains a number of sulfur-containing minerals. There is a significant proportion of rocks that are composed of sulfides. Various deposited metal sulfides ores are copper, lead, zinc, gold, nickel, iron, uranium, PO_4^{-3} ores, oil shales, coal seams and mineral sands. Sulfides eventually get exposed to oxygen during mining and give rise to acidic mine drainage. Tailing dams, waste rock dumps, open pit floors, open-pit faces, haul roads, quarries and other rock mines are all common places to expose sulfide minerals. In the presence of oxygenated groundwater or the atmosphere, the sulfides are oxidized to produce acidic water rich in sulfate, metalloids and heavy metals (Anju and Banerjee 2010). There are a few types of sulfide minerals that are relatively common, including pyrite (FeS_2). At mine sites, acidic water containing sulfate and metal(loids) is released mostly as a result of the weathering of these minerals. Consequently, acid mine drainage has emerged as a main environmental issue that has an impact on the mining sector as a whole (Bernd 2007).

9.2.2 Acidic Mine Waste

The formation of low pH waters is attributed to the oxidation of sulfur-bearing minerals. Mining activities tend to cause this type of event, affected by Acid Mine Drainage (AMD) from open-pit mined areas or after mine abandonment, the formation of Acidic Mining Lakes (AML) in surface depressions (İlay et al. 2019). Hazardous substances and metals and metalloids, Al, Cr, As, Mn, Ni, Cd and Zn are a few examples, have higher concentrations at low pH. Aquatic life downstream can be destroyed by AML drainage water. The dissolution of carbonate minerals in soil caused by acidic mining lakes causes landslides along with degraded water quality (Geller et al. 1998, Schultze and Geller 1996).

Acid mine drainage has an impact on aquatic species that live in water bodies. Acid mine drainage lowers the diversity of the environment, abundance and aquatic macroinvertebrates in streams. Polluted waters also affect many fish species (Letterman and Mitsch 1978). Pollution levels can alter the abundance of specific macroinvertebrates, as some species range widely, while others are confined to a limited area (Rasmussen and Lindegaard 1988).

9.2.3 Bauxite and Magnesite Mine

Mining has generated mine waste for several centuries. In mining tailings, metal concentrations range from 1 to 50 g kg^{-1}, depending on the metal (Monica et al. 2008). In the environment, metals are a significant source of toxicity for biota. The effects of these elements are felt by organisms like algae, microbes, yeast and fungi as well as in addition to ecosystems, activities and processes facilitated by microbes (Giller et al. 2009, Wang et al. 2010, Babich and Stotzky 1985, Baath 1989).

9.2.4 Smelter Contaminated Soils

Smelting and mining of zinc and lead ores have significant negative effects on the environment (Anju and Banerjee 2011, Dudka and Adriano 1997, Lambert et al. 1994). Health loss, pollution of water and soil, phytotoxicity, soil deterioration and environmentally harmful effects occur as a result of contact with Pb, Zn and Cd polluted smelter sites (Adriano et al. 1997, Pierzynski 1997). Instead of the level of biological contamination, the presence of heavy metals in the soil serves as an indicator of the level of action needed for soil cleanup or "bioavailability" of heavy metals. However, soil heavy metal bioavailability is linked to environmental risk. Heavy metals are assimilated by humans through the following multiple routes from the contaminated soils: (i) Exposure to Cd and Zn via plant absorption in the food chain, and (ii) accidental consumption of contaminated soil. Hand-to-mouth activity (*pica*) results in inadvertent consumption of soil. Exposure to metal(loid)s through incidental soil ingestion represents a significant source of non-dietary exposure to metals (Anju and Banerjee 2003, Chaney and Ryan 1994, Day et al. 1979, Duggan et al. 1985, Wixson and Davies 1994).

9.2.5 Fertilizer Products

Heavy metal(loid)s input from fertilizers, particularly phosphate (P) fertilizers, is believed to be the primary source of Cd. However, phosphate fertilizers with low Cd content can be used as fertilizers. PO_4^{3-} rocks with low Cd contents may therefore be utilized to manufacture fertilizer products. Several countries have introduced voluntary Cd limits on P fertilizers distributed by fertilizer manufacturers. By the year 2000, the Cd level of P fertilizer in New Zealand decreased by 280 mg Cd kg^{-1} P, compared to 340 milligrams Cd kilogram^{-1} P in the 1990s. The production of P fertilizers can be achieved by using PRs that have low Cd content; however, for practical and economic reasons, many countries continue to use sources with high Cd content (Bolan et al. 2003). Several chemical approaches have been researched to eliminate Cd from phosphoric acid followed by its conversion to fertilizers containing P. Two of these are the use of amines to extract wet phosphoric acids and the use of ion-exchange resins. Phosphate rocks are preheated in a calcination process through a stream or by a volatilization process to reduce Cd content in phosphate rocks. However, because calcination is expensive and reduces PRs' reactivity, it could not be a practical choice in the fertilizer sector; consequently, they are less suited for immediate usage as a phosphate source (Ando 1987).

9.2.6 Manure and Biosolids

Manure usage is gradually gaining prevalence globally, especially in metal enrichment in soils. The amount of metal(loid) in soils has multiplied due to frequent and extensive applications. Animal manure contributes 5247, 1821 and 225 mg of zinc, copper and nickel, correspondingly, to agricultural areas in England and Wales each year, accounting for 25 to 40% of overall contributions (Nicholson et al. 1999). Australia contains high Cd and Zn levels in vegetable soils when poultry manure was applied to agricultural soils Jinadasa et al. (1997). Copper concentrations in cow, chicken, pig and sheep dung in China were investigated by Xiong et al. (2010), who discovered the mean Cu concentrations to be higher than the national average in pig, poultry, sheep excrement and cattle, ranging from 699.6 to 31.8, 81.8 to 66.85 mg kg^{-1}, respectively. A large quantity of bioavailable cadmium and arsenic are found in poultry feces mixed with soil (Sims and Wolf 1994, McBride 1995, Haynes et al. 2009). Biosolids are commonly contaminated with heavy metals(loids) by contamination of Zn, Cu, Ni, Pb, Cd and Cr with industrial wastewater (Haynes et al. 2009).

When heavy metals like Zn, Ni and Pb are transported through sand and sandy loams with biosolids, Gove et al. (2001) discovered that drying and composting increase groundwater

Table 9.1. Metal(loid)s sources and toxicity to humans.

Metal and Metalloids	Toxicity to human	Sources	References
Arsenic (As)	Carcinogenesis, cardiac dysfunction, skin infection, lung damage, gastrointestinal problems	Paints, timber, pesticides industries, natural/ geogenic process, thermal, geothermal, smelting operation, burning fuels	Chandra et al. 2017, Lone et al. 2008, Gupta and Kumar 2017
Lead (Pb)	Mental disorders, kidney damage and affecting the nervous system	Metal products, electronic waste, petroleum additives, thermal power plants operated with coal, Bengal and ceramic industries	Chandra et al. 2017, Lone et al. 2008, Gupta and Kumar 2017
Cadmium (Cd)	Renal disorders, carcinogens and bone degeneration	Electronic industries, smelting and electroplating industries, paint industries	Chandra et al. 2017, Lone et al. 2008, Gupta and Kumar 2017
Copper (Cu)	Wilson disease, liver damage and insomnia	Electroplating, mine industries, timber, electronic waste, paint and pigment industries	Chandra et al. 2017, Lone et al. 2008, Gupta and Kumar 2017
Chromium (Cr)	Allergic reaction, dermal infection, carcinogen, DNA mutation, gastrointestinal hemorrhage	Leather industries, chromium salt industries, dye industries, pesticides, timber	Chandra et al. 2017, Lone et al. 2008, Gupta and Kumar 2017
Manganese (Mn)	The steel industry, municipal wastewater, fertilizers industries	Cardiovascular, respiratory and central nervous disorder	Chandra et al. 2017, Lone et al. 2008, Gupta and Kumar 2017
Mercury (Hg)	Mental retardation, deafness, blindness and kidney damage	E-waste, medical waste, thermal power plant and geothermal, fumigants	Chandra et al. 2017, Lone et al. 2008, Gupta and Kumar 2017
Nickel (Ni)	Cardiovascular damage, chronic asthma, nausea	Thermal power plants, smelting operations, e-waste, alloys, battery industries	Chandra et al. 2017, Lone et al. 2008

contamination with meta(loids). The biosolid treatment has little effect on Ni and Cr contents in soil (Illera et al. 2000). The occurrence of these metals in biosolids caused a significant rise in Zn, Cd, Cu and Pb levels. It has been noticed that grazing and immobilization of soil are harmful to the soil's microflora. It can rise the concentration of metal(oids) Zn, Pb and Cu whereas decreasing C and N in plants, animals and soil microbes (Haynes et al. 2009, Kao et al. 2006). Various metals and metalloid sources and their toxicity to human beings are shown in Table 9.1.

9.3 Heavy Metal(loid)s Reaction Mechanisms

Prior knowledge of the reaction mechanism between the binder substrate and metalloid is necessary for the development of sustainable treatment technology. The existing report confirms that the reaction mechanism plays a vital part in metalloid removal (Park et al. 2011). These detailed mechanisms will help in better understanding and development of technologies for real environmental applications. The systematic research on the metalloid's removal mechanisms is discussed below.

9.3.1 Adsorption

Adsorption, according to Sposito (1984), is the accumulation of a solute at the interface between a liquid solution and a solid. The process of forming a chemical connection with metal ions on the surface of adsorbents is known as adsorption. The adsorption is categorized dominantly into two sets, i.e., specific adsorption and non-specific (Bolan et al. 2014). By using functional groups, specific adsorption binds the solute to the adsorbents (Sposito 1984). Solutes are bound by non-specific adsorption through electrostatic attraction (Bolan et al. 2014). The adsorbent's properties depend on pH, as a rise in pH attributes to a –ve charge on the adsorbent's surface (Martínez-Villegas et al. 2004). Pb precipitates to form hydroxide ions and structures hydroxyl species which is more strongly bound in comparison to free metal ions (Martínez-Villegas et al. 2004). Multinuclear metal-hydroxyl species in aqueous forms can precipitate metal hydroxyl in homogenized solutions (Martínez-Villegas et al. 2004, Park et al. 2011).

Components of the soil, such as organic matter, silicate clay, manganese oxides, iron and aluminium, influence the metal(loid)s' adsorption process (Merdy et al. 2009). These soil components form inorganic and organic complexes with metal(oid)s (Bolan et al. 2003). Surface functional groups form semi-covalent bonds with dissolved ions (Bradl 2004). The rise in pH dissociates H^+ ions from functional groups like phenolic, carboxyl, carbonyl and hydroxyl, increasing the metal cation's affinity (Bolan et al. 2003). The soil's exterior surface, which has a number of hydroxyl groups, creates stable surface complexes with heavy metalloids (Hanlie et al. 2001). Heavy metalloids form two types of complexes with the soil: When water molecules are present between the soil's surface and the functional group in an outer-sphere complex, metalloids are directly attached to the functional groups of the soil (Hanlie et al. 2001). Complexes within the inner sphere are typically more stable than those outside (Bradl 2004). Natural ligands (fulvic, humic acid) and anthropogenic ligands (nitrile tri-acetic acid and EDTA) form complexes with heavy metals (Bradl 2004). Phytoextraction technology is used for the extraction of heavy metalloids by enhancing the mobility of heavy metal(loid)-EDTA or NTA complexes (Meers et al. 2009). Numerous variables, including the type of soil, affect the formation of metal(loid)-organic complexes, dominant cations, temperature, soil solution pH, and ionic strength (Luo et al. 2010). pH governs negative charge formation on surface functional groups (Yang et al. 2006). It has been noted that the soil's constituents also contribute to the formation of metalloid complexes (Bradl 2004). Fine particle soil has more active surface sites; therefore, it has a high retention potential for the metal(loid) (Bradl 2004). Adsorption of metalloids such as As(V) depends on the mineral surface characteristics like soil having low oxide content adsorption property has a very negligible effect of increasing pH (Smith et al. 1999).

9.3.2 Methylation/Demethylation

Toxic metal(loid)s are removed by methylation through biological methods by transforming (e.g., Hg) into methyl derivatives, which get removed further by the volatilization process (Frankenberger and Losi 1995). Se, Hg and As methylated derivatives are acquired from biological and chemical mechanisms. These mechanisms may reduce their toxicity by changing their volatility, solubility and mobility. Both biological and chemical mechanisms can result in the methylation of metal(loids) (Bolan et al. 2014). Biomethylation is a dominant process found in both aquatic and soil environments. Biomethylation reduces toxicity by excreting methylated compounds from cells, and are frequently volatile, e.g., organic arsenic. Researchers have categorized methylation into two groups: fission-methylation and trans-methylation (Thayer and Brinckman 1982). The fission of a chemical (methyl source) is known as fission methylation, not always having a methyl group in order to completely get rid of a compound like formic acid (Bolan et al. 2014). Subsequently, another substance that has been reduced to a methyl group captures the fission molecule. Unlike trans-methylation, which includes the exchange of methyl groups between sources (donors) (methyl acceptor). Microorganisms are active methylators in soil and sediments. For abiotic and biotic methylation in sediment and soil, organic matter serves as a source of methyl-donor. The methylation of Hg in sediments is influenced by organic matter and alternative electron acceptors (Martín-Doimeadios et al. 2004). A study has shown that methylated As species result in the breakdown of eliminating waste arsenicals or cellular organ arsenicals complex (Li et al. 2009).

Under anaerobic and aerobic conditions, methylation of Hg occurs (Martín-Doimeadios et al. 2004) and in undisturbed lake sediments under anaerobic conditions, significantly higher Hg methylation was observed (Martín-Doimeadios et al. 2004). Under these environments, Hg(II) ions are methylated biologically to produce monomethyl and dimethyl compounds, which are extremely lethal than any other form (Ullrich et al. 2001). Selenium removal from contaminated environments by biomethylation has also been reported (Adriano et al. 2004).

9.3.3 Oxidation/Reduction Mechanism

Microbes use oxidation/reduction reactions to decrease Cr, Se, Hg and As. Dissimilatory and assimilatory reactions are the two main classes into which the oxidation/reduction reactions are separated. In dissimilatory reactions, the metal(loid)s do not play any specific function in the growth of microbes (Bolan et al. 2014). They are found by accidental reductions linked to microbial oxidations to produce H_2, alcohols, simple organic acids and aromatic compounds (Holden and Adams 2003). In assimilatory reactions, the growth of microbes is promoted by metal(loid) to assist as the terminal e⁻ acceptor (Holden and Adams, 2003).

Bacteria reduce the toxicity of Cr(VI) and Hg(II) to less toxic forms by enzymatic reduction (Choppala et al. 2015). A study has shown that Se(VI) reduces to Se(0) from wastewater by anaerobic bacteria (Nejad et al. 2018). This remediation strategy has been proven as a successful approach for wastewater treatment. The oxidation mechanism's capability to transmute As(III) to As(V) has been found in archaebacterium *Sulfolobus acidocaldarius* (Lindström and Sehlin 1989). It has been reported that in an aqueous medium, in the presence of Fe(III), the oxidation rate of As(III) to As(V) is enhanced. As(V) are less toxic than As(III) and firmly bound with the inorganic soil components that result in immobilization and bioremediation through microbial oxidation (Lindström and Sehlin 1989). A similar reduction mechanism is followed by Cr(VI) to Cr(III) in the soil component. A study reported that *Bacillus* sp. isolated from the Cr-contaminated landfill reduces the potent Cr(VI) to a lesser toxic form Cr(III) (Bolan et al. 2003). In situations with a convenient source of e⁻ (Fe(II)), chromate (Cr(VI)) can be reduced to Cr(III), and microbial Cr(VI) reduction takes place in the presence of organic matter as an e⁻ donor (Bolan et al. 2014).

9.3.4 Precipitation

In polluted soils with basic concentrations and several anions present including phosphate, carbonate, sulfate and hydroxide, the precipitation mechanism of metal(loid)s removal has been discovered (Ok et al. 2010). Metal(loid) precipitation like Pb and Cu with carbonate and phosphate is an immobilization mechanism of bioremediation for elimination from soil or wastewater (McGowen et al. 2001). Studies show that phosphate reduced the discharge of Zn, Pb and Cd (McGowen et al. 2001). Similarly, additional research presented the precipitation of Cr(III) by the addition of lime by enhancing the soil pH (Bolan et al. 2003). The existence of iron oxyhydroxides causes changes in the surface chemicals present on the substrate and often leads to co-precipitation of metal(loid)s (Bolan et al. 2014, McGowen et al. 2001). The Pb(II) precipitates at pH 4 by the effect of ferric oxyhydroxides with hydroxide chloride [Pb(OH)Cl], chloride ($PbCl_2$) and carbonate ($PbCO_3$), while reacting with Mg/Al in an aqueous solution with hydroxides (Violante et al. 2007). Usually, phosphate compounds are added to the soil to prevent heavy metal(loid) leaching (Bolan et al. 2014). Stability of metallic phosphates is found in the following order Pb > Cu > Zn (Bolan et al. 2003).

9.3.5 Biological Transformation

Metal(loids)s solubility can be enhanced by microbial processes (Krebs et al. 1997). Microbes raise the bioavailability that results in immobilization (Park et al. 2011). Solubilization of the metalloids by microbes is grouped into two categories: Autotrophic (chemolithotrophic) and heterotrophic (chemoorganotrophic) (Krebs et al. 1997). Immobilization via microbes of metals could possibly be brought into the framework by the means of reduction, precipitation, biosorption, accumulation, sequestration and localization (Gadd 2010). Metal is removed through adsorption when metal(loid)s are present in microbial excreta and their derivative products (Park et al. 2011). Metallothioneins, the cysteine-rich polypeptides bind metal(loid)s. Microbial exopolymers, metal-thiolate clusters, cysteine-containing-glutamyl peptides and phytochelatins are associated with the reduction of metal-(loid) binding and detoxification (Cobbett and Goldsbrough 2002, Park et al. 2011). The bacterial species *Micrococcus luteus* and *Azotobacter* sp. immobilized Pb, i.e., 490 mg g^{-1} and on a dry weight basis about 310 mg Pb g^{-1} on (Tornabene and Edwards 1972). Other groups of researchers found that Sulfate-Reducing Bacteria (SRB) effectively remove Zn from the medium. SRBs use phosphogypsum as a sulfate source for terminal electron acceptors for energy production.

Stenotrophomonas maltophilia reduced the toxicity of Se(III) from the legume rhizosphere of *Astragalus bisulcatus* (Gregorio et al. 2005). The plant roots have a significant role in the chemistry of the metalloids and soil environment (Tao et al. 2004). Metal(loid)s removal from contaminated soils by plant roots is an emerging eco-friendly remediation technology (Ernst 1996). This bio-transformation technology in soil depends on pH changes, microbial activity, phytochelatins, metal binding by root exudates and plant uptake via roots. Plant root exudates form complexes with organic acid and metal(loid)s in soil (Koo et al. 2010).

9.4 Bioremediation Approaches of Metalloids

The hazardous metallic and metalloid contamination has appeared as a critical problem of unease with the development of industrial, and agricultural areas with a rising human population. Additionally responsible for the increasing amount of the metal(oids) are natural biogeochemical processes. Various kinds of bioremediation methods can be employed to remediate metalloids. Bioremediation, an ideal method at recovery, is a process that employs microorganisms, fungi and higher plants to decontaminate or degrade the pollutants in the environment through the processes of biodegradation, biotransformation and biodeterioration. The addition of fungi and bacteria enhances the bioremediation process to reduce pollutants to environmentally acceptable and legally permissible levels (Pointing 2001). When the crop is grown in afflicted areas, metallicoids can enter the body directly by drinking contaminated water or indirectly through the food chain.

9.4.1 Industrial Wastewater

Industrial effluents are the major sources contributing harmful contaminants to the environment. Many metalloid industries situated near the coastal areas discharge effluents to shore water bodies. Such harmful compounds generated in the surroundings also cause various harmful effects on living organisms throughout the food chain. Printing and dyeing industrial wastewater need to be treated before discharge. Dyeing, rubbing and desizing are the major sources of water contamination of effluents from textile industries. The residual dye-containing waters are characterized by visibly intense color, huge COD, suspended solids and alkaline pH (Anjaneya et al. 2011). Thus, effluents from these industries create a major concern for the environment. Bioremediation of textile wastewater has been done using potential organisms like *Brevibacillus chashiuensis*, *Bacillus subtilus*, *Pseudomonas* species, *B. cereus*, *Micrococcus* species and *B. mycoide* (Mahmood et al. 2013, Durve et al. 2012). Dubey et al. (2019) found that *Phormidium mucicola* has a maximum potential of biosorption of 86.12 and 94.63% removal abilities for zinc and copper. Hence, to some extent, with the aid of biodegradation, it is possible to mitigate the metal(oids) contamination in the environment because of the increasing growth of industries that pressurize society and all natural resources. Cyanobacteria have demonstrated a remarkable potential in industrial effluent treatment, aquatic and terrestrial habitat bioremediation, chemical effluent detoxifying, fertilizing ability, as a fuel alternative and as a protein source in the food industry. Some species of cyanobacteria viz. *Nodularia species*, *Cyanothece species*, *Synechococcus species* and *Oscillatoria species* have great biosorption and biodegradation abilities in industrial wastewater (Kumar et al. 2011).

9.4.2 Soil Contamination

Soil contamination is on the rise, which is catalyzed by the modern development of industry, chemical fertilizer uses on large-scale and pesticides, etc. The demand for food is increasing continuously, which is stressing the already contaminated soil for growing crops. Hence, there is a major need for the decontamination of soil, which can be brought about physio-chemically and biologically (Jin et al. 2021). But, increasing the quantity of various substances in the soil and waters because of the outbreak of industries has generated harrowing conditions for people and water bodies (Dixit et al. 2015). The plant-based techniques have an excellent capacity to degrade, extract and stabilize the contaminants in phytoremediation, which has become evident as a great alternative in respect of cost and is eco-friendly. Metals are introduced in underground water aquifers through soil by water flow, these metals are introduced into soil by anthropogenic activities (Shabir Hussain 2012).

Microbial remediation techniques make use of microorganisms with unique functions that help in reducing the pollutants by degrading them, by converting them into non-toxic substances with their metabolism under environmentally adapted conditions. Some of the limitations in microbial remediation are that the microbes mutate easily due to poor genetic stability and are not able to remove the pollutants completely. The inability to compete with the indigenous strain is a significant need that easily affects their performance (Rudakiya et al. 2019). About 72 species of acidophilic thermophilic species have been isolated by primary and secondary screening for resistance to heavy metal concentrations and their ability to biosorption (Umrania 2006). Phosphate-dissolving bacteria can detoxify metalloids (Saranya et al. 2018, Li et al. 2016). Plants convert pollutants into non-toxic forms (Jin et al. 2021). *Alfalfa* plants possess potentially reduced metalloids (Agnello et al. 2016).

9.5 Bioremediation Techniques for Heavy Metal(oid)s Removal Enhancement

Inorganic metals/metalloid pollutants like Cu, Hg, Zn, As, Cd, Mn and Se appear in the environment mainly as cations and anions rely only on the plant vascular system for their translocation and uptake (Dhankher et al. 2012). Inorganic pollutants are therefore changed (reduced/oxidized), transported within plants and volatilized (Se, Hg) in a few instances, but they cannot be treated as such. Many bioremediation techniques, such as microorganism-based techniques, have demonstrated their potential for degrading and detoxifying many organic as well as inorganic pollutants. In comparison to other traditional approaches, biological systems are barely resistant to environmental extremes; hence they have an edge over other approaches as they are less expensive (Cunningham and Ow 1997).

Plant-based remediation solutions, also known as phytoremediation, have sparked increased interest because they are potentially more cost-effective, have low adverse effects and are ecologically sound (Cunningham and Ow 1997). Metal ions are taken up from the root and delivered to the above-ground components through the shoot system during phytoremediation, where they concentrate. The components of the plant are harvested, and so is the metallic build-up, leading to the elimination of pollutants (i.e., metals) from the site (Nandakumar et al. 1995). Plants have demonstrated the capability to survive predominantly at higher levels of metal contaminates and organics toxicity by quickly converting them into less toxic metabolites in many circumstances. This can be accomplished through phytoextraction (the intake and recovery of metallic pollutants in the form of above-ground biomass), rhizofiltration (the filtering of metals in root systems) or phytostabilization (the stabilization of waste sites through erosion control and large-scale evapotranspiration), among other methods (Cunningham and Ow 1997). The phytoremediation procedures are not mutually exclusive, and they can be employed in tandem for greater effectiveness and efficiency. Phytostabilization, phytoaccumulation in harvest-worthy plant tissues (phytoextraction or rhizofiltration) in rare situations, and phytovolatilization are among the phytoremediation strategies available for inorganics. However, bioremediation techniques for

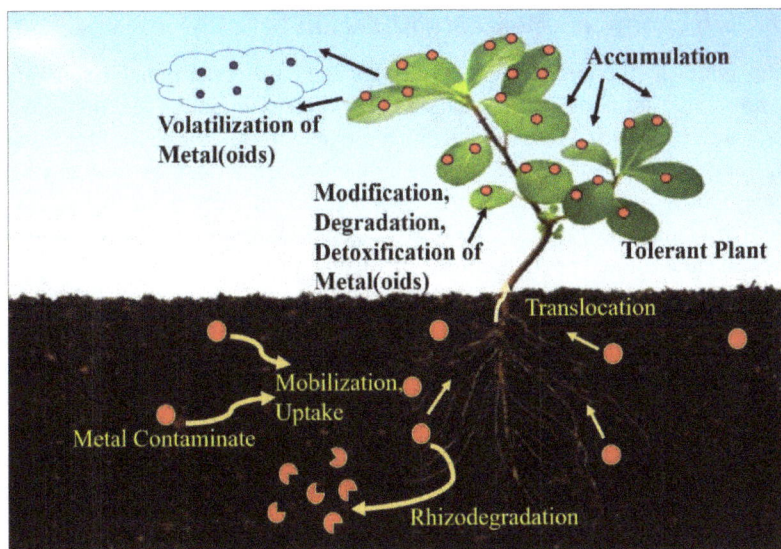

Figure 9.1. A description of the interactions between plants and metal contamination as well as potential outcomes for the metal contaminants (Ojuederie and Babalola 2017).

metal(oid) removal also have some shortcomings. These include the requirements for nutritional resources, certain climatic conditions, and appropriate soil properties for normal plant growth (Karami and Shamsuddin 2010). The most significant disadvantages of phytoextraction are the lengthy time required, which has hampered the widespread implementation of phytoremediation. An overview of plant-metal contaminants interaction and the possible fate of the metal contaminants is shown in Figure 9.1.

9.5.1 Valence Reduction/Oxidation

The mechanism involved in the reduction of Se, Hg, Cr, and As, among other metal(loid)s, was developed as a result of the oxidation/reduction processes carried out by beneficial microorganisms. Metal(loid) speciation and mobility are influenced by redox processes. As(III) gets converted to As(V) by microorganisms in sediments and soils (Bachate et al. 2012, Battaglia Brunet et al. 2002). As(V) has a strong attraction for inorganic soil elements, it gets immobilized following its oxidation. In well-drained soils, As(V) is the dominant form of As, whereas in poorly drained soils, As(III) predominates, however arsine gas (H_2As) [As(0)] and elemental arsenic can also be found. In most cases, the breakdown of organic materials by bacteria, the reduction is followed by the role of SO_4^- as the terminal e^- acceptor, and then reduction, mediates the reduction and methylation processes in sediments (Kim et al. 2002). In the instance of Cr, its mobility and bioavailability improve with its oxidation into Cr(VI). Oxidizing agents like Mn(IV), Fe(III) to a small extent, while the Cr(VI) - Cr(III) reduction is mediated by mechanisms that are both abiotic and biotic (Choppala et al. 2015). The settings having an accessible available electron source Fe(II), chromate Cr(VI) can get reduced into Cr(III). When organic fraction serves as an e^- donor, reduction via microbial Cr(VI) is boosted, while a marked enhancement of Cr(VI) reduction can be observed under acidic conditions (Choppala et al. 2015, Hsu et al. 2009). The metal(loid)s are usually reduced rather than oxidized in most biological systems. Chemical reductants like sulfide or hydroxylamine, or glutathione reductase biochemically reduce Se (Zhang et al. 2004). Microorganisms have a major bearing in converting the reactive Hg(II) species into its non-reactive counterpart Hg(0), which is susceptible to losses by volatilization. Mercuric reductases have been known to reduce Hg(II) into Hg(0), and the bacteria *Shewanella oneidensis,* which carries out dissimilatory metal(loid) reduction, has been demonstrated to reduce Hg(II) into Hg(0) when electron donors are present (Wiatrowski et al. 2006).

9.5.2 Immobilization or Phytostabilization

Immobilization refers to the vegetation's ability to maintain contaminated soils and sediments in place, as well as the immobilization of harmful pollutants in soils (Mukhopadhyay and Maiti 2010). *In-situ* inactivation or phyto immobilization are other terms for phytostabilization. Sorption, precipitation, complexation and metal valence reduction are strategies that can help plants to stabilize metals (Ghosh and Singh 2005). An *in-situ* remediation process is a suitable option for metal remediation (Jadia and Fulekar 2009). Root-zone microbiology and chemistry, as well as changes in the soil environment or contaminant chemistry, all, contribute to phytostabilization. Plant root exudates or CO_2 generation can modify the pH of the soil, that has a bearing on metal ion transport. Phytostabilization can affect metal solubility and mobility, as well as organic compound dissociation. Metals can be converted into an insoluble oxidation state from a soluble state in a plant-affected soil environment (Salt et al. 1995). Plants can also help to prevent metal-contaminated soil from eroding. Plants having elevated transpiration rates, like forage plants, grasses, reeds and sedges, can help with phytostabilization that could be used for metal remediation. The approach of utilizing trees, such as the densely rooted and perennial, in combination may be a good mix (Berti and Cunningham 2000).

9.5.3 Phytovolatilization

Toxic metals, including mercury, selenium and arsenic, are capable of being biomethylated to produce volatile substances that are discharged into the atmosphere. Phytovolatilization is the mechanism which is involved in the reduction of pollutants via the transpiration of plants. The plant absorbs the contaminant that is present in the water, passes through it, or undergoes transformation there, then is released into the atmosphere. Water passes through the plant's internal transport mechanism circulating from roots up to the leaves, where the inorganics get evaporated or volatilized and consequently released in the air enclosing the plant, potentially modifying the contaminant. Using phytovolatilization and phytoextraction to remove metals from commercial projects is a realistic option (Sakakibara et al. 2010). Tritium (3H), a radioactive isotope of hydrogen, has been successfully phytovolatilized; it decomposes into a stable form of helium, having a half-life of about 12 yr. HMs can be absorbed by several plants, including *Arabidopsis thaliana*, *Chara canescens* and *Brassica juncea*, converting them into their gaseous forms within plants, and then releasing them back into the environment (Ghosh and Singh 2005). Dimethylselenides and dimethyldiselenides are produced by plants (i.e., *Brassica juncea* and *Arabidopsis thaliana*), which are the volatile forms of volatile Se when grown on a high Se medium. Similarly, data from a study on heavy metal volatilization revealed that *P. vittata* is quite efficient at volatilizing Arsenic (As), as had been documented by its removal by almost 90% of total intake from As-affected soils in a greenhouse with subtropical conditions (Sakakibara et al. 2010). In contrast to the other ways of cleanup, after toxins have been removed via volatilization, they cannot be stopped from spreading to other areas. Similar occurrences of soil remediation based on volatilization have been recorded in many other publications (Tangahu et al. 2011, Conesa et al. 2012). Although it is well recognized that microbes dispense a significant function in the Se volatilization from soil systems (Karlson and Frankenberger 1989), it was investigated that plants can fulfil the same job. *B. juncea* has been recognized as a useful source for extracting Se from soils (Bauelos and Meek 1990, Baualos et al. 1993). The Se volatilization into methyl selenates has been hypothesized as a dominant mechanism for plant Se elimination (Zayed and Terry 1994, Terry et al. 1992). Non-volatile methyl selenate derivatives accumulate in the leaf of some plants, allowing them to extract Se from the soil. In the Se accumulator *Astragalus bisculatus*, the enzyme that serves in producing methyl selenocysteine was isolated and described (Neuhierl and Bock 1996).

9.5.4 Phytoextraction

Phytoextraction, better known as phytomining or phytoaccumulation, is the practice of raising a certain crop that has been known for collecting toxins within its shoot system and leaves (hyper accumulator or tolerant plant), harvesting them followed by elimination of contaminants from the affected areas. Unlike destructive degrading mechanisms, this process produces a consolidated plant and contaminate (mostly metallic) mass that must be disposed of or recycled. Correlated to landfilling and excavation, this technology is based on the concentration of pollutants which leaves a significantly smaller bulk of pollutants to be disposed off (Wani et al. 2012). Chelation is the process by which soil-borne metal pollutants are transported by roots to tissues. By translocation from roots to stems and leaves, metal(oids) are removed from the soil in Figure 9.2.

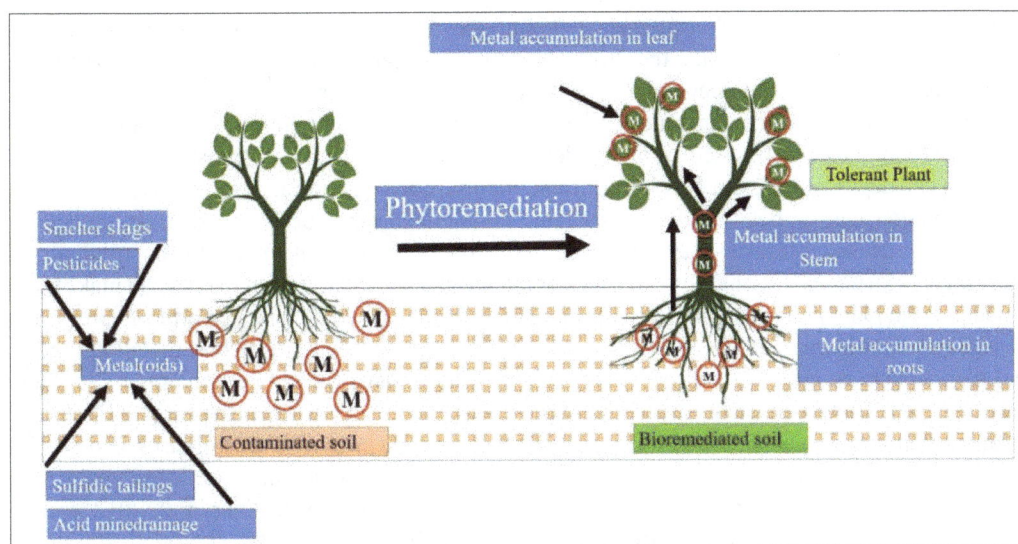

Figure 9.2. Metals uptake by Phytoextraction: Metal(oids) are eliminated from the soil by translocation into plant roots, stem, and leaves (Mishra et al. 2017).

A lot of studies have been undertaken on phytoextraction, including novel phytoextractors (Baker and Brooks 1989), for a better understanding of plant-metal absorption, metabolism, transport, excretion and tolerance (Krämer et al. 1996, Rauser 1995, Salt et al. 1999, Lasat et al. 1998) and genetic changes to improve plant metal accumulation and tolerance (Rauser 1995, Krämer et al. 1996, Kärenlampi et al. 2000, Clemens et al. 2002, Krämer 2005). An alternative option to advance this concept was to combine plants producing high biomass with chemical agents that boost the solubility of metals for plant uptake (Blaylock et al. 1997, Huang et al. 1997). Hyper-accumulators are plants that abnormally uptake metals. Over 400 varieties of plants have been found to improve the quality of soil and water (Lone et al. 2008). Since distinct plants have distinct capacities to absorb significant levels of contaminants, a variety of plants can be introduced for the phytoremediation process to see their effectiveness. One of the methods utilized to create a phytoremediation plant is the genetic engineering of improved hyperaccumulators. *Arabidopsis* sp., *Sedum alfredii* sp., and *Thlaspi* sp. are the hyper accumulators that have been examined the most by scientists (*Brassicaceae* and *Alyssum* are members of the *Brassicaceae* family) (Milner and Kochian 2008).

9.5.5 Modification of Rhizosphere

Rhizosphere bioremediation uses microbes accompanied by roots' remarkable abilities to break down organic contaminants and convert hazardous metals. As this plant-based approach is an *in-situ* photo restoration method, it is cost-effective, efficient and simple to implement in the field (Kumar and Fulekar 2018). The toxicity of metals could be reduced by volatilization, phytoextraction and the degradation process. Metals should usually be physically removed or immobilized, whereas organic substances can be destroyed. Rhizoremediation has been documented to be a virtuous technique along with various strategies like microbial augmentation as well as transgenic approaches (Kumar and Fulekar 2018).

9.5.5.1 Transgenic Technology in Bioremediation

Proteomic, metabolomic, transcriptomic, genomic and metagenomic approaches are being used to determine traits that maximize the utility of on-ground harvesting techniques and control the resistance, degradation and accumulation potential of plants and microorganisms to a diversity of inorganic and organic contaminates by recent advances in omics technologies. Transgenic and cisgenic techniques can be used to manipulate potential plant species to boost pollutant intake, transport and degradation, plant development and vitality, root formation and abiotic stress resilience. Organic matter can be detoxified and inorganic contaminants can be enriched using transgenic plants (Maestri and Marmiroli 2011). The production of genes that break down pollutants in prospective bioenergy systems aims to reduce organic contamination of plant tissues and make plant products from plant protection plantations more accessible. Selected metal transporters expressed in transgenic plants increase sulfur metabolism output. Chelators that detoxify metals, metallothioneins and phytochelatins, can also improve heavy metal uptake, transport and accumulation (Ruiz and Daniell 2009). Transgenic plants expressing three microbial reductases can also promote Hg and Se volatilization as well as arsenic accumulation in plant shoots. There are also several cases of microbial genes being successfully incorporated into plant tissue for better biodegradation. These genes code aid in transporting metals and the breakdown of organic pollutants (Dhankher et al. 2012, Iimura et al. 2007, Che et al. 2003, Bittsanszkya et al. 2005, Van Dillewijn et al. 2008, Doty et al. 2000). For instance, to decontaminate explosives (TNT and RDX), transgenic plants were created employing microbial pollutant-degrading genes (Iimura et al. 2007, Che et al. 2003, Bittsanszkya et al. 2005, Van Dillewijn et al. 2008). Plants that grow fast and generate a lot of biomasses, including jatropha, poplar and willow, could be used in two ways- for energy production and phytoremediation.

9.5.5.2 Designer Plant Approach

Organic pollutants are stored and built up in plant tissues, reducing the plant's lifespan while negatively impacting the environment by volatilizing through the leaves, that is one of the most critical limitations of phytoremediation. To address this issue, degrading microorganisms are put on plant tissues before their transfer to the polluted site, allowing pollutants to degrade in plant tissues (Aken et al. 2011). In addition, customizing plant-microbial interactions for specific applications is a new method for pollutant targeting in complex ecosystems (Abhilash et al. 2012).

Phytoremediation will need to be adopted in combination with other multipurpose remediation techniques in the future (i) conventional biotechnology and (ii) stressors, e.g., nutrient deficiency and location toxicity can be mitigated with integrated bioaugmentation employing advantageous microorganisms. Such multipurpose remediation systems, once applied, have the potential to change the remediation domain by providing environmental, economic as well as social advantages to all the stakeholders involved. Though all phytoremediation technologies are constrained by economic viability, plant species that provide additional benefits have a better chance of resolving this problem soon (Abhilash et al. 2012). The remediation of metal(loid)s by the combination of plants and microbial species are depicted in Table 9.2.

Table 9.2. Remediation of metal(loid)s by the combination of plants and microbial species.

Metals	Plant species	Microbial Species	Mechanism of removal	References
Zn	*Brassica juncea*	*Pseudomonas brassicacerarum, Rhizobium leguminosarum*	Organic acid secretion and metal chelation enhancement by phytochelatins	Adediran et al. 2015
	Cicuta virosa	*Rhodopseudomonas* sp., *Pseudomonas putida*	Siderophores and indole-3-acetic acid production	Nagata et al. 2015
Ni	*Alyssum pintodasilvae*	*Arthrobacter nicotinovorans*	Siderophores and organic acid production	Cabello-Conejo et al. 2014
	Helianthus annuus	*Pseudomonas libanensis* and *Claroideoglomus claroideum*	Solubilization of phosphate, ni phytostabilized by exopolysaccharide (EPS)	Ma et al. 2019
Al	*Miscanthus sinensis*	*Chaetomium cupreum*	Production of siderophore (oosporein)	Haruma et al. 2019
Cd	*Ocimum ratissimum*	*Arthrobacter* sp.	Production of EPS	Prapagdee and Khonsue 2015
Zn, Cu and Pb	*Clethra barbinervis*	*Clethra barbinervis, Rhizodermea veluwensis*	Melanin and siderophore production	Yamaji et al. 2016
Zn, Cd and Pb	*Sedum plumbizincicolaa*	*Endophytic bacterium E6S*	Organic acid production, Aminocyclopropane-1-carboxylate (ACC) and Indole-3-acetic acid (IAA) production, phosphate solubilization,	Ma et al. 2016
As, Cd, Cu, Pb and Zn	*Miscanthus sinensis*	*Pseudomonas koreensis AGB-1*	Aminocyclopropane-1-carboxylate (ACC) deaminase	Babu et al. 2015
Cr(VI), Fe, Mn, Zn, Cd, Cu and Ni	*Vetiveria zizanioides*	*Bacillus cereus*	Production of IAA, ACC, solubilize phosphate and production of ACC	Nayak et al. 2018
Cd and Pb	*Simplicillium chinense*	*Phragmites communis*	Pb bio sorption by EPS and Cd chelate formation	Jin et al. 2019
Cd, Zn and Cu	*Solanum nigrum*	*Pseudomonas* sp. *Lk9*	Organic acids and siderophore and biosurfactant production	Chen et al. 2014
Zn, Cd and Pb	*Salix dasyclados*	*Streptomyces* sp.	Siderophore production	Złoch et al. 2017

9.6 Case Studies on Bioremediation of Metal(oids)

9.6.1 Arsenic

Microorganisms have evolved arsenic defense systems because of arsenic's ubiquitous presence in the environment. The arsenic resilience operon's existence (ars) encodes enzymes degradation of arsenate in an exhaustive study undertaken in Leon, Spain (Mateos et al. 2006). It was noted that as arsenate gains entry into the cell via specialized (Pst) or nonspecific (Pit) phosphate transporters, its incorporation in phosphate-rich fluids or environments is lower. When *C. glutamicum* is cultivated in a low-phosphate medium or when the concentration of phosphate is reduced by precipitation, As(V) absorption increases, followed by disabling the three arsenite permease genes found in *C. glutamicum*'s genome, arsenite outflow was minimized. *C. glutamicum* ArsB1-B2, a double arsenite permease mutant, was particularly sensitive to arsenate and arsenite (Ordóñez et al. 2005). Ordóñez et al. (2005) successfully created *C. glutamicum* strains using omics approaches that may horde heavy metals outside the cell (as a biosorbent), similar to *E. coli* and *Ralstonia eutropha* (Kotrba et al. 1999) and *Ralstonia eutropha* (Valls et al. 2000).

9.6.2 Cadmium

In their analysis of 465 studies on the level of heavy metal contamination in Chinese agricultural soils (Zhang et al. 2015); the findings revealed that heavy metals polluted around 10.18% of arable soil, with cadmium (Cd^{2+}) having the highest contamination rate at 7.75%. Through the food chain and the soil-crop system, Cd^{2+} causes a health risk to humans (Nogawa and Kido 1996). According to research, Cd^{2+} causes lesions on bone, renal problems and pulmonary inadequacy in people (Sharma 1995, Chakravarty et al. 2010). As a result, steps should be taken to reduce or neutralize its negative effects in Cd-contaminated soil. To treat heavy metal-affected soils, microbial bioremediation is a promising approach. Based on its 16s rRNA sequence and metabolic profile, a study was undertaken in Wuhan, China (Xu et al. 2019). In which *Raoultella* sp. strain X13 was discovered to be a Cd-resistant bacterium that was isolated from a heavy metal-affected soil environment in southern China, demonstrated Cd^{2+} biosorption and bioaccumulation abilities and depicted tolerance of up to 8 mM Cd in LB medium. As a result, strain X13 has been proven to be used as a possible Cd^{2+} immobilizing substance in contaminated soil since it promotes plant development as well as Cd^{2+} fixation. In the case of Cd^{2+} pollution, many studies have shown that metal-resilient PGP microbes may promote plant development, improve heavy metal stability and raise plant resistance to heavy metals (Das et al. 2014).

9.6.3 Mercury

Mercury pollution has become a global problem, more precisely in the soil environment. It is a big problem in mercury mining locations, artisanal gold smelting areas and chlorine industrial plants, among other places, resulting in a lot of contamination of mercury (Driscoll et al. 2013).

The mercury mining regions of Wuchuan and Wanshan in China (Qiu et al. 2006, Li et al. 2008), Almadén, a mercury mining zone in the Amazon, are located in the heart of the region, these regions are mercury-prone areas (Higueras et al. 2006). In Dehua, Fujian Province, China, small-scale and artisanal gold mining activities occur, and an abandoned gold mining and smelting area (United Nations Environment Programme 2013) are all examples of mercury-polluted locations that cause serious ecological and environmental risks. Remediating mercury-polluted sites is an imminent but pressing task. In an investigation carried out in Beijing, China (Xun et al. 2017), *Cyrtomium macrophyllum* was naturally growing in the soil mercury region. *Cyrtomium macrophyllum* demonstrated remarkable mercury accumulation and translocation capabilities (Xun et al. 2017).

9.7 Future and Prospects

Rapidly increasing industrialization and rampant development have their share of negative impacts, such as contamination of soil and damage to soil health. The potential threat to living organisms due to the heavy metals contamination in soil and water cannot be overlooked. Many techniques or methods have served to reduce the risk, including many biological methods. The bioremediation method can be used ideally for the reduction of toxicity aroused by metal contaminants in soil. It is a safe and accommodating technique based on microbes present in soil essentially and does not cause harmful hazards in the habitat and organisms surviving in that region (Tarekegn et al. 2020). However, there are some limitations in the bioremediation methods, all microorganisms are not capable of removing toxic substances and this process hinders the effects of the microbes' activity (Dixit et al. 2015). Heavy metals are deleterious to living beings as it is followed by accumulation in tissues of the body that persist in the organs like kidneys, liver and bones for a longer time (Kapahi and Sachdeva 2019).

Conventional techniques for treatment have many limitations and would be replaced by low cost, efficient and eco-friendly methods like nature base solutions. To overcome the shortcomings, the need of the hour is to utilize advanced techniques like bioremediation that either chemically converts these compounds into less harmful products or degrades them together. Bioremediation techniques have indefinite use, like the production of biodiesel, bioethanol, CO_2 fixation, heavy metals pollution controls and so on. And it is also favorable to treat wastewater and polluted land because the contaminated site is highly nutrient-rich for some organisms like *Ideonella sakaiensis*, a plastic waste degrading bacterium, *Sulfurospirillum arsenophilum*, *Bacillus arsenicoselenatis*, *Chrysiogenes arsenatis* and Archaea (*Pyrobaculum arsenaticum*, *Pyrobaculum aerophilum*) used for arsenic pollution control of polluted water (/Kudo et al. 2013). Species like *Brassica napus* can even control soil heavy metals pollution by cadmium, cobalt and nickel (Boros-Lajszner et al. 2021).

Bioremediation is not a new technique to treat pollution load, but it is a slow process. But from the future prospective, highly efficient bioremediation techniques need to be developed by advanced research with the help of bioinformatics, biostatics analyses and other omics approaches helping to understand the metabolism of microorganisms efficiently. Certain significant aspects that should be taken care of include: (1) net reduction of pollution in actual and laboratory conditions, that requires better equipment handling, patience and regular monitoring; (2) studying the possible contributing factors (abiotic and biotic); and (3) the remediation technology should be cost-effective, reliable and rapid. Moreover, the endeavor should be made to integrate the phytoremediation method with bio-energy for the two-fold utilization of plants for phytoremediation and bio-fuel generation on polluted lands. These methods should be beneficial to the phytoremediation of polluted regions and concurrently generate sustainable power that can balance the prices that bear on this kind of methodologies (Mosa et al. 2016).

9.8 Conclusion

Being an eco-friendly technology for heavy metals remediation and organic contaminants, phytoremediation has a lot of promise. Rhizospheric bacteria and plants have shown the capability of detoxifying and converting organic pollutants into harmless compounds that may be eliminated from the soil without accumulating. The above-ground biomass can carry toxic metals, and plants can then detoxify, translocate, accumulate and recover heavy metals like lead. Although phytoremediation has excellent potential to be applied to contaminant removal from water, sediment and soil, it has not been widely commercialized or implemented on a broad basis. Field implementation of phytoremediation has still not been studied A lack of comprehensive insight into the uptake mechanism of metals from the soil system to the roots has hampered the commercialization of phytoremediation for heavy metals and metalloids. Several recent studies have focused on transcriptomic and proteomic techniques for metal accumulation in plants and elucidating heavy metal transport. This chapter summarizes the efficient, environmentally sound and cost-efficient methods for removing metal(oids) from the environment and protecting the ecology.

Acknowledgment

All the authors listed in the chapter have contributed substantially, directly and intellectually to the work and have approved its publication. The UGC, GOI, New Delhi, is sincerely appreciative of its support of the NFSC fellowship on behalf of one of the authors, Kumar M.

References

Abhilash, P. C., J. R. Powell, H. B. Singh and B. K. Singh. 2012. Plant-microbe interactions: novel applications for exploitation in multipurpose remediation technologies. Trends Biotechnol. 30: 416–420.

Adediran, G. A., B. T. Ngwenya, J. F. W. Mosselmans, K. V. Heal and B. A. Harvie. 2015. Mechanisms behind bacteria induced plant growth promotion and Zn accumulation in *Brassica juncea*. J. Hazard. Mater. 283: 490–499.

Adlane, B., Z. Xu, X. Xu, L. Liang, J. Han and G. Qiu. 2020. Evaluation of the potential risks of heavy metal contamination in rice paddy soils around an abandoned Hg mine area in Southwest China. Acta Geochim. 39: 85–95.

Adriano, D. C., J. Albright, F. W. Whick, I. K. Islandar and C. Sherony. 1997. Remediation of soil contaminated with metals and radionuclide-contaminated soils. pp. 27–46. *In*: I. K. Islandar and D. C. Adriano (Eds.). Remediation of Soils Contaminated with Metals. Northwood, UK. SciRev-Review the scientific review process.

Adriano, D. C. 2001. Bioavailability of trace metals. *In*: Trace Elements in Terrestrial Environments Springer, New York.

Adriano, D. C., W. W. Wenzel, J. Vangronsveld and N. S. Bolan. 2004. Role of assisted natural remediation in environmental cleanup. Geoderma. 2-4: 121–142.

Agnello, A. C., M. Bagard, E. D. van Hullebusch, G. Esposito and D. Huguenot. 2016. Comparative bioremediation of heavy metals and petroleum hydrocarbons co-contaminated soil by natural attenuation, phytoremediation, bioaugmentation and bioaugmentation-assisted phytoremediation. Sci. Total Environ. 563: 693–703.

Alexander, M. 2000. Aging, bioavailability, and overestimation of risk from environmental pollutants. Environ. Sci. Technol. 34: 4259–4265.

Alloway, B. J. 2013. Sources of heavy metals and metalloids in soils. *In*: B. Alloway (Eds.). Heavy Metals in Soils: Environmental Pollution Springer, Dordrecht.

Ando, J. 1987. Thermal phosphate. pp. 93–124. *In*: F. T. Nielsson (Eds.). Manual of Fertilizer Processing, Fertilizer Science and Technology Series. New York, Marcel Dekker Inc.

Anjaneya, O., S. Y. Souche, M. Santoshkumar and T. B. Karegoudar. 2011. Decolorization of sulfonated azo dye Metanil Yellow by newly isolated bacterial strains: *Bacillus* sp. strain AK1 and *Lysinibacillus* sp. strain AK2. J. Hazard. Mater. 190(1): 351–358.

Anju, M. and D. K. Banerjee. 2003. Heavy metal levels and solid phase speciation in street dusts of Delhi, India. Environ. Pollut. 123: 95–105.

Anju, M. and D. K. Banerjee. 2010. Comparison of two sequential extraction procedures for heavy metal partitioning in mine tailings. Chemosphere. 78: 1393–1402.

Anju, M. and D. K. Banerjee. 2011. Associations of cadmium, zinc, and lead in soils from a lead and zinc mining area as studied by single and sequential extractions. Environ. Monit. Assess. 176: 67–85.

Baath, E. 1989. Effects of heavy metals in soil on microbial processes and populations (a review). Water Air Soil Pollut. 47: 335–379.

Babich, H. and G. Stotzky. 1985. Heavy metal toxicity to microbe-mediated ecologic processes: a review and potential application to regulatory policies. Environ. Res. 36: 111–37.

Babu, A. G., P. J. Shea, D. Sudhakar, I. BooJung and B. T. Oh. 2015. Potential use of *Pseudomonas koreensis* AGB-1 in association with *Miscanthus sinensis* to remediate heavy metal(loid)-contaminated mining site soil. J. Environ. Manage. 151: 160–166.

Bachate, S. P., Khapare, R. M. Kisan and M. K. Kodam. 2012. Oxidation of arsenite by two β-proteobacteria isolated from soil. Appl. Microbiol. Biotechnol. 93: 2135–2145.

Baker, A. and R. Brooks. 1989. Terrestrial higher plants which hyper accumulate metallic elements—a review of their distribution, ecology and phytochemistry. Biorecovery. 1: 81–126.

Battaglia-Brunet, F., M. C. Dictor, F. Garrido, C. Crouzet, D. Morin and K. Dekeyser. 2002. An arsenic (III)-oxidizing bacterial population: selection, characterization, and performance in reactors. J. Appl. Microbiol. 93: 656–6.

Bauelos, G. and D. Meek. 1990. Accumulation of selenium in plants grown on selenium-treated soil. J. Environ. Qual. 19: 772.

Bernd, L. 2007. Characterization, Treatment and Environmental Impacts. In Mine Wastes.: Springer, Berlin, Heidelberg.

Baualos, G. S., G. Cardon, B. Mackey, J. Ben-Asher, L. Wu, P. Beuselinck, S. Akohoue and S. Zambrzuski. 1993. Plant and environment interactions, boron and selenium removal in boron-laden soils by four sprinkler irrigated plant species. J. Environ. Quali. 22: 786–792.

Bittsanszkya. A., T. Kömives, G. Gullner, G. Gyulai, J. Kiss, L. Heszky, L. Radimszky and H. Rennenberg. 2005. Ability of transgenic poplars with elevated glutathione content to tolerate zinc (2+) stress. Environ. Int. 31: 251–254.

Blaylock, M. J., D. E. Salt, S. Dushenkov, O. Zakharova, C. Gussman, Y. Kapulnik, B. D. Ensley and I Raskin. 1997. Enhanced accumulation of Pb in Indian mustard by soil-applied chelating agents. Environ. Sci. Technol. 31: 860–865.

Bolan, N., A. Kunhikrishnan, R. Thangarajan, J. Kumpiene, J. Park and T. Makino. 2014. Remediation of heavy metal(loid)s contaminated soils-To mobilize or to immobilize. J. Hazard. Mater. 266: 141–166.

Bolan, S., C. Adriano and R. Naidu. 2003. Role of phosphorus mobilization and bioavailability of heavy metals in the soil-plant system. Rev. Environ. Contam. Toxicol. 177: 1–44.

Boros-Lajszner, E., J. Wyszkowska, A. Borowik and J. Kucharski 2021. The response of the soil microbiome to contamination with cadmium, cobalt and nickel in soil sown with *Brassica napus*. Minerals. 11: 498.

Bradl, H. 2004. Adsorption of heavy metal ions on soils and soils constituents. J. Colloid Interface Sci. 277: 1–18.

Cabello-Conejo, M. I., C. Becerra-Castro, A. Prieto-Fernández, C. Monterroso, A. Saavedra-Ferro, M. Mench and P. S. Kidd. 2014. Rhizobacterial inoculants can improve nickel phytoextraction by the hyperaccumulator *Alyssum pintodasilvae*. Plant and Soil. 379: 35–50.

Chakravarty, P., N. S. Sarma and H. P. Sarma. 2010. Biosorption of cadmium(II) from aqueous solution using heartwood powder of *Areca catechu*. Chem. Eng. Sci. 162: 949–9.

Chandra, R., V. Kumar and S. Tripathi. 2017. Phytoremediation of industrial pollutants and life cycle assessment. pp. 14–35. *In:* R. Chandra and DNK, KV [eds.]. Phytoremediation of Environmental Pollutants.: CRC Press (Taylor & Francis Group), USA.

Chaney, R. L. and J. A. Ryan. 1994. Risk based standards for arsenic, lead and cadmium in urban soils. In Summary of information and methods developed to estimate standards for Cd, Pb and As in urban soils. Germany.

Che, D., R. B. Meagher, A. C. P. Heaton, A. Lima, C. L. Rugh and S. A. Merkle. 2003. Expression of mercuric ion reductase in Eastern cottonwood (*Populus deltoides*) confers mercuric ion reduction and resistance. Plant Biotechnol. J. 1: 311–319.

Chen, L., S. Luo, X. Li, Y. Wan, J. Chen and C. Liu. 2014. Interaction of Cd-hyperaccumulator *Solanum nigrum* L. and functional endophyte *Pseudomonas* sp. Lk9 on soil heavy metals uptake. Soil Biol. Biochem. 68: 300–308.

Choppala, G., N. Bolan, A. Kunhikrishnan, W. Skinner and B. Seshadri. 2015. Concomitant reduction and immobilization of chromium in relation to its bioavailability in soils. Environ. Sci. Pollut. Res. 22: 8969–8978.

Clemens, S., M. G. Palmgren and U. Krämer. 2002. A long way ahead: understanding and engineering plant metal accumulation. Trends Plant Sci. 7: 309–315.

Cobbett, C. and P. Goldsbrough. 2002. Phytochelatins and metallothioneins: roles in heavy metal detoxification and homeostasis. Annu. Rev. Plant Biol. 53: 159–182.

Conesa, H. M., M. W. Evangelou, B. H. Robinson, S. R. Conesa, H. M. Evangelou, M. W. Robinson, B. H. and R. Schulin. 2012. Critical view of current state of phytotechnologies to remediate soils: still a promising tool? Sci. World J. 1–10.

Cunningham, S. and D. Ow. 1997. Promises and prospects of root zone of crops. Phytoremediation. Plant Physiol. 110: 715–719.

Cunningham, S. and W. Berti. 2000. Phytoremediation of toxic metals: using plants to clean up the environment. Phytostabilization of Metals. pp. 71–88.

Das, S., J. S. Jean, S. Kar, M. L. Chou and C. Y. Chen. 2014. Screening of plant growth-promoting traits in arsenic-resistant bacteria isolated from agricultural soil and their potential implication for arsenic bioremediation. J. Hazard. Mater. 272: 112–120.

Day, J. P., J. E. Fergusson and T. M. Chee. 1979. Solubility and potential toxicity of lead in urban street dust. Bull. Environ. Contam. Toxicol. 23: 497–502.

Dhankher, O. P., E. A. Pilon-Smits, R. B. Meagher and S. Doty. 2012. Biotechnological approaches for phytoremediation. Plant Biotechnology and Agriculture, 309–328.

Dixit, R., M. D. Wasiullah, K. Pandiyan, U. Singh and A. Sahu. 2015. Bioremediation of heavy metals from soil and aquatic environment: an overview of principles and criteria of fundamental processes. Sustainabiliity. 7: 2189–2212.

Doty, S. L., T. Q. Shang, A. M. Wilson, J. Tangen, A. D. Westergreen, L. A. Newman, S. E. Strand and M. P. Gordon. 2000. Enhanced metabolism of halogenated hydrocarbons in transgenic plants containing mammalian cytochrome P450 2E1. PNAS. 6287–6291.

Douay, F., C. Pruvot, H. Roussel, H. Ciesielski, H. Fourrier and N. Proix. 2008. Contamination of urban soils in an area of northern france polluted by dust emissions of two smelters. Water Air Soil Pollut. 188: 247–260.

Driscoll, C. T., Mason, H. M. Chan, D. Jacob and N. Pirrone. 2013. Mercury as a global pollutant: 543 sources, pathways, and effects. Environ. Sci. Technol. 47: 4967–4983.

Dubey, S. K., P. Vyas, P. Tiwari, A. J. Viswas and S. P. Bajpai. 2019. Bioremediation of industrial effluent using cyanobacterial species: *Phormidium mucicola* and *Anabaena aequalis*. Annu. Res. Rev. Biol., 1–8.

Dudka, S. and D. C. Adriano. 1997. Environmental impacts of metal ore mining and processing: a review. J. Environ. Qual. 26: 590–602.

Duggan, M. J., M. J. Inskip, S. A., Rundle and J. S. Moorcroft. 1985. Lead in playground dust and on the hands of schoolchildren. Sci. Total Environ. 44: 65–79.

Durve, A., S. Naphade, M. Bhot, J. Varghese and N. Chandra. 2012. Characterisation of metal and xenobiotic resistance in bacteria isolated from textile effluent. Adv. Appl. Sci. Res. 3: 2801–2806.

Ernst, W. H. O. 1996. Bioavailability of heavy metals and decontamination of soils by plants. J. Appl. Geochem. 11: 163–167.

Frankenberger, W. T. and M. E. Losi. 1995. Applications of bioremediation in the cleanup of heavy metals and metalloids. In bioremediation: science and applications. SSSA Special Publications, 173–210.

Gadd, G. 2010. Metals, minerals and microbes: geomicrobiology and bioremediation. Micro Soc. 156: 609–643.

Geller, W., H. Klapper and M. Schultze. 1998. Natural and anthropogenic sulfuric acidification of lakes. *In*: W. Geller, H. Klapper and W. Salomons [eds.]. Acidic Mining Lakes: Acid Mine Drainage, Limnology and Reclamation. Berlin, Heidelberg: Springer.

Ghosh, M. and S. P. Singh. 2005. A review on phytoremediation of heavy metals and utilization of it's by products. Appl. Ecol. Environ. Sci. 3: 1–18.

Giller, E., E. Witter and S. P. McGrath. 2009. Heavy metals and soil microbes. Soil Biol. Biochem. 41: 2031–2037.

Gove, L., C. M. Cooke, F. A. Nicholson and A. J. Beck. 2001. Movement of water and heavy metals (Zn, Cu, Pb and Ni) through sand and sandy loam amended with biosolids under steady-state hydrological conditions. Bioresour. Technol. 78: 171–9.

Gregorio, S. D., S. Lampis and G. Vallini. 2005. Selenite precipitation by a rhizospheric strain of *Stenotrophomonas* sp. isolated from the root system of *Astragalus bisulcatus*: a biotechnological perspective. Environ. Int. 31: 233–241.

Gupta, P. and V. Kumar. 2017. Value added phytoremediation of metal stressed soils using phosphate solubilizing microbial consortium. World J. Microbiol. Biotechnol. 9: 33.

Hanlie, H., F. Zhengyi and M. Xinmin. 2001. The adsorption of [Au(Hs)2]—On Kaolinite Surfaces: Quantum Chemistry Calculations. Canad Mineral. 39: 1591–1596.

Haruma, T., K. Yamaji, K. Ogawa, H. Masuya, Y. Sekine and N. Kozai. 2019. Root-endophytic *Chaetomium cupreum* chemically enhances aluminium tolerance in *Miscanthus sinensis* via increasing the aluminium detoxicants, chlorogenic acid and oosporein. PLoS One. 14: 0212644.

Haynes, R. J., G. Murtaza and R. Naidu. 2009. Inorganic and organic constituents and contaminants of biosolids: implications for land application. Adv. Agron. 104: 165–267.

Higueras, P., R. Oyarzun, J. Lillo, J. Sánchez-Hernandez, J. Molina, J. Esbri and S. Lorenzo. 2006. The Almadén district (Spain): anatomy of one of the world's largest Hg-contaminated sites. Sci. Total Environ. 356: 112–124.

Holden, J. F. and M. Adams. 2003. Microbe–metal interactions in marine hydrothermal environments. Curr. Opin. Chem. Biol. 7: 160–165.

Hsu, N. H., S. L. Wang, Y. C. Lin, G. D. Sheng and J. F. Lee. 2009. Reduction of Cr(VI) by crop-residue derived black carbon. Environ. Sci. Technol. 43: 8801–8806.

Huang, J. W., J. Chen and W. R. Berti. 1997. Phytoremediation of lead-contaminated soils: role of synthetic chelates in lead phytoextraction. Environ. Sci. Technol. 31: 800–805.

Iimura,Y., M. Yoshizumi, T. Sonoki, M. Uesugi, K. Tatsumi, K. Horiuchi, S. Kajita and Y. Katayama. 2007. Hybrid aspen with a transgene for fungal manganese peroxidase is a potential contributor to phytoremediation of the environment contaminated with bisphenol. J. Wood Sci. 53: 541–544.

İlay, R., A. Baba and Y. Kavdır. 2019. Removal of metals and metalloids from acidic mining lake (AML) using olive oil solid waste (OSW). Int. J. Environ. Sci. Technol. 16: 4047–4058.

Illera, V., I. Walter, P. Souza and V. Cala. 2000. Short-term effects of biosolid and municipal solid waste applications on heavy metals distribution in a degraded soil under a semi-arid environment. Sci. Total Environ. 255: 29–44.

Jadia, C. D. and M. H. Fulekar. 2009. Phytoremediation of heavy metals: recent techniques. Afr. J. Biotechnol. 8: 921–928.

Jin, T., C. Shi, P. Wang, J. Liu and L. Zhan. 2021. A review of bioremediation techniques for heavy metals pollution in soil. In IOP Conference Series: Environ. Earth Sci. IOP Publishing. 012012.

Jin, Z., S. Deng, Y. Wen, Y. Jin, L. Pan and Y. Zhang, T. Black, K. C. Jones, H. Zhang and D. Zhang. 2019. Application of *Simplicillium chinense* for Cd and Pb biosorption and enhancing heavy metal phytoremediation of soils. Sci. Total Environ. Sci. 697: 134148.

Jinadasa, K. B. P. N., P. J. Milha, C. A. Hawkins, P. S. Cornish, P. A. Williams and C. J. Kaldo. 1997. Heavy Metals in the Environment. Richmond, Australia: University of Western Sydney, Hawkesbury Campus, Faculty of Agriculture and Horticulture.

Kao, P. H. C. C. Huang and Z. Y. Hseu. 2006. Response of microbial activities to heavy metals in a neutral loamy soil treated with biosolid. Chemosphere. 64: 63–70.

Kapahi, M. and S. Sachdeva. 2019. Bioremediation options for heavy metal pollution. J. Health Pollut. 24: 191203.

Karami, A. and Z. H. Shamsuddin. 2010. Phytoremediation of Heavy Metals with Several Efficiency Enhancer Methods. 9: 3689–3698.

Kärenlampi, S., H. Schat, J. Vangronsveld, J. A. C. Verkleji, D. van Der Lelei, M. Mergeay and A. I. Tervahauta 2000. Genetic engineering in the improvement of plants for phytoremediation of metal polluted soils. Environ. Pollut. 107: 22.

Karlson, U. and W. J. Frankenberger. 1989. Accelerated rates of selenium volatilization from California soils. Soil Sci. Soc. Am. J. 53: 749–753.

Kim, J. G., J. B. Dixon and C. C. Chusuei. 2002. Oxidation of chromium(III) to (VI) by manganese oxides. Soil Sci Soc Am J. 66: 306–315.

Kirkham, M. B. 2006. Cadmium in plants on polluted soils: effects of soil factors, hyperaccumulation, and amendments soil factors, hyperaccumulation, and amendments. Geoderma. 137: 19–32.

Koo, B. J., W. Chen, A. C. Chang, A. L. Page, T. C. Granato and R. H. Dowdy. 2010. A root exudates based approach to assess the long-term phytoavailability of metals in biosolids-amended soils. Environ. Pollut. 158: 2582–2588.

Kotrba, P., L. Dolecková, V. de Lorenzo and T. Ruml. 1999. Enhanced bioaccumulation of heavy metal ions by bacterial cells due to surface display of short metal binding peptides. Appl. Environ. Microbiol. 65: 1092–1098.

Krämer, U., J. D. Cotter-Howells, J. M. Charnock, A. J. M. Baker and J. A. C. Smith. 1996. Free histidine as a metal chelator in plants that accumulate nickel. Nature. 379: 635–638.

Krämer, U. 2005. Phytoremediation: novel approaches to cleaning up polluted soils. Curr. Opin. Biotechnol. 16: 133–141.

Krebs, W., C. Brombacher, P. P. R. Bosshard Bachofen and H. Brandl. 1997. Microbial recovery of metals from solids. FEMS Microbiol. Rev. 20: 605–617.

Kudo, K., N. Yamaguchi, T. Makino, T. Ohtsuka, K. Kimura, D. T. Dong and S. Amachi. 2013. Release of arsenic from soil by a novel dissimilatory arsenate reducing bacterium, *Anaeromyxobacter* sp. Strain PSR-1. Appl. Environ. Microbiol. 79: 4635–4642.

Kumar, M. and R. Singh. 2017. Phytoremediation: a green technology for remediation of metal-contaminated sites. pp. 306–328. *In*: R. N. Bharagava (Ed.). Environmental Pollutants and their Bioremediation Approaches. CRC Press, Boca Raton, Florida.

Kumar, P. and M. H. Fulekar. 2018. Research article rhizosphere bioremediation of heavy metals (copper and lead) by *Cenchrus ciliaris*. J. Environ. Sci. 12: 166–176.

Kumar, S., J. Dubey, S. Mehra, P. Tiwari and A. J. Bishwas. 2011. Potential use of cyanobacterial species in bioremediation of industrial effluents. Afr. J. Biotechnol. 10: 1125–1132.

Lambert, M., G. Pierzynski, L. Erickson and J. Schnoor. 1994. Remediation of lead, zinc, and cadmium contaminate. Environ. Sci. Technol. 7: 91–102.

Lasat, M. M., A. J. Baker and L. V. Kochian. 1998. Kochian Altered Zn compartmentation in the root symplasm and stimulated Zn absorption into the leaf as mechanisms involved in Zn hyperaccumulation in *Thlaspi caerulescens*. Plant Physiol. 118: 875–883.

Letterman, R. D. and W. J. Mitsch. 1978. Impact of mine drainage on a mountain stream in Pennsylvania. Environ. Pollut. 17: 53–73.

Li, P., X. Feng, G. Qiu, L. Shang and S. Wang. 2008. Mercury exposure in the population from 568 Wuchuan mercury mining area, Guizhou, China. Sci. Total Environ. 395: 72–79.

Li, R. Y., Y. Ago, W. J. Liu, N. Mitani, J. Feldmann, S. P. McGrath, J. F. Ma and F. J. Zhao. 2009. The rice Aquaporin Lsi1 mediates uptake of methylated Arsenic species. Plant Physiology. 150: 2071–2080.

Li, X., W. Peng, Y. Jia, L. Lu and W. Fan. 2016. Bioremediation of lead contaminated soil with *Rhodobacter sphaeroides*. Chemosphere. 156: 228–235.

Lindström, E. B. and H. M. Sehlin. 1989. High efficiency of plating of the thermophilic sulfur-dependent archaebacterium *Sulfolobus acidocaldarius*. Appl. and Environ. Microbiol. 55: 3020–3021.

Liu, Y., C. J. Lin, W. Yuan, Z. Lu and X. Feng. 2021. Translocation and distribution of mercury in biomasses from subtropical forest ecosystems: evidence from stable mercury isotopes. Acta Geochim. 40: 42–50.

Lone, M. I., Z. He, P. J. Stoffella and X. Yang. 2008. Phytoremediation of heavy metal polluted soils and water: progresses and perspectives. J. Zhejiang Univ. Sci. 9: 210–220.

Luo, Z., A. Wadhawan and E. J. Bouwer. 2010. Sorption behavior of nine chromium (III) organic complexes in soil. Int. J. Environ. Sci. Technol. 7: 1–10.

Ma, Y., C. Zhang, R. S. Oliveira, H. Freitas and Y. Luo. 2016. Bioaugmentation with endophytic bacterium E6S homologous to *Achromobacter piechaudii* enhances metal rhizoaccumulation in host *Sedum plumbizincicola*. Front. Plant Sci. 7: 1–9.

Ma, Y., M. Rajkumar, R. S. Oliveira, C. Zhang and H. Freitas. 2019. Potential of plant beneficial bacteria and arbuscular mycorrhizal fungi in phytoremediation of metal-contaminated saline soils. J. Hazard. Mater. 379: 120813.

Maestri, E. and M. Marmiroli. 2011. Genetic and molecular aspects of metal tolerance and hyperaccumualtion. Metal Toxicity in Plants: Perception, Signalling and Remediation, 41–61.

Mahmood, R., F. Sharif, S. Ali and M. U. Hayyat. 2013. Bioremediation of textile effluent by indigenous bacterial consortia and its effects on Zea mays L. CV C1415. J. Anim. Plant Sci. 23: 1193–1199.

Makino, T. K., K. Sugahara, Y. Sakurai, H. Takano, T. Kamiya and K. Sasaki. 2006. Remediation of cadmium contamination in paddy soils by washing with chemicals: selection of washing chemicals. Environ. Pollut. 144: 2–10.

Martín-Doimeadios, R. R., E. Tessier, D. Amouroux, R. Guyoneaud, R. Duran, P. Caumette and O. F. X. Donard. 2004. Mercury methylation/demethylation and volatilization pathways in estuarine sediment slurries using species-specific enriched stable isotopes. Mar. Chem. 90: 107–123.

Martínez-Villegas, N., L. M. Flores-Vélez and O. Domınguez. 2004. Sorption of lead in soil as a function of pH: a study case in México. Chemosphere. 57: 1537–1542.

Mateos, L. M., E. Ordóñez, M. Letek and J. A. Gil. 2006. *Corynebacterium glutamicum* as a model bacterium for the bioremediation of arsenic. Int. J. Microbiol. 3: 207–215.

Mavropoulos, E., A. M., Rossi, A. M. Costa, C. A. C. Perez, J. C. Moreira and M. Saldanha. 2002. Studies on the mechanisms of lead immobilization by hydroxyapatite. Environ. Sci. Technol. 36: 1625–1629.

McBride, M. B. 1995. Toxic metal accumulation from agricultural use of sludge Are USEPA regulations protective. J. Environ. Qual. 24: 5–18.

McGowen, S. L., N. T. Basta and G. O. Brown. 2001. Use of diammonium phosphate to reduce heavy metal solubility and transport in smelter-contaminated soil. J. Environ. Qual. 30: 493–500.

Mclaughlin, M. J., K. G. Tiller, R. Naidu and D. P. Stevens. 1996. Review: the behaviour and environmental impact of contaminants in fertilizers. Aust. J. Soil Res. 34: 1–54.

Meers, E., M. Qadir, P. De Caritat, F. M. G. Tack, G. Du Laing and M. H. Zia. 2009. EDTA-assisted Pb phytoextraction. Chemosphere. 74(10): 1279–1291.

Merdy, P., T. Gharbi and Y. Lucas. 2009. Pb, Cu and Cr interactions with soil: Sorption experiments and modelling. Colloids Surf. A: Physicochem. Eng. Asp. 347: 192–199.

Milner. M. J. and V. Kochian. 2008. Investigating heavy-metal hyperaccumulation using *Thlaspi caerulescens* as a model system. Ann. Bot. 102: 3–13.

Mishra, J., R. Singh and N. K. Arora. 2017 Alleviation of heavy metal stress in plants and remediation of soil by rhizosphere microorganisms. Front. Microbiol. 8: 1706.

Monica, M. O., N. W. Julia and M. M. Raina. 2008. Characterization of a bacterial community in an abandoned semiarid lead-zinc mine tailing site. Appl. Environ. Microbiol. 74: 3899–3907.

Mosa, K. A., I. Saadoun, K. Kumar, M. Helmy and O. P. Dhankhe. 2016. Potential biotechnological strategies for the cleanup of heavy metals and metalloids. Front. Plant Sci. 7: 303.

Mukhopadhyay, S. and S. K. Maiti. 2010. Phytoremediation of metal mine waste. Appl. Ecol. Environ. Sci. 8: 207–222.

Nagata, S., K. Yamaji, N. Nomura and H. Ishimoto. 2015. Root endophytes enhance stress-tolerance of *Cicuta virosa* L. growing in a mining pond of eastern Japan. Plant Species Biol. 30: 116–125.

Nandakumar, P., V. Dushenkov, H. Motto and I. Raskin. 1995. Phytoextraction: The use of plants to remove heavy metals from soils. Environ. Sci. Technol. 29: 1232–1238.

Nayak, A. K., S. S. Panda, A. Basu and N. K. Dhal. 2018. Enhancement of toxic Cr (VI), Fe, and other heavy metals phytoremediation by the synergistic combination of native *Bacillus cereus* strain and *Vetiveria zizanioides* L. Int. J. Phytoremediation. 20: 682–691.

Nejad, Z. D., M. C. Jung and K. H. Kim. 2018. Remediation of soils contaminated with heavy metals with an emphasis on immobilization technology. Environ. Geochem. Health. 40: 927–953.

Neuhierl, B. and A. Bock. 1996. On the mechanism of selenium tolerance in selenium-accumulating plants: purification and characterization of a specific selenocysteine methyltransferase from cultured cells of *Astragalus bisculatus*. Eur. J. Biochem. 239: 235–238.

Nicholson, F. A., B. J. Chambers, J. R. Williams and R. J. Unwin. 1999. Heavy metal contents of livestock feeds and animal manures in England and Wales. Bioresour. Technol. 70: 23–31.

Nogawa, K. and T. Kido. 1996. Itai-Itai disease and health effects of cadmium. Toxicology of Metals, 353–369.

Ojuederie, O. B. and O. O. Babalola. 2017. Microbial and plant-assisted bioremediation of heavy metal polluted environments: a review. Int. J. Environ. Res. Public Health. 14: 1504.

Ok, Y. S., S. E. Oh, M. Ahmad, S. Hyun, K. R. Kim and D. H. Moon. 2010. Effects of natural and calcined oyster shells on Cd and Pb immobilization in contaminated soils. Environ. Earth Sci. 61: 1301–8.

Ordóñez, E., M. Letek, N. Valbuena, J. A. Gil and L. M. Mateos. 2005. Analysis of genes involved in arsenic resistance in *Corynebacterium glutamicum* ATCC 13032. Appl. Environ. Microbiol. 71: 6206–6215.

Park, J. H., D. Lamb, P. Paneerselvam, G. Choppala, N. Bolan and J. W. Chung. 2011. Role of organic amendments on enhanced bioremediation of heavy metal(loid). J. Hazard. Mater. 185: 549–574.

Pérez, A. L. and K. A. Anderson. 2009. DGT estimates cadmium accumulation in wheat and potato from phosphate fertilizer applications. Sci. Total Environ. 407: 5096–103.

Pierzynski, G. M. 1997. Strategies for remediating trace-element contaminated sites. pp. 67–84. *In*: I. K. Iskandar and D. C. Adriano [Eds.]. Remediation of Soils Contaminated with Metals. Advances in Environmental Science. Middlesex, UK. Science Reviews.

Pointing, S. B. 2001. Feasibility of bioremediation by white-rot fungi. Appl. Microbiol. Biotechnol. 57: 20–33.

Prapagdee, B. and N. Khonsue. 2015. Bacterial-assisted cadmium phytoremediation by *Ocimum gratissimum* L. in polluted agricultural soil: a field trial experiment. Int. J. Environ. Sci. Technol. 12: 3843–3852.

Qiu, G., W. Feng, S. Wang and L. Shang. 2006. Environmental contamination of mercury from 621 Hg-mining areas in Wuchuan, northeastern Guizhou, China. Environ. Pollut. 142: 549–558.

Rasmussen, K. and C. Lindegaard. 1988. Effects of iron compounds on macroinvertebrate communities in a Danish lowland river system. Water Res. 22: 1101–1108.

Rauser, W. 1995. Phytochelatins and related peptides. Plant Physiology. 109: 1141–1149.

Roberts, A. H. C., R. D. Longhurst and M. W. Brown. 1994. Cadmium status of soils, plants, and grazing animals in New Zealand. New Zealand J. Agric. Res. 37: 119–129.

Rudakiya, D., Y. Patel, U. Chhaya and A. Gupte. 2019. Carbon nanotubes in agriculture: production, potential, and prospects. pp. 121–130. *In*: D. Panpatte and Y. Jhala [Eds.]. Nanotechnology for Agriculture. Singapore: Springer.

Ruiz, O. N. and H. Daniell. 2009. Genetic engineering to enhance mercury phytoremediation. Curr. Opin. Biotechnol. 20: 213–219.

Sakakibara, M., A. Watanabe, M. Inoue, S. Sano and T. Kaise. 2010. Phytoextraction and phytovolatilization of arsenic from As-contaminated soils by *Pteris vittata*. In Proceedings of the Annual International Conference on Soils, Sediments and Water. 26.

Salt, D. E., M. Blaylock, N. P. Kumar, V. Dushenkov, B. D. Ensley, I. Chet and I. Raskin. 1995. Phytoremediation: a novel strategy for the removal of toxic metals from the environment using plants. Biotechnol. 13: 468–474.

Salt, D. E., R. C. Prince, A. J. Baker, I. Raskin and I. J. Pickering. 1999. Zinc ligands in the metal hyperaccumulator *Thlaspi caerulescens* as determined using X-ray absorption spectroscopy. Environ. Sci. Technol. 33: 713–717.

Sanchez, A. G., A. Alastuey and X. Querol. 1999. Heavy metal adsorption by different minerals: application to the remediation of polluted soils. Sci. Total Environ. 242: 179–188.

Sangita, P. I, A. U. Kakde and R. C. Maggirwar. 2012. Biodegradation of tannery effluent by using tannery effluent isolate. Int. Multidiscip. Res. J. 2: 43–44.

Saranya, K., A. Sundaramanickam, S. Shekhar, M. Meena, R. S. Sathishkumar and T. Balasubramanian. 2018. Biosorption of multi-heavy metals by coral associated phosphate solubilising bacteria *Cronobacter muytjensii* KSCAS. J. Environ. Manage. 222: 396–401.

Schultze, M. and W. Geller. 1996. The acid lakes of lignite mining district of the former german democratic republic. pp. 89–105. *In*: R. Reuther [ed.]. Geochemical Approaches to Environmental Engineering of Metals. Environmental Science.

Shabir Hussain, W. 2012. Phytoremediation: Curing soil problems with crops. Afr. J. Agric. Res.7: 3991–4002.

Sharma, Y. C. 1995. Economic treatment of Cadmium(II)-Rich hazardous waste by indigenous material. J. Colloid Interface Sci. 173: 66–70.

Shen, F., R. Liao, A. Ali, A. Mahar, D. Guo and R. Li. 2017. Spatial distribution and risk assessment of heavy metals in soil near a Pb/Zn smelter in Feng County, China. Ecotoxicol. Environ. Saf. 139: 254–262.

Shukla, A., S. Srivastava and S. F. D'Souza. 2018. An integrative approach toward biosensing and bioremediation of metals and metalloids. Int. J. Environ. Sci. Technol. 15: 2701–2712.

Sims, J. T. and D. C. Wolf. 1994. Poultry waste management: agricultural and environmental issues. Adv. Agron. 52: 1–83.

Smith. E., R. Naidu and A. M. Alston. 1999. Chemistry of arsenic in soils: i. sorption of arsenate and arsenite by four Australian Soils. J. Environ. Qual. 28: 1719–1726.

Sposito. 1984. The Surface Chemistry of Soils, Oxford University Press New York.

Tangahu, B. V., S. R. S. Abdullah, H. Basri, M. Idris, N. Anuar and M. Mukhlisin. 2011. A review on heavy metals (As, Pb, and Hg) uptake by plants through phytoremediation. Int. J. Chem. Eng., 1–31

Tao, S., W. X. Liu, Y. J. Chen, F. L. Xu, R. W. Dawson, B. G. Li, J. Cao, X. J. Wang, J. Y. Hu and J. Y. Fang. 2004. Evaluation of factors influencing root-induced changes of copper fractionation in rhizosphere of a calcareous soil. Environ. Pollut. 129: 5–12.

Tarekegn, M. M., F. Z. Salilih, A. I. Ishetu and Y. Fatih. 2020. Microbes used as a tool for bioremediation of heavy metal from the environment. Cogent. Food. Agric. 6: 1783174.

Terry N., C. Carlson, T. K. Raab and Adel, M. Zayed. 1992. Rates of selenium volatilization among crop species. ASA. 21: 341–344.

Thayer, J. S. and F. E. Brinckman. 1982. The biological methylation of metals and metalloids. Adv. Organomet. Chem. 20: 313–356.

Tornabene, T. G. and H. W. Edwards. 1972. Microbial uptake of lead. Science. 176: 1334–1335.

Ullrich, S. M., T. W. Tanton and S. A. Abdrashitova. 2001. Mercury in the aquatic environment: a review of factors affecting methylation. Crit. Rev. Environ. Sci. Technol. 31: 241–293.

Umrania, V. 2006. Bioremediation of toxic heavy metals using acidothermophilic autotrophes. Bioresour. Technol. 97: 1237–1242.

United Nations Environment Programme. 2013. Global Mercury Assessment Sources, Emissions, Releases, and Environmental Transport. Geneva, Switzerland: United Nations Environment Programme Chemicals Branch.

Valls, M., S. Atrian, V. de Lorenzo and L. A. Fernández. 2000. Engineering a mouse metallothionein on the cell surface of *Ralstonia eutropha* CH34 for immobilization of heavy metals in soil. Nat. Biotechnol. 18: 661–665.

Aken, B. V., R. Tehrani and J. L. Schnoor. 2011. Endophyte-assisted phytoremediation of explosives in poplar trees by *Methylobacterium populi* BJ001 T. In Endophytes of Forest Trees (pp. 217–234). Springer, Dordrecht.

Van Dillewijn, P., J. L. Couselo, E. Corredoira, A. Delgado, A. Ballester and J. L. Ramo. 2008 Bioremediation of 2, 4, 6-trinitrotoluene by bacterial nitroreductase expressing transgenic aspen. Environ. Sci. Technol. 42: 7405–7410.

Violante, A., S. D. Gaudio, M. Pigna, M. Ricciardella and D. Banerjee. 2007. Coprecipitation of arsenate with metal oxides. 2. Nature, Mineralogy, and Reactivity of Iron(III) Precipitates. Environ. Sci. Technol. 41: 8275–8280.

Wang, F., J. Yao, Y. Si, H. Chen, M. Russel and K. Chen. 2010. Short-time effect of heavy metals upon microbial community activity community activity. J. Hazard. Mater. 15: 510–6.

Wani, S. H., G. S. Sanghera, H. Athokpam, J. Nongmaithem, R. Nongthongbam, B. S. Naorem and H. S. Athokpam. 2012. Phytoremediation: curing soil problems with crops. Afr. J. Agric. Res. 7: 3991–4002.

Wiatrowski, H. A., P. M. Ward and T. Barkay. 2006. Novel reduction of mercury (II) by mercury-sensitive dissimilatory metal-reducing bacteria. Environ. Sci. Technol. 40: 6690–6696.

Wixson, G. and B. Davies. 1994. Guidelines for lead in soil: proposal of the society of environmental geochemistry and health. Environ. Sci. Technol. 28: 26A–31A.

Xiong, X., L. Yanxia, L. Wei, L. Chunye H. Wei and Y. Ming. 2010. Copper content in animal manures and potential risk of soil copper pollution with animal manure use in agriculture. Resour. Conserv. Recycl. 54: 985–990.

Xu, S., Y. Xing, S. Liu, Q. Huang and W. Chen. 2019. Chen Role of novel bacterial *Raoultella* sp. strain X13 in plant growth promotion and cadmium bioremediation in soil. Appl. Microbiol. Biotechnol. 103: 3887–3897.

Xun, Y., L. Feng, Y. Li and H. Dong. 2017. Mercury accumulation plant *Cyrtomium macrophyllum* and its potential for phytoremediation of mercury polluted sites. Chemosphere. 189: 161–170.

Yamaji, K., Y. Watanabe, H. Masuya, A. Shigeto, H. Yui and T. Haruma. 2016. Root fungal endophytes enhance heavy-metal stress tolerance of *Clethra barbinervis* growing naturally at mining sites via growth enhancement, promotion of nutrient uptake and decrease of heavy-metal concentration. PLoS One. 11: 1–15.

Yang, J. Y., X. E. Yang, Z. L. He, T. Q. Li, J. L. Shentu and P. J. Stoffella. 2006. Effects of pH, organic acids, and inorganic ions on lead desorption from soils. Environ. Pollut. 143: 9–15.

Ok, Y. S., S. E. Oh, M. Ahmad, S. Hyun, K. R. Kim and D. H. Moon. 2010. Effects of natural and calcined oyster shells on Cd and Pb immobilization in contaminated soils. Environ. Earth Sci. 61: 1301–8.

Zayed, A. M. and N. Terry. 1994. Selenium volatilization in roots and shoots: effects of shoot removal and sulfate level. J. Plant Physiol. 143: 8–14.

Zhang, M. Y., A. Bourbouloux, O. Cagnac, C. V. Srikanth, D. Rentsch, A. K. Bachhawat and S. Delrot. 2004. A novel family of transporters mediating the transport of glutathione derivatives in plants. Plant Physiol. 134: 482–491.

Zhang, X., T. Zhong and L. Liu. 2015. Ouyang X. Impact of soil heavy metal pollution on food safety in China. PLoS One. 10: 1–14.

Zhuang, X., J. Chen, H. Shim and Z. Bai. 2007. New advances in plant growth-promoting rhizobacteria for bioremediation. Environ. Int. 33: 406–13.

Złoch, M., C. T. Kowalkowski, J. Tyburski and K. Hrynkiewicz. 2017. Modeling of phytoextraction efficiency of microbially stimulated *Salix dasyclados* L. in the soils with different speciation of heavy metals Int. J. Phytoremediation. 19: 1150–1164.

CHAPTER 10

Metal(loid)s Toxicity and Bacteria Mediated Bioremediation

Sushant Sunder,[1] *Anshul Gupta,*[1] *Mehak Singla,*[1]
Rohit Ruhal[2] *and Rashmi Kataria*[1],*

10.1 Introduction

Pollutants are impurities that enter the natural environment beyond permissible limits and have a notable negative impact on the inhabitants. These chemical compounds could be new to the environment or present naturally, such as metal pollution in mining regions, volcanic eruptions, air fallout, sea salt particles, etc. Industrialization and aggressive utilization of natural resources have produced a number of contaminants or pollutants that are either biodegradable or non-biodegradable in nature. Contaminants including toxic metals, heavy metals, insecticides, herbicides and radioactive compounds are among the pollutants that have caused significant environmental and health problems. It has been observed that pollutants are being continuously released from industrial waste and anthropogenic sources to the environment, and this waste circulates from the physical environment to plants, plants to animals, and so on enter the food chain to reach humans.

10.2 Toxic Metals

The metabolic processes in plants, animals and humans are impacted by different toxic metals. Copper (Cu), Cobalt (Co), Zinc (Zn), Nickel (Ni) and Chromium (Cr) are examples of potentially hazardous metals, and some of these serve both as micronutrients and essential components of redox reactions. These metals are important for regulation, electrostatic interactions and cofactor's stability in many compounds and reactions. Trace components, also known as heavy metals, are necessary for the smooth functioning of the human body metabolism; however, their harmful effects cannot be neglected if present in excessive quantities. The high-density metallic elements are toxic in nature, even in relatively smaller quantities. The detrimental effects of these toxic metals are discussed in Table 10.1. Any amount of lead (Pb) may have a negative impact on the body. In some cases, lighter

[1] Department of Biotechnology, Delhi Technological University (DTU), Bawana Road, Delhi 110042.
[2] Department of Biological Sciences. 1250 W Wisconsin Avenue, Marquette University, Milwaukee, Wisconsin 53233, USA.
* Corresponding author: rashmikataria@dtu.ac.in

Table 10.1. Effects of toxic metals on human health.

Toxic Metals	Impact on Human Health	References
As	• Bladder cancer, skin cancer, leukaemia and lung cancer • Neurological disorders throughout puberty and adulthood • Diabetes and heart related diseases • Pregnancy-related foetal death and premature birth	García-Esquinas et al. 2013, Lin et al. 2013, Heck et al. 2014
Cd	• Inhalation issues, pulmonary edema, and mucous membrane degradation • Vomiting and diarrhoea • Severe impacts on bone and kidney	Nogawa et al. 2004, Nordberg 2004
Co	• Pulmonary fibrosis, coughing, lungs inflammation and reduced pulmonary functioning • Plasma cells cancer, damage in recurrent laryngeal nerves and brachial plexus neuropathy • Loss of visual and hearing capacity	Sheikh 2016, Leyssens et al. 2017
Cr	• Asthma, chronic bronchitis, chronic irritation, chronic pharyngitis, chronic rhinitis, congestion and hyperaemia, polyps of the upper respiratory tract, tracheobronchitis and nasal mucosa ulceration with probable septal perforation • Irritant and allergic contact dermatitis • Respiratory system malignancies such as lung cancer, nose cancer and sinus cancer • Insufficient oxygen and blood flow supply to kidney, sudden kidney failure • Derangement of liver cells, necrosis, damage in lymph cells, increase in Kupffer cells and histiocytic infiltration • Anaemia and haematocrit, leucocytosis, reticulocyte counts and plasma haemoglobin	Brutti et al. 2013, Abdel-Gadir et al. 2016, Ray 2016, Lim et al. 2017
Cu	• Inflammation in mouth, eyes and nose, headaches, drowsiness, vomiting and stomach cramps • Severe damage to liver and kidney • Wilson's illness, brain damage, damage to myelin sheath, improper functioning of kidney and copper deposition in the cornea	Kodama et al. 2012, Okereafor et al. 2020
Hg	• Neurological disorders and behavioural problems such as tremors, sleeping disorders, loss of appetite, droopy eyelids, double vision, cramps and movement issues • Red blood cell accumulation (competes with iron for haemoglobin binding) and suppression of myelin formation in developing foetus • Immune, enzyme and genetic alterations • Young's syndrome (Azoospermia sinopulmonary infections)	Hamada et al. 2013, Andreoli and Sprovieri 2017, Al Osman et al. 2019
Mn	• Risk of Parkinson disease • Abrupt functioning of cardiovascular system • Memory loss, tremors, neurological disorders, mortality of infants. • Sexual incompatibility in males	Spangler and Spangler 2009, Meeker et al. 2010
Ni	• Respiratory issues, vomiting, dizziness, nausea, headache, coughing • Lowering of oxygenated blood in lungs, improper functioning of lungs, increase chances of lung cancer, chronic bronchitis • Inflammatory responses in skins due to allergies • Oncogenic haematotoxicity, congenital malfunction to foetus, toxic to immune system	Das et al. 2008, Zdrojewicz et al. 2016
Pb	• Renal damage, inflammation, neurological disorders, fatigue, sleep disorder, hallucinations, vertigo, hypertension, arthritis, birth defects, mental retardation • Abrupt functioning of Central Nervous System (CNS) due to disruption of intracellular second messenger systems	Okereafor et al. 2020
Zn	• Respiratory disease caused by inhaling zinc smoke, epigastric pains, increased risk of prostate cancer and lethargy • Imbalance of copper in body • Inflammation and corrosiveness in gastrointestinal tract, nausea, vomiting and interstitial nephritis	Plum et al. 2010, Okereafor et al. 2020

metals such as beryllium may be hazardous. However, the necessary elements or metals, such as iron (Fe), could be poisonous. Metals in certain oxidation phases may also be toxic, such as Cr (III) is an essential micronutrient, whilst the other form Cr (VI) is a potential carcinogen. The consequences of heavy metal toxicity include decreased fitness, problems with reproduction, cancers and mortality. Organometallic forms, such as methyl Hg (mercury) and tetraethyl Pb, may be extremely hazardous, while organometallic derivatives, such as cobaltocenium cation, are less toxic. Radium, which is a radioactive heavy metal, mimics the property of calcium as it gets absorbed in bones and, along with Pb and Hg, causes several health complications. However, there are exceptions in cases of Barium (Ba) and Aluminum (Al), which are rapidly eliminated by the kidneys.

10.3 Effects of Toxic Metals

In the environment, toxic metals are present significantly in effluents, waste streams and sediments, affecting bacterial diversity, population size, and overall activity (Xie et al. 2013). Toxicity induced by metals is frequently documented as an environmental health challenge which causes bioaccumulation of toxic metals in the food chain. The severe consequences are seen in humans, animals and plants (Aycicek et al. 2008). The impacts of poisonous trace metals are reliant on a number of variables, including the element's dietary content, absorption by the body, maintaining homeostasis for these toxic metallic elements, and the number of animals affected by these elements (Rajaganapathy et al. 2011). Furthermore, the oxidation state of a hazardous metal influences its toxicological and biological impacts on the environment. The functional and structural changes in DNA, RNA and proteins, along with challenges of oxidative phosphorylation and osmotic imbalances, are widely observed due to metal toxicity (Yao et al. 2008). Toxic metal contamination has occurred as a severe health issue due to industrial waste streams and agricultural residues. Humans are typically exposed not only orally but also via cutaneous contact, inhalation and ingestion. The severity of a single compound's harmful effects (including immunotoxin) may be altered by interactions with one or more other heavy metals or xenobiotics. As a result, the synergistic interactions of insecticides and heavy metals may lead to different health problems.

10.3.1 Effects of Toxic Metals on Human Health

A large fraction of the population in most developing nations faces the challenge of toxic metal contamination in food items (D'Souza and Peretiatko 2002, Cheng 2003, Meharg 2004). These toxic metals promote serious health complications in the human body because their introduction causes the malfunctioning of crucial cellular processes by displacing important metals from their correct positions (Vieira et al. 2011). People who are more exposed to trace element-polluted areas or those who work in metal mines are continuously being jeopardized with rising health issues (Table 10.1). Metal poisoning in humans may cause slow development in children, dementia in adults, nervous system related problems, renal illnesses, liver diseases, insomnia , depression and visual abnormalities (Jan et al. 2011).

Toxic metal transmission in humans is rather complicated. Lead (Pb) is taken up by the body mainly through the stomach and respiratory tracts before it enters the circulatory system. Phosphate enters the respiratory tract and digestive tract in the form of soluble salts, protein complexes and ions with a concentration of more than 95% insoluble Pb accumulating for bones build-up (Okereafor et al. 2020). It is also a strong pro-organizational element that alters and damages the kidney, liver, reproductive organs, neurological functioning and digestive system. The urinary system, the immunological networking and the cells performing fundamental physiological processes and gene expression (Mahurpawar 2015) are also affected. A longer duration of exposure to Pb results in brain damage, short-term memory loss, adapting skills and coordination complexities in children. Prenatal Pb exposure may lead to a decrease in immunity and weight of new born babies. It is one of the main reasons why certain babies are associated with asthma and allergies. Studies have concluded that lead (Pb) could influence behavioural inhibition systems, resulting in an increase of aggression, and it could also promote tooth decay.

The rate of consumption is also a determining step for toxicity in human health. Metals such as copper, nickel and zinc are essential trace elements for the body but are harmful if consumed in excess. Prolonged exposure to copper is linked to abnormalities that severely affect capillaries, kidneys and liver functions. It also leads to central nervous system irritation along with depression. System dysfunction impedes growth and development. Excess zinc is related to infertility. Exposure to nickel among humans enhances the risk of developing respiratory cancer nasopharyngeal carcinoma (Barta et al. 2019). Excessive mercury harms the neurological system and affects muscle functioning, body balance, partial blindness and abnormalities in infants (Baby et al. 2010). According to some studies, mercury may harm both the foetal and embryonic neural systems at concentrations lower than the WHO's recommended limits, resulting in learning difficulties, poor memory and reduced attention spans.

Cadmium (Cd) poisoning damages the liver, brain, lungs, placenta, kidneys and bones. The severity depends on the amount of exposure (Sobha et al. 2007). Extreme exposure has been related to mortality and pulmonary oedema. Emphysema, bronchiolitis and alveolitis are the other consequences of exposure to Cd. Cadmium is associated with many clinical problems, including heart failure, malignancies, cataract development and lung-related issues. The toxicity symptoms of arsenic, mercury and lead, depend on the form in which these are consumed. Arsenic promotes protein coagulation with coenzymes and inhibits adenosine triphosphate (ATP) synthesis. Arsenic oxides are carcinogenic and serve violent mortality in higher concentrations. Arsenic is associated with several disorders, such as Guillain-Barre syndrome, which is known as an anti-immune disorder. In this condition, the patient's Peripheral Nervous System (PNS) is attacked by the human immune system, which results in nerve inflammation and muscle weakness.

10.3.2 Effects of Toxic Metals on Plants

A minimum concentration of metal ions is essential for optimal plant growth. The absence of the same could result in growth deficiencies, whereas excessive access to Cd, As, Hg and Pb are highly toxic in plants (Okereafor et al. 2020). There are many parameters such as temperature, pH of the soil and plant growth environment, which decide the plant's metal absorption (Nagajyoti et al. 2010). Excessive levels of hazardous metals affect plant growth and development (Musilova et al. 2016). Different metals like Mn, Pb, Cd, Cr and Co cause slow development in maize plants (Mujtaba Munir et al. 2020). If exposed to a high concentration, toxic metals may cause oxidative stress, cell structural deformation and slow photosynthetic activities in the plants. Zn and Cd phytotoxicity is shown in different plant species such as *Brassica juncea* by delayed growth and development, alteration in metabolism and inductive oxidative damage (Okereafor et al. 2020). Cd and Zn at extremely high concentrations may cause oscillations in the catalytic competence of enzymes in pea plants (Romero-Puertas et al. 2004). Zinc poisoning has also been observed in plants causing slow development of roots and shoots and chlorosis in younger leaves. There have been reports of agricultural yield reductions owing to Ni toxicity, which inhibited specific enzymatic activity (amylase, protease and ribonuclease), reducing seed germination. Membrane stability, nitrate reductase and carbonic anhydrase were also impacted by Ni (Yusuf et al. 2012).

Cu is a micronutrient for plants, which is essential for ATP production and CO_2 absorption. Cu is found in several proteins, including plastocyanin which is important in the photo system, Cytochrome Oxidase and Electron Transport Chain (Demirevska-Kepova et al. 2004). When it is present in excess, it may cause stress, growth retardation and leaf chlorosis in plants. Plants suffer from oxidative stress because of elevated Cu levels, which disrupts metabolic pathways and damages macromolecules. It is also observed that copper poisoning affects *Alyssum montanum*, whereas copper and cadmium cause poor germination, small seedling length and roots development in *Solanum melongena* (Al Khateeb and Al-Qwasemeh 2014).

Mercury (Hg) is found widely in higher aquatic plant species in several forms, including HgS, Hg^{2+}, methyl-Hg and Hg^0 (Wang and Greger 2004, Kamal et al. 2004, Malar et al. 2015). Higher quantities of Hg^{2+} are known to be highly phytotoxic to plant cells, causing apparent lesions and physiological problems, which affects the closing of stomata in leaves and blockage of transportation of water in plants (Zhou et al. 2007). It is also observed that high amounts of Hg^{2+} have been found to interfere with the plant mitochondrial function, resulting in the alteration of cell membrane components, including lipids, and affecting the cellular metabolism (Messer et al. 2005, Cargnelutti et al. 2006).

Lead impacts negatively on plant shape, growth, photosynthesis and metabolic activities by interfering with essential enzymes, limiting seed germination (Zulfiqar et al. 2019). Another problem caused by increased Pb concentrations is oxidative stress. In plants, it promotes the formation of Reactive Oxygen Species (ROS) (Reddy et al. 2005). Plants which act as accumulators are highly resistant to heavy metals in their habitat and can tolerate high concentrations. These plants can use a different mechanism, such as Exclusion, which limits the excessive transportation of metals in the plant. A constant concentration of different metals can be maintained in the shoots. Secondly, Inclusion, a mechanism that prompts the plants to absorb the maximum metals from the soil. Thirdly, Bioaccumulation, which is the build-up of the toxic metal concentrations in the plants. Table 10.2 summarizes the different toxic metals and their impacts on plants.

10.3.3 Effects on Animals

Mutation, carcinogenicity, teratogenicity, poor physical condition and decreased reproduction are the most common adverse consequences associated with prolonged exposure to heavy metals in. pets (dogs and cats, in particular). Animal species such as pigs, hens and dairy cows have shown that elements like Pb and Cd could have subtoxic effects due to their prolonged consumption through dietary amounts. Both Pb and Cd get accumulated in organs like the liver and kidneys, while lead (Pb) could also be deposited in the bones of dairy animals. Modest consumption of the livers and kidneys from Pb-exposed animals could be harmful. Consumption of these organs from Cd-exposed animals should be avoided. It is quite evident that some toxic metals such as lead, mercury and selenium with an inappropriate level of some essential trace elements like chromium are observed in some wild plant species from Northeast India (Kennady et al. 2018). With the help of cutting-edge technology such as plasma spectroscopy, toxicity behavioural studies can be concluded from bones, hairs and cellular and surface ultrastructure features. Behavioural investigations revealed numerous symptoms associated with specific elemental anomalies, such as lack of appetite, constipation, salivation, photophobia, etc.

Since metals are not degraded by bacteria, and the presence of pollutants in rivers disturbs the balance of the aquatic environment and causes a massive reduction in marine life (Woo et al. 2009, Ay et al. 2009). Cellular intoxication, interference in the fish's metabolic processes, and death at the cellular level are a few histopathological changes observed due to toxic metals. The continual flow of water via gills and food sources continuously exposes fish to waterborne and particulate hazardous metals. The Reactive Oxygen Species (ROS) are often generated by toxic metals and cause degradation of cellular components, including protein, DNA, enzymes and fatty acids in the aquatic organisms. These toxic elements interact with the large surface area of fish gills to produce various abnormalities, including epithelial lifting issues, leucocytic necrosis, etc. (Mehana et al. 2020).

10.4 Bacteria Mediated Bioremediation

Bacteria use the environment and interactions with other organisms to attain carbon sources and nutrients for microbial cell maintenance. Bacterial bioremediation involves the use of aerobic and anaerobic bacteria to reduce environmental pollutants (Nagajyoti et al. 2010). The availability of

Table 10.2. Effects of toxic metals on plants growth and development.

Toxic Metals	Plants	Impacts on Plant	References
As	Rice (*Oryza Sativa*)	Smaller leaves, decreased germination capacity and lower seedling height.	Abedin et al. 2002, Okereafor et al. 2020
	Tomatoes (*Lycopersicon esculentum*)	Drop in fruit yield and lower yield of fresh leaves.	
	Canola oil (*Brassica napus*)	Restricted growth, chlorosis and wilting.	
Cd	Wheat (*Triticum* sp.)	Lower seed germination capacity, low nutritional value of plant.	Wang et al. 2007, Yourtchi and Bayat 2013
	Garlic (*Allium sativum*)	Lower shoot induction, Cadmium build-up.	
	Maize (*Zea mays*)	Lower root and shoot induction.	
Co	Tomato (*Lycopersicon esculentum*)	Reduced abstraction of nutrition from plants.	Jayakumar et al. 2007, 2008, Okereafor et al. 2020
	Mung Bean (*Vigna radiata*)	A decrease in the antioxidant quality of the plant, as well as loss of carbohydrates, sugar, amino acids and protein content.	
	Radish (*Raphanus sativus*)	Reduced root and shoot induction, as well as decreased leaf area, also a decrease in antioxidant enzyme activity, chlorophyll content, sugars, starch, amino acids and protein content.	
Cr	Wheat (*Triticum* sp.)	Root and shoot development stunted.	Ray 2017, Okereafor et al. 2020
	Tomato (*Lycopersicon esculentum*)	Plant nutrient acquisition reduced.	
	Onion (*Allium cepa*)	Germination inhibition, as well as a decrease in plant biomass.	
Cu	Bean (*Phaseolus vulgaris*)	Build-up of Cu in plant results in malformation and reduction in roots length.	Sheldon and Menzies 2005, Okereafor et al. 2020
	Black bindweed (*Polygonum convolvulus*)	Increased plant mortality, decreased production of plant biomass and seeds.	
	Rhode grass (*Chloris gayana*)	Root growth slowed.	
Hg	Rice (*Oryza sativa*)	Deprived shoot induction, reduction in flowering capacity, lower yield of biomass production, enhanced bioaccumulation.	Du et al. 2005, Shekar et al. 2011
	Tomato (*Lycopersicon esculentum*)	Lower germination capacity, attenuated shoot growth, decrease in the percentage of germination, reduced plant height, decreased fruit weight, chlorosis.	
Mn	Broad bean (*Vicia faba*)	Chlorosis due to manganese build-up in the shoot and root.	Doncheva et al. 2005, Asrar et al. 2005, Arya and Roy, 2011
	Spearmint (*Mentha spicata*)	Decrease in the amount of chlorophyll and carotenoid.	
	Pea (*Pisum sativum*)	Decreased relative growth rate and photosynthetic activity.	
	Tomato (*Lycopersicon esculentum*)	Slower plant development; decreased chlorophyll content.	
Ni	Pigeon pea (*Cajanus cajan*)	Inhibition in photosynthesis, decreased enzyme activity affecting the Calvin cycle and CO_2 fixation.	Okereafor et al. 2020
	Rye grass (*Lolium perenne*)	Reduced nutrient uptake by the plant, decreased shoot production, chlorosis.	
	Wheat (*Triticum* sp.)	Plant nutrient uptake is reduced.	
	Rice (*Oryza sativa*)	Root growth is slowed	

nutrients is an important factor for microbial growth that affects pollutant degradation. Several microbial mediated approaches are used. Biostimulation is the technique that involves a selective enrichment of existing soil microbe for efficient bioremediation; bioaugmentation involves the addition of a microbial strain for biodegradation of specific contaminant; bioaccumulation consists of the uptake of pollutants in the microbial cells for storage and utilization in metabolic activity (Mehana et al. 2020). Biofilm biosorption involves the removal of contaminants via surface adsorption. Bacterial bioremediation has the benefit of being resistant to the presence of particular pollutants.

10.4.1 Microbial Mechanism for Bioremediation of Toxic Metals

Heavy metals are extremely unsafe for plants, animals and human health. Microbes have developed defense mechanisms against pollutants by developing biological processes to use heavy metals. The detoxifying pathways for contaminants include extracellular barrier exclusion, binding of toxic metals, extracellular polymeric substances, intracellular sequestration (confinement of these toxic metals in the cytosol), active transport, efflux (microbial efflux machinery to transport away the toxic elements) and enzyme mediated detoxification (toxic metals are transformed into less or non-toxic substances). Thus, microbial biodegradation represents an affordable, efficient and economical approach for the bioremediation of pollutants from the environment.

Microbes aided remediation tries to harness various microbial activities and the metabolic potentials of important bacteria depending on the specific site conditions. Microbe-mediated remediation of sediments polluted by heavy and toxic metals involves multiple approaches, including bioleaching, biosurfactants, bioaccumulation, biosorption, bioprecipitation, biotransformation and bio volatilization, as shown in Figure 10.1.

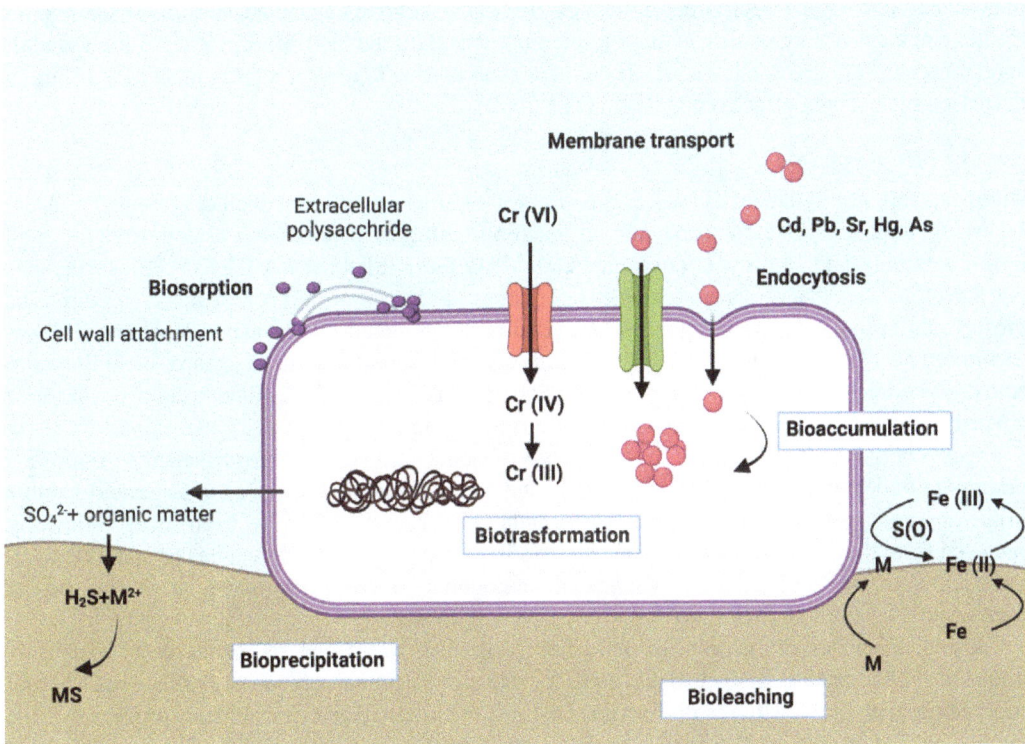

Figure 10.1. Microbial bioremediation of toxic metals (created by www.biorender).

10.4.1.1 Bioleaching

The bioleaching approach by acidophilic bacteria facilitates the solubility of heavy metals from sediments and contaminated sites. It predominantly affects the concentrations of toxic metals, including iron and sulfur, in sediments. Thus, groups of bacteria oxidize iron and sulfur compounds include: Firmicutes (e.g., *Alicyclobacillus* sp., *Sulfobacillus* sp.), Nitrospirae (e.g., *Leptospirillum* sp.), Proteobacteria (e.g., *Acidithiobacillus* sp., *Acidiphilium* sp., *Acidiferrobacter* sp., *Ferrovum* sp.), Actinobacteria, archaea (Crenarchaeota). These microorganisms are known as bioleaching microorganisms (Akcil et al. 2015). They can establish an acidic environment by oxidizing minerals, and therefore dissolve toxic metals ions into an aqueous medium (Akcil et al. 2015).

Chemolitho-autotrophic bacteria *Leptospirillum* oxidized Fe^{+2} under acidic conditions (pH = 1.5 – 1.8) in a sample adulterated with uranium (U_{92}) (Bertrand et al. 2015), whereas *Thiobacillus* is found to be capable of generating energy by oxidation of sulfur and thiosulfate (Xin et al. 2009). Fungal strain *Aspergillus niger* is also found to be efficient for leaching the elements and removing pollutants from sediment (Zeng et al. 2015).

The efficiency of bioleaching depends on various parameters, including abiotic stresses, characteristics of ions present in sediments, pH, and size of sediments. The biotic factors like the presence of microbial communities, metabolic pathways used for the process, and adaptability to minerals also play a significant role. As and Fe are the most leached metals from sediments. With these, a small percentage of Cu is also leachable. The bioleaching process is not suitable for Hg. The remediating capabilities of indigenous heterotrophic bacterial isolates of *Bacillus* have been observed in the leaching of toxic metals from polluted sites (Štyriaková et al. 2016).

These methods are viable both financially and environmentally. According to different research studies, bioleaching of toxic metals is more effective than the conventional leaching methods (Liu et al. 2003, Deng et al. 2013). However, several significant disadvantages restrict the utilization of this technique for an effective detoxification process. Excessive metal concentrations usually limit the development or functioning of susceptible microbes (Collinet and Morin 1990). A high load of solid materials and organic materials in polluted sources also hinders microorganisms, resulting in decreased bioleaching effectiveness (Cho et al. 2002).

10.4.1.2 Biosurfactant

Biosurfactants are synthesized by various classes of microorganisms, including fungi, and are known to be a substitute for conventional leaching owing to factors such as low toxicity, high biodegradability and significant environmental friendliness (Shekhar et al. 2015). Biosurfactants are the molecules that serve as biological chelating agents for a number of heavy metals. The biosurfactant-producing microorganisms are used at heavy metal-contaminated locations (Pacwa-Płociniczak et al. 2011). Biosurfactants are composed of different functional groups, including fatty acids, glycolipids and phospholipids (Pacwa-Płociniczak et al. 2011). Rhamnolipid is the most studied biosurfactant for heavy metal elimination (Mulligan et al. 1999).

Z-glycolipid in *Burkholderia* sp. is capable of being utilized as a biosurfactant to remove a mixture of toxic metals like Pb, As, Cu, Cd, Zn and As from polluted soil. Due to large acid-soluble properties and high chelation with biosurfactants, Mn, Zn and Cd have been more efficiently removed from the soil. The biosurfactant, rhamnolipid, demonstrated efficient heavy metal removal from polluted river sediments in exchangeable, carbonate bound or iron-manganese oxide–bound fractions (Chen et al. 2017).

Using biosurfactants to remove pollutants from soil, waste and sediments is an extremely interesting approach. It may also be used for the pre-treatment of contaminated areas before proceeding with stabilization/solidification, natural attenuation or electrokinetic processes. This method has the edge over other bioprocesses due to its ability to operate effectively even at high pH up to 11, while other bioprocesses (such as bioleaching and biosorption) work effectively only at low pH levels. Biosurfactants with excellent degradability and biocompatibility provide a considerable advantage with high environmental acceptability (Mulligan 2005). However, the exorbitantly high cost of biosurfactant production and the poor productivity of biosurfactants serve as a barrier to their commercial applications in bioremediation. Further, endeavors are needed in this area to recycle and reuse biosurfactants in order to decrease total costs via the development of cost-effective recovery techniques.

10.4.1.3 Bioaccumulation

Bioaccumulation is a complicated and dynamic procedure that requires the absorption and deposition of heavy metal ions in microbial intracellular components. The metabolism-dependent heavy metal transporting mechanisms determine its effectiveness. Heavy metals move across cell membranes via ion channels, protein pathways and carrier-mediated movement (Mishra and Malik 2013). The lipid bilayer allows for passive diffusion and could even enable toxic metals to enter the cell through endocytosis. Heavy metal active transportation is observed in a number of microbial species, including *pseudomonas, Micrococcus, Bacillus* and *Aspergillus*. However, this technique is not very useful for bioremediation since heavy metal deposited in the cell may mediate harmful effects on microbial metabolism. The efflux pathway releases heavy metals into the environment after a certain level of build up. Despite these restrictions, bioaccumulation has been utilized as a complete bioremediation method when the requirements of a minimum growing medium with low harmful impact on cells during treatment are required. In one study, the removal of Hg via bioaccumulation was shown to be more efficient than biosorption when using a live seaweed *Ulva lactuca* (Henriques et al. 2015).

10.4.1.4 Biosorption

Microorganisms, including bacteria, are used to extract toxic metals from sites like contaminated sewage, soil and sediments (Bano et al. 2018). Biosorption is a metabolism-independent, passive absorption mechanism that binds heavy metals to the cell membrane. It comprises both the physical and chemical bonding such as electrostatic, covalent, exopolysaccharides, ion-exchange, Van der Waal's force and microprecipitation with different functional groups (Montazer-Rahmati et al. 2011). Several factors, including acidity, temperature, ionic strength of the environment, permeability, origin, pre-treatment of biosorbents, amount and speciation of heavy metals, affect the adsorption process (Zhu et al. 2013, Fomina and Gadd 2014). The bacteria *Ochrobactrum* MT180101 showed high biosorption efficiency for copper chelated with other compounds (Sun et al. 2021). The affinity for metal binding is attributed to the presence of functional groups on the microbial cell that includes alkanes, amides, amines, as well as negatively charged exopolysaccharides (EPS). Furthermore, EPS modifications like acetylation, carboxymethylation, and methylation may increase the affinity for metal ions (Gupta and Sar 2020).

10.4.1.5 Bioprecipitation

The process of altering soluble toxic metals ions into insoluble species like hydroxides, carbonates, phosphates and sulfide groups by microorganisms is known as bioprecipitation. In this approach, microbe-assisted precipitation is not contingent on the activity of microbes, as it could be present in live and dead cells. Moreover, it may result in the precipitation of toxic metals that are directly linked to the cells. Ambient factors such as pH levels and redox potentials influence the efficiency of bioprecipitation. Sulfate-reducing bacteria produce hydrogen sulfide in an anaerobic environment using organic substances as a nucleophile, and they can precipitate metal ions. Biological oxidation of soluble ferrous iron under oxidative circumstances produces Fe (III) hydroxides that co-precipitate additional ions such as sulfide, Cd, and U indirectly (Kaplan et al. 2016, Rinklebe and Shaheen 2017). The leaching capability and the availability of toxic metals in sediments are greatly reduced by these insoluble complexes. By oxidation, the sulfide group is resolubilized into the aqueous phase. Hence, it is a critical process to track fluctuations in environmental redox potential and microorganisms' activity.

10.4.1.6 Biotransformation

Microbes interact with toxic metals and alter their form to a reasonably less toxic form via multiple biological reactions such as condensation, hydrolysis, formation of new carbon bonds, isomerization and introduction of functional groups, etc. (Table 10.2) (Guo et al. 2019). In *Bacillus* species converts Cr (VI) to Cr (III) under physical conditions like pH (7–9), temperature (between 30°C and 40°C), and Cr (VI) levels (50–250 mg L^{-1}) (Lei et al. 2019). Under optimal conditions of pH 7 and at 37°C, this species completely reduced the Cr (VI) concentration of 120 mg L^{-1} in just 48 hr. By using acetate as a carbon source, a heterogenous anaerobic colony containing *Anaerolineaceae*, *Spirochaeta* and *Spirochaetaceae* exhibited the capability to the reduction of Cr (VI) and V (V) with high efficiency of 97 and 99.1%, sequentially (Wu et al. 2019). In the presence of hematite and dissolved organic compounds, *Geobacter sulfurreducens* showed the capacity to eliminate Cr (VI).

Microorganisms use a sequence of methylation reactions for the conversion of As to volatile forms. Microorganisms transform arsenic trioxide (As_2O_3) into volatile toxic tri-methyl arsine $(CH_3)_3As$ and increase the mobility and release of As into the environment. *Acinetobacter* sp. and *Micrococcus* sp. are likewise found to be converting poisonous As (III) into nontoxic and less soluble As (III) and lessen its toxicity. In methylation of Hg, bacteria convert mercuric ions into methyl mercuric, which increases the bioavailability of Hg through food sources in the aquatic environment. The biochemical processes of methylation of Hg and SO_4^{2-} in contaminated sites are significantly associated with reducing the pollutant levels in sediment pore water (Hines et al. 2012).

10.5 Biofilm in Toxic Metal Removal

An aggregation of microbes consisting of one or more strains bonded to a substratum and encapsulated in a self-synthesized composite including water, proteins, carbohydrates and extra cellular DNA, is known as a biofilm (Jia et al. 2013). It is expected that diverse microbial species found in biofilm communities, each with its unique metabolic breakdown mechanism, can degrade numerous pollutants collectively or individually (Gieg et al. 2014). As they fight for nutrients and oxygen, biofilm-producing bacteria are adapted to thrive in adverse circumstances. It is considered that biofilm-mediated remediation is an economical and environment-friendly alternative strategy of removing toxic chemicals. Biofilms are useful for bioremediation because they accumulate, precipitate, and eliminate many environmental pollutants. Indigenous populations in severely polluted regions create bacterial biofilms to help them survive, flourish and cope with the harsh environment. Gene expression in biofilms varies and is unique, analogous to free-floating planktonic cells. Several genes play an important role in the degradation of pollutants through different metabolic pathways. Chemotaxis and motility in bacteria are very important in biofilms formation (Horemans et al. 2013). Microorganisms use various motility behaviours such as chemotaxis, swimming and quorum sensing to coordinate the movement of microbe toward pollutants and improve biodegradation (Pratt and Kolter 1999).

Under normal environmental conditions, bacteria surviving in the form of biofilm enclosed in a polymeric substance are known as exopolysaccharides (EPS) and offer a favorable assembly to biofilm-producing microorganisms in bioremediation (Lacal et al. 2013, More et al. 2014). The EPS is composed of polysaccharides, amino acids, fatty acids, nucleic acids, proteins and humic compounds (Flemming and Wingender 2001). The composition of these EPS is variable in different species and depends upon growth conditions, surface for attachment for biofilms and type of physical stress (Jung et al. 2013). Biofilms form filamentous and mushroom-like structures that grow rapidly in flowing and stagnant water, correspondingly (Miqueleto et al. 2010). In the presence of predatory protozoa, bacteria frequently adapt to produce biofilms in the form of vast inedible microcolonies that allow survival and persistence under hostile environments (Edwards et al. 2000). The biofilm matrix protects microorganisms against adverse environmental conditions, mechanical stress, pH stress, antibiotics, antimicrobial compounds, dehydration, predators, solvents, UV radiations and elevated doses of dangerous chemicals and pollutants better than planktonic cells (Davey and O'toole 2000, Matz and Kjelleberg 2005).

Biofilm-producing bacteria are efficient to bioremediate as they are imprisoned in an EPS that also immobilizes pollutants during breakdown (Mah and O'Toole 2001). Due to the low concentration of oxygen towards the center, all the microbes, including aerobes and anaerobes, heterotrophs (nitrifying bacteria) and sulfate reducers are present in close proximity in the three-dimensional network of EPS, facilitating quicker degradation of various contaminants in natural and artificial systems (Sutherland 2001). Toxic metals are removed from the aqueous environment using EPS from cyanobacteria as a biosorbent (de Philippis et al. 2011). Several enzymes are present in biofilms EPS, which detoxify heavy metals and organic compounds. Due to the presence of many negatively charged functional groups in EPS, it acts as a trap for metals and metalloids, allowing the formation of chelates with toxic metals and organic contaminants and thus facilitating their elimination (Li and Yu 2014). Different metals like zinc, lead, nickel, magnesium, cadmium, iron, manganese and copper are known to bind to EPS (Pal and Paul 2008). Nutrient restriction may lead to enhancement in EPS and copper production and allow to absorb more pollutants from the polluted site. The EPS of phosphorus-accumulating bacteria in biofilms serves as a reservoir and aids in the bioremediation and recovery of phosphorus from wastewater (Zhang et al. 2013).

Biofilms may be observed in natural environments such as soil, plants, sediments, streams, ponds, rivers and man-made environments. Fungus, bacteria, protozoa and algae make up biofilms that form on the water's surface.

During bloom periods, fragile structures termed flocs are formed. The floc-activity of activated sludge plants is used to treat municipal wastewater. Slow-sand filters are used to separate organic chemicals and metals from the natural environment by using biofilms (Burmølle et al. 2006). *Planctomycetes* found in seaweed biofilms in coastal habitats can be used to remove nitrogen from wastewater by anammox reactions by converting ammonia to dinitrogen anaerobically (Kartal et al. 2010).

Biofilms are extensively used for risk assessment or as an indication for monitoring and assessing toxic metal pollution. There may be a change in biofilm structural and physiological functions in the presence of toxic compounds. Microbial biofilms could accumulate contaminants, and contaminants offer an easy attachment site for biofilms. They are applied as an indicator system (Bengtsson and Øvreås 2010). Microbes are the first organisms in aquatic habitats to interact with minerals and pollutants, and as a result, biofilms may be used as a risk assessment tool in aqueous bodies (Fuchs et al. 1997). Several biofilm marker properties, including biomass production, microbial species, photosynthetic machinery, pigment formation and enzyme activity, may be used in monitoring environmental pollution. The microbial species present in river biofilms vary depending on the season and as well as the level of pollution (Peacock et al. 2004). Contaminated sites with various toxic metals such as zinc and cadmium may also affect the microbial species; hence, it could be related to the level of contamination (Brümmer et al. 2000). Biofilm sampling has been shown to provide a more accurate assessment of heavy metal pollution in aquatic microbial populations (Bricheux et al. 2013). The algal biofilm could be used as an indicator of pollution due to changes in biomass in the presence of heavy metals (Ancion et al. 2013). The pigment-formation property of the biofilm may be altered after exposure to hazardous chemicals, which serves as a biomarker (Navarro et al. 2002, Dorigo et al. 2004, Dewez et al. 2008).

10.6 Bioremediation Using Extremophiles

During the past several decades, the methods for sustainable bioremediation of toxic metals have been explored using extremophilic bacteria. Acidophilic bacteria capable of sliving in extremely acidic environments are used as host cultures for the bioremediation of toxic chemicals (Gumulya et al. 2018, Saavedra et al. 2020). *Acidothiobacillus* species, a most ubiquitous acidophilic and chemolithotrophic bacterium, have been used to develop bioremediation techniques.

Acidothiobacillus ferroxidans, for instance, has demonstrated the ability to achieve industrial-scale bioleaching (Zhang et al. 2018). *Acidocella aromatica* and *Acidiphilium symbioticum* demonstrated the capability to reduce vanadium ions and biosorption of cadmium cations under extreme pH levels (Okibe et al. 2016).

A more effective method for toxic metal removal could be accomplished by using a microbe's consortium. The bioremediation of these pollutants by using an acid-isolated acidophilic microbe consortia was performed on a contaminated sediment site. An acidophilic microbial composed of *Acidothiobacillus thiooxidans*, *Leptospirillum ferrooxidans* and *Acidiphilium cryptum* demonstrated effectiveness in extracting over 90% of Cu^{2+}, Cd^{2+}, Hg^{2+} and Zn^{2+} (Beolchini et al. 2009) (Table 10.3).

Halophilic bacteria provide significant benefits in the remediation of hazardous contaminants in extreme saline environments. As marine bacteria can survive at high salinity, bioremediation using marine bacteria may be a viable alternative for cleaning seawater composed of toxic metals. *Vibrio harveyi* has shown a significant potential to acquire cadmium ions. The adsorption capacity was observed to be up to 23.3 mg Cd^{2+}/g dry cells (Abd-Elnaby et al. 2011). The other marine bacteria, *Enterobacter cloaceae*, could make complexes with Cd, Cu and Co from mixed-salts solutions (Iyer et al. 2005). In conjunction with marine bacteria, thermophilic microbes have high biosorption capability, indicating these microbes have a high potential for removing contaminants from polluted environments (Özdemir et al. 2013).

Another approach is to study new extremophilic bacterial enzymes: extremozymes, which have peculiar structure-function properties such as stability at high temperatures, severe pH, high ionic strength, in the presence of organic solvents and toxic metals (Cabrera and Blamey 2018). Extremophilic bacteria such as *Metallosphaera sedula*, *Leptospirillum ferriphilum* and *Sulfolobus solfataricus* have been sequenced, and segments containing the Hg-resistance gene merA have been discovered.

Table 10.3. Efficiency of extremophiles in bioremediation of toxic metals.

Toxic Metal	Bioremediation Mechanism	Extremophiles	Initial Conc.	Efficiency	References
As (III)	Bioleaching	*Acidothiobacillus ferrooxidans BY-3*	-	35.9%	Chen et al. 2011
U (VI)	Bioleaching	*At. ferrooxidans*	100 mg L^{-1}	50%	Romero-González et al. 2016
Cu (II)	Bioprecipitation	*Acidothiobacillus ferrivorans*	50 mM	> 99%	Jameson et al. 2010
V(V)	Bioreduction	*Acidocella aromatica*	1 mM	70%	Okibe et al. 2016
Cd (II)	Bio-accumulation	*Vibrio harveyi*	30–60 mg L^{-1}	84%	Chakravarty and Banerjee 2012
	Biosorption	*Acidiphilium symbioticum H8*	250 mg L^{-1}	248.62mg Cd (II)/g biomass	Abd-Elnaby et al. 2011
		Enterobacter cloaceae	100 mg L^{-1}	65%	Iyer et al. 2005
		Geobacillus thermantarcticus	50 mg L^{-1}	85.4%	Özdemir et al. 2013
		Anoxybacillus amylolyticus	50 mg L^{-1}	74.1%	Özdemir et al. 2013
Cr (VI)	Bioreduction	*Pyrobaculum islandicum*	600 μM	100%	Kashefi and Lovley 2000

As a result, extremophiles could be used in the elimination of toxic metals from toxic locations and sludges. However, more research into advancing technology for investigating microbial surroundings and gaining insight into the pathways analysis which influence microbial activity and metal degradation metabolic pathways under severe environments is necessary.

10.7 Recombinant DNA Technology for Bioremediation

Genetic engineering methods such as recombinant DNA technology, which include genetic transfer between bacteria, are used to manipulate the genetic makeup of organisms, which are known as Genetically Modified Microorganisms (GMM) or Genetically Engineered Microorganisms (GEM). Microbial Metal Resistance genes (MMRg) are useful genetic strategies for modifying bacteria. Several MMRg-based bioremediation technologies have been proposed (Zheng et al. 2019). Genetically modified microorganisms are being used for the effective removal of toxic metals from the environment. Customized microbial genes in genetically modified organisms provide novel metabolic pathways that improve the efficiency of bioremediation methods (Holliger and Zehnder 1996). In genetically modified microorganisms, the expression of genes is controlled, which is more important for the conversion of toxic metals to fewer toxic species/forms (Bondarenko et al. 2008). The microbial potential has been successfully utilized in different investigations.

The genome sequence of many bacterial communities involved in bioremediation has been done (Rahman et al. 2017, Yang et al. 2017). The genetic makeup of *Pseudomonas* sp. KT2440 (6.2 MB) has been analyzed, revealing the presence of genes encoding a wide range of enzymes and proteins and efflux pumps, each of which plays a critical role in the deterioration of a number of chemicals. Several additional investigations have revealed that microorganisms are engaged in the bioremediation of toxic metals, dyes and other chemicals, depending on their genome (Belda et al. 2016, Dangi et al. 2017).

A genetically engineered *E. coli* strain could successfully remove mercury from the contaminated area, including water or soil (Sharma and Shukla 2020). Transgenic bacteria with metallothioneins and polyphosphate kinase genes are appropriate for mercury bioremediation. Similarly, Cd-contaminated industrial effluent was reported to be treated using genetically modified *Ralstonia metallidurans* and *Caulobacter* spp. (Patel et al. 2010, Azad et al. 2014). Arsenic (As) may be removed from contaminated soil by transgenic bacteria expressing the ArsM gene through volatilization (Liu et al. 2019). Bioaccumulation has also been observed in genetically modified strains of *E. coli* with enhanced expression of the ArsR gene (Kostal et al. 2004). Nickel is one of the most refractory pollutants that could be accumulated by genetically engineered *E. coli* strain from an aqueous solution (Pacwa-Płociniczak et al. 2011). In another study, the *merB* and *merG* genes were added to the mercury-resistant *Cupriavidus metallidurans* strain MSR33 to regulate Hg biodegradation (Rojas et al. 2011).

Even though genetic engineering has devised a number of variants and bacterial species capable of degrading contaminants, there are many barriers. A major concern towards developed strains and microorganism's species is their low bioremediation efficiency. In microbiological ecology, the use of Stable Isotope Probing (SIP) and related methods have shown that *Rhodococcus* and *Pseudomonas*, which grow faster, are often used as biodegradation hosts but are much less effective in various natural conditions (Tahri et al. 2013). The main issue with this effective bioremediation state is keeping the ground conditions for engineered microbes under control. *P. fluorescens* HK44 has been actively monitored for the optimum ground conditions for bioremediation in the ecosystem (Ripp et al. 2000). As a result, when it comes to pollution clean-up, GEMs do not appear without the risks of their introduction into the environment. The adverse field conditions for the designed microorganisms constitute a significant issue in bioremediation. In naturalistic settings, bacteria such as *E. coli* (Bondarenko et al. 2008), *B. subtilis* (Ivask et al. 2011), and *P. putida* (Wu et al. 2006) have been used to focus on ways in which the molecular significance is primarily restricted. The need for adaptation of created bacterial strains to meet the new challenge is a crucial characteristic of open biotechnological applications. The most pressing issue is developing genetically modified bacteria with an acceptable degree of environmental assurance for field release in bioremediation. The efforts to evaluate the performance of modified bacteria under severe environmental conditions include endurance and horizontal gene transfer capacity, that may influence the native microflora of the environment. To avoid this, field bacteria are specially developed for *in vitro* bioremediation. It is unclear if the intentional discharge of genetically engineered bacteria for bioremediation has any negative impact on native microorganisms (Sayler and Ripp 2000). As a result, the survival of genetically modified microorganisms in a hostile environment remains a major concern.

10.8 Conclusions

Several studies have been conducted for the bioremediation of toxic metals from the environment. Microbial mediated bioremediation is found to be a sustainable approach. However, certain areas need to be more focused on and explored so that they can be applied for commercial use. As bioremediation in most studies is successful in the laboratory and controlled conditions, hence there is a need to explore the mechanism to carry out bioremediation under natural environment without controlled conditions. Besides this, GMOs are being used due to their high capability to perform bioremediation in comparison to wild-type microbial strains. However, there are several risks associated with this technology, such as ecological disturbance and horizontal gene transfer, which cause the limited use of GMOs in bioremediation. Extremophiles need to be explored more, and there might be unculturable microbes present in the environment, hence the need to develop the process to culture these unexplored microbes. The process development for scale-up is required for more commercial applications.

Acknowledgments

This work is supported by Grant no. BT/RLF/Re-entry/40/2017 from the Department of Biotechnology (DBT), Ministry of Science and Technology and SERB project file no: EEQ/2020/000614 Govt. of India. We acknowledge the CSIR pool scientific scheme Grant no. 9103, Govt. of India. Pictures are created by using www.biorender.com.

References

Abdel-Gadir, A., R. Berber, J. B. Porter, P. D. Quinn, D. Suri, P. Kellman et al. 2016. Detection of metallic cobalt and chromium liver deposition following failed hip replacement using T2* and R2 magnetic resonance. J. Cardiovasc. Magn. Reson. 18: 29.

Abd-Elnaby, H., G. M. Abou-Elela and N. A. El-Sersy. 2011. Cadmium resisting bacteria in Alexandria Eastern Harbor (Egypt) and optimization of cadmium bioaccumulation by *Vibrio harveyi*. Afr. J. Biotech. 10: 3412–3423.

Abedin, Md. J., J. Cotter-Howells and A. A. Meharg. 2002. Arsenic uptake and accumulation in rice (*Oryza sativa* L.) irrigated with contaminated water. Plant Soil. 240: 311–319.

Akcil, A., C. Erust, S. Ozdemiroglu, V. Fonti and F. Beolchini. 2015. A review of approaches and techniques used in aquatic contaminated sediments: metal removal and stabilization by chemical and biotechnological processes. J. Clean. Prod. 86: 24–36.

Al Khateeb, W. and H. Al-Qwasemeh. 2014. Cadmium, copper and zinc toxicity effects on growth, proline content and genetic stability of *Solanum nigrum* L., a crop wild relative for tomato; comparative study. Physiol. Mol. Biol. Plants. 20: 31–39.

Al Osman, M., F. Yang and I. Y. Massey. 2019. Exposure routes and health effects of heavy metals on children. Biomet.: Int. J. Role. Met. Ion. Biol. Biochem. Med. 32: 563–573.

Ancion, P. Y., G. Lear, A. Dopheide and G. D. Lewis. 2013. Metal concentrations in stream biofilm and sediments and their potential to explain biofilm microbial community structure. Environ. Pollut. 173: 117–124.

Andreoli, V. and F. Sprovieri. 2017. Genetic aspects of susceptibility to mercury toxicity: an overview. Int. J. Environ. Res. Public Health. 14: 93.

Arya, S. K. and B. K. Roy. 2011. Manganese induced changes in growth, chlorophyll content and antioxidants activity in seedlings of broad bean (*Vicia faba* L.). J. Environ. Biol. 32: 707–711.

Asrar, Z., R. Khavari-Nejad and H. Heidari. 2005. Excess manganese effects on pigments of Mentha spicata at flowering stage. Arch. Agron. Soil Sci. 51: 101–107.

Ay, T. A., iran, O. O. Fawole, S. O. Adewoye and M. A. Ogundiran. 2009. Bioconcentration of metals in the body muscle and gut of *Clarias gariepinus* exposed to sublethal concentrations of soap and detergent effluent. J. Cell Anim. Biol. 3: 113–118.

Aycicek, M., O. Kaplan and M. Yaman. 2008. Effect of cadmium on germination, seedling growth and metal contents of sunflower (*Helianthus annus* L.). Asian J. Chem. 20: 2663–2672.

Azad, M. A. K., L. Amin and N. M. Sidik. 2014. Genetically engineered organisms for bioremediation of pollutants in contaminated sites. Chin. Sci. Bull. 59: 703–714.

Baby, J., J. S. Raj, E. T. Biby, P. Sankarganesh, M. V. Jeevitha, S. U. Ajisha et al. 2010. Toxic effect of heavy metals on aquatic environment. Int. J. Biol. Chem. Sci. 4.

Bano, A., J. Hussain, A. Akbar, K. Mehmood, M. Anwar, M. S. Hasni et al. 2018. Biosorption of heavy metals by obligate halophilic fungi. Chemosphere. 199: 218–222.

Barta, J. A., C. A. Powell and J. P. Wisnivesky. 2019. Global epidemiology of lung cancer. Annals Glob. Health. 85(1): 8.

Belda, E., R. G. A. van Heck, M. José Lopez-Sanchez, S. Cruveiller, V. Barbe, C. Fraser et al. 2016. The revisited genome of *Pseudomonas putida* KT2440 enlightens its value as a robust metabolic chassis. Environ. Microbiol. 18: 3403–3424.

Bengtsson, M. M. and L. Øvreås. 2010. Planctomycetes dominate biofilms on surfaces of the kelp *Laminaria hyperborea*. BMC Microbiol. 10: 261.

Beolchini, F., A. Dell'Anno, L. de Propris, S. Ubaldini, F. Cerrone and R. Danovaro. 2009. Auto- and heterotrophic acidophilic bacteria enhance the bioremediation efficiency of sediments contaminated by heavy metals. Chemosphere. 74: 1321–1326.

Bertrand, J. C., P. Caumette, P. Lebaron, R. Matheron, P. Normand and T. Sime-Ngando. 2015. Environmental microbiology: fundamentals and applications. Springer Netherlands.

Bondarenko, O., T. Rõlova, A. Kahru and A. Ivask. 2008. Bioavailability of Cd, Zn and Hg in soil to nine recombinant luminescent metal sensor bacteria. Sensors. 8(11): 6899–6923.

Bricheux, G., G. le Moal, C. Hennequin, G. Coffe, F. Donnadieu, C. Portelli et al. 2013. Characterization and evolution of natural aquatic biofilm communities exposed *in vitro* to herbicides. Ecotoxicol. Environ. Saf. 88: 126–134.

Brümmer, I. H. M., W. Fehr and I. Wagner-Döbler. 2000. Biofilm community structure in polluted rivers: abundance of dominant phylogenetic groups over a complete annual cycle. Appl. Environ. Microbiol. 66: 3078–3082.

Brutti, C. S., R. R. Bonamigo, T. Cappelletti, G. M. Martins-Costa and A. P. S. Menegat. 2013. Occupational and non-occupational allergic contact dermatitis and quality of life: a prospective study. Anal. Bras. de Derma. 88: 670–671.

Burmølle, M., J. S. Webb, D. Rao, L. H. Hansen, S. J. Sørensen and S. Kjelleberg. 2006. Enhanced biofilm formation and increased resistance to antimicrobial agents and bacterial invasion are caused by synergistic interactions in multispecies biofilms. Appl. Environ. Microbiol. 72: 3916–3923.

Cabrera, M. Á. and J. M. Blamey. 2018. Biotechnological applications of archaeal enzymes from extreme environments. Biol. Res. 51: 1–15.

Cargnelutti, D., L. A. Tabaldi, R. M. Spanevello, G. de Oliveira Jucoski, V. Battisti, M. Redin et al. 2006. Mercury toxicity induces oxidative stress in growing cucumber seedlings. Chemosphere. 65: 999–1006.

Chakravarty, R. and P. C. Banerjee. 2012. Mechanism of cadmium binding on the cell wall of an acidophilic bacterium. Bioresour. Technol. 108: 176–183.

Chen, P., L. Yan, F. Leng, W. Nan, X. Yue, Y. Zheng et al. 2011. Bioleaching of realgar by *Acidithiobacillus ferrooxidans* using ferrous iron and elemental sulfur as the sole and mixed energy sources. Bioresour. Technol. 102: 3260–3267.

Chen, W., Y. Qu, Z. Xu, F. He, Z. Chen, S. Huang et al. 2017. Heavy metal (Cu, Cd, Pb, Cr) washing from river sediment using biosurfactant rhamnolipid. Environ. Sci. Pollut. Res. 24: 16344–16350.

Cheng, S. 2003. Heavy metal pollution in China: origin, pattern and control. Environ. Sci. Pollut. Res. 10: 192–198.

Cho, K. S., H. W. Ryu, I. S. Lee and H. M. Choi. 2002. Effect of solids concentration on bacterial leaching of heavy metals from sewage sludge. J. Air Waste Manag. Association. 52: 237–243.

Collinet, M. N. and D. Morin. 1990. Characterization of arsenopyrite oxidizing *Thiobacillus*. Tolerance to arsenite, arsenate, ferrous and ferric iron. Antonie van Leeuwenhoek. 57: 237–244.

Dangi, A. K., K. K. Dubey and P. Shukla. 2017. Strategies to Improve *Saccharomyces cerevisiae*: technological advancements and evolutionary engineering. Indian J. Microbiol. 57: 378–386.

Das, K. K., S. N. Das and S. A. Dhundasi. 2008. Nickel, its adverse health effects & oxidative stress. Indian J. Med. Res. 128: 412.

Davey, M. E. and G. A. O'toole. 2000. Microbial biofilms: from ecology to molecular genetics. Microbiol. Mol. Biol. Rev. 64: 847–867.

Demirevska-Kepova, K., L. Simova-Stoilova, Z. Stoyanova, R. Hölzer and U. Feller. 2004. Biochemical changes in barley plants after excessive supply of copper and manganese. Environ. Exp. Bot. 52: 253–266.

Deng, X., L. Chai, Z. Yang, C. Tang, Y. Wang and Y. Shi. 2013. Bioleaching mechanism of heavy metals in the mixture of contaminated soil and slag by using indigenous *Penicillium chrysogenum* strain F1. J. Hazard. Mater. 248-249: 107–114.

de Philippis, R., G. Colica and E. Micheletti. 2011. Exopolysaccharide-producing cyanobacteria in heavy metal removal from water: molecular basis and practical applicability of the biosorption process. Appl. Microbiol. Biotechnol. 92: 697–708.

Dewez, D., O. Didur, J. Vincent-Héroux and R. Popovic. 2008. Validation of photosynthetic-fluorescence parameters as biomarkers for isoproturon toxic effect on alga *Scenedesmus obliquus*. Environ. Pollut. 151: 93–100.

Doncheva, S., K. Georgieva, V. Vassileva, Z. Stoyanova, N. Popov and G. Ignatov. 2005. Effects of succinate on manganese toxicity in pea plants. J. Plant. Nutr. 28: 47–62.

Dorigo, U., X. Bourrain, A. Bérard and C. Leboulanger. 2004. Seasonal changes in the sensitivity of river microalgae to atrazine and isoproturon along a contamination gradient. Sci. Total. Environ. 318: 101–114.

D'Souza, C. and R. Peretiatko. 2002. The nexus between industrialization and environment: a case study of Indian enterprises. Environ. Manag. H. 13: 80–97.

Du, X., Y.-G. Zhu, W.-J. Liu and X.-S. Zhao. 2005. Uptake of mercury (Hg) by seedlings of rice (*Oryza sativa* L.) grown in solution culture and interactions with arsenate uptake. Environ. Exp. Bot. 54: 1–7.

Edwards, K. J., P. L. Bond, T. M. Gihring and J. F. Banfield. 2000. An archaeal iron-oxidizing extreme acidophile important in acid mine drainage. Science. 287: 1796–1799.

Flemming, H. C. and J. Wingender. 2001. Relevance of microbial extracellular polymeric substances (EPSs) - Part I: Structural and ecological aspects. Water Sci. Technol. IWA Publishing. 43(6): 1–8.

Fomina, M. and G. M. Gadd. 2014. Biosorption: current perspectives on concept, definition and application. Bioresour. Technol. 160: 3–14.

Fuchs, S., T. Haritopoulou, M. Schäfer and M. Wilhelmi. 1997. Heavy metals in freshwater ecosystems introduced by urban rainwater runoff—Monitoring of suspended solids, river sediments and biofilms. Water. Sci. Technol. 36: 277–282.

García-Esquinas, E., M. Pollán, J. G. Umans, K. A. Francesconi, W. Goessler, E. Guallar et al. 2013. Arsenic exposure and cancer mortality in a US-based prospective cohort: the strong heart study. Caner Epidemiol. Pren. Biomark. 22: 1944–1953.

Gieg, L. M., S. J. Fowler and C. Berdugo-Clavijo. 2014. Syntrophic biodegradation of hydrocarbon contaminants. Curr. Opin. Biotechnol. 27: 21–29.

Gumulya, Y., N. J. Boxall, H. N. Khaleque, V. Santala, R. P. Carlson and A. H. Kaksonen. 2018. In a quest for engineering acidophiles for biomining applications: challenges and opportunities. Genes. 9: 116.

Guo, T., L. Li, W. Zhai, B. Xu, X. Yin, Y. He et al. 2019. Distribution of arsenic and its biotransformation genes in sediments from the East China Sea. Environ. Pollut. 253: 949–958.

Gupta, A. and P. Sar. 2020. Characterization and application of an anaerobic, iron and sulfate reducing bacterial culture in enhanced bioremediation of acid mine drainage impacted soil. J. Environ. Sci. Health. - Part A Toxic/ Hazard. Subst. Environ. Eng. 55: 464–482.

Hamada, A. J., S. C. Esteves and A. Agarwal. 2013. A comprehensive review of genetics and genetic testing in azoospermia. Clinics. 68: 39–60.

Heck, J. E., A. S. Park, J. Qiu, M. Cockburn and B. Ritz. 2014. Risk of leukemia in relation to exposure to ambient air toxics in pregnancy and early childhood. Int. J. Hyg. Environ. Health. 217: 662–668.

Henriques, B., L. S. Rocha, C. B. Lopes, P. Figueira, R. J. R. Monteiro, A. C. Duarte et al. 2015. Study on bioaccumulation and biosorption of mercury by living marine macroalgae: prospecting for a new remediation biotechnology applied to saline waters. Chem. Eng. J. 281: 759–770.

Hines, M. E., E. N. Poitras, S. Covelli, J. Faganeli, A. Emili, S. Žižek et al. 2012. Mercury methylation and demethylation in Hg-contaminated lagoon sediments (Marano and Grado Lagoon, Italy). Estuar. Coast Shelf Sci. 113: 85–95.

Holliger, C. and A. J. B. Zehnder. 1996. Anaerobic biodegradation of hydrocarbons. Curr. Opin. Biotechnol. 7: 326–330.

Horemans, B., P. Breugelmans, J. Hofkens, E. Smolders and D. Springael. 2013. Environmental dissolved organic matter governs biofilm formation and subsequent linuron degradation activity of a linuron-degrading bacterial consortium. Appl. Environ. Microbiol. 79: 4534–4542.

Ivask, A., H. C. Dubourguier, L. Põllumaa and A. Kahru. 2011. Bioavailability of Cd in 110 polluted topsoils to recombinant bioluminescent sensor bacteria: effect of soil particulate matter. J. Soils Sediments. 11: 231–237.

Iyer, A., K. Mody and B. Jha. 2005. Biosorption of heavy metals by a marine bacterium. Mar. Pollut. Bull. 50: 340–343.

Jan, A. T., A. Ali and Q. M. R. Haq. 2011. Glutathione as an antioxidant in inorganic mercury induced nephrotoxicity. J. Postgrad. Med. 57: 72.

Jameson, E., O. F. Rowe, K. B. Hallberg and D. B. Johnson. 2010. Sulfidogenesis and selective precipitation of metals at low pH mediated by *Acidithiobacillus* spp. and acidophilic sulfate-reducing bacteria. Hydrometall. 104: 488–493.

Jayakumar, K., C. A. Jaleel and P. Vijayarengan. 2007. Changes in growth, biochemical constituents, and antioxidant potentials in radish (*Raphanus sativus* L.) under Cobalt Stress. Turk. J. Biol. 31: 127–136.

Jayakumar, K., C. Jaleel and M. Azooz. 2008. Phytochemical changes in green gram (Vigna radiata) under cobalt stress. Glob. J. Mol. Sci. 3(2): 46–49.

Jia, Y., H. Huang, M. Zhong, F. H. Wang, L. M. Zhang and Y. G. Zhu. 2013. Microbial arsenic methylation in soil and rice rhizosphere. Environ. Sci. Technol. 47: 3141–3148.

Jung, J. H., N. Y. Choi and S. Y. Lee. 2013. Biofilm formation and exopolysaccharide (EPS) production by *Cronobacter sakazakii* depending on environmental conditions. Food. Microbiol. 34: 70–80.

Kamal, M., A. E. Ghaly, N. Mahmoud and R. Côté. 2004. Phytoaccumulation of heavy metals by aquatic plants. Environ. Int. 29: 1029–1039.

Kaplan, D. I., R. Kukkadapu, J. C. Seaman, B. W. Arey, A. C. Dohnalkova, S. Buettner et al. 2016. Iron mineralogy and uranium-binding environment in the rhizosphere of a wetland soil. Sci. Total Environ. 569-570: 53–64.

Kartal, B., J. G. Kuenen and M. C. M. van Loosdrecht. 2010. Sewage treatment with anammox. Science. 328: 702–703.

Kashefi, K. and D. R. Lovley. 2000. Reduction of Fe(III), Mn(IV), and toxic metals at 100°C by *Pyrobaculum islandicum*. Appl. Environ. Microbiol. 66: 1050–1056.

Kennady, V., R. Verma and V. Chaudhiry. 2018. Detrimental impacts of heavy metals on animal reproduction: a review. J. Entomol. Zool. Stud. 06: 27–30.

Kodama, H., C. Fujisawa and W. Bhadhprasit. 2012. Inherited copper transport disorders: biochemical mechanisms, diagnosis, and treatment. Curr. Drug Metab. 13: 237–250.

Kostal, J., R. Yang, C. H. Wu, A. Mulchandani and W. Chen. 2004. Enhanced arsenic accumulation in engineered bacterial cells expressing ArsR. Appl. Environ. Microbiol. 70: 4582–4587.

Lacal, J., J. A. Reyes-Darias, C. García-Fontana, J. L. Ramos and T. Krell. 2013. Tactic responses to pollutants and their potential to increase biodegradation efficiency. J. Appl. Microbiol. 114: 923–933.

Lei, P., H. Zhong, D. Duan and K. Pan. 2019. A review on mercury biogeochemistry in mangrove sediments: hotspots of methylmercury production? Sci. Total. Environ. 680: 140–150.

Leyssens, L., B. Vinck, C. Van Der Straeten, F. Wuyts and L. Maes. 2017. Cobalt toxicity in humans—a review of the potential sources and systemic health effects. Toxicol. 387: 43–56.

Li, W. W. and H. Q. Yu. 2014. Insight into the roles of microbial extracellular polymer substances in metal biosorption. Bioresour. Technol. 160: 15–23.

Liu, H. L., C. W. Chiu and Y. C. Cheng. 2003. The effects of metabolites from the indigenous *Acidithiobacillus thiooxidans* and temperature on the bioleaching of cadmium from soil. Biotechnol. Bioeng. 83: 638–645.

Liu, S., Y. Zheng, Y. Ma, A. Sarwar, X. Zhao, T. Luo et al. 2019. Evaluation and proteomic analysis of lead adsorption by lactic acid bacteria. Int. J. Mol. Sci. 20: 5540.

Lim, H. W., S. A. B. Collins, J. S. Resneck, J. L. Bolognia, J. A. Hodge, T. A. Rohrer et al. 2017. The burden of skin disease in the United States. J. Am. Acad. Dermatol. 76: 958–972.

Lin, H.-J., T.-I. Sung, C.-Y. Chen and H.-R. Guo. 2013. Arsenic levels in drinking water and mortality of liver cancer in Taiwan. J. Hazard. Mater. 262: 1132–1138.

Mah, T. F. C. and G. A. O'Toole. 2001. Mechanisms of biofilm resistance to antimicrobial agents. Trends Microbiol. 9: 34–39.

Mahurpawar, M. 2015. Effects of heavy metals on human health. Int. J. Res. -Granth. 3: 1–7.

Malar, S., S. V. Sahi, P. J. C. Favas and P. Venkatachalam. 2015. Assessment of mercury heavy metal toxicity-induced physiochemical and molecular changes in *Sesbania grandiflora* L. Int. J. Environ. Sci. Technol. 12: 3273–3282.

Matz, C. and S. Kjelleberg. 2005. Off the hook—How bacteria survive protozoan grazing. Trends. Microbiol. 13: 302–307.

Meeker, J. D., M. G. Rossano, B. Protas, V. Padmanahban, M. P. Diamond, E. Puscheck et al. 2010. Environmental exposure to metals and male reproductive hormones: circulating testosterone is inversely associated with blood molybdenum. Fertil. Steril. 93: 130–140.

Mehana, E.-S. E., A. F. Khafaga, S. S. Elblehi, M. E. Abd El-Hack, M. A. E. Naiel, M. Bin-Jumah, S. I. Othman and A. A. Allam. 2020. Biomonitoring of heavy metal pollution using *Acanthocephalans parasite* in ecosystem: An updated overview. Animals 10(5): 811.

Meharg, A. A. 2004. Arsenic in rice—understanding a new disaster for South-East Asia. Trends. Plant Sci. 9: 415–417.

Messer, R. L. W., P. E. Lockwood, W. Y. Tseng, K. Edwards, M. Shaw, G. B. Caughman et al. 2005. Mercury (II) alters mitochondrial activity of monocytes at sublethal doses via oxidative stress mechanisms. J. Biomed. Mater. Res. Part B: Appl. Biomater. 75: 257–263.

Miqueleto, A. P., C. C. Dolosic, E. Pozzi, E. Foresti and M. Zaiat. 2010. Influence of carbon sources and C/N ratio on EPS production in anaerobic sequencing batch biofilm reactors for wastewater treatment. Bioresour. Technol. 101: 1324–1330.

Mishra, A. and A. Malik. 2013. Recent advances in microbial metal bioaccumulation. Crit. Rev. Environ. Sci. Technol. 43: 1162–1222.

Montazer-Rahmati, M. M., P. Rabbani, A. Abdolali and A. R. Keshtkar. 2011. Kinetics and equilibrium studies on biosorption of cadmium, lead, and nickel ions from aqueous solutions by intact and chemically modified brown algae. J. Hazard. Mater. 185: 401–407.

More, T. T., J. S. S. Yadav, S. Yan, R. D. Tyagi and R. Y. Surampalli. 2014. Extracellular polymeric substances of bacteria and their potential environmental applications. J. Environ. Manag. 144: 1–25.

Mujtaba Munir, M. A., G. Liu, B. Yousaf, M. U. Ali, Q. Abbas and H. Ullah. 2020. Synergistic effects of biochar and processed fly ash on bioavailability, transformation and accumulation of heavy metals by maize (*Zea mays* L.) in coal-mining contaminated soil. Chemosphere. 240: 124845.

Mulligan, C. N., R. N. Yong, B. F. Gibbs, S. James and H. P.J. Bennett. 1999. Metal removal from contaminated soil and sediments by the biosurfactant surfactin. Environ. Sci. Technol. 33: 3812–3820.

Mulligan, C. N. 2005. Environmental applications for biosurfactants. Environ. Pollut. 133: 183–198.

Musilova, J., J. Arvay, A. Vollmannova, T. Toth and J. Tomas. 2016. Environmental contamination by heavy metals in region with previous mining activity. Bull. Environ. Contam. Toxicol. 97: 569–575.

Nagajyoti, P. C., K. D. Lee and T. V. M. Sreekanth. 2010. Heavy metals, occurrence and toxicity for plants: a review. Environ. Chem. Lett. 8: 199–216.

Navarro, E., H. Guasch and S. Sabater. 2002. Use of microbenthic algal communities in ecotoxicological tests for the assessment of water quality: the Ter river case study. J. Appl. Phycol. 14: 41–48.

Nogawa, K., E. Kobayashi, Y. Okubo and Y. Suwazono. 2004. Environmental cadmium exposure, adverse effects and preventive measures in Japan. Biometals. 17: 581–587.

Nordberg, G. F. 2004. Cadmium and health in the 21st century—historical remarks and trends for the future. Biometals. 17: 485–489.

Okereafor, U., M. Makhatha, L. Mekuto, N. Uche-Okereafor, T. Sebola and V. Mavumengwana. 2020. Toxic metal implications on agricultural soils, plants, animals, aquatic life and human health. Int. J. Environ. Res. Public Health. 17: 2204.

Okibe, N., M. Maki, D. Nakayama and K. Sasaki. 2016. Microbial recovery of vanadium by the acidophilic bacterium, *Acidocella aromatica*. Biotechnol. Lett. 38: 1475–1481.

Özdemir, S., E. Klnç, A. Poli and B. Nicolaus. 2013. Biosorption of heavy metals (Cd2+, Cu2+, Co 2+, and Mn2+) by thermophilic bacteria, *Geobacillus thermantarcticus* and *Anoxybacillus amylolyticus*: equilibrium and kinetic studies. Bioremediat. J. 17: 86–96.

Pacwa-Płociniczak, M., G. A. Płaza, Z. Piotrowska-Seget and S. S. Cameotra. 2011. Environmental applications of biosurfactants: recent advances. Int. J. Mol. Sci. 12: 633–654.

Pal, A. and A. K. Paul. 2008. Microbial extracellular polymeric substances: central elements in heavy metal bioremediation. Indian J. Microbio. 48: 49–64.

Patel, J., Q. Zhang, R. M. L. McKay, R. Vincent and Z. Xu. 2010. Genetic engineering of *caulobacter crescentus* for removal of cadmium from water. Appl. Biochem. Biotechnol. 160: 232–243.

Peacock, A. D., Y. J. Chang, J. D. Istok, L. Krumholz, R. Geyer, B. Kinsall et al. 2004. Utilization of microbial biofilms as monitors of bioremediation. Microb. Ecol. 47: 284–292.

Plum, L. M., L. Rink and H. Haase. 2010. The essential toxin: impact of zinc on human health. Int. J. Environ. Res. Public Health. 7: 1342–1365.

Pratt, L. A. and R. Kolter. 1999. Genetic analyses of bacterial biofilm formation. Curr. Opin. Microbiol. 2: 598–603.

Rahman, S. F., R. S. Kantor, R. Huddy, B. C. Thomas, A. W. van Zyl, S. T. L. Harrison et al. 2017. Genome-resolved metagenomics of a bioremediation system for degradation of thiocyanate in mine water containing suspended solid tailings. Microbiol. Open. 6: e00446.

Rajaganapathy, V., F. Xavier, D. Sreekumar and P. K. Mandal. 2011. Heavy metal contamination in soil, water and fodder and their presence in livestock and products: a review. J. Environ. Sci. Technol. 4: 234–249.

Ray, R. R. 2016. Adverse hematological effects of hexavalent chromium: an overview. Interdiscip. Toxicol. 9: 55–65.

Ray, R. R. 2017. Review article. Adverse hematological effects of hexavalent chromium: an overview. Interdiscip. Toxicol. 9: 55–65.

Reddy, A. M., S. G. Kumar, G. Jyothsnakumari, S. Thimmanaik and C. Sudhakar. 2005. Lead induced changes in antioxidant metabolism of horsegram (*Macrotyloma uniflorum* (Lam.) Verdc.) and bengalgram (*Cicer arietinum* L.). Chemosphere. 60: 97–104.

Rinklebe, J. and S. M. Shaheen. 2017. Redox chemistry of nickel in soils and sedimentsA review. Chemosphere. 179: 265–278.

Ripp, S., D. E. Nivens, Y. Ahn, C. Werner, J. Jarrell IV, J. P. Easter et al. 2000. Controlled field release of a bioluminescent genetically engineered microorganism for bioremediation process monitoring and control. Environ. Sci. Technol. 34: 846–853.

Rojas, L. A., C. Yáñez, M. González, S. Lobos, K. Smalla and M. Seeger. 2011. Characterization of the metabolically modified heavy metal-resistant *Cupriavidus metallidurans* strain MSR33 generated for mercury bioremediation. PLoS ONE. 6(3): e17555.

Romero-González, M., B. C. Nwaobi, J. M. Hufton and D. J. Gilmour. 2016. *Ex-situ* bioremediation of U(VI) from contaminated mine water using *Acidithiobacillus ferrooxidans* Strains. Front. Environ. Sci. 4: 39.

Romero-Puertas, M. C., M. Rodríguez-Serrano, F. J. Corpas, M. Gómez, L. a. D. Río and L. M. Sandalio. 2004. Cadmium-induced subcellular accumulation of O2·− and H2O2 in pea leaves. Plant Cell Environ. 27: 1122–1134.

Saavedra, A., P. Aguirre and J. C. Gentina. 2020. Biooxidation of Iron by *Acidithiobacillus ferrooxidans* in the presence of D-Galactose: understanding its influence on the production of EPS and cell tolerance to high concentrations of iron. Front. Microbiol. 11: 759.

Sayler, G. S. and S. Ripp. 2000. Field applications of genetically engineered microorganisms for bioremediation processes. Curr. Opin. Biotechnol. 11: 286–289.

Sharma, B. and P. Shukla. 2020. Futuristic avenues of metabolic engineering techniques in bioremediation. Biotech. Appl. Biochem. https://doi.org/10.1002/bab.2080.

Sheikh, I. 2016. Cobalt Poisoning: a comprehensive review of the literature. J. Med. Toxicol. Clin. Forensic Med. 2.

Shekar, C. C., D. Sammaiah, T. Shasthree and K. J. Reddy. 2011. Effect of mercury on tomato growth and yield attributes. Int. J. Pharm. Bio. Sci. 2.

Shekhar, S., A. Sundaramanickam and T. Balasubramanian. 2015. Biosurfactant producing microbes and their potential applications: a review. Crit. Rev. Environ. Sci. Technol. 45: 1522–1554.

Sheldon, A. and N. W. Menzies. 2005. The effect of copper toxicity on growth and morphology of Rhodes grass (*Chlorisgayana*) in solution culture. Plant Soil, 278.

Sobha, K., A. Poornima, P. Harini and K. Veeraiah. 2007. A study on biochemical changes in the fresh water fish, Catla catla (Hamilton) exposed to the heavy metal toxicant cadmium chloride. Kathmandu Univ. J. Sci. Eng. Technol. 3: 1–11.

Spangler, A. H. and J. G. Spangler. 2009. Groundwater manganese and infant mortality rate by county in North Carolina: an ecological analysis. EcoHealth. 6: 596–600.

Štyriaková, I., I. Štyriak, A. Balestrazzi, C. Calvio, M. Faè and D. Štyriaková. 2016. Metal leaching and reductive dissolution of iron from contaminated soil and sediment samples by indigenous bacteria and *Bacillus* isolates. Soi. Sediment Contam. 25: 519–535.

Sun, W., B. Zhu, F. Yang, M. Dai, S. Sehar, C. Peng et al. 2021. Optimization of biosurfactant production from Pseudomonas sp. CQ2 and its application for remediation of heavy metal contaminated soil. Chemosphere 265: 129090.

Sutherland, I. W. 2001. The biofilm matrix—An immobilized but dynamic microbial environment. Trends Microbiol. 9: 222–227.

Tahri, N., W. Bahafid, H. Sayel and N. el Ghachtouli. 2013. Biodegradation: involved microorganisms and genetically engineered microorganism. Ch. 11. pp. 289–320. *In:* R. Chamy and F. Rosenkranz [eds.]. Biodegradation - Life of Science. IntechOpen.

Vieira, C., S. Morais, S. Ramos, C. Delerue-Matos and M. B. P. P. Oliveira. 2011. Mercury, cadmium, lead and arsenic levels in three pelagic fish species from the Atlantic Ocean: Intra- and inter-specific variability and human health risks for consumption. Food Chem. Toxicol. 49: 923–932.

Wang, Y. and M. Greger. 2004. Clonal differences in mercury tolerance, accumulation, and distribution in willow. J. Environ. Qual. 33: 1779–1785.

Wang, M., J. Zou, X. Duan, W. Jiang and D. Liu. 2007. Cadmium accumulation and its effects on metal uptake in maize (*Zea mays* L.). Bioresour. Technol. 98: 82–88.

Woo, S., S. Yum, H.-S. Park, T.-K. Lee and J.-C. Ryu. 2009. Effects of heavy metals on antioxidants and stress-responsive gene expression in Javanese medaka (*Oryzias javanicus*). Comp. Biochem. Physiol. Part C: Toxicol. Pharmacol. 149: 289–299.

Wu, C. H., T. K. Wood, A. Mulchandani and W. Chen. 2006. Engineering plant-microbe symbiosis for rhizoremediation of heavy metals. Appl. Environ. Microbiol. 72: 1129–1134.

Wu, M., Y. Li, J. Li, Y. Wang, H. Xu and Y. Zhao. 2019. Bioreduction of hexavalent chromium using a novel strain CRB-7 immobilized on multiple materials. J. Hazard. Mater. 368: 412–420.

Xie, X., W. Zhu, N. Liu and J. Liu. 2013. Bacterial community composition in reclaimed and unreclaimed tailings of Dexing copper mine, China. Afr. J. Biotechnol. 12(30): 4841–4849.

Xin, B., D. Zhang, X. Zhang, Y. Xia, F. Wu, S. Chen et al. 2009. Bioleaching mechanism of Co and Li from spent lithium-ion battery by the mixed culture of acidophilic sulfur-oxidizing and iron-oxidizing bacteria. Bioresour. Technol. 100: 6163–6169.

Yang, S., M. Yu and J. Chen. 2017. Draft genome analysis of *Dietzia* sp. 111N12-1, isolated from the South China Sea with bioremediation activity. Braz. J. Microbiol. 48: 393–394.

Yao, J., L. Tian, Y. Wang, A. Djah, F. Wang, H. Chen et al. 2008. Microcalorimetric study the toxic effect of hexavalent chromium on microbial activity of Wuhan brown sandy soil: An *in vitro* approach. Ecotoxicol. Environ. Saf. 69: 289–295.

Yourtchi, M. S. and H. R. Bayat. 2013. Effect of cadmium toxicity on growth, cadmium accumulation and macronutrient content of durum wheat (Dena CV.). Int. J. Agric. Crop Sci. (IJACS). 6: 1099–1103.

Yusuf, M., Q. Fariduddin, P. Varshney and A. Ahmad. 2012. Salicylic acid minimizes nickel and/or salinity-induced toxicity in Indian mustard (*Brassica juncea*) through an improved antioxidant system. Environ. Sci. Pollut. Res. 19: 8–18.

Zdrojewicz, Z., E. Popowicz and J. Winiarski. 2016. Nickel—role in human organism and toxic effects. Pol. Merk. Lek. 41: 115–118.

Zeng, X., S. Wei, L. Sun, D. A. Jacques, J. Tang, M. Lian et al. 2015. Bioleaching of heavy metals from contaminated sediments by the *Aspergillus niger* strain SY1. J. Soil Sediment 15: 1029–1038.

Zhang, H. L., W. Fang, Y. P. Wang, G. P. Sheng, R. J. Zeng, W. W. Li et al. 2013. Phosphorus removal in an enhanced biological phosphorus removal process: roles of extracellular polymeric substances. Environ. Sci. Technol. 47: 11482–11489.

Zhang, S., L. Yan, W. Xing, P. Chen, Y. Zhang and W. Wang. 2018. *Acidithiobacillus ferrooxidans* and its potential application. Extremophiles. 22: 563–579.

Zheng, X., L. Chen, M. Chen, J. Chen and X. Li. 2019. Functional metagenomics to mine soil microbiome for novel cadmium resistance genetic determinants. Pedosphere. 29: 298–310.

Zhou, Z. S., S. Q. Huang, K. Guo, S. K. Mehta, P. C. Zhang and Z. M. Yang. 2007. Metabolic adaptations to mercury-induced oxidative stress in roots of *Medicago sativa* L. J. Inorg. Biochem. 101: 1–9.

Zhu, X., H. Yu, H. Jia, Q. Wu, J. Liu and X. Li. 2013. Solid phase extraction of trace copper in water samples via modified corn silk as a novel biosorbent with detection by flame atomic absorption spectrometry. Anal. Methods. 5: 4460–4466.

Zulfiqar, U., M. Farooq, S. Hussain, M. Maqsood, M. Hussain, M. Ishfaq, M. Ahmad and M. Z. Anjum. 2019. Lead toxicity in plants: impacts and remediation. J. Environ. Manag. 250: 109557.

Chapter 11

Lead Induced Toxicity, Detoxification and Bioremediation

Shalini Dhiman,[1] *Arun Dev Singh,*[1] *Isha Madaan,*[2,7]
Raman Tikoria,[3] *Driti Kapoor,*[4] *Priyanka Sharma,*[5]
Nitika Kapoor,[6] *Geetika Sirhindi,*[2] *Puja Ohri*[3] and
Renu Bhardwaj[1,*]

11.1 Introduction

Lead is a hazardous element that comes from various technogenic sources like ammunition, batteries-based industries, bangle manufacturing, building material, ceramic ware, cosmetics, gasoline, glassware, plastic pipes, paints with Pb pigments, petrochemicals, radiation protection, leaded fuels, bullets, fishing sinkers, mining, smelting, electroplating of the metallic ores (Dotaniya et al. 2020). All the possible routes of Pb exposure beyond its permissible limits lead to Pb toxicity in soil, water, air, humans, animals as well as in plant systems. Pb inside soil disrupts soil properties and soil ecosystem mainly by reducing nutrients availability to plants, which in turn effects soil-forming processes, microbiota, soil health, crop quality and productivity. However, Pb uptake beyond permissible limits inside plant systems causes many abnormal morphological and physiological symptoms like disturbance in photosynthesis, water potential, nutrient uptake, respiration and causes nuclear damage. Moreover, plants also show tolerance and detoxification mechanism against heavy metal stress such as compartmentalization, metallothionine, phytochelatins, Pb-immobilization by plants roots exudates, organic acid, etc. (Mitra et al. 2020).

[1] Department of Botanical and Environmental Sciences, Guru Nanak Dev University, Amritsar, Punjab, India, 143005.

[2] Department of Botany, Punjabi University, Patiala, India, 147002.

[3] Department of Zoology, Guru Nanak Dev University, Amritsar, India, 143005.

[4] Department of Botany, School of Bioengineering and Biosciences, Lovely Professional University, Phagwara, India, 144402.

[5] School of Bioengineering Sciences & Research, MIT-ADT Loni Kalbhor, Pune, Maharashtra, India, 412201.

[6] PG Department of Botany, Hans Raj Mahila Maha Vidyalaya, Jalandhar, Punjab, India, 144623.

[7] Government College of Education, Jalandhar, Punjab, India, 144001.

* Corresponding author: renubhardwaj82@gmail.com

Harmless removal of Pb from Pb-contaminated sites is the need of this modern century. Such safe disposal and proper remediation of Pb from polluted sites have been done using various modern tools and techniques such as bioremediation, especially phytoremediation. Some plants are very capable of remediating contaminants from contaminated sites, and such processes are collectively also known as phytoremediation (Jagetiya and Kumar 2020). Soils polluted with Pb are remediated by other living forms such as fungi, algae, microbes, etc. A considerable number of studies have been successfully carried out on Pb-bioremediation because these techniques are acceptable, safer, cheaper and more ecofriendly than any other techniques for the removal of Pb.

11.2 Lead Toxicity in Soil Plant System

Various heavy metals act as pollutants and create serious threats to plants' health by altering their metabolism, growth rate and other physiology. Among them, lead (Pb) is the second most dangerous heavy metal due to its potential toxicity to plants (Shahid et al. 2011). In the soil-plant system, Pb comes from fumes of automobiles, factories, storage batteries, mining, metal plating and finishing operations. Once it gets established in the soil, it is more easily accumulated in it (Hadi and Aziz 2015).

Plant nutrient dynamics is an important factor for determining the growth and yield of the crop. In healthy soil, a good amount of plant biomass and yield is produced. Large numbers of organic acid (having low molecular weight) are secreted by the plants into the soil. Most of the microorganisms get their raw food material from them and enhance the nutrient mineralization processes. Studies showed that approximately 40–65% of the photosynthesis products were released by the living plants as root exudates. These exudates are constituted of amino acids, plant growth hormones, etc. (Dotaniya et al. 2020). The toxic levels of Pb inside soil negatively impact the normal root exudation processes and decline their secretion. These conditions are very critical in the presence of higher concentrations of Pb as the fertility of soil decreases up to a level from where land will not produce even a single blade of grass. Pb creates oxidative stress on plants and generates reactive oxygen species, which are phytotoxic. The most common symptoms of Pb toxicity are loss of chlorophyll content, disruption in respiration and photosynthesis, etc. It has been reported that younger plants and leafy crops are more susceptible to Pb toxicity as compared to older and less leafy crops.

Similarly, peri-urban vegetables also tend to be more susceptible to lead toxicity. Once established in the food chain, Pb toxicity greatly affects human health, especifically infants. Such infants show symptoms like stunted growth, mental retardation, hair loss and organ and metabolic process failure. Different plant species get affected to varying degrees of toxicity due to Pb. Some plants have their own excluder strategies through roots to reduce the effect of Pb toxicity. Some plants have a higher potential for tolerating Pb toxicity. These plants are referred to as hyperaccumulators, and are also used in phytoremediation (Dotaniya et al. 2017, Dotaniya et al. 2018a).

11.2.1 Lead Toxicity in Soil

Lead is found naturally in the crust layer of the earth, having levels less than 50 mg kg^{-1} (Arias et al. 2010). But anthropogenic activities may lead to change this level very frequently. These anthropogenic activities may result in the accumulation of Pb onto the surface layer of the soil. Further, its content declines with an increase in depth (Cecchi et al. 2008). According to ATSDR (2020), the Pb level in soil was increased by one thousand-fold in the past three centuries because of the availability of more advanced techniques that can detect Pb in very small fractions (ATSDR 2020). In soil, Pb may be present in many forms, including free metal ion, complex forms (devise complexes with inorganic constituents like carbonate, bicarbonate, sulfate and chloride ion) or maybe found in its organic forms such as fulvic acid, amino acid and humic acid. Alternatively, adsorption of Pb occurs on the surface of particles like iron oxides, biological and organic matter, etc. (Vega et al. 2010, Sammut et al. 2010). However, due to strong binding affinities of Pb with colloidal/ organic material, it created the possibility that only a tiny fraction of Pb is soluble in soil, and therefore less accessible for plant uptake (Punamiya et al. 2010). An investigatory report given by Kumpiene et al. (2017) showed that lead is a versatile element present in the soil with a background amount of 27 mg kg⁻¹. Lead-contaminated soil is divided into five categories depending on the level of contamination, i.e., extremely low (< 150 ppm), low (150–400 ppm), moderate (400–1000 ppm), high (1000–2000 ppm) and extremely high (> 2000 ppm). The degree of Pb contamination also varies from one season to another, and in different mediums (Patel et al. 2010). Reports suggested that a high level of lead is reported in some places where anthropogenic activities are prominent. For example—at smelting sites, lead concentration in soil ranged from 10 to 7100 mg kg⁻¹ (Chlopecka et al. 1996), road dust (105–110 mg kg⁻¹) (Bi et al. 2018), mining site (132–45016 mg kg⁻¹) (Higueras et al. 2017). Further, the industrial site contained lead at a level of 42–131 mg kg⁻¹ (He et al. 2017); in previous garden soil, it was 1020–1030 mg kg⁻¹ (Egendorf et al. 2018). Similarly, Pb in flooded soil was in between 105–115 mg kg⁻¹ (Antić-Mladenović et al. 2017), and in the soil of shooting ranges was 32,500–33,500 mg kg⁻¹ (Mariussen et al. 2018).

11.2.1.1 *How Lead Toxicity Affects Soil Ecosystem*

The presence of Pb in excessive concentrations in the soil affects the soil ecosystem in several ways:

11.2.1.1.1 Reducing Nutrient Availability to Plants

Nutrients inside the soil ecosystem play a vital role in the growth of plants and animals. An increase in the amount of lead in the soil decreased the other metal ions' availability to the plants. Soil contaminated with Pb may generate abiotic stress to the living flora, which further reduces the uptake of other macronutrients ions like potassium, calcium and magnesium and micronutrients like zinc, copper, nickel as well as nitrogen uptake (Wu et al. 2011). For example, enhancement of Pb level reduces the nickel ion concentration in spinach in Madhya Pradesh, India. Further, elevated concentrations of Pb in black soil also affected its enzymatic properties (Pipalde and Dotaniya 2018).

11.2.1.1.2 Effect on soil Forming Processes

There are various pedogenic processes like oxidation-reduction, mineralization-immobilization, dissolution-precipitation, sorption-desorption, etc., which are responsible for the formation of soil. These reactions are involved in regulating the dynamics of various nutrient cycles and ultimately, soil biodiversity by creating competition among the nutrients. Once elevation occurs in the concentration of Pb, it leads to disruption in these reactions and affects the dynamics, ultimately pedogenesis (Dotaniya and Pipalde 2018).

11.2.1.1.3 Effect on Soil Health and Microbiota

Lead toxicity affects the microbial count in the soil ecosystem by direct or indirect effects. The decomposition rate of organic material was suppressed due to the presence of moderate or excessive concentrations of metal ions inside the soil (Bahar et al. 2012). According to reports, approximately 8–16 times higher concentration of lead was discovered in crops developed on polluted sites (smelter plant) in France (Douay et al. 2013). Further, enhancement in the level of Pb leads to a decline in the biodiversity, richness and population count of microorganisms (Banat et al. 2010). If Pb concentration was reached to 100 mg/kg, it reduced the rate of carbon mineralization as well as enzymatic activities such as dehydroascorbate reductase. Reports suggested that lead toxicity reduces the amount of nitrogen fixation organisms, thus affecting the rate of biological nitrogen fixation (Dotaniya et al. 2020).

11.2.1.1.4 Crop Quality and Productivity

Crop loss and fall in quality are among the most typical consequences of Pb toxicity. Lead accumulation in crops reduce the food grain quality, which has detritus effects on human health after consuming these crops. Vegetables and their parts are more prone to the accumulation of toxic metals as compared to edible crops. Further, crops grown on land continuously supplied by sewage water also have a high level of lead, which ultimately leads to a bad color of the crops (Bhupal Raj et al. 2009). Onion crops quality was found inferior in the soil irrigated by lead-contaminated water in Ratlam, India (Meena et al. 2020).

Severe loss of yield in crops was observed in soybean, gram, fenugreek and garlic. These crops got affected by Pb contaminated water, which decreased farmers' output (Panwar et al. 2010).

11.2.1.2 Factors Affecting Pb Toxicity in Soil

Lead toxicity in the soil is affected by several factors. These factors might be topographic, climatic or due to human activities. Some of these factors are:

11.2.1.2.1 Soil Properties

Soil properties like pH, porosity, amount of organic matter, cation exchange capacity, clay content, soil structure and the presence/absence of other ions inside soils are the key factors that ultimately determine lead toxicity (Meena et al. 2020). If clay content is high in the soil, it forms complexes with humus, which ultimately reduces the mobility of Pb (Dotaniya et al. 2016). These complexes also serve as sources of carbon for soil microorganisms, enhancing their growth and variety. Further, the organic matter containing the carboxyl group binds with Pb and forms complexes, which ultimately decrease its availability (Dotaniya et al. 2020). Solubility and accumulation of Pb in the soil are influenced by the presence of phosphates and carbonates. Additionally, the availability of lead to plants is typically low in the pH range from 5 to 7 (Blaylock et al. 1997). The presence of several metal ions like zinc, copper, nickel and chromium also had adverse effects on Pb accumulation in soil (Orroño et al. 2012). Rhizosphere containing soil fauna has a vital role in Pb availability, as found in the case of *Thlaspi caerulescens* and *Lantana camera* (Jusselme et al. 2012).

11.2.1.2.2 Concentration of other Elements

An increase in the level of nickel ion to the level of 100 mg kg^{-1} leads to a reduction in Pb toxicity (Pipalde and Dotaniya 2018). By introducing acid-neutralizing material into the soil, the mobility of Pb can be reduced. Water stress might be helpful in lowering Pb concentration as on adding water, Pb concentration is diluted (Dotaniya et al. 2018b). The toxicity of soil also depends upon the forms of Pb. For example, Pb is present in oxyanion complexes in soil and water streams (Dotaniya et al. 2018c). Similarly, an increased amount of phosphatic fertilizers could be helpful in reducing the Pb accumulation (Lenka et al. 2016).

11.2.1.2.3 Industries

Agricultural soil/lands that are situated closer to industrial areas are more prone to lead contamination because of leakage of lead-containing pollutants (Saha et al. 2013). Pb level was found greater in the soil surrounding the Coimbatore area near an electroplating and paint industry. Similarly, in Tamil Nadu's Dindigul district, the cement factory is the major source of lead poisoning (Meena et al. 2020). Lead contamination into the soil is also caused by industries that make daily life products such as Pb-based paints, solder, ceramics and pesticides. Industries that are involved in mining smelting and tailing release high amounts of lead into nearby soil (Mitra et al. 2020, Anju and Banerjee 2011).

11.2.1.2.4 Urbanization

Lead increase in urban soils might be due to increasing industrialization. Combustion of leaded petrol results in vehicular emission containing tetraethyl Pb, which contributes significantly to lead in urban areas. As per USEPA standards, the threshold limit of Pb in the soil is 400 mg/L, and in portable water, it is 0.01 mg/L (Bureau of Indian Standard) (Mitra et al. 2020).

11.2.2 *Lead Uptake and Toxicity in Plant System*

Pb is available in different exchangeable and non-exchangeable forms in soil. Among these, Pb (II) is the only form that can be absorbed by plants. Higher concentrations of Pb in soil may lead to its entry inside the plant system. Soil pH is also one of the essential factors which affect the bioavailability of Pb to plants. Pb is adsorbed by plant roots at low pH, as high pH favors the formation of stable and insoluble covalent Pb compounds such as acetates, carbonates and hydroxides (Shahid et al. 2017). The most common mechanism by which Pb enters the plant body is through roots. However, Pb is translocated through the apoplast pathway and Ca channels on the plasma membrane to above ground parts of the plant. Nevertheless, the level of Pb that reaches the shoot region is very low as most of it is sequestered through chelation with the mucilage and galacturonic acids of the cell walls of roots. Even lower concentrations of Pb that enter the plant system causes significant disturbances in plant fundamental processes.

Pb being a toxic heavy metal, has no biological function in plant primary as well secondary processes. Excessive amounts of Pb in growing environments may negatively influence plant activities from seed germination to crop yield. Pb contamination disrupts seed germination by inhibiting activities of various enzymes such as amylases, proteases thereby leading to retarded seedling development. Decreased root, shoot length, leaf area and impaired photosynthesis are other symptoms reflected in plants grown in Pb prone areas.

Pb pollution inhibited root growth, root hair differentiation, water and essential divalent cations absorption such as Mg, Ca, Mn, Fe, etc., by roots, which resulted in damage to normal machinery of plants (Rucińska-Sobkowiak 2016). Pb may also directly interact with many functional groups such as -OH, -COOH, -SH, CO, CHO, etc., thereby, leading to conformational changes in primary biomolecules, i.e., lipids, proteins, carbohydrates and nucleic acids. In addition to this, the production of Reactive Oxygen Species (ROS) under Pb stress has also been reported in many research studies. These ROS may oxidize vital plant metabolites and hence, hamper the overall efficiency of various metabolic processes such as photosynthesis, respiration, transpiration, etc. (Ghori et al. 2019, Hasanuzzaman et al. 2020).

Pb is also shown to inhibit the action of enzymes involved in the production of chlorophyll synthesis and upregulates the activity of chlorophyllase, chlorophyll degrading enzyme which results in decreased chlorophyll content in plants (Yang et al. 2020). In certain cases, the replacement of the central Mg atom of chlorophyll by Pb has also been found. Other processes which are disturbed include hill reaction, Calvin cycle and grana stacking in the chloroplast. All these factors lead to impaired photosynthesis, one of the most indispensable processes influencing crop productivity (Santos et al. 2015). The toxic effects of Pb on plant health have been summarized in Figure 11.1. A large amount of nuclear damage resulting from Pb toxicity has also been reported in different plants, including polyploidy, replication errors leading to single stranded DNA formation, incomplete cell cycle and chromosome stickiness (Pizzaia et al. 2019). In addition to these, modifications in the level of water potential, membrane integrity and hormonal signaling may also occur due to Pb-induced induced plant damage. All these alterations assertively result in dwindled crop productivity and hence create a serious threat to the world's health and economy.

11.3 Lead Tolerance and Detoxification Mechanism Inside Plant System

Pb enters the environment as a result of a number of anthropogenic activities. Plants being sessile organisms, cannot change their habitat in response to such conditions. Therefore, they adapt to such an environment by stimulating intracellular mechanisms that ensure the normal survival of plants. These mechanisms may include Pb accumulation in plant parts without any visible symptoms, sequestration of Pb into vacuoles, exudation from roots into soils in the insoluble precipitated form that cannot be reabsorbed. All such mechanisms are referred to as tolerance and detoxification mechanisms (Figure 11.2). Usually, entry of Pb stimulate either or all the following detoxification processes:

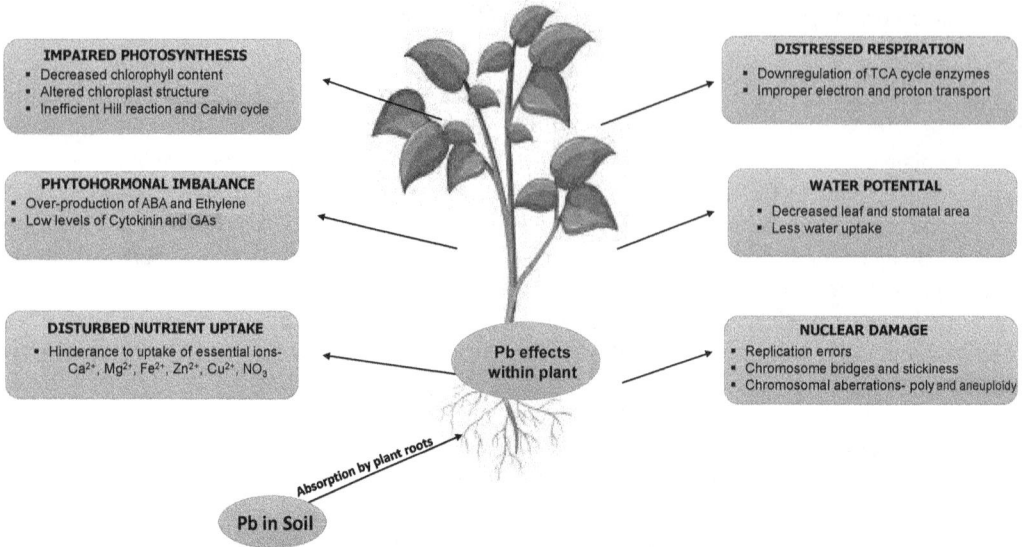

Figure 11.1. Uptake of lead from soil and its toxic effects within the plant.

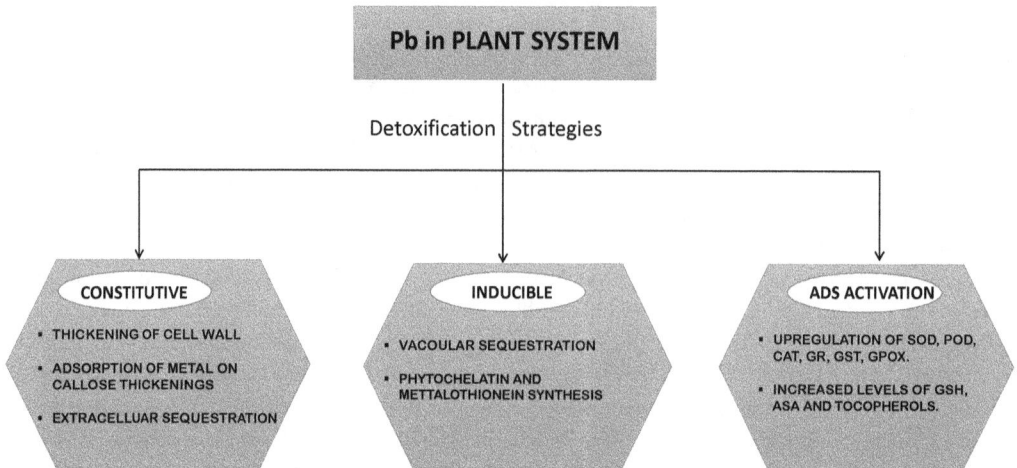

Figure 11.2. Lead detoxification mechanisms within a plant system.

11.3.1 Constitutive Mechanisms

Such mechanisms act as the first level of plant defense against mild Pb stress. This method of Pb detoxification involves the adsorption of Pb on the cellular compartments, mainly on thickened cell walls. Entry of Pb stimulates the synthesis of polysaccharides, callose and pectin, which causes widening of the cell wall and hence restricts entry of Pb into the cell (Rucińska-Sobkowiak et al. 2013).

It also provides new sites for the Pb adsorption and enhances its sequestration extracellularly.

11.3.2 Inducible Mechanism

These mechanisms consume a large amount of energy and are usually activated under conditions of high Pb toxicity. It involves the task of different transporters and ATPases for the efflux of symplastic Pb into vacuoles. Along with this, the synthesis of Pb binding ligands such as phytochelatins (PCs), metallothioneins (MTs) is also induced. PCs and MTs aid in Pb detoxification by its sequestration within cytoplasm before it enters vacuoles and chloroplast, thereby minimizing the toxic effects of Pb on the plant cell (Pourrut et al. 2013).

11.3.3 Activation of Antioxidative Defense System (ADS)

Harmful effects of Pb-induced ROS generation and oxidative stress are reciprocated by ADS activation, which involves both enzymatic and non-enzymatic antioxidants. The activities of several enzymes, i.e., superoxide dismutase (SOD), guaicol peroxidase (POD), catalase (CAT), Glutathione Reductase (GR), Glutathione-S-Transferase (GST), glutathione peroxidase (GPOX), polyphenol peroxidase (PPO), etc., is found to be upregulated in many plant species such as wheat, rice, sunflower, mung bean, black grams and maize (Kandziora-Ciupa et al. 2013). Levels of non-enzymatic antioxidants, including glutathione, tocopherols and ascorbic acid, have also been reported to get enhanced significantly as their functional groups help in the sequestration of Pb by forming stable metal-chelating complexes.

These detoxification mechanisms help in the survival of plants in Pb stressed habitats. However, it is the need of the hour to cease human activities which introduce Pb into the environment as under adverse conditions, even these processes may fail and hence lead to plant death.

11.4 Lead Bioremediation

Extensive reports indicate that continuous industrial, mining, urban development and overpopulation are releasing toxic metal lead (Pb) persistently in soil flora beyond the permissible threshold value. Being a non-biodegradable and biologically non-functional element, Pb is accountable for its toxicity in rhizospheric flora, environment and all-living entities, resulting in consequential health issues to human beings. Therefore, there is an urgent need to explore efficient and ecological friendly sustainable techniques to minimize heavy metal contaminants like Pb, which are released in soil, water, etc. The conventional physico-chemical methods such as landfill, soil incineration, chemical precipitation, electrokinetic system, ion exchange, ultrafiltration, adsorption, etc., are effective, but these approaches are very expensive and also generate secondary waste, that are the major constraints for soil fertility, loss of biota and rhizospheric microbial flora (Akmal and Jianming, 2009, Li et al. 2014, Akhtar et al. 2017).

Hence, advancements in bioengineering are allowing more competitive approaches to grow efficient, cost-effective, environment-friendly and sustainable methods for the remediation/ detoxification of non-biodegradable toxic Pb metal contaminants from the environment into non-hazardous forms (Su 2014, Kushwaha et al. 2018). The latest researches are signifying a sustainable bioremediation strategy for remediation of lead contaminated soil repositories either via phytoremediation methods (using plants) or microbial remediation (via microbes) (Wuana and Okieimen 2011, Kushwaha et al. 2018).

Bioremediation is a process where highly toxic, harmful pollutants such as Pb are transformed/ degraded to a non-reactive oxidative state or reduced to minerals via *in-* and *ex-situ* practices (Kumar et al. 2021). Plants via the process of phytoremediation, recently played a promising role to extract, immobilize, stabilize and transform the toxic ionization state of lead (Pb) from contaminated sites through different processes viz. phytoextraction, phytostabilization, accumulation of toxic compounds and rhizofiltration (Chibuike and Obiora 2014, Rigoletto et al. 2020) (Figure 11.3). Currently, phytoremediation techniques have emerged as techniques of great interest due to their cost-effectiveness, efficiency and environmental sustainability (Prakash et al. 2013).

Similarly, microbial communities also engage in different defensive strategies to combat fatal toxicity by influencing the activity of Pb through siderophore, ion chelator production, biosorption, extracellular sequestration, efflux, compartmentalization, etc. Thus, microbes and plants are promising resolutions for sustainable development by restoring the contaminated sites of soil, water and the environment. Next divergent bioremediation strategies by using microbes, plants, and algae for the detoxification, stabilization and removal of lead from contaminated areas are described.

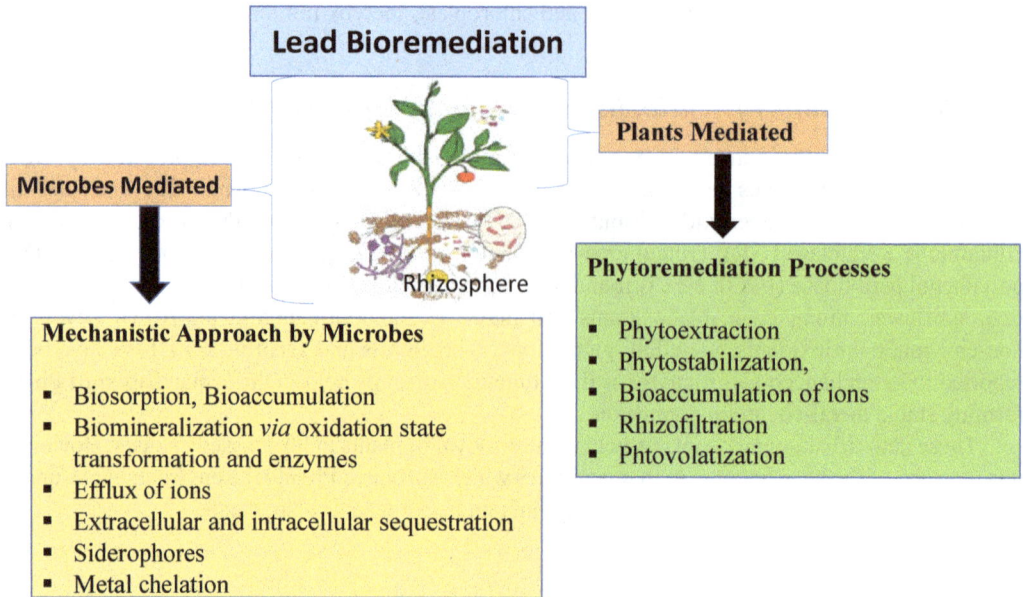

Figure 11.3. Overview of bioremediation strategies to reduce lead toxicity.

11.4.1 Bacteria Assisted Remediation of Pb

Microorganisms are prevalent in the environment and have evolved unique systems for determining the fate of hazardous heavy metals. Microbes like bacteria, algae, archaebacteria, fungi, etc., mediate crucial interactions with ions of metals without unveiling negative effects on their structure, action and metabolism (Rahman and Singh 2020). However, some toxic metals affect the activities and survival of microbe cells above the threshold limit by imitating their biological function, biomolecular denaturation and accelerating ROS production (Prabhakaran et al. 2016). But, due to the behavior of microorganisms like varying phenotypic expressions, they combat metal toxicity and show a metal-resistant function (Yin et al. 2019). Recently, a lot of research has been conducted to unravel the action of microorganisms to diminish the concentration of Pb via biosorption, oxidation state transformation, efflux, sequestration, metal chelation, use of enzymes, synthesis of exopolysaccharides, intracellular bioaccumulation, etc. Microbial assisted remediation of Pb shows key advantages due to their biochemical features like specificity, viability for genetic modulation and suitability for *in situ* conditions (Singh and Prasad 2015, Ullah et al. 2015).

Biosorption is a technique where metal ions bind to the peptidoglycan cell wall of bacteria to confine their entry into the cell and enhance the extracellular sequestration of toxic ions. It is observed that *Pseudomonas aeruginosa* biosorb lead, on account of functionally charged groups available at the surface viz. sulfonate, amine, carboxyl, hydroxyl, phosphonate, etc. (Gabr et al. 2008, Kushwaha et al. 2018). However, efflux is a process to maintain the homeostasis of ions intracellularly under the action of membrane transport proteins. Several researches revealed that many ATPase allied transmembrane transporter proteins facilitated lead resistance in microbes, e.g., *Staphylococcus aureus, R. metallidurans, Cupriavidus metallidurans* in the efflux of Pb (Borremans et al. 2001, Hynninen et al. 2009). The toxicity of HMs to organisms relies on their bioavailability; thus, microorganisms boost the sequestration of toxic metals at the membrane level. Researchers have found that exopolysaccharide (EPS) help in bioaccumulation of lead in *Pseudomonas aeruginosa, Marinobacter* sp., *Acinetobacter junii*, etc. (Bramhachari et al. 2007, De and Ramaiah 2007, Kushwaha et al. 2017).

In the environment, the most common microbes are bacteria. During remediation of heavy metals via bioaccumulation, sequestration and adsorption action, bacteria onset these mechanisms by the functional groups present/occurring on their surface to adsorb metal ions (Yin et al. 2016). In soil, bacteria restricted Pb concentration via fostering insoluble lead complexes with hydroxide, sulfide and carbonates, transforming the active form into stable insoluble state.

Further, sequestration of toxic ions is mainly through exopolysaccharide, an organic polysaccharide with smaller proteins and lipids. Many microorganisms like *Xanthomonas, Bacillus, Agrobacterium, Alcaligenes, Pseudomonas* spp., etc., have been identified as genera of EPS-producing organisms to achieve heavy metal remediation by utilizing the charged property of EPS, where they are incorporated with abundant anionic functional groups (Tayang and Songachan 2021). This mechanism is critical in the process of biomineralization, metal ions biosorption and bioaccumulation (Thakare et al. 2021). Similarly, Chen et al. (2015) described *Bacillus thuringiensis* as a potential biosorbent for Pb (II) transformation. Further, *Bacillus cereus* could transform Pb into Pb hydroxyapatite via enzymatic action. An experiment conducted by scientists revealed that the bacterium *Streptomyces* and *Staphylococcus* showed a prominent binding affinity for lead and other metals. Therefore, they can be effectively used for the biosorption of lead (Sahmoune 2018).

Li et al. (2017) findings revealed that the bacterial strains of *Pseudomonas* sps., can efficiently absorb Pb (II) from wastewater sites. For the first time, Borremans et al. (2001) found a lead-resistance strain, i.e., CH34 in *R. metallidurans*, which enhances uptake and efflux mechanism by the *pbr* operon system. Later many studies have shown the involvement of specific genes expression for resistance to Pb by metallothionein proteins, specifically in *P. aeruginosa* (Kumari and Das 2019). Kang et al. (2016) confirmed microbial (bacterial) remediation of Pb-contaminated soils due to the function of precipitation, sequestration or variation in the oxidation state of Pb. They revealed the synergistic effect of bacterial consortium (*E. cloacae, Sporosarcina, Viridibacillus arenosi*, and *Enterobacter cloacae*) on a mixture of Pb along with other heavy metals against single strain culture. These bacteria are accountable for the transformation of HMs by enzymes production (Huang et al. 2009). It was observed that *Bacillus iodinium, Klebsiella aerogenes* and *Bacillus pumilus* precipitate Pb (II) into PbS 9 (Govarthanan et al. 2013).

It has been noted that c-type cytochromes and porin–cytochrome proteins in outer membrane proteins in the microbes are involved in declining contaminants toxicity (Shi et al. 2016). Several studies have shown that microbes transformed the state of metal by changing the valence status of metals via redox-mediated processes (Dixit et al. 2015, Shi et al. 2016). The bacterial organisms such as *Bacillus* sps., *A. eutrophus, Pseudomonas* sps., produce siderophores enabling extraction of Pb from soil (Naik and Dubey 2017, Kalita and Joshi 2017). Similarly, in another report, the positive interaction among siderophores and metal Pb and Ni was revealed by the bacteria *P. aeruginosa* (Braud et al. 2009, 2010). The *Pteris vittata* plant exhibited rapid growth in the Pb-contaminated area, which was enabled due to *Pseudomonas* spp., resistance against metal Pb via the process of extracellular sequestration (Manzoor et al. 2019). Therefore, these examples of evidence show the significance of siderophore-producing bacteria, which cause extraction and mobilization of Pb from contaminated soil.

11.4.2 Plants Assisted Remediation of Pb

Phytoremediation, or the use of plants to clean up contaminated soil, is a well-known practice that is both environmentally beneficial and cost-effective (Ali et al. 2013). Various phytoremediation processes like phytoextraction, rhizodegradation/rhizofiltration, phytovolatilization, phytodegradation/ phytoaccumulation, phytostabilization and phytorestoration are included in the phytoremediation of polluted soil. This technique is also performed to eliminate heavy metals through immobilization or detoxification (Ali et al. 2013). During the process of phytoremediation, heavy metals are accumulated through the cultivation of hyper accumulator plants as these plants have great potential to absorb heavy metals and then gather them in the aboveground plant parts (Aliyu and Adamu 2014). Heavy metals are then degraded within the plants through internal (via metabolic aspects) or external (through the release of some chemicals in the rhizosphere by plant roots) breakdown through the process of phytodegradation or phytotransformation. Hyperaccumulators plants have the highest metal accumulation ability but little biomass efficiency, whereas non-hyperaccumulators plants have high biomass efficacy and lesser accumulation potential (Ojuederie and Babalola 2017).

Rhizofiltration is one of the techniques used to clean up different metal-contaminated water resources through the precipitation of metals in the plant roots (Zulfiqar et al. 2019). Numerous plant cultivars like *Nicotiana tabaccum, Zea mays, Brassica juncea, Helianthus annus,* and *Spinacia oleracea* have been studied to clean Pb from the terrestrial and aquatic systems (Camargo et al. 2003). Phytostabilization is another *in-situ* plant-mediated remediation of metals via inhibiting their mobility and bioavailability (Kunito et al. 2001, Ali et al. 2013). *Miscanthus floridulus* showed greater phytoremediation potential for Pb metal (Cheng et al. 2016). Co-plantation of *Pteris vittata* and the *Ricinus communis* exhibited significant phytoextraction potential for the remediation of Pb from the soil through an elevation in the yield of *P. vittata* after Pb uptake (Yang et al. 2017).

Pb content was observed to decline from 218 to 32 mg kg^{-1} in stem and 7232 to 1196 mg kg^{-1} in *Lolium italicum* and *Festuca arundinaceae* roots system, respectively due to the presence of compost (Kushwaha et al. 2018). *Ludwigia stolonifera* exhibited significant efficiency of 99% for the removal of Pb (Saleh et al. 2019). *Cyamopsis tetragonoloba* and *Sesamum indicum* showed significant Pb phytoremediation potential with higher Pb accumulation in the root. It was further reported that these plant cultivars can be grown on marginally contaminated soils and could be practised for phytostabilization (Amin et al. 2018). *Glycine max* exhibited good potential for Pb removal from the soil, and the content of organic carbon of soil was also reduced in the remediated soil (Aransiola et al. 2013). Biochar prepared from rice husk and pinewood showed remarkable ability to remove the Pb through the process of adsorption (Liu and Zhang 2009, Raj and Maiti 2020). *Alyssum maritimum* showed good phytoremediation potential by accumulating Pb in various plant organs, with the highest in the leaf followed by the root and stem. It has also been suggested that this plant can be involved in the family of hyperaccumulator plants, with a high potential for remediating Pb from soil (Solgi et al. 2020). *Athyrium wardii* was reported to show significant Pb phytostabilization ability via accumulating most of the Pb in roots (Zhao et al. 2016). *Salix purpurea* exhibited higher phytostabilization as compared to *Salix viminalis*; however, both the species were shown to develop lengthy roots under Pb stress (Sylvain et al. 2016). Overall, Pb content in different essential food crops should be evaluated cautiously to mitigate the Pb triggered health impacts. Various plant species that can be practised for Pb phytoremediation are presented in Table 11.1.

11.4.3 Algae and Fungi Assisted Remediation of Pb

Plant-associated algae and fungi play an important function in the remediation of Pb from the soil by promoting the Pb accumulation in plants and plant growth in contaminated soil. Recent studies reported that plant endophytes enhance their phytoremediation potential (Waranusantigul et al. 2011, Deng et al. 2011, Gupta et al. 2013). A study on Arbuscular Mycorrhizal (AM) fungi *viz. Glomus deserticola* by Arias et al. (2010) revealed its Pb accumulation potential in *Prosopis* sp. (mesquite plants). The Transmission Electron Microscopy (TEM) revealed *G. deserticola*'s presence within the roots of *Prosopis* sp., which improved metal tolerance and accumulation in plants. AM fungi reduce the Pb toxicity by inducing its accumulation in the plant's roots, stems and leaves. Further, phytoremediated Pb was found to be accumulated in the xylem and phloem cells of mesquite plants. Another fungal endophyte, *Mucor* sp., CBRF59 helps in the remediation of Pb from contaminated soil that has been recovered from rape roots growing in a heavy metal contaminated environment. Inoculation of plant roots with mycelia of *Mucor* sp., CBRF59 significantly enhanced the Pb availability of soil by 77% from contaminated soils (Deng et al. 2011). AM fungi are also observed to reduce oxidative stress and improve plants growth in heavy metal contaminated sites. *Glomus mosseae* inoculated Vetiver grass has increased chlorophyll content and decreased levels of GSH, thereby increasing its ability to withstand metal-induced stress better. *G. mosseae* enhanced the accumulation and translocation of Pb in Vetiver grass shoots (Punamiya et al. 2010).

Table 11.1. Role of various plant species for the remediation of Pb.

Sr. No.	Plant species	Effect on Pb accumulation	References
1.	*Spinacia oleracea*	Application of *S. oleracea* with tartaric acid markedly increased the uptake of Pb and its translocation from the root to stem through the process of phytoextraction.	Khan et al. 2016
2.	*Schoenoplectus californicus*	*S. californicus* growing in marshy environments showed increased Pb absorption by largely retaining it in their roots.	Arreghini et al. 2017
3.	*Sesuvium portulacastrum*	Enhancement in Pb levels in the shoot of *S. portulacastrum* compared to *Brassica juncea* and Pb accumulation in its upper portions was observed to be 3.4 mg g^{-1} of dry weight.	Zaier et al. 2014
4.	*Lolium multiflorum*	Significant phytoremediation potential was shown by *L. multiflorum* through the process of phytoextraction. The contents of heavy metals transported from roots to the upper parts were greater than that remained in the soil.	Salama et al. 2016
5.	*Achillea wilhelmsii, Erodium cicutarium, Nonnea persica* and *Mentha longifolia*	These plant species were most appropriate for phytostabilizing Pb and had good potential for phytoremediation.	Mahdavian et al. 2017
6.	*Coronopus didymus*	The roots of *C. didymus* showed maximum Pb accumulation compared to the shoot and *C. didymus* was recognized as a good phytoremediation candidate in Pb-polluted soils.	Sidhu et al. 2018
7.	*Chrysanthemum indicum*	Greater remediation efficiency was shown by *C. indicum* through increased clean-up of the Pb contaminated soils with maximum concentration in the root followed by the shoot and flower.	Mani et al. 2015
8.	*Phragmites australis*	95% of the Pb was removed from the Pb contaminated water by using *P. australis* through the process of phytostabilization.	Bello et al. 2018
9.	*Daucus carota*	Pb accumulation was observed to be increased in the *D. carota* via chelate application and can be used as a hyperaccumulator plant for Pb-phytoextraction and phytostabilization from polluted soils.	Babaeian et al. 2016

Different strategies adopted by microorganisms to survive in heavy metal contaminated soils are extrusion of metal ions by using metal efflux pumps, biotransformation of ions, intra/extracellular metal sequestration, enzymatic usage, exopolysaccharide (EPS) generation, and metallothionein and bio surfactants synthesis, etc. (Dixit et al. 2015, Igiri et al. 2018). Microorganisms can further detoxify the metal ions by several different methods, including ion exchange, electrostatic interaction, precipitation, surface complexation, etc. (Yang et al. 2015). Microorganisms have negatively charged groups on their cell surface that facilitate them to bind to cationic metal ions (Gavrilescu 2004).

Fungal hyphae remediate the heavy metal contaminated soils by intracellular sequestration of toxic metal ions. Chitin, lipids, mineral ions, N-polysaccharide, polyphosphates and proteins are major constituents of the cell wall of fungi. Fungal hyphae and their spores can eradicate the heavy metal ions from the soil by ATPase pump-mediated uptake, extracellular and intracellular precipitation and change in the oxidation state of metal ions. On the fungi cell wall outer surface, there is the presence of various metal ions binding ligands/functional groups that enhanced the rate of binding of toxic metals to hyphae, thereby reducing the availability of these toxic metals to plants. The foremost metal-binding ligands present on the surface of the cell wall are the hydroxyl group, carboxyl group, phosphoryl group, sulfate group, sulfite group, ester, amine group, carboxylate group and sulfanyl group. Out of these functional groups, the amine group is most involved in metal absorption as it can bind with both cationic as well as anionic metal ions by surface complexation and electrostatic interaction, respectively (Gupta et al. 2015, Xie et al. 2016).

Various reports exhibited that active as well as inactive fungal cells, play an essential role in the adsorption of inorganic metal ions (Srivastava and Thakur 2006, Tiwari et al. 2013, Igiri et al. 2018). Lakkireddy and Kües (2017) reported that *Coprinopsis atramentaria* could accumulate 94.7% of 800 mg L^{-1} of Pb^{2+}. As a result, it is being identified as a good heavy metal ion accumulator for mycoremediation. Some of the fungal biomasses *viz. Aspergillus niger, Rhizopus oryzae, Penicillium chrysogenum* and *Saccharomyces cerevisiae* are also effective in converting the most hazardous oxidation state of heavy metal to less toxic/non-toxic oxidation state of heavy metals (Park et al. 2005). Inoculum of Arbuscular Mycorrhizal Fungi (AMF) is found to enhance the Pb remediation efficiency of Japanese clover (*Kummerowia striata* (Thunb.)) and barnyard grass (*Echinochloa crus-galli* L.) (Chen et al. 2005). Fungi like *Aspergillus* sp. and *Coprinopsis* sp. are largely used as biosorbents for eliminating toxic metals with great potential for metal absorption and recovery (Akar et al. 2005, Dursun et al. 2003).

Biosurfactants produced by fungi also play a major role in cleaning heavy metals from contaminated soil. Luna et al. (2016) reported that anionic biosurfactant from *Candida sphaerica* has 79% removal efficiency for Pb from heavy metal contaminated soil. The biosurfactant was found to be effective in removing the exchangeable, oxide, carbonate and organic fractions of heavy metals by forming complexes with metal ions.

Yeast biosurfactants are also found to be effective in cleaning heavy metals and petroleum derivatives from contaminated soils by reducing soil permeability. The crude biosurfactant significantly reduced the concentration of Pb and other heavy metals from the test sample of soil. Biosurfactants being amphoteric in nature, not only help in eliminating heavy metals but can also be applied to remove hydrophobic organic compounds. They reduce the interfacial tension and solubilize hydrocarbons in the aqueous phase or capture the oil droplets within their micelles. On the other hand, anionic nature biosurfactants capture the metal ions through electrostatic interactions or complexation (Rufino et al. 2011). In recent years, biosurfactants have received a lot of interest for their biodegradable nature, low toxicity and diversity. Several yeast strains such as *S. cerevisiae, Rhodotorula pilimanae, Hansenula polymorpha, Yarrowia lipolytica* and *Rhodotorula mucilage* have been utilized to convert more toxic forms of heavy metals to non-toxic ones (Ksheminska et al. 2008, Chatterjee et al. 2012).

Phycoremediation is an important aspect that deals with removing or degrading heavy metals from contaminated sites with the help of algal biomasses. Features that make dead algae biomass an ideal candidate for the removal of heavy metals include the presence of sulfate and carboxylic acid functional groups on the cell wall that facilitate metal adsorption and large surface area/volume ratios. In comparison to other microbial biosorbents, algae are autotrophic, require little nutrients and produce large amounts of biomass. Heavy metal removal has been achieved using these biosorbents with a high sorption capacity. Algal biomass bioremediates the heavy metal contaminated effluent either through adsorption or by integrating inside the cells (Abbas et al. 2014, Chabukdhara et al. 2017, He and Chen 2014).

When compared to other microbial biosorbents, algal biomasses had biosorption effectiveness of 15.3–84.6%. Ion exchange techniques are used to accomplish this (Mustapha and Halimoon 2015). For successful heavy metals cleanup from the polluted area, algal biomass has been immobilized using various chemical pretreatments, which lead to the formation of stable cellular aggregates with appropriate size, efficient mechanical strength, rigidity, porosity and increased biomass concentration (Laxman and More 2002). Red marine algae *Jania ruben* L. was found to be effectively bioadsorb Pb, which was further confirmed by thermal analysis (Hanbali et al. 2014). Various reports are available in literature regarding the use of algal biomasses in the decontamination of toxic metals. Goher et al. (2016) reported that dead cells of *Chlorella vulgaris* could be used for the removal of copper (Cu^{2+}), cadmium (Cd^{2+}) and lead (Pb^{2+}) ions from an aqueous solution under several conditions of biosorbent dosage, pH and contact time. The biomass of *C. vulgaris* removed cadmium (Cd^{2+}), copper (Cu^{2+}), and lead (Pb^{2+}) at the rate of 95.5, 97.7 and 99.4%, respectively, constituting a combined solution of 50 mg dm^{-3} of each metal ion. Thus applying appropriate microbial inoculum under specific conditions of pH, temperature and dosage could aid plants to remove heavy metals like Pb from contaminated soil efficiently.

Remediation of toxic contaminants from various substrates by employing a microbial agent, fungus, is also known as mycoremediation. Fungi have a capacity to recover HMs owing to filamentous structures that exhibit a charged group on the cell wall. In addition to functional groups, they also display features of metal transporters by enzymatic activity, vacuolar sequestration and antioxidant systems (Kumar 2017, Vacar et al. 2021). Fawzy et al. (2017) screened many fungal species like *Emercilla quadrillineata*, *Rhizopus stolonifier*, *Aspergillus niger*, etc., and observed an effective resistance even at higher concentrations of Pb in contaminated soil. Similarly, other fungal isolates such as *Rhizophagus irregularis* and *Funneliformis mosseae* showed promising high biomass of *Helianthus annuus* against Pb ions application (Hassan et al. 2013). Fungi, particularly *Aspergillus niger*, revealed a detoxification mechanism in the environment by the processes of compartmentalized sequestration, biosorption and chelation with organic acids. Thus, such pathways can be used for decreasing lead levels through immobilization or mobilization (Bellion et al. 2006, Iram et al. 2015).

Some species of fungi like *Metarhizium* and *Paecilomyces* can convert metallic lead into chloropyromorphite in a lead mining site. This transformation occurred because of the organic acid secreted by the fungi, which caused the precipitation of Pb (Rhee et al. 2012, Rigoletto et al. 2020). Alongwith this, Povedano-Priego et al. (2017) confirmed that biomineralization of lead phosphate in decaying wood caused tolerance to isolated fungi. They noted that *Penicillium* and *Asperigillus* strains showed more tolerance to heavy metals. A summary of various algae and fungi involved in the remediation of Pb from contaminated sites is presented in Table 11.2.

Table 11.2. Groups of microbes involved in phycoremediation/mycoremediation of Pb contaminated sites.

Microbes Group	Bioremediators	Sorption efficiency (mg g^{-1})	References
Fungi	*Aspergillus niger*	34.4	Dursun et al. 2003
	Saccharomyces cerevisiae	80	Farhan and Khadom 2015
	Phanerochaete chrysosporium	88.16	Iqbal and Edyvean 2004
	Botrytis cinerea	107.1	Akar et al. 2005
	Aspergillus terreus	59.67	Joshi et al. 2011
Algae	*Codium vermilara*	63.3	Romera et al. 2007
	Cladophora sp.	13.7	Lee and Chang 2011
	Spirogyra sp.	38.2	Lee and Chang 2011
	Spirogyra sp.	140	Gupta and Rastogi 2008
	Asparagopsis armata	63.7	Romera et al. 2007
	Cystoseira barbata	196.7	Yalçın et al. 2012

11.5 Conclusion

Heavy metal Pb has a large number of applications in industries such as lead used as lead acetate in sweeteners, installations of drinkable water, paints, additives used in gasoline and many more that further increase the probability of Pb release, its exposure and penetration in all existing organisms of our ecosystem. Beyond permissive limits, Pb accumulation inside the organisms creates toxic hazards and severe morphological and physiological implications and finally decreases the efficiency of the organism. Thus, the removal of Pb-based hazardous products should be firmly perused in order to improve and avoid environmental as well as health risks. An essential criteria for attaining Pb based contaminant-free environment require the development of highly efficient bioremediation technologies such as phytoremediation, mycoremediation, phycoremediation and microbial remediation, including research in the field of bioremediation. Additionally, advanced transgenic plants and microbes, as well as modern genetic strategies widen the application of bioremediation and finally lead to lesser environmental disturbances and production of a much safer and healthy ecosystem.

References

Abbas, H. S., M. I. Ismail, M. T. Mostafa and H. A. Sulaymon. 2014. Biosorption of heavy metals: a review. J. Chem. Sci. Technol. 3: 74–102.

Akar, T., S. Tunali and I. Kiran. 2005. *Botrytis cinerea* as a new fungal biosorbent for removal of Pb (II) from aqueous solutions. Biochem. Eng. J. 25(3): 227–235.

Akhtar, N., S. Kha, I. Malook, S. U. Rehman and M. Jamil. 2017. Pb-induced changes in roots of two cultivated rice cultivars grown in lead-contaminated soil mediated by smoke. Environ. Sci. Pollut. Res. 24: 21298–21310.

Akmal, M. and X. Jianming. 2009. Microbial biomass and bacterial community changes by Pb contamination in acidic soil. J. Agric. Biol. Sci. 1: 30–37.

Ali, H., E. Khan and M. A. Sajad. 2013. Phytoremediation of heavy metals—concepts and applications. Chemosphere. 91(7): 869–881.

Aliyu, H. G. and H. M. Adamu. 2014. The potential of maize as phytoremediation tool of heavy metals. Eur. Sci. J. 10(6).

Amin, H., B. A. Arain, T. M. Jahangir, M. S. Abbasi and F. Amin. 2018. Accumulation and distribution of lead (Pb) in plant tissues of guar (*Cyamopsis tetragonoloba* L.) and sesame (*Sesamum indicum* L.): profitable phytoremediation with biofuel crops. Geol. Ecol. Landscapes. 2(1): 51–60.

Anju, M. and D. K. Banerjee. 2011. Associations of cadmium, zinc, and lead in soils from a lead and zinc mining area as studied by single and sequential extractions. Environ. Monit. Assess. 176(1): 67–85.

Antić-Mladenović, S., T. Frohne, M. Kresović, H. J. Stärk, Z. Tomić, V. Ličina and J. Rinklebe. 2017. Biogeochemistry of Ni and Pb in a periodically flooded arable soil: fractionation and redox-induced (im) mobilization. Environ. Manag. Today. 186: 141–50.

Aransiola, S. A., U. J. J. Ijah and O. P. Abioye. 2013. Phytoremediation of lead polluted soil by *Glycine max* L. Appl. Environ. Soil Sci. 2013.

Arias, J. A., R. Jose, P. Videa, J. T. Ellzey, M. Ren, M. N. Viveros and J. L. Gardea-Torresdey. 2010. Effects of *Glomus deserticola* inoculation on *prosopis*: enhancing chromium and lead uptake and translocation as confirmed by X-Ray mapping, ICP-OES and TEM techniques. Environ. Exp. Bot. 68(2): 139–148.

Arreghini, S., L. de Cabo, R. Serafini and A. F. de Iorio. 2017. Effect of the combined addition of Zn and Pb on partitioning in sediments and their accumulation by the emergent macrophyte *Schoenoplectus californicus*. Environ. Sci. Pollut. Res. 24(9): 8098–8107.

ATSDR. 2020. "Toxicological Profile for Lead." Atlanta, Georgia.

Babaeian, E., M. Homaee and R. Rahnemaie. 2016. Chelate-enhanced phytoextraction and phytostabilization of lead-contaminated soils by carrot (*Daucus carota*). Arch. Agron. Soil Sci. 62(3): 339–358.

Bahar, M. D., M. M. Megharaj and R. Naidu. 2012. Arsenic bioremediation potential of a new arsenite-oxidizing bacterium *Stenotrophomonas* Sp. MM-7 Isolated from Soil. Biodegradat. 23(6): 803–12.

Banat, I. M., A. Franzetti, I. Gandolfi, G. Bestetti, M. G. Martinotti, L. Fracchia, T. J. Smyth and R. Marchant. 2010. Microbial biosurfactants production, applications and future potential. Appl. Microbiol. Biotechnol. 87(2): 427–44.

Bellion, M., M. Courbot, C. Jacob, D. Blaudez and M. Chalot. 2006. Extracellular and cellular mechanisms sustaining metal tolerance in ectomycorrhizal fungi. FEMS. Microbiol. Lett. 254(2): 173–181.

Bello, A. O., B. S. Tawabini, A. B. Khalil, C. R. Boland and T. A. Saleh. 2018. Phytoremediation of cadmium-, lead- and nickel-contaminated water by *Phragmites australis* in hydroponic systems. Ecol. Eng. 120: 126–133.

Bhupal Raj, G., M. V. Singh, M. C. Patnaik and K. M. Khadke. 2009. Four decades of research on micro and secondary nutrients and pollutant elements in soils of Andhra Pradesh. Bhopal, India: Indian Institute of Soil Science.

Bi, C., Y. Zhou, Z. Chen, J. Jia and X. Bao. 2018. Heavy metals and lead isotopes in soils, road dust and leafy vegetables and health risks via vegetable consumption in the industrial areas of Shanghai, China. Sci. Total Environ. 619-620: 1349–57.

Blaylock, M. J., D. E. Salt, S. Dushenkov, O. Zakharova, C. Gussman, Y. Kapulnik, B. D. Ensley and I. Raskin. 1997. Enhanced accumulation of Pb in Indian mustard by soil-applied chelating agents. Environ. Sci. Technol. 31(3): 860–65.

Borremans, B., J. L. Hobman, A. Provoost, N. L. Brown and D. van der Lelie. 2001. Cloning and functional analysis of the pbr lead resistance determinant of *Ralstonia metallidurans*. CH34. J. Bacteriol. 183: 5651–5658.

Bramhachari, P. V., P. B. Kavi Kishor, R. Ramadevi, R. Kumar, B. R. Rao and S. K. Dubey. 2007. Isolation and characterization of mucous exopolysaccharide produced by *Vibrio furnissii* VB0S3. J. Microbiol. Biotechnol. 17: 44–51.

Braud, A., K. Jezequel, S. Bazot and T. Lebeau. 2009. Enhanced phytoextraction of an agricultural Cr and Pb-contaminated soil by bioaugmentation with sideropho reproducing bacteria. Chemosphere. 74: 280–286.

Braud, A., V. Geoffroy, F. Hoegy, G. L. A. Mislin and I. J. Schalk. 2010. The siderophores pyoverdine and pyochelin are involved in *Pseudomonas aeruginosa* resistence against metals: another biological function of these two siderophores. Environ. Microbiol. Rep. 2: 419–425.

Camargo, F. A. O., B. C. Okeke, F. M. Bento and W. T. Frankenberger. 2003. *In vitro* reduction of hexavalent chromium by a cell-free extract of *Bacillus* sp. ES 29 stimulated by Cu 2+. Appl. Microbiol. Biotechnol. 62(5): 569–573.

Cecchi, M., C. Dumat, A. Alric, B. Felix-Faure, P. Pradere and M. Guiresse. 2008. Multi-metal contamination of a calcic cambisol by fallout from a lead-recycling plant. Geoderma. 144(1-2): 287–98.

Chabukdhara, M., S. K. Gupta and M. Gogoi. 2017. Phycoremediation of heavy metals coupled with generation of bioenergy. In Algal biofuels pp. 163–188. Springer, Cham.

Chatterjee, S., N. C. Chatterjee and S. Dutta. 2012. Bioreduction of chromium (VI) to chromium (III) by a novel yeast strain *Rhodotorula mucilaginosa* (MTCC 9315). Afr. J. Biotechnol. 11(83): 14920–14929.

Chen, X., C. Wu, J. Tang and S. Hu. 2005. *Arbuscular mycorrhizae* enhance metal lead uptake and growth of host plants under a sand culture experiment. Chemosphere. 60(5): 665–671.

Chen, Z., X. Pan, H. Chen, Z. Lin and X. Guan. 2015. Investigation of lead (II) uptake by *Bacillus thuringiensis* 016. World J. Microbiol. Biotechnol. 31: 1729–1736.

Cheng, S. F., C. Y. Huang, K. L. Chen, S. C. Lin and Y. C. Lin. 2016. Phytoattenuation of lead-contaminated agricultural land using *Miscanthus floridulus*—An *in situ* case study. Desalin. Water Treat. 57(17): 7773–7779.

Chibuike, G. U. and S. C. Obiora. 2014. Heavy metal polluted soils: effect on plants and bioremediation methods. Appl. Environ. Soil Sci. 2014.

Chlopecka, A., J. R. Bacon, M. J. Wilson and J. Kay. 1996. Forms of cadmium, lead, and zinc in contaminated soils from Southwest Poland. J. Environ. Qual. 25(1): 69–79.

De, J. and N. Ramaiah. 2007. Characterization of marine bacteria highly resistant to mercury exhibiting multiple resistances to toxic chemicals. Ecol. Indic. 7: 511–520.

Deng, Z., L. Cao, H. Huang, X. Jiang, W. Wang, Y. Shi and R. Zhang. 2011. Characterization of Cd-and Pb-resistant fungal endophyte *Mucor* sp. CBRF59 isolated from rapes (*Brassica chinensis*) in a metal-contaminated soil. J. Hazard. Mater. 185(2-3): 717–724.

Dixit, R., D. Malaviya, K. Pandiyan, U. B. Singh, A. Sahu, R. Shukla, B. P. Singh, J. P. Rai, P. K. Sharma and H. Lade. 2015. Bioremediation of HMs from soil and aquatic environment: an overview of principles and criteria of fundamental processes. Sustainability. 7(2): 2189–2212.

Dotaniya, M. K., V. D. Meena, K. Kumar, B. P. Meena, S. L. Jat, M. Lata, A. Ram, C. K. Dotaniya and M. S. Chari. 2016. Impact of biosolids on agriculture and biodiversity: environmental impact on biodiversity. Today and Tomorrow's Printer and Publisher, 11–20.

Dotaniya, M. L., V. D. Meena, S. Rajendiran, M. V. Coumar, J. K. Saha, S. Kundu and A. K. Patra. 2017. Geo-accumulation indices of heavy metals in soil and groundwater of Kanpur, India under long term irrigation of Tannery Effluent. Bull. Environ. Contam. Toxicol. 98(5): 706–11.

Dotaniya, M. L. and J. S. Pipalde. 2018. Soil enzymatic activities as influenced by lead and nickel concentrations in a vertisol of Central India. Bull. Environ. Contam. Toxicol. 101(3): 380–85.

Dotaniya, M. L., V. D. Meena, J. K. Saha, S. Rajendiran, A. K. Patra, C. K. Dotaniya, H. M. Meena, K. Kumar and B. P. Meena. 2018a. Environmental impact measurements: tool and techniques. pp. 1–31. *In:* L. M. T. Martínez, O. V. Kharissova and B. I. Kharisov [eds.]. Handbook of Ecomaterials. Cham: Springer International Publishing, Singapore.

Dotaniya, M. L., N. R. Panwar, V. D. Meena, C. K. Dotaniya, K. L. Regar, M. Lata and J. K. Saha. 2018b. Bioremediation of metal contaminated soil for sustainable crop production. pp. 143–73. *In:* V. S. Meena [ed.]. Role of Rhizospheric Microbes in Soil. Springer, Singapore.

Dotaniya, M. L., S. Rajendiran, V. D. Meena, M. V. Coumar, J. K. Saha, S. Kundu and A. K. Patra. 2018c. Impact of long-term application of sewage on soil and crop quality in vertisols of central India. Bull. Environ. Contam. Toxicol. 101(6): 779–86.

Dotaniya, M. L., C. K. Dotaniya, P. Solanki, V. D. Meena and R. K. Doutaniya. 2020. Lead contamination and its dynamics in soil–plant system. pp. 83–98. *In:* D. K. Gupta, S. Chatterjee and C. Walther [eds.]. Lead in Plants and the Environment. Springer Nature, Switzerland.

Douay, F., A. Pelfrêne, J. Planque, H. Fourrier, A. Richard, H. Roussel and B. Girondelot. 2013. Assessment of potential health risk for inhabitants living near a former lead smelter. part 1: metal concentrations in soils, agricultural crops, and homegrown vegetables. Environ. Monit. Assess. 185(5): 3665–80.

Dursun, A. Y., G. Uslu, Y. Cuci and Z. Aksu. 2003. Bioaccumulation of copper (II), lead (II) and chromium (VI) by growing *Aspergillus niger*. Process Biochem. 38(12): 1647–1651.

Egendorf, S. P., Z. Cheng, M. Deeb, V. Flores, A. Paltseva, D. Walsh, P. Groffman and H. W. Mielke. 2018. Constructed soils for mitigating lead (Pb) exposure and promoting urban community gardening: The New York City clean soil bank pilot study. Landsc. Urban Plan. 175: 184–94.

Farhan, S. N. and A. A. Khadom. 2015. Biosorption of heavy metals from aqueous solutions by *Saccharomyces cerevisiae*. Int. J. Ind. Chem. 6(2): 119–130.

Fawzy, E. M., F. F. Abdel-Motaal and S. A. El-Zayat. 2017. Biosorption of heavy metals onto different eco-friendly substrates. J. Bioremediat. Biodegrad. 8: 394.

Gabr, R. M., S. H. A. Hassan and A. A. M. Shoreit. 2008. Biosorption of lead and nickel by living and non-living cells of *Pseudomonas aeruginosa* ASU 6a. Int. Biodeteriorat. Biodegrad. 62: 195–203.

Gavrilescu, M. 2004. Removal of heavy metals from the environment by biosorption. Eng. Life Sci. 4(3): 219–232.

Ghori, N. H., T. Ghori, M. Q. Hayat, S. R. Imadi, A. Gul, V. Altay and M. Ozturk. 2019. Heavy metal stress and responses in plants. Int. J. Environ. Sci. Technol. 16(3): 1807–1828.

Goher, M. E., A. E. M. AM, A. M. Abdel-Satar, M. H. Ali, A. E. Hussian and A. Napiórkowska-Krzebietke. 2016. Biosorption of some toxic metals from aqueous solution using non-living algal cells of *Chlorella vulgaris*. J. Elementol. 21(3).

Govarthanan, M., K. J. Lee, M. Cho, J. S. Kim, S. Kamala-Kannan and B. T. Oh. 2013. Significance of autochthonous *Bacillus* sp. KK1 on biomineralization of lead in mine tailings. Chemosphere. 90: 2267–2272.

Gupta, D. K., H. G. Huang and F. J. Corpas. 2013. Lead tolerance in plants: strategies for phytoremediation. Environ. Sci. Pollut. Res. 20(4): 2150–2161.

Gupta, V. K. and A. Rastogi. 2008. Biosorption of lead from aqueous solutions by green algae *Spirogyra* species: kinetics and equilibrium studies. J. Hazard. Mater. 152(1): 407–414.

Gupta, V. K., A. Nayak and S. Agarwal. 2015. Bioadsorbents for remediation of heavy metals: current status and their future prospects. Environ. Eng. Res 20(1): 1–18.

Hadi, F. and T. Aziz. 2015. A mini review on lead (Pb) toxicity in Plants. J. Biol. Life Sci. 6(2): 2157–6076.

Hanbali, M., H. Holail and H. Hammud. 2014. Remediation of lead by pretreated red algae: adsorption isotherm, kinetic, column modeling and simulation studies. Green Chem. Lett. Rev. 7(4): 342–358.

Hasanuzzaman, M., M. H. M. Bhuyan, F. Zulfiqar, A. Raza, S. M. Mohsin, J. A. Mahmud, M. Fujita and V. Fotopoulos. 2020. Reactive oxygen species and antioxidant defense in plants under abiotic stress: revisiting the crucial role of a universal defense regulator. Antioxidants. 9(8): 681.

Hassan, S. E., M. Hijri and M. St-Arnaud. 2013. Effect of arbuscular mycorrhizal fungi on trace metal uptake by sunflower plants grown on cadmium contaminated soil. N. Biotechnol. 30(6): 780–7. doi: 10.1016/j.nbt.2013.07.002s.

He, J. and J. P. Chen. 2014. A comprehensive review on biosorption of heavy metals by algal biomass: materials, performances, chemistry, and modeling simulation tools. Bioresour. Technol. 160: 67–78.

He, K., Z. Sun, Y. Hu, X. Zeng, Z. Yu and H. Cheng. 2017. Comparison of soil heavy metal pollution caused by e-waste recycling activities and traditional industrial operations. Environ. Sci. Pollut. Res. 24: 9387–9398.

Higueras, P., J. M. Esbrí, E. García-Ordiales, B. González-Corrochano, M. A. López-Berdonces, E. M. García-Noguero, J. Alonso-Azcárate and A. Martínez-Coronado. 2017. Potentially harmful elements in soils and holm-oak trees (*Quercus Ilex* L.) Growing in mining sites at the Valle de Alcudia Pb-Zn District (Spain)–some clues on plant metal uptake. J. Geochem. Explor. 182: 166–79.

Huang, S. H., P. Bing, Z. H. Yang, L. Chai and L. Zhou. 2009. Chromium accumulation, microorganism population and enzyme activities in soils around chromium containing slag heap of steel alloy factory. Trans. Nonferrous Metals Soc. China. 19: 241–248.

Hynninen, A., T. Touze, L. Pitkanen, D. Mengin-Lecreulx and M. Virta. 2009. An efflux transporter PbrA and a phosphatase PbrB cooperate in a lead-resistance mechanism in bacteria. Mol. Microbiol. 74: 384–394.

Igiri, B. E., S. I. Okoduwa, G. O. Idoko, E. P. Akabuogu, A. O. Adeyi and I. K. Ejiogu. 2018. Toxicity and bioremediation of heavy metals contaminated ecosystem from tannery wastewater: a review. J. Toxicol. 2018.

Iqbal, M. and R. G. J. Edyvean. 2004. Biosorption of lead, copper and zinc ions on loofa sponge immobilized biomass of *Phanerochaete chrysosporium*. Minerals Eng. 17(2): 217–223.

Iram, S., R. Shabbir, H. Zafar and M. Javaid. 2015. Biosorption and Bioaccumulation of copper and lead by heavy metal-resistant fungal isolates. Arab. J. Sci. Eng. 40: 1867–1873.

Jagetiya, B. and S. Kumar. 2020. Phytoremediation of lead: a review. Lead Plants Environ., 171–202.

Joshi, P. K., A. Swarup, S. Maheshwari, R. Kumar and N. Singh. 2011. Bioremediation of heavy metals in liquid media through fungi isolated from contaminated sources. Indian J. Microbiol. 51(4): 482–487.

Jusselme, M. D., F. Poly, E. Miambi, P. Mora, M. Blouin, A. Pando and C. Rouland-Lefèvre. 2012. Effect of earthworms on plant *Lantana Camara* Pb-uptake and on bacterial communities in root-adhering soil. Sci. Total Environ. 416: 200–207.

Kalita, D. and S. R. Joshi. 2017. Study on bioremediation of lead by exopolysaccharide producing metallophilic bacterium isolated from extreme habitat. Biotechnol. Rep. 16: 48–57.

Kandziora-Ciupa, M., R. Ciepał, A. Nadgórska-Socha and G. Barczyk. 2013. A comparative study of heavy metal accumulation and antioxidant responses in *Vaccinium myrtillus* L. leaves in polluted and non-polluted areas. Environ. Sci. Pollut. Res. 20(7): 4920–4932.

Kang, C. H., Y. J. Kwon and J. S. So. 2016. Bioremediation of heavy metals by using bacterial mixtures. Ecol. Eng. 89: 64–69. doi: 10.1016/j.ecoleng.2016.01.023.

Khan, I., M. Iqbal, M. Y. Ashraf, M. A. Ashraf and S. Ali. 2016. Organic chelants-mediated enhanced lead (Pb) uptake and accumulation is associated with higher activity of enzymatic antioxidants in spinach (*Spinacea oleracea* L.). J. Hazard. Mater. 317: 352–361.

Ksheminska, H., D. Fedorovych, T. Honchar, M. Ivash and M. Gonchar. 2008. Yeast tolerance to chromium depends on extracellular chromate reduction and Cr (III) chelation. Food Technol. Biotechnol. 46(4): 419–426.

Kumar, A., H. Touseef, C. Susmita, D. K. Maurya, M. Danish and S. A. Farooqui. 2021. Microbial remediation and detoxification of heavy metals by plants and microbes. pp. 589–614. *In*: M. Shah, S. Rodriguez-Couto and K. Mehta [eds.]. The Future of Effluent Treatment Plants, Elsevier, SBN 9780128229569.

Kumar, V. V. 2017. Mycoremediation: a step toward cleaner environment. pp. 171–187. *In*: R. Prasad [ed.]. Mycoremediation and Environmental Sustainability. Springer International Publishing: Cham, Switzerland.

Kumari, S. and S. Das. 2019. Expression of metallothionein encoding gene bmtA in biofilm-forming marine bacterium *Pseudomonas aeruginosa* N6P6 and understanding its involvement in Pb (II) resistance and bioremediation. Environ. Sci. Pollut. Res. (28): 28763–28774.

Kumpiene, J., L. Giagnoni, B. Marschner, S. Denys, M. Mench, K. Adriaensen, J. Vangronsveld, M. Puschenreiter and G. Renella. 2017. Assessment of methods for determining bioavailability of trace elements in soils: a review. Pedosphere. 27(3): 389–406.

Kunito, T., K. Saeki, K. Nagaoka, H. Oyaizu and S. Matsumoto. 2001. Characterization of copper-resistant bacterial community in rhizosphere of highly copper-contaminated soil. Eur. J. Soil Biol. 37(2): 95–102.

Kushwaha, A., N. Hans, S. Kumar and R. Rani. 2018. A critical review on speciation, mobilization and toxicity of lead in soil-microbe-plant system and bioremediation strategies. Ecotoxicol. Environ. Saf. 147: 1035–1045.

Kushwaha, A., R. Rani, S. Kumar, T. Thomas, A. A. David and M. Ahmed. 2017. A new insight to adsorption and accumulation of high lead concentration by exopolymer and whole cells of lead-resistant bacterium *Acinetobacter junii* L. Pb1 isolated from coal mine dump. Environ. Sci. Pollut. Res. 24: 10652–10661.

Lakkireddy, K. and U. Kües. 2017. Bulk isolation of basidiospores from wild mushrooms by electrostatic attraction with low risk of microbial contaminations. AMB Express. 7(1): 1–22.

Laxman, R. S. and S. More. 2002. Reduction of hexavalent chromium by *Streptomyces griseus*. Minerals Eng. 15(11): 831–837.

Lee, Y. C. and S. P. Chang. 2011. The biosorption of heavy metals from aqueous solution by *Spirogyra* and *Cladophora filamentous* macroalgae. Bioresour. Technol. 102(9): 5297–5304.

Lenka, S., S. Rajendiran and V. Coumar. 2016. Impact of fertilizers use on environmental quality. *In*: National Seminar on Environmental Concern for Fertilizer Use in Future. Krishi Viswavidyalaya, Kalyani.

Li, X., L. Zhang and G. Wang. 2014. Genomic evidence reveals the extreme diversity and wide distribution of the arsenic-related genes in Burkholderiales. PLoS One. 9(3): e92236.

Li, X., D. Meng, J. Li, H. Yin, H. Liu and X. Liu. 2017. Response of soil microbial communities and microbial interactions to long-term heavy metal contamination. Environ. Pollut. 231: 908–917.

Liu, Z. and F. S. Zhang. 2009. Removal of lead from water using biochars prepared from hydrothermal liquefaction of biomass. J. Hazard. Mater. 167: 933–939.

Luna, J. M., R. D. Rufino and L. A. Sarubbo. 2016. Biosurfactant from *Candida sphaerica* UCP0995 exhibiting heavy metal remediation properties. Process Saf. Environ. Prot. 102: 558–566.

Mahdavian, K., S. M. Ghaderian and M. Torkzadeh-Mahani. 2017. Accumulation and phytoremediation of Pb, Zn, and Ag by plants growing on Koshk lead–zinc mining area, Iran. J. Soils Sediments 17(5): 1310–1320.

Mani, D., C. Kumar, N. K. Patel and D. Sivakumar. 2015. Enhanced clean-up of lead-contaminated alluvial soil through *Chrysanthemum indicum* L. Int. J. Environ. Sci. Technol. 12(4): 1211–1222.

Manzoor, M., A. Rathinasabapathi, L. M. De Oliveira, E. da Silva, F. Deng, C. Rensing, M. Arshad, G. Iram, P. Xiang and L. Q. Ma. 2019. Metal tolerance of arsenic-resistant bacteria and their ability to promote plant growth of *Pteris vittata* in Pb-contaminated soil. Sci. Total Environ. 660: 18–24. https://doi.org/10.1016/j.scitotenv.2019.01.013.

Mariussen, E., I. V. Johnsen and A. E. Strømseng. 2018. Application of sorbents in different soil types from small arms shooting ranges for immobilization of Lead (Pb), Copper (Cu), Zinc (Zn), and Antimony (Sb). J. Soil Sediment. 18(4): 1558–68.

Meena, V., M. L. Dotaniya, J. K. Saha, H. Das and A. K. Patra. 2020. Impact of lead contamination on agroecosystem and human health. pp. 67–82. *In:* D. K. Gupta, S. Chatterjee and C. Walther [eds.]. Lead in Plants and the Environment. Switzerland: Springer Nature.

Mitra, A., S. Chatterjee, A. V. Voronina, C. Walther and D. K. Gupta. 2020. Lead toxicity in plants: a review. pp. 99–116. *In:* D. K. Gupta, S. Chatterjee and C. Walther [eds.]. Lead in Plants and the Environment. Switzerland: Springer Nature.

Mustapha, M. U. and N. Halimoon. 2015. Microorganisms and biosorption of heavy metals in the environment: a review paper. J. Microb. Biochem. Technol. 7(5): 253–256.

Naik, M. M. and S. K. Dubey. 2017. Lead resistant bacteria: lead resistance mechanisms, their applications in lead bioremediation and biomonitoring. Ecotoxicol. Environ. Saf. 98: 1–7.

Ojuederie, O. B. and O. O. Babalola. 2017. Microbial and plant-assisted bioremediation of heavy metal polluted environments: a review. Int. J. Environ. Res. Public Health. 14(12): 1504.

Orroño, Daniela I., V. Schindler and R. S. Lavado. 2012. Heavy metal availability in *Pelargonium hortorum* rhizosphere: interactions, uptake and plant accumulation. J. Plant Nutr. 35(9): 1374–86.

Panwar, N. R., J. K. Saha, T. Adhikari, S. Kundu, A. K. Iswas, A. Rathore, S. Ramana, S. Srivastava and A. Subba Rao. 2010. Soil and water pollution in India: some case studies. IISS Technical Bulletin, Indian Institute of Soil Science, Nabi Bagh, Bhopal, pp. 1–40.

Park, D., Y. S. Yun, J. H. Jo and J. M. Park. 2005. Mechanism of hexavalent chromium removal by dead fungal biomass of *Aspergillus niger*. Water Res. 39(4): 533–540.

Patel, K. S., B. Ambade, S. Sharma, D. Sahu, N. K. Jaiswal, S. Gupta, R. K. Dewangan, S. Nava, F. Lucarelli, B. Blazhev, R. Stefanova and J. Hoinkis. 2010. Lead environmental pollution in Central India. *In:* B. Ramov [ed.]. New Trends in Technologies. IntechOpen.

Pipalde, J. S. and M. L. Dotaniya. 2018. Interactive effects of lead and nickel contamination on nickel mobility dynamics in spinach. International Journal of Environmental Research 12(5): 553–560.

Pizzaia, D., M. L. Nogueira, M. Mondin, M. E. A. Carvalho, F. A. Piotto, M. F. Rosario and R. A. Azevedo. 2019. Cadmium toxicity and its relationship with disturbances in the cytoskeleton, cell cycle and chromosome stability. Ecotoxicol. 28(9): 1046–1055.

Pourrut, B., M. Shahid, F. Douay, C. Dumat and E. Pinelli. 2013. Molecular mechanisms involved in lead uptake, toxicity and detoxification in higher plants. Heavy Metal Stress in Plants, 121–147.

Povedano-Priego, C., I. Martin-Sanchez, F. Jroundi, I. Sanchez-Castro and M. L. Merroun. 2017. Fungal biomineralization of lead phosphates on the surface of lead metal. Miner. Eng. 106: 46–54.

Prabhakaran, P., M. A. Ashraf and W. S Aqma. 2016. Microbial stress response to heavy metals in the environment. RSC Adv. 6(111): 109862–109877.

Prakash, D., P. Gabani, A. K. Chandel, Z. Ronen and O. V. Singh. 2013. Bioremediation: a genuine technology to remediate radionuclides from the environment. Microbiol. Biotechnol. 6: 349–360. 10.1111/1751-7915.12059.

Punamiya, P., R. Datta, D. Sarkar, S. Barber, M. Patel and P. Das. 2010. Symbiotic role of *Glomus Mosseae* in phytoextraction of lead in vetiver grass [*Chrysopogon Zizanioides* (L.)]. J. Hazard. Mater. 177(1–3): 465–74.

Rahman, Z. and V. P. Singh. 2020. Bioremediation of toxic heavy metals (THMs) contaminated sites: concepts, applications and challenges. Environ. Sci. Pollut. Res. 27(22): 27563–27581. doi: 10.1007/s11356-020-08903-0.

Raj, D. and S. K. Maiti. 2020. Sources, bioaccumulation, health risks and remediation of potentially toxic metal(loid)s (As, Cd, Cr, Pb and Hg): an epitomised review. Environ. Monitor. Assess. 192(2): 1–20.

Rhee, Y. J., S. Hillier and G. M. Gadd. 2012. Lead transformation to pyromorphite by fungi. Curr. Biol. 22: 237–241.

Rigoletto, M., P. Calza, E. Gaggero, M. Malandrino and D. Fabbri. 2020. Bioremediation methods for the recovery of lead-contaminated soils: a review. App. Sci. 10(10): 3528. https://doi.org/10.3390/app10103528.

Romera, E., F. González, A. Ballester, M. L. Blázquez and J. A. Munoz. 2007. Comparative study of biosorption of heavy metals using different types of algae. Bioresour. Technol. 98(17): 3344–3353.

Rucińska-Sobkowiak, R. 2016. Water relations in plants subjected to heavy metal stresses. Acta Physiologiae Plantarum. 38(11): 1–13.

Rucińska-Sobkowiak, R., G. Nowaczyk, M. Krzesłowska, I. Rabęda and S. Jurga. 2013. Water status and water diffusion transport in lupine roots exposed to lead. Environ. Exp. Botany. 87: 100–109.

Rufino, R. D., G. I. B. Rodrigues, G. M. Campos-Takaki, L. A. Sarubbo and S. R. M. Ferreira. 2011. Application of a yeast biosurfactant in the removal of heavy metals and hydrophobic contaminant in a soil used as slurry barrier. Applied and Environmental Soil Science, 2011.

Saha, J. K., N. Panwar and M. V. Singh. 2013. Risk assessment of heavy metals in soil of a susceptible agro-ecological system amended with municipal solid waste compost. J. Indian Soc. Soil Sci. 61: 15–22.

Sahmoune, M. N. 2018. Performance of *Streptomyces rimosus* biomass in biosorption of heavy metals from aqueous solutions. Microchem. J. 141: 87–95.

Salama, A. K., K. A. Osman and N. A. R. Gouda. 2016. Remediation of lead and cadmium-contaminated soils. Int. J. Phytoremediation. 18(4): 364–367.

Saleh, H. M., R. F. Aglan and H. H. Mahmoud. 2019. Ludwigia stolonifera for remediation of toxic metals from simulated wastewater. Chem. Ecol. 35(2): 164–178.

Sammut, M. L., Y. Noack, J. Rose, J. L. Hazemann, O. Proux, M. Depoux, A. Ziebel and E. Fiani. 2010. Speciation of Cd and Pb in dust emitted from sinter plant. Chemosphere. 78(4): 445–50.

Santos, C. L., B. Pourrut and J. M. P. Oliveira. 2015. The use of comet assay in plant toxicology: recent advances. Frontiers in Genetics. 6: 216.

Shahid, M., C. Dumat, S. Khalid, E. Schreck, T. Xiong and N. K. Niazi. 2017. Foliar heavy metal uptake, toxicity and detoxification in plants: a comparison of foliar and root metal uptake. Journal of Hazardous Materials. 325: 36–58.

Shahid, M., E. Pinelli, B. Pourrut, J. Silvestre and C. Dumat. 2011. Lead-induced genotoxicity to vicia faba l. roots in relation with metal cell uptake and initial speciation. Ecotoxicol. Environ. Saf. 74(1): 78–84.

Shi, L., H. Dong, G. Reguera, H. Beyenal, A. Lu, J. Liu, H. Q. Yu and J. K. Fredrickson. 2016. Extracellular electron transfer mechanisms between microorganisms and minerals. Nat. Rev. Microbiol. 14: 651–662. https://doi.org/10.1038/nrmicro.2016.93.

Sidhu, G. P. S., A. S. Bali, H. P. Singh, D. R. Batish and R. K. Kohli. 2018. Phytoremediation of lead by a wild, non-edible Pb accumulator *Coronopus didymus* (L.) Brassicaceae. Int. J. Phytoremediation. 20(5): 483–489.

Singh, A. and S. M. Prasad. 2015. Remediation of heavy metal contaminated ecosystem: an overview on technology advancement. Int. J. Environ. Sci. Technol. 12 (1): 353–366. doi:10.1007/s13762-014-0542-y.

Solgi, E., R. Yazdanyar and M. Taghizadeh. 2020. Assessment of phytoremediation potential of *Alyssum Maritimum* in remediation of lead-contaminated soils. J. School Pub. Health Inst. Public Health Res. 17(4): 363–372.

Srivastava, S., and I. S. Thakur. 2006. Isolation and process parameter optimization of *Aspergillus* sp. for removal of chromium from tannery effluent. Bioresource Technology. 97(10): 1167–1173.

Su, C. 2014. A review on heavy metal contamination in the soil worldwide: situation, impact and remediation techniques. Environ. Skeptics and Critics. 3: 24–38.

Sylvain, B., M. H. Mikael, M. Florie, J. Emmanuel, S. Marilyne, B. Sylvain and M. Domenico. 2016. Phytostabilization of As, Sb and Pb by two willow species (*S. viminalis* and *S. purpurea*) on former mine technosols. Catena. 136: 44–52.

Tayang A. and L. S. Songachan. 2021. Microbial bioremediation of heavy metals. Current. Sci. 120(6): 1013–1025.

Thakare, M., H. Sarma, S. Datar, A. Roy, P. Pawar, K. Gupta, S. Pandit and R. Prasad. 2021. Understanding the holistic approach to plant-microbe remediation technologies for removing heavy metals and radionuclides from soil. Cur. Res. Biotech. 3: 84–98.

Tiwari, S., S. N. Singh and S. K. Garg. 2013. Microbially enhanced phytoextraction of heavy-metal fly-ash amended soil. Communications in Soil Science and Plant Analysis. 44(21): 3161–3176.

Ullah, A., S. Heng, M. F. H. Munis, S. Fahad and X. Yang. 2015. Phytoremediation of heavy metals assisted by plant growth promoting (PGP) bacteria: a review. Environ. Exp. Bot. 117: 28–40.

Vacar, C. L., E. Covaci, S. Chakraborty, B. Li, D. C. Weindorf, T. Frentiu, M. Parvu and D. Podar. 2021. Heavy metal-resistant filamentous fungi as potential mercury bioremediators. J. Fungi. 7: 386.

Vega, F. A., M. L. Andrade and E. F. Covelo. 2010. Influence of soil properties on the sorption and retention of cadmium, copper and lead, separately and together, by 20 soil horizons: comparison of linear regression and tree regression analyses. J. Hazard. Mater. 174 (1–3): 522–33.

Waranusantigul, P., H. Lee, M. Kruatrachue, P. Pokethitiyook and C. Auesukaree. 2011. Isolation and characterization of lead-tolerant *Ochrobactrum intermedium* and its role in enhancing lead accumulation by *Eucalyptus camaldulensis*. Chemosphere. 85(4): 584–590.

Wu, Y., Q. Liang and Q. Tang. 2011. Effect of Pb on growth, accumulation and quality component of tea plant. Procedia Eng. 18: 214–19.

Wuana, R. A. and F. E. Okieimen. 2011. Heavy metals in contaminated soils: a review of sources, Chemistry, Risks and Best Available Strategies for Remediation. ISRN Ecol. 2011: 1–20.

Xie, Y., J. Fan, W. Zhu, E. Amombo, Y. Lou, L. Chen and J. Fu. 2016. Effect of heavy metals pollution on soil microbial diversity and bermudagrass genetic variation. Frontiers in Plant Science, 7: 755.

Yalçın, S., S. Sezer and R. Apak. 2012. Characterization and lead (II), cadmium (II), nickel (II) biosorption of dried marine brown macro algae *Cystoseira barbata*. Environmental Science and Pollution Research. 19(8): 3118–3125.

Yang, J., J. Yang and J. Huang. 2017. Role of co-planting and chitosan in phytoextraction of As and heavy metals by *Pteris vittata* and castor bean–a field case. Ecol. Eng. 109: 35–40.

Yang, T., M. L. Chen and J. H. Wang. 2015. Genetic and chemical modification of cells for selective separation and analysis of heavy metals of biological or environmental significance. TrAC Trends in Analytical Chemistry. 66: 90–102.

Yang, Y., L. Zhang, X. Huang, Y. Zhou, Q. Quan, Y. Li and X. Zhu. 2020. Response of photosynthesis to different concentrations of heavy metals in *Davidia involucrata*. PloS one. 15(3): 0228563.

Yin, K., M. Lv, Q. Wang, Y. Wu, C. Liao, W. Zhang and L. Chen. 2016. Simultaneous bioremediation and biodetection of mercury ion through surface display of carboxylesterase E2 from *Pseudomonas aeruginosa* PA1. Water Res. 103: 383–390.

Yin, K., Q. Wang, M. Lv and L. Chen. 2019. Microorganism remediation strategies towards heavy metals. Chem. Eng. J. 360: 1553–1563.

Zaier, H., T. Ghnaya, R. Ghabriche, W. Chmingui, A. Lakhdar, S. Lutts and C. Abdelly. 2014. EDTA-enhanced phytoremediation of lead-contaminated soil by the halophyte *Sesuvium portulacastrum*. Environ. Sci. Pollut. Res. 21(12): 7607–7615.

Zhao, L., T. Li, X. Zhang, G. Chen, Z. Zheng and H. Yu. 2016. Pb Uptake and phytostabilization potential of the mining ecotype of *Athyrium wardii* (Hook.) grown in Pb-Contaminated soil. Clean–Soil Air Water. 44(9): 1184–1190.

Zulfiqar, U., M. Farooq, S. Hussain, M. Maqsood, M. Hussain, M. Ishfaq, M. Ahmad and M. Z. Anjum. 2019. Lead toxicity in plants: impacts and remediation. J. Environ. Manag. 250: 109557.

CHAPTER 12

Microalgal Bioremediation of Heavy Metals:
An Integrated Low-cost Sustainable Approach

Anubha Kaushik,[1] *Sharma Mona,*[2,3,*] *Randhir Bharti*[1]
and *Sujata*[3]

12.1 Introduction

Microalgae consisting of green algae and cyanobacteria have a remarkable range of applications, including bioremediation of pollutants and production of biofuel, bioactive compounds and biofertilizers. Microalgae have a short life cycle, and can be easily cultured in the laboratory. Microalgae may be grown either in the conventional suspended form or in the attached form. The suspended form has relatively lower efficiency for bioremediation, whereas, in the attached form, they tend to form a biofilm that is more effectual in wastewater treatment and bioremediation (Hasan et al. 2021). With a broad range of use for various heavy metals in different sectors, their concentrations in wastewaters are increasing at an alarming rate in the wastewaters and solid wastes, which ultimately enter the aquatic and terrestrial ecosystems, pollute the environment and health is also affected (Briffa et al. 2020). There is a need to eliminate toxic metals from the environment to safe levels for which bioremediation has emerged as an environmentally sound and low-cost approach. The part of microalgae in the bioremediation of various toxic metals, various processes and mechanisms, and the possibilities of adopting a biorefinery approach for wastewater treatment with microalgae have been described in the chapter to make it an economically and environmentally sustainable method.

[1] University School of Environment Management, GGS Indraprastha University, New Delhi-110078, India.
[2] Department of Environmental Studies, Central University of Haryana, Mahendergarh-123031, Haryana, India.
[3] Department of Environmental Science & Engineering, Guru Jambheshwar University of Science & Technology, Hisar-125001, Haryana, India.
* Corresponding author: drmonasharma1@gmail.com, mona@cuh.ac.in

12.1.1 Heavy Metals Pollution in Water and their Impacts

Heavy metals are explained as metals with high atomic weight and high density. Metallic chemicals and metalloids (selenium, arsenic) are highly toxic to human health, while some are typically non-toxic, like gold and silver (Tchounwou et al. 2012). While toxic metals need to be removed, precious metals must be essentially recovered from the wastewaters. There are some essential heavy metals likes Cu, Fe, Mg, Zn, while non-essential heavy metals are Cr, Ni, Pb, As, Hg (Kumar et al. 2020). Heavy metal accumulation in worldwide rivers and lakes is higher than the permissible limits as per standards of the WHO (World Health Organization) and has been increasing over the past 4–5 decades (Zhou et al. 2020). This increased metal pollution is mainly attributed to industrial waste and mining activities, while the geochemical origin is the natural causal factor (Zeitoun et al. 2014). Chemicals loaded with heavy metals are used in industries like metal plating, battery production, tanneries, petroleum refineries and mining; wastewaters from these industries include large amounts of heavy metals. Metals are mobilized in the environment, where they alter the biogeochemical processes and bioaccumulate in the food chain (Waldichuk 1994). Plants, animals and microorganisms are all affected by metal polluted water and soil. Human health is adversely affected because metals tend to get bioaccumulated in the food chain and human beings occupy higher levels in the food web. Concentrations of various metals such as Fe, Ni, Cd and Co which are known to have serious health implications, were found to be exceeding the permissible limits in river Yamuna flowing across the state of Haryana (Kaushik et al. 2001) and river Ghaggar (Kaushik et al. 2000) and these metals were mainly related to anthropogenic sources. Table 12.1 shows the health impacts of some non-essential heavy metals and metalloids as reported by researchers.

Table 12.1. Health impacts and sources of non-essential heavy metals.

Heavy metals	Major sources	Health impacts	References
Arsenic	Industrial waste, agricultural runoff, coal mining, fossil fuel burning, etc.	Psychological effect, decreased mental performance, cardiovascular disease, kidney damage, carcinogenesis, effect on reproduction, bone demineralization	Rehman et al. 2018
Cadmium	Mine drainage water, waste from the processing of ores, agricultural runoff	Kidney damage, osteoporosis, cardiovascular disease, cancer, lungs damage, obesity and diabetes	Fatima et al. 2019
Lead	Paint industry, batteries, unregulated cosmetics and medicine	Affects brain and nervous system, high risk of blood pressure, kidney damage, effects pregnant woman or exposing the fetus	Demayo et al. 1982, García-Lestón 2010
Nickel	Batteries, paper industry, fossil fuels, mining, volcanoes, forest fires	Cardiovascular, kidney damage, lung fibrosis, lung and nasal cancer and respiratory manifestations	Genchi et al. 2020
Mercury	Industrial effluents, coal combustion, battery and fluorescent lamp production, cement production, Hg mining, and biofuel burning	Acute chemical bronchitis and pneumonitis, Minamata disease	Zhang et al. 2007

12.1.2 Metal Remediation Approaches

With increasing awareness about the harmful impacts of various heavy metals on ecosystems and organisms' health, there is a greater emphasis on the elimination and retrieval of the metals from polluted waters and soils by adopting various techniques (physical, chemical, biological) available, as shown in Figure 12.1.

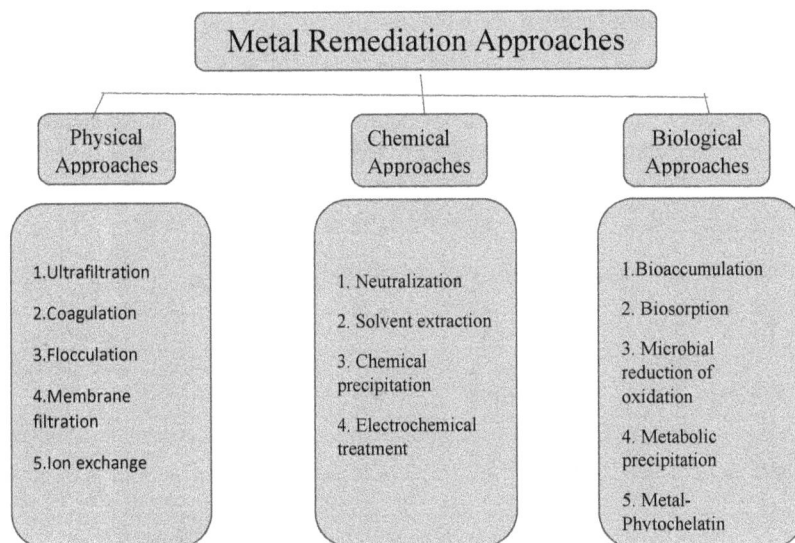

Figure 12.1. Various approaches for heavy metal remediation.

By using physical methods, almost all the pollutants can be removed, but these methods have some limitations. Physical methods based on the distribution of the practical size of pollutants need further processing and have a comparatively high cost of application. Although, chemical methods of metal remediation are highly effective in these methods formation of byproducts increases further downstream processing steps (Mona et al. 2008). The biological methods for metal remediation are less costly and do not create any secondary pollution (Selvi et al. 2019).

12.1.3 Bioremediation using Microalgae—Merits and Potential

Bioremediation is a technique to exude and modify harmful pollutants (heavy metals) into less harmful substances and/or eliminate toxic elements from the polluted environment (Eccles 1999). Microalgae are found to be very potent in the bioremediation of different heavy metals from wastewaters. Microalgae have several advantages like small size, simple structure, easy handling, high photosynthetic activity, short life cycle, simple nutrient requirements, high adaptability and tolerance to different types of stress conditions, which increase their potential for applications in bioremediation. Therefore, there has been great interest in using microalgae in the phytoremediation of toxic heavy metals. The cell wall of microalgae shows more binding affinity, the richness of binding sites and wide surface area, all of which favor effective biosorption of the metals (Cameron et al. 2018). Moreover, microalgae show good biosorption capacity as living or dead cells, free or immobilized cells. Besides metal elimination capacity and being eco-friendly, bioremediation of heavy metals has added significance, such as the development of value-added products.

Chlorella vulgaris and the *Chlorella salina* (marine alga) have been shown to remove 14 to 100% of heavy metals viz. Fe, Mn, Ni, Zn, Cu, Co and Cr from wastewaters along with other pollutants such as TDS, pH, COD, BOD, calcium, magnesium, ammonia, nitrate, phosphate, sulfate, sodium, potassium (El-Sheekh et al. 2016). While several microalgae species have shown very good metal tolerance, they also show tolerance to certain toxic dyes and have additional merits of being able to produce biohydrogen (Mona and Kaushik 2015a). Many cyanobacterial species show excellent co-tolerance to metals and salts (Kiran et al. 2008). The metal-salt co-adapted *Lyngbya* and *Gloeocapsa* strains were found to show better Cr removal capability in the presence of salts (Kiran et al. 2007a). This indicates that the indigenous strains of microalgae may be more effective in bioremediation when they have been exposed for a long time to different pollutants.

12.2 Microalgal Remediation of Metals

12.2.1 Metal Tolerance and Resistance in Microalgae

Many algae are found to grow in metal-polluted environments. These algae tolerate high concentrations of heavy metals. Different groups of algae have varying levels of metal tolerance and resistance (Table 12.2), and they may be genetically and physiologically fixed (Gaur et al. 2001). The concentration and nature of metals define the toxicity of heavy metals.

Table 12.2. Metal Tolerance by different Microalgae.

Heavy metal	Microalgae	References
Zinc	*Chlorophyceae, Anacystis nidulans, Navicula, Caloneis, Pinnularia*	Reed and Gadd 1989, Gaur et al. 2001
Copper	*Cyanophyceae, Caloneis, Eunotia, Cyanidium calarium, Chlorella vugaris*	Hall 2002, Priyadarshini et al. 2019
Cadmium	*Chlorophyceae, Cyanophyceae, Nostoc linckia*	Brinza et al. 2009, Mona et al. 2011b
Mercury	*Chlorella, Pseudochloroccum typicum, Phormidium ambiguum*	Gaur et al. 2001, Priyadarshini et al. 2019
Nickel	*Cyanidium calarium, Euglena gracilis*	García-García et al. 2018
Lead	*Pseudochlorococcum typicum, Scenedesmus quadricauda, Cladophora aglomerata*	Priyadarshini et al. 2019
Chromium	*Scenedesmus dimorphus, Lyngbya putealis Nostoc calcicola, Chroococcum*	Toranzo et al. 2020, Kiran et al. 2007a, Kamra et al. 2007, Mona et al. 2011a

Some metals like Cu, Ni, Mn, Fe are essential for the growth and conditioning of microorganisms as they are required for the cells (Toranzo et al. 2020). While metals like Pb, Cd, Hg and As are harmful at very low concentrations, essential metals are also toxic for microbial activity beyond a threshold level (Hall 2002). Essential metals are displaced by non-essential metals because of their high ionic force from their area of bindings and make complex bindings with working sites of microbial cell walls.

12.2.2 Biosorption and Bioaccumulation

Metal removal by microalgae may take place through bioaccumulation or biosorption. In biosorption, the biomass matrix functions as a sorbent. These activities present a low-cost, sustainable and reversible solution for the remediation of various contaminants by rapid binding on the functional groups of the biomass surface. It is independent of cellular metabolism. The biosorption process has the edge over other regular treatment methods due to minimal use of chemicals, operation receiving ambient conditions, low cost, high efficiency and little biological sludge. Further, auxiliary nutrients are not required, and there is a potential for renewal of the biomass (sorbent) and regaining of metals (Kratochvil and Volesky 1998). A critical review on biosorption of metals by algae demonstrated excellent biosorptive properties of various groups of algae, particularly microalgae. Biosorption can take place in pH range of 3–9 and temperature (4–45°C). Biosorbent particle size is found to be more suitable between 1–2 mm, and due to the small size, the equilibrium states of adsorption and desorption are attained fast (Michalak et al. 2013).

Biosorption is a more reliable method for remediate of heavy metals (Sweetly et al. 2014). The conventional adsorption method mediated through activated carbon is one of the efficient processes for the extraction of contaminants like heavy metals, but there are disadvantages of this method like non-recyclable and less cost-efficient (Naimabadi et al. 2020). The algae biosorbents have been used for regaining of heavy metals. Thus, their use in metal bioremediation enhanced the attention

Table 12.3. Microalgal biosorption of heavy metals.

Name of Microalgae	Metal (s)	Removal efficiency (%)	References
Nostoc linckia	Chromium, Cobalt	58–60	Mona and Kaushik 2015b, Mona et al. 2011b
Maugeotia genuflexa	Arsenic	24	Ubando et al. 2021
Laurencia obtuse	Chromium, Cobalt, Cadmium	98.6, 98.2, and 98.0	Hamdy 2000
Nostoc, Gloeocapsa	Chromium	90–95	Sharma et al. 2016

of researchers across the world, leading to a huge amount of literature on the subject. Table 12.3 shows some reports on metal removal by various microalgae species.

Bioaccumulation is a procedure wherein heavy metal is controlled metabolically, producing energy and altering heavy metal concentration (Arunakumara and Zhang 2008). Bioaccumulation of metals by living cells depends on both the intra and extracellular processes, and passive gaining is restricted (Fomina and Gadd 2014). *Cladophora herpestica*, a green alga that grows abundantly in the Maruit Lake surface, has shown accumulation of residual nutrients from the atmospheric and aquatic environment in introduction to heavy metal ions (Dahlia and Hassan 2017). Some microalgae have also been reported to heavy metals bioaccumulation. Bioaccumulation of heavy metals (Zn, Fe, Cu, Cd,Als) was shown by *Chlorella vulgaris, Phacus curvicauda, Euglena acus* and *Oscillatoria bornettia* (Abrihire and Kadiri 2011). Amongst these species, *Oscillatoria* had a high concentration of metal factor for Zn (0.306), Fe (0.302), Cu (0.091), Cd (0.276), while *Phacus* and *Euglena* had relatively higher concentration factor for Al (0.439).

12.2.3 Factors Influencing Metal Remediation by Microalgae

In each type of environment, many factors present unique influences on metal remediation from water. The biosorption method is regulated by many operating factors, including pH, temperature, organic molecules, salinity, primary metal concentration, contact time and co-pollutants. There have been extensive studies showing the influence of such factors on the metal biosorption process (García-García et al. 2018).

(a) **pH:** Solution pH regulates the biosorption process and affects solution chemistry and functional group activity in the biosorbents and competition with other co-pollutants (Vijayaraghavan and Yun 2008). The optimum pH for maximum biosorption of the metals shows great variations. *Lyngbya putealis* was found to show the highest Cr (VI) biosorption at acidic (pH 3.0). The isoelectric point for algal biomass being acidic pH (3.0), there is hydronation of some functional groups, and the occurrence of hydronium ions near the binding sites lead to greater binding of Cr (VI) to the algal surface. Cr (VI), which exists as $HCrO^{4-}$, $Cr_2O_7^{2-}$, in solution form at optimum sorption pH, tends to bind to the protonated active sites of the biosorbent (Kiran et al. 2007a). Extracellular polymeric substance (EPA) of *Gloeocapsa calcarean* and *Nostoc punctiforme* remove maximum Cr (VI) at pH 2 (Mona et al. 2008). Cr (VI) and Co (II) are removed by *Nostoc* at pH 2 and 3.5, respectively (Mona et al. 2013).

(b) **Temperature:** Temperature plays a crucial part in the heavy metal removal process from water. With increasing temperature, the flow of adsorbate diffusion on the biosorbent surface, solubility of heavy metals, enzymatic activity and metabolism increase, thus generally escalating the removal process (Igiri et al. 2018). Numerous studies on the impact of temperature on heavy metals remediation are available, but the response of different microalgae to temperature shows inconsistent effects on remediation. Kumar et al. (2015) reported varying impacts of temperature on removal of heavy metals; some indicated a rise in metal remediation with an increase in temperature, while others showed a decline at high temperature. Aksu (2002) reported that the dry biomass of *Chorella vulgaris* shows improved adsorption of Ni at increased temperature.

The temperature has the highest effect on metal remediation, but it has less impact than the impact of pH (Lau et al. 1999). Therefore, it is important to find the optimum temperature for each microalgal system before designing the biosorption operation for different metals.

(c) **Organic matter:** Organic matter concentration influences the biosorption method, and is dependent on the type of metal and microalgal species used in bioremediation. For example, biosorption of copper decreases from 3.2–2.3 mg L^{-1} and arsenic from 2.2–0.0 mg L^{-1} in *Chlorella*, whereas in the case of *Scendesmus almeriensis*, removal declines from 2.1 to 1.6 mg L^{-1} for copper and 2.3 to 1.7 mg L^{-1} for arsenic in the existence of organic matter (Saavedra et al. 2019). Since most wastewaters have some organic load, and organic matter tends to chelate the metals-forming complexes, therefore it is paramount to comprehend the effect of organic matter on the overall metal bioremediation protocol.

(d) **Carbon dioxide:** Microalgae need carbon dioxide for their photosynthesis as raw material, and carbon dioxide is emitted by several industries. Microalgae can be grown by using this waste gas and converting the same into useful biomass, which can then be used as a biosorbent to clean metal contaminated waste streams. *Oedogonium* sp. shows rapid uptake of a few heavy metals (particularly Cd, Al, Ni and Zn), and its biomass productivity gets increased when CO_2 concentration is higher than normal conditions and escalates with productivity (Roberts et al. 2013).

(e) **Nutrients:** Algal growth is largely affected by the availability of nutrients. There are some dominant species like *Chlamydomonas, Spirogyra, Euglena* and *Dinoflagellates* that use carbon and other nutrients from the wastewater for their maturation and photosynthetic activity, increasing the efficiency of metal remediation microorganisms (Sayara et al. 2021).

Many other factors like initial metal concentration, types of metals such as tertiary and quaternary multi-metal systems and contact time are also important factors for metal removal (Kiran and Kaushik 2012).

12.2.4 Role of Exopolymers

Biofilms are formed on the surface where the algae get attached due to the secretion of exopolymer substances that are essentially composed of proteins, lipids, nucleic acids, polysaccharides, lipopolysaccharides and glycoproteins (Zeraatkar et al. 2016). The structure of exopolymers varies depending upon the type of microorganism, age of biofilms and on environmental conditions. Exopolysaccharides are surface-active bio-agents originating from algae, fungi, bacteria, cyanobacteria and help in sequestering toxic metals from an aqueous solution (Sharma et al. 2009, Mona and Kaushik 2015b). Chelating agents produced by exopolysaccharides for the elimination of positively charged metal ions from water (Potnis et al. 2021).

The yield of exopolysaccharides by algae increases when disclosed to a higher concentration of metal ions, which seems to play a crucial part in metal biosorption (Sharma et al. 2009). The bioremediation of heavy metals by exopolysaccharides of the algae is gaining importance for future use in bioremediation programs of wastewaters containing these metals.

12.2.5 Bioremediation Mechanisms

The remediation of pollutants (heavy metals) in algae occurs in two ways; first is fast passive biosorption that takes place where the pollutants assimilate on the biomass surface within a lesser period, and this step is not based on the metabolism of cells, and the second step is slow active sorption of pollutants (heavy metals) into the cytoplasm of algal biomass, and this step is completely based on metabolism of biomass (Monteiro et al. 2012).

Various analytical techniques have been used by researchers, which mainly include atomic "Scanning Electron Microscope (SEM) or Transmission Electron Microscope (TEM) coupled with energy-dispersive X-ray spectroscopy, X-ray spectroscopy, Infrared spectroscopy or Fourier-transform infrared, spectroscopy (IR or FTIR), X-ray Absorption Spectroscopy (XAS) and Nuclear Magnetic Resonance (NMR) to gain insight into the process and mechanisms of biosorption process (Fomina and Gadd 2014, Michalak et al. 2013, Kiran et al. 2016)." It can be confirmed that surface adsorption of the metals is found in the case of Cr (VI) bound at specific binding sites on the algal biosorbent based on SEM and FTIR spectral analysis (Kiran et al. 2008).

To understand the procedure of the biosorption and optimize the processes, modeling and simulation of the experimental data are generally done for which several models have been developed (Volesky 2003a). The two regularly used models are Langmuir and Freundlich models, in which the adsorption mechanism is demonstrated as a batch equilibrium isotherm curve to contrast pollutant uptake proportion of different bio-sorbent and affinities for the metals (Mona et al. 2011a). These equilibrium sorption models provide some normal information about a given process. As the biosorbent adsorbs the metal to the equilibrium point, the value of metal uptake (qe) by the bio-sorbent is plotted against the equilibrium (final) metal concentration (C). The Langmuir isotherm assume a finite number of equal adsorption locations and the absence of lateral interrelation between adsorbed species. The most regularly multilayer adsorption model is the Brunauer–Emmett–Teller (BET) isotherm, which presume that the Langmuir equation is applied to every layer (Vijayaraghavan and Yun 2008). Equilibrium data is right to Langmuir isotherm, and a linear plot is received from this isotherm (Figure 12.2); and, the value of the Langmuir constant is calculated:

$$q_e = Q_o b C_e + b C_e$$

where;
qe = adsorption of metal (mg g^{-1}); Ce = residual metal (mg L^{-1})

Qo (mg g^{-1}) and b (L mg^{-1}) are Langmuir constants exhibit the adsorption range and adsorption energy (Mona et al. 2015a).

Freundlich isotherm considers a heterogeneous base of the adsorbent, and the equation is

$$Log\ q_e = Log\ Kf + n\ Log\ C_e$$

"where;
qe = metal adsorbed (mg g^{-1}); Ce = residual metal ion concentration (mg L^{-1}); n = Freundlich exponent; Kf = Freundlich constant indicating adsorbent capacity (mg g^{-1} dry weight)."

And a linear plot of Log qe versus Log Ce explains the applicability of this isotherm (Figure 12.2) for the biosorbent (Gadd 2009).

There is another model, Brunauer Emmer and Teller (BET); in this model the 1st layer of sorbent gets absorbed on the upper layer with the energy approximate to the heat of adsorption for single layer sorption, and the next layer has same energy.

$$\frac{C_e}{q_e} = \frac{1}{q_m KL} + \frac{C_e}{q_m}$$

"where;

C_e = the equilibrium concentration of the adsorbate and qe is the adsorption capacity adsorbed at equilibrium, qm is maximum adsorption capacity

KL = the Langmuir adsorption constant"

Response Surface Methodology (RSM) is one of the most widely used approaches adopted for optimization of the parameters to get the maximum metal removal response from the biosorption process (Mona et al. 2011a). Box-Behnken Design (BBD) and Central Composite Design (CCD) are two models of Response Surface Methodology (RSM). Box-Behnken model (BBM) of RSM is used

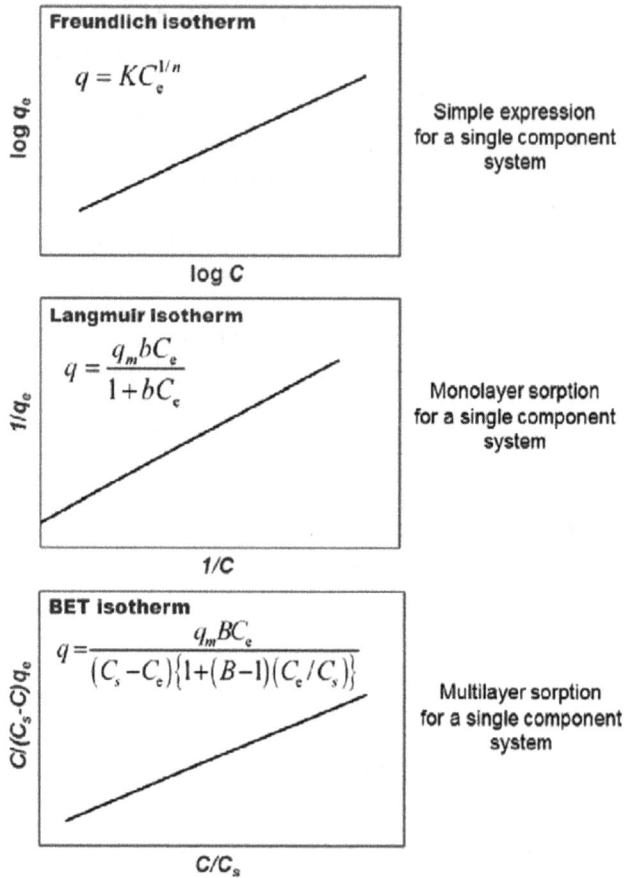

Figure 12.2. Equations and graphical representation of Freundlich, Langmuir and Brunuaer–Emmett–Teller (BET) isotherms (adopted from Gadd 2009, Park et al. 2011).

in the study with maximum and minimum levels for each parameter (initial metal concentration, pH, temperature). The CCD methodology is used for the prediction of the impact of different parameters on the metal remediation process (Podstawczyk et al. 2015). To regulate the adsorption equilibrium and kinetic behavior of chromium in an aqueous solution, batch mode experiments were performed for *Lyngbya*, and maximum metal adsorption capacity was determined. The impact of other parameters like initial metal ion concentration (10–100 mg/L), pH (2–6) and temperature (25–45°C) on chromium adsorption were also noted , using Box–Behnken design (BBD) model of RSM and the optimized conditions (50–60 mg/L initial Cr concentration, 2–3 pH, 45°C temperature) were computed, when 82% of the metal was extracted from the solution by the alginate immobilized alga (Kiran et al. 2007b).

The biosorption range of algal cells directly depends on the presence of different functional groups viz. OH^-, PO_3O_2, NH_2, $COOH$, SH, etc., on the cell of algae. These groups provide a negative charge to the cell (Kaplan 2013), which has a strong binding capacity for the positively charged heavy metals present in the water. These functional groups form a complex or get linked with cell wall components (teichoic acids, peptidoglycan, polysaccharides and proteins), directly or indirectly provide metal-binding sites (Volesky and Holan 1995). Biosorption of Cd in *Chlamydomonas reinhardtti* occurs in a unique manner, where Cd^{2+} ions form a complex with carboxylic groups of the algal cells (Adhiya et al. 2002). Covalent bonding and electrostatic attraction are directly responsible for the adsorption of Ni and Zn on the *Chaetophora elegans* (Andrade et al. 2005). Bioremediation of aluminum (Al) follows a different mechanism of sorption on algal cells as it binds to the algal cell in a polynuclear form; after binding of aluminum ions, these ions prevent the other metal ions from binding on the binding surface of the biomass (Bottero et al. 1980).

12.2.6 Metal Removal in Continuous Packed Bed-Reactors

For the industrial application of microalgal bioremediation of metals, the operation should be carried out in a continuous mode, where packed bed reactors are found useful. For the dynamic and static mode studies, *Lyngbya putealis* HH-15 cyanobacteria extracted from a metal-polluted surface was used as a biosorbent of Cr (VI) from aqueous solutions (Kiran and Kaushik 2008). Through regularly flow column experiments, the data effect of an initial concentration (5–20 mgL^{-1}), flow rate (1–3 mL min^{-1}) and bed height (5–10 cm) on breakthrough time and adsorption capacity of the immobilized alga was developed into Bohart-Adams model (Singh et al. 2012). The chromium elimination efficiency and regeneration capacity of this biosorbent recommend its application use in industrial activities, and the data generated strongly suggested potential for more increase in the adsorption process. There is another discovery for metal sorption of a column packed with Spirogyra granules, which could be successfully used up in many cycles of sorption and removing Cu (II) and Pb (II), respectively (Singh et al. 2012). The excellent metal bonding ability of algae has been proved by the assessment of maximum sorption capacity based on isotherm studies by various researchers. Continuous flow studies in packed bed columns appear highly effective and economically suitable than the metal absorption by batch operation. Thomas mass transfer model, Adam-Bohart advection–dispersion-reaction equation and bed-depth-service-time model have been established for understanding the breakthrough curve. Some fresh approaches, such as artificial neural networking, may prove still more useful for elucidating breakthrough curves and metal sorption in multi-metal systems (Kumar et al. 2016).

12.2.7 Pretreatment and Immobilization Approaches for Improved Bioremediation

Microalgal biosorption of metals can be improved by adopting physical and chemical pretreatments that provide extra binding sites on the cell surface by changing the cell surface structure. Physical pretreatment (heating, boiling, freezing, crushing and drying) enhanced metal ion biosorption (Ahluwalia et al. 2007). These kinds of pretreatments increase the cell wall surface area and favor improved biosorption of the metals (Uzunoglu et al. 2014, Kiran et al. 2016). General algal pretreatments include treating with calcium chloride, formaldehyde, glutaldehyde, NaOH and HCl. The impact of temperature and acid treatment on the absorption of tetravalent chromium [Cr (IV)] by the *Chlamydomuns reinhardtii* was observed and noted that biosorption capacity is higher than the untreated biomass (Zeraatkar et al. 2016). Biofilm generation on natural or artificial packing, engagements of biomass inside matrices, adsorption and binding of cells to a surface are considered as the prime immobilization technique used in the metal elimination process (Nasirpour et al. 2017).

Saccharomyces cerevisiae immobilized polyacrylamide was used to bioaccumulate Cu^{2+}, Cd^{2+}, Co^{2+} (Duncan et al. 1997). Sequestration of Cr and Co by exopolysaccharides (EPS) of freshwater microalga has been demonstrated with high biosorption capacities 14.3mg Cr g^{-1} EPS and 17.9 mg Co g^{-1} EPS (Mona and Kaushik 2015b).

12.3 Challenges and Future Opportunities

Bioremediation of heavy metals, particularly the biosorption technique, has been one of the most detailed studied subjects from the past six decades, with more than 13,000 research papers in reviewed journals. A vital understanding of the complex biosorption mechanism has made it possible to quantify the process based on equilibrium and kinetics studies and optimize the process by manipulating various operational parameters using modeling approaches based on RSM. Though testing of the process in pilot projects and at the industrial-scale is still in the initial stage, the

BV SORBEX, a company owned by the pioneer researcher in the field of biosorption, McGill University and Bohumil Volesky from Canada, has applied the laboratory process of biosorption to an industrial level (Volesky 2003b).

However, there are some limitations in the application of the biosorption method on an industrial scale. This limitation is due to less durability and low mechanical resistance of the biomass, which create problems in the process and other factors, including regeneration of the sorbent and its subsequent declination. Thus the success of the sorption process depends on the recovery and reuse of the sorbent. It is further required that the biosorption process be integrated into other microalgae-based applications, including energy recovery in the form of biofuels like biohydrogen or biodiesel and value-added biomaterial production to make the overall procedure more economically feasible and sustainable. Next new directions for integrating various microalgae-based technologies have been described with a view to make it a competitive metal bioremediation technology with additional applications to raise its sustainability. Various new opportunities for such combined technologies and the biorefinery approach for microalgal wastewater treatment and bioremediation are discussed next.

12.4 Combining Microalgal Technologies as Low-cost and Sustainable Approach

Schenk et al. (2008) emphasized on exploiting the biological pathways for improving the efficiency of the bioremediation potential of algae. Wastewaters have been used for biomass production of microalgae during the bioremediation process and utilized for yielding carotenoids (Kalra et al. 2020). Microalgae can grow about 10% of their total volume during every 2 d. The biomass of *Chlorella vulgaris* and *Scenedesmus obliquus* have the biochemical composition of 30–35% proteins, 6–8% lipids, 39–30% carbohydrates (Bhatia et al. 2017) and are used in pisciculture feeds. The algal biomass derived from wastewater bioremediation sludge can also be used as a biostimulant for crop plants (Kalra et al. 2020). The biomass of algae can be processed with the torrefaction process to produce 72.5 ± 1.7% charcoal yields.

Further, microalgae can successfully hold carbon in the form of gas from the environment. The heterotrophic algae can also intake carbon in the form of organic carbon from acetate, glycerol and glucose, whereas autotrophic algae take up inorganic carbon. Carbonates soluble in water act as a fount of carbon dioxide that enters the algal cell by diffusion at pH 5.0–7.0 (Picardo et al. 2013, Sydney et al. 2014). Bicarbonate is transformed to CO_2 within the microalgal cell before being turned into energy-rich molecules by enzyme ribulose bisphosphate carboxylase/oxygenase (Gonçalves et al. 2017).

Wastewater contains heavy metals in addition to phosphate, nitrogen and carbon. Microalgal assisted wastewater treatment is regarded as one of the most effective heavy metals elimination methods. However, the metal ion bioconcentration capacity is influenced by various factors, such as temperature, pH, availability of nutrients, cell number, cell size and cell structure (Kosek et al. 2016).

12.4.1 Integrating Biofuel Production and Bioremediation using Microalgae

To make the overall process of microalga-based bioremediation of metals an economically sustainable technology, it has been proposed to integrate the process with various other applications. It has been reported that even the waste biomass of cyanobacteria from biohydrogen rectors can be used successfully for metal bioremediation (Mona et al. 2011c). The twofold uses of the cyanobacterium for constant hydrogen production and elimination of some contaminants from textile industry wastewater by spent algal biomass as an integrated laboratory-scale system were demonstrated by Kaushik et al. (2011). Waste algal biomass of *Nostoc linckia* tested at multi-stages (5, 15, 25 d) from hydrogen-producing photobioreactors when tested for metal bio removal capability show very good adsorption capacity (15–18 mg/g) for Cr (VI) and Cd metals, and the efficiency is higher when the spent biomass is obtained at a relatively earlier stage (Mona et al. 2013). Such studies suggest that careful planning can lead to the highly successful integration of the algae-based bioremediation-cum-bioenergy production.

These studies explore the application of waste algal biomass from biofuel producing systems for metal bioremediation and use microalgal biomass produced in wastewater for biofuel production.

12.4.2 Using Microalgae for Bioremediation Adopting Biorefinery Approach

Mass cultivation of algae can be done in a cost-effective way by growing them in wastewaters that generally contain carbonaceous matter, nitrates and phosphates nutrients required for algal growth. The cost of cultivating 1 kg microalgae using synthetic fertilizers in the culture medium or freshwater is very high. While growing in such wastewaters, microalgae utilize all the inorganic and organic nutrients (TP, TN, NH_4^+, NO^{3-}, TOC). These nutrients are removed from the wastewater, and COD and BOD are also decreased. When algal production is carried out by raceway ponds, the cost decreases of biofuel production from wastewater treatment increases with added benefits of less impacts of environment (Park et al. 2011). *Chlorella* sp. extracted from wastewater showed high lipid production efficiency in a bioreactor under heterotrophic and photoautotrophic conditions (Viswanath and Bux 2012). Under heterotrophic conditions, it produced 3.6-fold more biomass than photoautotrophic growth conditions and enhanced lipid production by 4.4-fold. Microalgal heterotrophic growth is an efficient way for producing high biomass and biochemicals of algae, which could lower the cost of producing microalgal biomass. Since various microalgal species also have an innate ability to produce certain peptides having functional groups such as amine, sulfate, phosphate, carboxyl and hydroxyl, that can be attached to heavy metals; these biomolecules take part in conclusive to the metal ions compactly (Gong et al. 2005). The dead microalgal cells also show biosorption function for bioremediation of heavy ccc metals. As demonstrated by Kaushik et al. (2011), even the spent biomass of algae from bioreactors shows very good metal biosorption capability; thus, there are enormous opportunities that need to be explored and investigated systematically for adopting a biorefinery approach (Figure 12.3).

There are several methods to produce different types of fuels from algal biomass using wastewater as a biorefinery approach (Craggs et al. 2011), while several value-added biomaterials, including pigments and carbohydrates, can also be obtained from the microalgae along with remediation of the wastewater (Markou and Nerantzis 2013, Preeti et al. 2015).

Biodiesel production from microalgae occurs via transesterification, i.e., the conversion of lipids into fatty acid methyl esters. The biodiesel formation from microalgae completed the sequential steps: biomass production, lipid extraction from biomass, lipids transesterification and biodiesel purification.

Bioethanol production from algal biomass or algal meals takes place through saccharification and fermentation. The carbohydrates in algal biomass, such as starch, cellulose, glucose and hemicellulose, get converted to bioethanol through fermentation. *Porphyridium cruentum, Spirogyra* sp., *Nannochloropsis oculate* and *Chlorella* sp. are some of the most used microalgae for carbohydrate synthesis (Markou and Nerantzis 2013).

Biogas and biomethane are created by fermented bacteria during the controlled, anaerobic breakdown of microalgal biomass (Li et al. 2019). Methane fermentation involves a series of biochemical reactions carried out by specific microbes, including hydrolysis, acidogenesis and methanogenesis (Jankowska et al. 2017). Biobutanol can be made from green waste left behind after extracting microalgae oil. It has a higher energy density than biomethanol or bioethanol and a chemical structure like gasoline. It can also be used as a solvent in industries, in addition to being a biofuel (Yeong et al. 2018). Microalgal strains which have high starch and sugar content, such as *Chlorella vulgaris, Tetraselmis subcordiformis* and *Scenedesmus obliquus*, are favored for biobutanol synthesis.

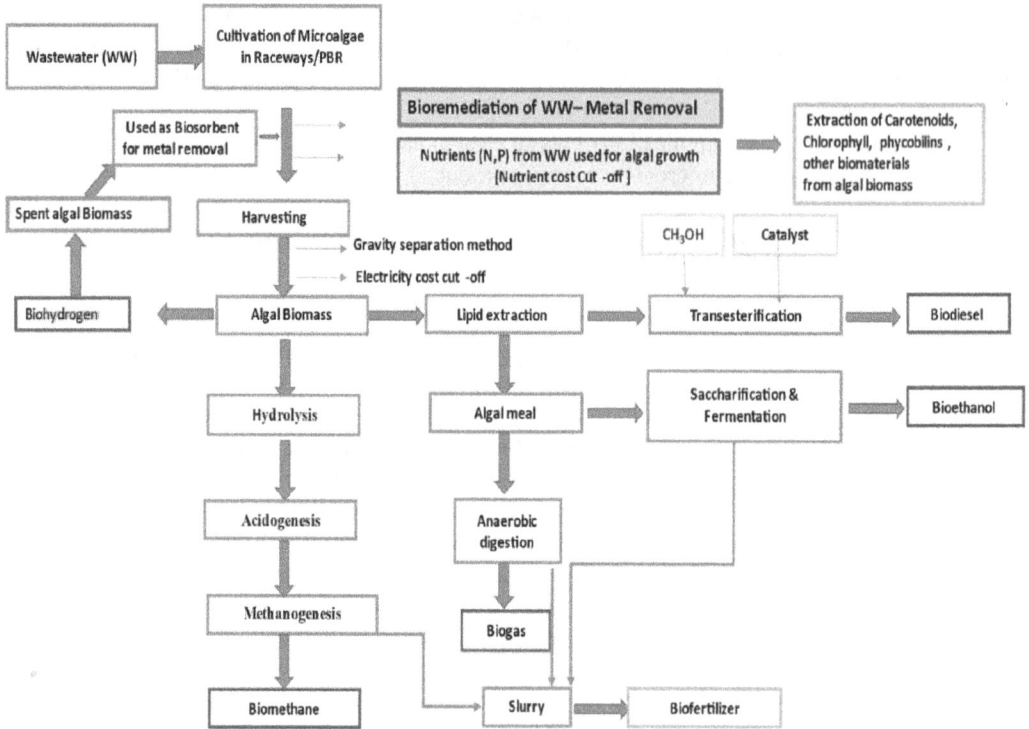

Figure 12.3. Wastewater bioremediation with biorefinery approach using microalgae.

12.5 Conclusion

Extensive studies on the treatment of using wastewater and metal removal have demonstrated the enormous potential of various species of both BGA (cyanobacteria) and green algae (Chlorophyceae). Batch mode studies have established that it is possible to optimize various operational parameters for high metal removal while using the RSM approach, and applying different statistical models further facilitates the selection of the best combinations of parameters for maximum metal bioremediation. Both adsorption and absorption processes have been demonstrated; however, the former has been found to be more effective and easier. Moreover, the mechanism and kinetics of biosorption have been well understood for different systems. Besides this, methods have been developed to regenerate the biosorbent and recover the metals to make the whole process more economical and sustainable. Immobilization of algal biomass has been found to impart greater structural stability and protection to the algae in the presence of toxic pollutants, and immobilized algae are preferred more in continuous mode studies using packed bed reactors. It has been found that pretreatment of algal biomass is useful in improving the biosorption capacity of algal. However, industrial-scale application of the biosorption process must compete with other conventional technologies. In order to make the process economically and environmentally more sustainable, integration of the bioremediation process with other processes leading to the production of biofuels and biomaterials using the biorefinery approach has been proposed. Thus, there are great opportunities for the future linking various algal technologies to derive multiple benefits during bioremediation, making microalgal metal bioremediation a low-cost sustainable approach.

References

Abrihire, O. and M. O. Kadiri. 2011. Bioaccumulation of heavy metals using microalgae. Asian J. Microbiol. Biotechol. Environ. Sci. 13: 91–94.

Adhiya, J., X. Cai, R. T. Sayre and S. J. Traina. 2002. Binding of aqueous cadmium by the lyophilized biomass of *Chlamydomonas reinhardtii*. Colloid. Surf. A: Physicochem Eng. Asp. 210: 1–11.

Ahluwalia, S. S. and D. Goyal. 2007. Microbial and plant derived biomass for removal of heavy metals from wastewater. Bioresour. Technol. 2243–2257.

Aksu, Z. 2002. Determination of the equilibrium, kinetic and thermodynamic parameters of the batch biosorption of nickel (II) ions onto *Chlorella vulgaris*. Process Biochem. 38(1): 89–99.

Andrade, A. D., M. C. E. Rollemberg and J. A. Nóbrega. 2005. Proton and metal binding capacity of the green freshwater alga *Chaetophora elegans*. Process Biochem. 40: 1931–1936.

Arunakumara, K. K. I. U. and X. Zhang. 2008. Heavy metal bioaccumulation and toxicity with special reference to microalgae. J. Ocean Uni. China. 7: 60–64.

Bhatia, D., N. R. Sharma, J. Singh and R. S. Kanwar. 2017. Biological methods for textile dye removal from wastewater: a review. Crit. Rev. Environ. Sci. Technol. 47: 1836–1876.

Bottero, J. Y., J. M. Cases, F. Fiessenger and J. Poirier. 1980. Studies of hydrolysed Aluminium Chloride solutions. Nature of Aluminium species and composition of aqueous solutions. J. Phys. Chem. 84: 2933–2939.

Briffa, J., E. Sinagra and R. Blundell. 2020. Heavy metal pollution in the environment and their toxicological effects on humans. Heliyon. 6: e04691.

Brinza, L., C. A. Nygård, M. J. Dring, M. Gavrilescu and L. G. Benning. 2009. Cadmium tolerance and adsorption by the marine brown alga *Fucus vesiculosus* from the Irish Sea and the Bothnian Sea. Bioresour. Technol. 100(5): 1727–1733.

Cameron, H., M. T. Mata and C. Riquelme. 2018. The effect of heavy metals on the viability of *Tetraselmis marina* AC16-MESO and an evaluation of the potential use of this microalga in bioremediation. Peer. 6: e5295.

Craggs, R. J., S. Heubeck, T. J. Lundquist and J. R. Benemann. 2011. Algal biofuels from wastewater treatment high rate algal ponds. Water Sci. Technol. 63: 660–665.

Dahlia, M. and I. Hassan. 2017. Heavy metals bioaccumulation by the green alga *Cladophera herpestica* in Lake Mariut, Alexandria, Egypt. J. Pollut. 1: 2.

Demayo, A., M. C. Taylor, K. W. Taylor, P. V. Hodson and P. B. Hammond. 1982. Toxic effects of lead and lead compounds on human health, aquatic life, wildlife plants, and livestock. Crit. Rev. Environ. Sci. Technol. 12(4): 257–305.

Duncan, J. R., D. Brady and B. Wilhelmi. 1997. Immobilization of yeast and algal cells for bioremediation of heavy metals. In Biorem. Proto. 91–97.

Eccles, H. 1999. Treatment of metal-contaminated wastes: why select a biological process. Trends Biotechnol. 17: 462–465.

El-Sheekh, M. M., A.A. Farghl, H. R. Galal and H. S. Bayoumi. 2016. Bioremediation of different types of polluted water using microalgae. Rendiconti Lincei, 27: 401–410.

Fatima, G., A. M. Raza, N. Hadi, N. Nigam and A. A. Mahdi. 2019. Cadmium in human diseases: it's more than just a mere metal. Indian J. Clin. Biochem. 34: 371–378.

Fomina, M. and G. M. Gadd. 2014. Biosorption: current perspectives on concept, definition and application. Bioresour. Technol. 160: 3–14.

Gadd, G. M. 2009. Biosorption: critical review of scientific rationale, environmental importance and significance for pollution treatment. Journal of Chemical Technology & Biotechnology: International Research in Process. Environmental & Clean Technology. 84(1): 13–28.

García-García, J. D., K. A. Peña-Sanabria, R. Sánchez-Thomas and R. Moreno-Sánchez. 2018. Nickel accumulation by the green algae-like *Euglena gracilis*. J. Hazard. Mater. 343: 10–18.

García-Lestón, J., J. Méndez, E. Pásaro and B. Laffon. 2010. Genotoxic effects of lead: an updated review. Environ. Int. 36(6): 623–636.

Gaur, J. P. and L. C. Rai. 2001. Heavy metal tolerance in algae. Algal adaptation to environmental stresses. Springer, Berlin, Heidelberg, 363–388.

Genchi, G., A. Carocci, G. Lauria, M. S. Sinicropi and A. Catalano. 2020. Nickel: human health and environmental toxicology. Int. J. Environ. Res. Public Health, 17: 679.

Gonçalves A. L., J. C. Pires and M. Simões. 2017. A review on the use of microalgal consortia for wastewater treatment. Algal Res. 24: 403–415.

Gong R., Y. Ding, H. Liu, Q. Chen and Z. Liu. 2005. Lead biosorption and desorption by intact and pretreated Spirulina maxima biomass. Chemosphere. 58: 125–130.

Hall, J. Á. 2002. Cellular mechanisms for heavy metal detoxification and tolerance. J. Exp. Bot. 53: 1–11.

Hamdy, A. A. 2000. Biosorption of heavy metals by marine algae. Current Microbiol. 41: 232–238.

Hasan, H. A., S. N. Hatika, A. Bakar and M. S. Takriff. 2021. Microalgae biofilms for the treatment of wastewater. Microalgae. Academic Press, 2021. 381–407.

Igiri, B. E., S. I. Okoduwa, G. O. Idoko, E. P. Akabuogu, A. O. Adeyi and I. K. Ejiogu. 2018. Toxicity and bioremediation of heavy metals contaminated ecosystem from tannery wastewater: a review. Journal of Toxicology.

Jankowska, E., K. S. Ashish and O. P. Piotr. 2017. Biogas from microalgae: review on microalgae's cultivation, harvesting and pretreatment for anaerobic digestion. Renew. Sustain. Energy Rev. 75: 692–709.

Kalra, R., S. Gaur and M. Goel. 2020. Microalgae bioremediation: a perspective towards wastewater treatment along with industrial carotenoids production. J. Water Process Eng. 101794.

Kamra, A., A. Kaushik, K. Bala and N. Rani. 2007. Biosorption of Cr(VI) by immobilized biomass of two indigenous strains of cyanobacteria isolated from metal contaminated soil. J. Hazard. Mater. 148: 383–386.

Kaplan, D. 2013. Absorption and adsorption of heavy metals by microalgae. Handbook of microalgal culture: applied Phycology and Biotechnology. 2: 602–611.

Kaushik, A., S. Jain, J. Dawra and M. S. Bishnoi. 2000. Heavy metal pollution of river Ghaggar in Haryana. Indian J. Environ. Toxicol. 10: 63–66.

Kaushik, A., S. Jain, J. Dawra and C. P. Kaushik. 2001. Heavy metal pollution of river Yamuna in the industrially developing state of Haryana. J. Environ. Biol. 43: 164–168.

Kaushik, A., S. Mona and C. P. Kaushik. 2011. Integrating photobiological hydrogen production with dye-metal bioremoval from simulated textile wastewater. Bioresour. Technol. 102: 9957–9964.

Kiran, B., A. Kaushik and C. P. Kaushik. 2007a. Biosorption of Cr (VI) by native isolate of *L. putealis* (HH-15) in the presence of salts. J. Hazard. Mater. 141: 662–667.

Kiran, B., A. Kaushik and C. P. Kaushik. 2007b. Response surface methodological approach for optimizing removal of Cr (VI) from aq. sol. using immobilized cyanobacterium. Chem. Eng. J. 126: 147–153.

Kiran, B. and A. Kaushik. 2008. Cyanobacterial biosorption of Cr (VI): application of two parameter and Bohart Adams models for batch and column studies. Chem. Eng. J. 144: 391–399.

Kiran, B., A. Kaushik and C. P. Kaushik. 2008. Metal–salt co-tolerance and metal removal by indigenous cyanobacterial strains. Process Biochemistry. 43.6: 598–604.

Kiran, B. and A. Kaushik. 2012. Equilibrium sorption study of Cr (VI) from multi-metal systems in aq. sol. by *Lyngbya putealis.* Ecol. Eng. 38: 93–96.

Kiran, B., N. Rani and A. Kaushik. 2016. FTIR spectroscopy and scanning electron microscopic analysis of pretreated biosorbent to observe the effect on Cr (VI) remediation. Int. J. Phytoremediat. 18: 1067–1074.

Kosek, K., Z. Polkowska, B. Żyszka and J. Lipok. 2016. Phytoplankton communities of polar regions diversity depending on enviro. conditions and chemical anthropopressure. J. Environ. Manag. 171: 243–259.

Kratochvil, D. and B. Volesky. 1998. Biosorption of Cu from Ferruginous Wastewater by Algal Bio-mass. Water Res. 32: 2760–2768.

Kumar, D., L. K. Pandey and J. P. Gaur. 2016. Metal sorption by algal biomass: from batch to continuous system. Algal Res. 18: 95–109.

Kumar, K. S., H. U. Dahms, E. J. Won, J. S. Lee and K. H. Shin. 2015. Microalgae–a promising tool for heavy metal remediation. Ecotoxicol. Enviro Saf. 113: 329–352.

Kumar, V., A. Sharma and A. Cerda. 2020. Heavy Metals in the Environment: Impact, Assessment, and Remediation Elsevier.

Lau, P. S., H. Y. Lee, C. C. K. Tsang, N. F. Y. Tam and Y. S. Wong. 1999. Effect of metal interference, pH and temperature on Cu and Ni biosorption by *C. Vulgaris* and *C. Miniata*. Environ. Technol. 20: 953–96.

Li, Y., Y. Chen and J. Wu. 2019. Enhancement of methane production in anaerobic digestion process: A review. Applied Energy 240: 120–137.

Markou, G. and E. Nerantzis. 2013. Microalgae for high-value compounds and biofuels production: a review with focus on cultivation under stress conditions. Biotechnol. Adv. 31: 1532–1542.

Michalak, I., K. Chojnacka and A. Witek-Krowiak. 2013. State of the art for the biosorption process—a review. Appl. Biochem. Biotechnol. 170: 1389–1416.

Mona, S., A. Kaushik and C. P. Kaushik. 2011a. Biosorption of Chromium (VI) by spent cyanobacterial biomass from a hydrogen fermentor using box-behnken model. Int. Biodeterior. Biodegrad. 65: 656–663.

Mona, S., A. Kaushik and C. P. Kaushik. 2011b. Sequestration of Co (II) from aqueous solution using immobilized biomass of *Nostoc linckia* waste from a hydrogen bioreactor. Desalin. 276: 408–415.

Mona, S., A. Kaushik and C. P. Kaushik. 2011c. Hydrogen production and metal-dye bioremoval by a *N. linckia* strain isolated from textile mill oxidation pond. Bioresour. Technol. 102: 3200–3205.

Mona, S., A. Kaushik and C. P. Kaushik. 2013. Prolonged hydrogen production by Nostoc in photobioreactor and multi-stage use of the biological waste for column biosorption of some dyes and metals. Biomass Bioenergy 54: 27–35.

Mona, S. and A. Kaushik. 2015a. Screening metal-dye-tolerant photoautotrophic microbes from textile wastewaters for biohydrogen production J. Appl. Phycol. 27: 1185–1194.

Mona, S. and A. Kaushik. 2015b. Cr and Cb sequestration using exopolysaccharides produced by freshwater cyanobacterium *Nostoc linckia.* Ecol. Eng. 82: 121–125.

Monteiro, C., P. Castro and F. Malcata. 2012. Metal uptake by microalgae: underlying mechanisms and practical applications. Biotechnol. Progress. 28: 299–311.

Naimabadi, A., A. Gholami and A. M. Ramezani. 2020. Determination of heavy metals and health risk assessment in indoor dust from different functional areas in Neyshabur, Iran. Indoor Built Environ. 1420326X20963378.

Nasirpour, N., S. M. Zamir and S. A. Shojaosadati. 2017. Immobilization techniques for microbial bioremediation of toxic metals. In Handbook of Metal-microbe Interactions and Biomed. 357–370.

Park, J. B, K. R. J. Craggs and A. N. Shilton. 2011. Wastewater treatment high rate algal ponds for biofuel production. Bioresour. Technol. 102: 35–42.

Picardo, M. C., J. L. de Medeiros, F. A. Ofélia de Queiroz and R. M. Chaloub. 2013. Effects of CO_2 enrichment and nutrients supply intermittency on batch cultures of *Isochrysis galbana*. Bioresour. Technol. 143: 242–250.

Podstawczyk, D., A. Witek-Krowiak, A. Dawiec and A. Bhatnagar. 2015. Biosorption of copper (II) ions by flax meal: empirical modeling and process optimization by response surface methodology (RSM) and artificial neural network (ANN) simulation. Ecol. Eng. 83: 364–379.

Potnis, A. A., P. S. Raghavan and H. Rajaram. 2021. Overview on cyanobacterial exopolysaccharides and biofilms: role in bioremediation. Reviews in Environmental Science and Bio/Technology. 20.3: 781–794.

Preeti, R., A. Kaushik and C. P. Kaushik. 2015. Potential of cyanobacterial consortium of hot springs for high value pigment production and dye decolorization. Inter. J. Adv. Sci. Technol. 53: 11–317.

Priyadarshini, E., S. S. Priyadarshini and N. Pradhan. 2019. Heavy metal resistance in algae and its application for metal nanoparticle synthesis. Appl. Microbiol. Biotechnol. 103: 3297–3316.

Reed, R. H. and G. M. Gadd. 1989. Metal tolerance in eukaryotic and prokaryotic algae. Heavy metal tolerance in plants: Evolutionary Aspects. 105–118.

Rehman, K., F. Fatima, I. Waheed and M. S. H. Akash. 2018. Prevalence of exposure of heavy metals and their impact on health consequences. J. Cell. Biochem. 119: 157–184.

Roberts, D. A., R. de Nys and N. A. Paul. 2013. The effect of CO_2 on algal growth in industrial waste water for bioenergy and bioremediation applications. PloS one. 8: e81631.

Saavedra, R., R. Muñoz, M. E. Taboada and S. Bolado. 2019. Influence of organic matter and CO_2 supply on bioremediation of heavy metals by *C. vulgaris* and *S. almeriensis* in a multi metallic matrix. Ecotoxicol. Environ. Saf. 182: 109393.

Sayara, T., S. Khayat, J. Saleh and P. Van Der Steen. 2021. Evaluation of the effect of reaction time on nutrients removal from Sec. effluent of wastewater: Field demonstrations using algal-bacterial photobioreactors. Saudi J. Biol. Sci. 28: 504–511.

Schenk, P. M., S. R. Thomas-Hall, E. Stephens, U. C. Marx, J. H. Mussgnug, C. Posten, O. Kruse and B. Hankamer. 2008. Second generation biofuels: high-efficiency microalgae for biodiesel production. Bioenergy Res. 1: 20–43.

Selvi, A., A. Rajasekar, J. Theerthagiri, A. Ananthaselvam, K. Sathishkumar, J. Madhavan and P. K. Rahman. 2019. Integrated remediation processes toward heavy metal removal/recovery from various environments-a review. Frontiers Environ. Sci. 7: 66.

Sharma, M., A. Kaushik, K. Bala and A. Kamra. 2008. Sequestration of chromium by exopolysaccharides of *Nostoc* and *Gloeocapsa* from dilute aqueous solutions. Journal of Hazardous Materials 157(2-3): 315–318.

Sharma, M., N. Rani, A. Kamra, A. Kaushik and K. Bala. 2009. Growth, exopolymer production and metal bioremoval by *Nostoc punctiforme* in Na and Cr (VI) spiked medium. J. Environ. Res. Dev. 4(2).

Sharma, S., S. Rana, A. Thakkar, A. Baldi, R. S. R. Murthy and R. K. Sharma. 2016. Physical, chemical and phytoremediation technique for removal of heavy metals. J. Heavy Met. Toxic. Dis. 1: 1–15.

Singh, A., D. Kumar and J. P. Gaur. 2012. Continuous metal removal from solution and industrial effluents using Spirogyra biomass-packed column reactor. Water Res. 46(3): 779–788.

Sweetly, K., J. Sangeetha and B. Suganthi. 2014. Biosorption of heavy metal lead from aq. Sol by non- living biomass of *Sargassum myriocystum*. Int. J. Appl. Innov. Eng. Manag. 3: 39–45.

Sydney, E. B., A. C. Novak, J. C. de Carvalho and C. R. Soccol. 2014. Respirometric balance and carbon fixation of industrially important algae. In Biofuels from algae. Elsevier. 67–84.

Tchounwou, P. B., C. G. Yedjou, A. K. Patlolla and D. J. Sutton. 2012. Heavy metal toxicity and the environment. Molecular, Clinical and Environmental Toxicology. 133–164.

Toranzo, R., G. Ferraro, M. V. Beligni, G. L. Perez, D. Castiglioni, D. Pasquevich and C. Bagnato. 2020. Natural and acquired mechanisms of tolerance to Cr in a *S. dimorphus* strain. Algal Res. 52: 102100.

Ubando, A. T., A. D. M. Africa, M. C. Maniquiz-Redillas, A. B. Culaba, W. H. Chen and J. S. Chang. 2021. Microalgal biosorption of heavy metals: a comprehensive bibliometric review. J. Hazard. Mater. 402: 123431.

Uzunoğlu, D., N. Gürel, N. Özkaya and A. Özer. 2014. The single batch biosorption of copper (II) ions on *S. acinarum*. Desalin. Water Treat. 52(7-9): 1514–1523.

Vijayaraghavan, K. and Y. S. Yun. 2008. Bacterial biosorbents and biosorption. Biotechnol. Adv. 26(3): 266–291.

Viswanath, B. and F. Bux. 2012. Biodiesel production potential of wastewater microalgae *Chlorella* sp. Under photoautotrophic and Heterotrophic growth conditions. Br. J. Eng. Technol. 1: 251–264.

Volesky, B. and Z. R. Holan. 1995. Biosorption of heavy metals. Biotechnol. Progress. 11(3): 235–250.

Volesky, B. 2003a. Biosorption process stimulation tools. Hydrometall. 71(1-2): 179–190.

Volesky, B. 2003b. Sorption and biosorption. St. Lambert, Quebee: BV Soebex.

Waldichuk, M. 1974. Some biological concerns in heavy metals pollution. Pollution and physiology of marine Organisms. 1: 1–59.

Yeong, T. K., K. Jiao, X. Zeng, L. Lin, S. Pan and M. K. Danquah. 2018. Microalgae for biobutanol production– Techno. Evaluation and value proposition. Algal Res. 31: 367–376.

Zeitoun, M. Moustafa and E. E. Mehana. 2014. Impact of water pollution with heavy metals on fish health: overview and updates. Global Vet. 12: 219–231.

Zeraatkar, A. K., H. Ahmadzadeh, A. F. Talebi, N. R. Moheimani and M. P. McHenry. 2016. Potential use of algae for heavy metal bioremed., a critical review. J. Environ. Manag. 181: 817–831.

Zhang, L. and M. H. Wong. 2007. Environmental Hg contamination in China: sources and impacts. Environ. Int. 33(1): 108–121.

Zhou, Q., N. Yang, Y. Li, B. Ren, X. Ding, H. Bian and X. Yao. 2020. Total concentrations and sources of heavy metal pollution in global river and lake water bodies from 1972 to 2017. Global Ecology and Conservation. 22: e00925.

CHAPTER **13**

Bioremediation of Heavy Metals from Aquatic Environments

Fariha Latif,[1,*] *Shahena Perveen,*[1] *Sana Aziz,*[2]
Rehana Iqbal,[1] and *Muhammad Mudassar Shahzad*[3]

13.1 Introduction

The severe contamination of freshwater resources, particularly in developing countries, is causing serious health hazards and environmental issues. Though aquatic pollution is not a novel phenomenon, but the pace of the industrial revolution and urbanization have aggravated its negative impacts on the aquatic environment. Predominant anthropogenic sources of aquatic contamination include mining operations, untreated industrial effluents, domestic sewage, waste dump leachates and combustion emissions. These sources are adding a number of pollutants including metallic ions, polycyclic aromatic hydrocarbons, dioxins, polychlorinated biphenyls and other xenobiotics in the natural aquatic ecosystems. Metallic ions toxicity affects the physiology and ecology of aquatic organisms due to their specific properties of long persistence, bioaccumulation and biomagnification in the food chain. Currently, all areas of the environment are being polluted by several chemical and biological pollutants and the aquatic ecosystems are serving as a major repository for them. Among these pollutants, metallic ions pollution is widespread and has become an issue of great concern as metallic ions are indiscriminately discharged into freshwaters, deteriorating their quality and ultimately affecting the inhabitant fish fauna.

13.1.1 Heavy Metals

Metalloids and metals having a density of more than 4000 kg m^{-3} are called heavy metals. Most of them are mostly poisonous at different concentrations (Shah et al. 2018). The concentration can be higher, lower or moderate. Examples of heavy metals are Zinc (Zn), Copper (Cu), Nickel (Ni), Strontium (Sr), Molybdenum (Mo), Titanium (Ti), Lead (Pb), Cobalt (Co), Chromium (Cr), Vanadium (V,) Mercury (Hg) and Tin (Sn) (Satyanarayana et al. 2019).

[1] Institute of Zoology, Bahauddin Zakariya University, Multan, Pakistan.
[2] Department of Zoology, Wildlife and Fisheries, University of Agriculture, Faisalabad, Pakistan.
[3] Department of Zoology, Division of Science & Technology, University of Education, Lahore, Pakistan.
* Corresponding author: farihalatif@bzu.edu.pk

Water and soil are now becoming polluted due to the release of garbage. Untreated wastewater is added to water, and it contains various metals that create pollution (Rudakiya and Pawar 2013). Heavy metals have a negative effect on the food web and food chain. They can also impact numerous activities of cells (Vardhan et al. 2019).

13.1.1.1 Metals Sources

There are two kinds of metal sources in aquatic environments:

  i. Natural sources

 ii. Anthropogenic sources

i. Natural Sources

Natural sources include the weathering of magmatic, sedimentary and metamorphic rocks in the rock cycle that add heavy metals to the atmosphere and aquatic reservoirs.

ii. Anthropogenic Sources

Heavy metals are mainly introduced into groundwater as well as surface water by agricultural and industrial activities, landfilling, mining and transportation. Various origins of pollution are caused by metallic contaminants. Sometimes single or in mixture form, metals are discharged into the dumping areas. Different organic pollutants consist of various metal pollutants. The pesticides contain a number metals such as zinc, arsenic and copper. The formation of stains or dyes, alloys, pigments and batteries also require metals in their composition (Roane et al. 2015). The high level of metal concentrations and their use causes various negative impacts on humans, plants and animals.

13.1.1.2 Heavy Metal Toxicity

Heavy metals generally exist in the form of carbonates, hydroxides, oxides, sulfides, sulfates, phosphates, silicates and organic compounds. They also exist in their metallic, elemental form, but are mobilized by the action of humans (anthropogenic activity extraction, smelting) or natural phenomenon (weathering and leaching). The World Health Organization reported that heavy metals including cadmium, arsenic, chromium, cobalt, copper, mercury, nickel, manganese, titanium, lead and tin are toxic (Index 2018). Various heavy metals cause cellular toxicity through different mechanisms as depicted in Figure 13.1.

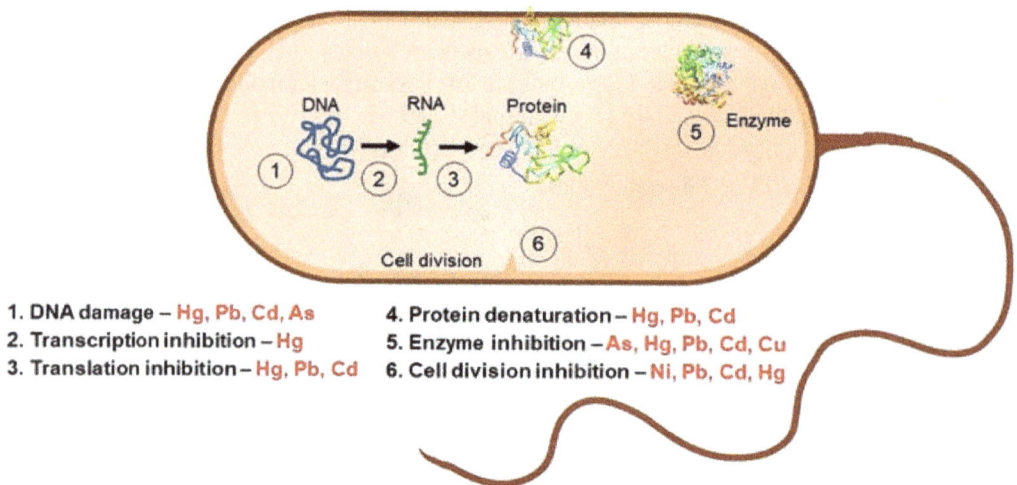

1. DNA damage – Hg, Pb, Cd, As
2. Transcription inhibition – Hg
3. Translation inhibition – Hg, Pb, Cd

4. Protein denaturation – Hg, Pb, Cd
5. Enzyme inhibition – As, Hg, Pb, Cd, Cu
6. Cell division inhibition – Ni, Pb, Cd, Hg

Figure 13.1. Toxicity mechanisms of some heavy metals.

13.2 Bioremediation

The method in which plants, microorganisms, bacteria and fungi are used to eliminate pollutants from the environment is called bioremediation. The purpose of "bioremediation" methods is to decrease the level of the contaminant to a minimum and within safe limits viz. within the range set by different environmental protection agencies (Pointing 2001). Bioremediation can also be increased by adding different materials such as nitrogen and carbon sources (Rudakiya et al. 2019a). The fungi and bacteria can augment the bioremediation process (Zaki et al. 2014).

The bioremediation process can reduce negative impacts on the natural environment. It is a less costly process. The total remediation of contaminants is attainable on-site without the requirement of excavation (Vidali 2001). It needs a small amount of energy and protects the soil. It can be made more reliable to the community (Zhang and Qiao 2002). Aerobic soil is essential for the fungal system to work. There are many drawbacks to the bioremediation process. This method can be used for those compounds that are biodegradable. The by-products of the bioremediation process are less hazardous than the original compounds. These processes are highly specified and complex. It requires a longer duration for their work (Vidali 2001).

The bioremediation of heavy metals is done by using the following processes (Figure 13.2):

13.2.1 Biosorption

13.2.2 Bioaccumulation

13.2.3 Bioleaching

13.2.4 Biotransformation

13.2.5 Biomineralization

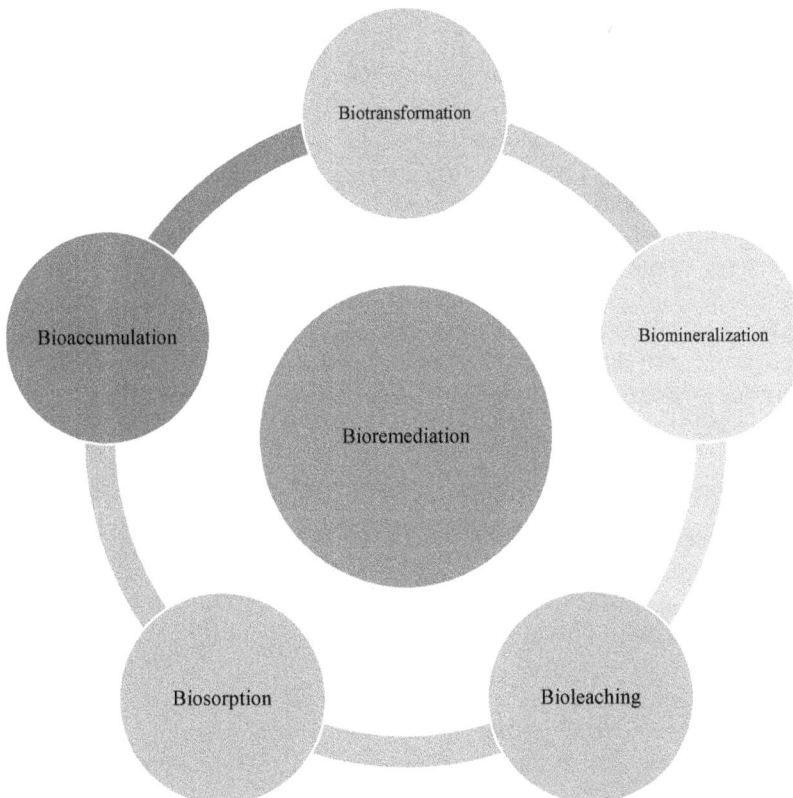

Figure 13.2. Various processes in the bioremediation.

13.2.1 Biosorption

The capacity of biological materials to concentrate heavy metals from wastewater via metabolically mediated or physico-chemical pathways is called biosorption (Fourest and Roux 1992). The biological material can bind heavy metals from an aqueous solution. It has the characteristics of some substances to concentrate and bind some ions from metal-containing water (Volesky 2007). The best biosorbents that adsorb metals are bacteria, fungi, yeasts and algae. Biosorption is an alternative rapid method for the removal of pollutants by using non-active and nonliving microbes. The advantage of this process are as follows:

➢ No additional nutrient requirement

➢ biosorbent can be regenerated

➢ metal recovery

➢ low cost

➢ high efficiency

➢ decreasing amount of biological sludge

The process of biosorption is affected by various physico-chemical parameters such as temperature, pH, metal ion's initial concentration, biosorbent concentration and speed of mixing of biosorbent and solution. The entire biomass of biosorbent can be treated physically and chemically before use and can be made economical by recycling and reusing the biosorbent material after removing the metals. Various bioreactors can be used in biosorption for the removal of metal ions from large volumes of water or effluents (Kanamarlapudi et al. 2018). When the physico-chemical interaction of metal ions occurs with cellular compounds heavy metals can be adsorbed.

13.2.2 Bioaccumulation

This is an active metabolic process that relies on the import-storage system. Using transporter proteins, the system moves heavy metal ions through the lipid bilayer and into the cytoplasm or intracellular regions. The heavy metals that these entities sequester might exist in particulate forms, insoluble forms and their by-products. The metal ions are sequestered by metal-binding entities such as proteins and peptide ligands. The process in which ions of metals accumulate in the living cell is called bioaccumulation. It is a complex and irreversible process that depends on cell metabolism and is mainly done by microorganism biomass cultivation. The metabolic process starts in the organism that activates the intracellular transport system (Srichandan et al. 2014, Gu et al. 2018). A few examples include; the yeast *Pichia stipites* can bio-accumulate Cr^{3+} and Cu^{2+} with the highest absorbance ability of 9.10 mg g^{-1} and 15.85 mg g^{-1}, respectively (Ojha et al. 2015). *Aspergillus niger* can eliminate Pb and Cu with the greatest uptake of 34.4 mg g^{-1} and 15.6 mg g^{-1}, respectively (Srichandan et al. 2015).

13.2.3 Bioleaching

The bioleaching process is also called bio-hydrometallurgy or microbial leaching. The process in which solubilization of metals occurs from some ores that are not soluble is called bioleaching. Microorganisms are also used in this process as reducing agents (Mishra et al. 2005). These methods include membrane separation, solvent extraction, adsorption, ion exchange, selective precipitation and electrowinning.

Microbial bioleaching is used for the solubilization of various metals viz. cobalt, nickel copper, zinc and uranium (Rohwerder et al. 2003). For example, in the bioleaching process of arsenopyrite, the sulfides of arsenic and iron are dissolved. The endured gold is recuperated by the procedure known as cyanidation.

The natural process of bioleaching has its limitations therefore, researchers have broadened this process to treat solid wastes artificially to remove or solubilize metals. These solid wastes are released from different industries and mining processes causing various health hazards to animal diversity as well as human health. Different microbes are used in the bioleaching process. Examples of some microbes are archaea, fungi and acidophilic bacteria (Brandl et al. 2001, Natarajan and Ting 2014). All three types of acidophilic bacteria such as mesophiles, moderate thermophiles and thermophiles are used in this process. Fungi can also be used in this process. Some of the fungi that are used in the bioleaching process include *Penicillium chrysogenum, Penicillium simplicissimum, Aspergillus niger* and *Aspergillus flavus*. The fungal bioleaching process requires a pH from 3.0 to 7.0 and a temperature ranging from 25 to 35°C.

13.2.4 Biotransformation

This is the method by which a chemical compound's structure is changed, resulting in the production of a molecule with comparatively higher polarity. In other words, through the process of metal-microbe interaction, metal and organic molecules are changed from a harmful state to a form that is substantially less toxic. This technique basically enables microbes to acclimatize to changing environments.

Microorganisms control trace element transformation (microbial or biotransformation) through a number of mechanisms such as oxidation, reduction, methylation, demethylation, complex formation and biosorption. Microbial transformation is important in the behavior and fate of toxic elements in soils and sediments, particularly Arsenic (As), Chromium (Cr), mercury (Hg), and Selenium (Se). Biotransformation processes can change the speciation and redox state of these elements, controlling their solubility and mobility (Kunhikrishnan et al. 2017). These processes are critical for trace element bioavailability, mobility, ecotoxicity and environmental health. Microbial cells have a high surface-to-volume ratio, a rapid rate of growth, as well as metabolic activity, and is simple to maintain sterile conditions for microbes. They are thus ideal organisms for the process of biotransformation. Condensation, hydrolysis, the creation of new carbon bonds, isomerization, the addition of functional groups, oxidation, reduction and methylation are all methods that can be used to carry out this process. These processes might cause metals to volatilize, thus decreasing their ability (Tayang and Songachan 2021).

13.2.5 Biomineralization

A natural process of mineral production is the biomineralization of heavy metals. Minerals like phosphates, oxides, sulfates, silicates and carbonates are naturally synthesized in this process, which involves a variety of mechanisms in living things. Mineral production depends on the presence of highly variable and reactive surfaces, such as cell walls and extra organic layers with varying levels of hydration, content and structure. Additionally, there are organic ligands that deprotonate and impart a net negative charge on the microbial surface as pH rises, including amine, carboxyl, hydroxyl, phosphoryl and sulfur.

Positively charged potential hazardous metals precipitate unevenly into more stable and compact mineral products. Phosphate precipitation, carbonate precipitation, oxalate precipitation and complexation can all result in the immobilization or complexation of metals (Tayang and Songachan 2021).

13.3 Types of Bioremediations

The process in which incomplete and occasionally complete detoxification of pollutants takes place by small microorganisms is called biodegradation (Gouma et al. 2014). The more specified term is bio-mineralization in which bacteria, fungi and plants release different acids and convert them into the unsolvable form of metal (Rudakiya et al. 2019). The following kinds of bioremediations are used to eliminate toxic heavy metals, metalloids and non-metals from different areas:

13.3.1 Bacterial remediation

13.3.2 Fungal remediation

13.3.3 Algal remediation

13.3.4 Phyto remediation

13.3.1 Bacterial Bioremediation

There are different mechanisms used by bacteria to remediate or eliminate metals (Rudakiya and Pawar 2017). Bacteria use the processes of metal binding, biosorption, biotransformation and biosorption to eradicate metals from aquatic environments (Rudakiya 2018). The process in which metals are oxidized and reduced to change into nontoxic forms is called biotransformation. The following processes occur during biotransformation outside the cell:

➢ Reduction

➢ Oxidation

➢ Methylation.

➢ Dealkylation

Bacteria consist of carbohydrates, protein and fatty acids. These components are very essential for the binding of metals. Enzymes can also be used for the transformation of metals. For example, chromate reductase. It forms a species of metals that are not usually soluble (Rudakiya et al. 2020).

13.3.2 Fungal Bioremediation

The fungi are especially important organisms to treat metalloids and metals based on different properties like biochemical, ecological and morphological (Rudakiya and Gupte 2017). Fungi can be used to treat metals throughout the various phases of their life cycle through different mechanisms (Baldrian 2003). These components of the cell wall contain various functional groups viz. hydroxyl, phosphate, carboxyl and amine. They are involved in the process of 'metal chelation' (Couto et al. 2004).

Many other microorganisms make different substances like Extracellular Polymeric Substances (EPS) in the environment where they exist (Guibaud et al. 2005). These substances can be used in binding, reduction of heavy metals and tolerance limits. EPS are anionic and acidic in nature. The anionic characteristics of EPS transmit "electrostatic interaction" with heavy metals however, the acidic property of Extracellular Polymeric Substances is due to different functional groups such as amino acids, phenolic, hydroxyl, carboxyl and uronic acids. These groups interact with metal ions that have a positive charge (Gutnick and Bach 2000). Heavy metals like Cu, Pb and Cd have been separated by *A. niger* (Kapoor et al. 1999).

Interaction among metals and fungi: Fungi can be used for the bioremediation of heavy metals due to their various properties such as rapid growth, easily culturable due to their small reproduction cycle and can be modified both genetically and morphologically (Dhankhar and Hooda 2011). The elimination of toxic substances by fungi is called mycoremediation. Fungi can use various mechanisms for heavy metal bioremediation, especially the biosorption process. The dead and living cells of fungi can be used for fungal biosorption. The dead cells of fungi have a higher capacity to remove heavy metals (Volesky 2007).

Fungi can use two mechanisms for the removal of metals (Veglio and Beolchini 1997). The first method is the primary method which includes the binding of heavy metals to the outer surface

Table 13.1. Bioremediations of various heavy metals by fungi.

Sr. No.	Metal	Fungi	References
1	Hg	*Aspergillus niger*	Acosta-Rodríguez et al. 2018
		Aspergillus flavus	Kurniati et al. 2014
		P. canescens	Say et al. 2004
		R. arrhizus	Tobin et al. 1984
2	Cr	*A. niger*	Acosta-Rodríguez et al. 2018
		Rhizopus nigricans	Bai and Abraham 2001
3	Pb	*Penicillium* sp.	Siegel et al. 1983
		Rhizopus arrhizus	Fourest and Roux 1992
4	Cd	*Aspergillus niger*	Das et al. 2013
		Penicillium spinulosum	Townslcy and Ross 1986
5	Zn	*A. niger*	Acosta-Rodríguez et al. 2018
6	As	*Aspergillus nidulans*	Maheshwari and Murugesan 2009
		Aspergillus niger	Acosta-Rodríguez et al. 2018
7	Co	*Trichoderma, Apecilomyces*	Townslcy and Ross 1986
		Penicillium, Pythium, Rhizopus, Aspergillus niger	Acosta-Rodríguez et al. 2018

and the second method involves the absorption of metals by the metabolism called bioaccumulation (Kadukova and Vircikova 2005). Some of the studies that are based on the biosorption process are detailed in Table 13.1.

These methods are surface binding, the complexion by various functional groups including hydroxyl, phosphate, carboxyl, and amine groups and ion exchange reaction (Wang and Chen 2009). The fungus cell wall makes a connection with ions of metals (Figure 13.3). The extracellular methods use cell wall binding, precipitation and chelation (Bellion et al. 2006). The recycling of fungi in bioremediation is an expensive process (Tsezos 1984). Living cells of fungi are killed by various methods like chemical drying, mechanical drying, autoclaving and vacuum drying.

Factors Affecting the Fungal Biosorption: The removal of 'heavy metals' by fungi depends on various environmental factors and types of metals (Hassen et al. 1998). Different parameters such as oxygen level and glucose affect the absorption of metals (Javaid et al. 2011). The parameters like concentration, contact time, initial metal ion concentration, temperature, pH and biomass have been noted to affect the fungal biosorption process (Kapoor and Viraraghavan 1995).

Fungal Bioleaching: Fungal bioleaching has been used in mining areas and for the treatment of waste material containing metals (Bosecker 2001). Fungi can also be used for sorting of heavy metals from low-grade ores (Chaudhary et al. 2014). *Penicillium verruculosum* and *Aspergillus* were used to solubilize iron *Penicillium* and *Aspergillus* leached Ni, Fe and Co (Valix et al. 2001).

Fungal Bio-immobilization: The interaction between metals and fungus causes a notable decrease in the mobility of heavy metals. This immobility has been achieved by different methods such as bio-precipitation, bio-reduction/bio-oxidation, biosorption and bioaccumulation. The fungus, *Rhizopus arrhizus* can be used for the elimination of Cu^{2+}, Fe^{3+} and Cd^{2+} by the fungal bio-immobilization process (Lewis and Kiff 1998).

Fungal Biomineralization: Biologically Controlled Mineralization (BCM) and Biologically Induced Mineralization (BIM) are types of biomineralization. In BCM, the microorganisms are used to control the growth and nucleation of the biominerals, however in BIM organisms modify the native environment to produce optimum conditions for the mineral's precipitation (Gadd 2010).

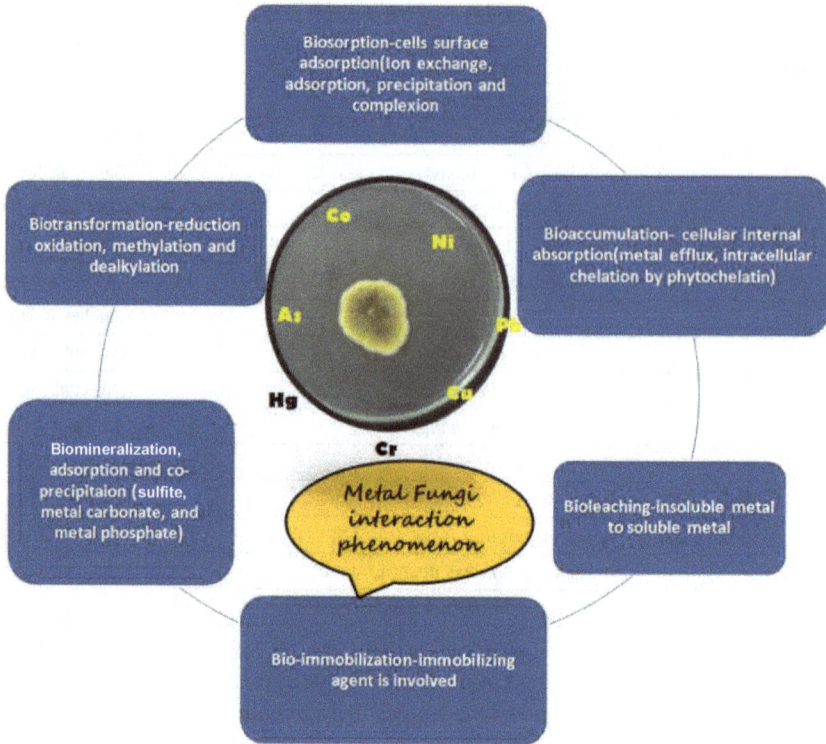

Figure 13.3. Metals and fungal interactions in bioremediation.

13.3.3 Algal Bioremediation

Microorganisms such as fungi, yeast, bacteria and microalgae can be used for the bioremediation of metals. Microalgae have been widely used due to their various properties viz. simple structure and high photosynthetic capacity. Microalgae can grow well in any given environment. Microalgae can be used in the phycoremediation of poisonous heavy metals (Cameron et al. 2018). The dead and living cells of microalgae can be used as biosorbents which is an eco-friendly process. The microalgae can be used for the recovery of gold and silver (Birungi and Chirwa 2015, Jaafari and Yaghmaeian 2019).

Mechanisms of Heavy Metal Removal by Microalgae

Heavy metals including manganese (Mn), zinc (Zn), Boron (B), Cobalt (Co), copper (Cu), iron (Fe) and Molybdenum (Mo) can be used by microalgae. These can be consumed for cell and enzymatic metabolism (Figure 13.4). Metals like lead, chromium, cadmium and arsenic are poisonous to microalgae. Some cyanobacteria including *Phormidium*, *Spirogyra*, *Anabaena* and *Oscillatoria* can grow in environments where heavy metals are present. These species have a tolerance ability to metals stress (Balaji et al. 2016).

Microalgae have adopted different techniques for their protection from heavy metals toxicity including gene regulation, exclusion, chelation and immobilization (Gómez-Jacinto et al. 2015). They can form complexes of heavy metals and proteins without altering their activity (Priatni et al. 2017).

The elimination of heavy metals by using microalgae is acquired by two processes. Biosorption (extracellular passive adsorption) is the first stage and bioaccumulation (intracellular positive diffusion and accumulation) is the second stage. The first stage is rapid while the second stage is slow. The microalgae cell wall contains lipids, proteins and polysaccharides. The polysaccharides

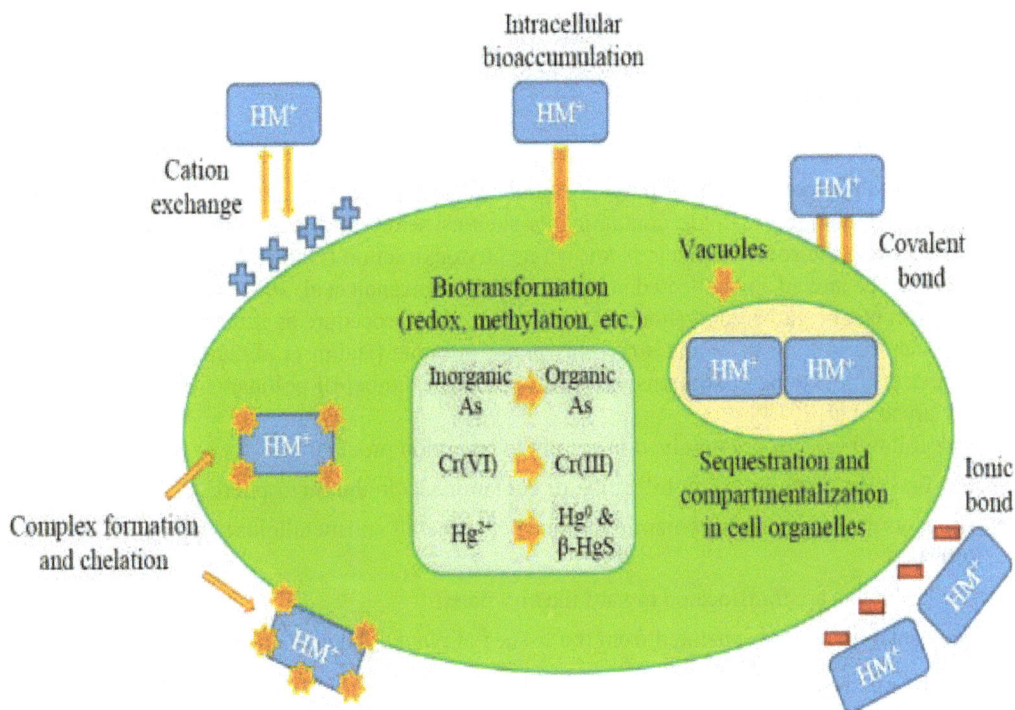

Figure 13.4. Mechanism of heavy metal removal by microalgae.

are alginate and cellulose. Its cell wall also provides various functional groups including imidazole, phosphate, sulfonate, thiol, amino, carboxyl and hydroxyl. These functional groups can bind with heavy metals (Priatni et al. 2017). They also contain a large amount of alcohol and carboxyl groups that attract the cations and anions of heavy metals (Pradhan et al. 2019).

The spectroscopy including NMR (Nuclear Magnetic Resonance) and FTIR (Fourier transform infrared) can be used to find out different functional groups of microalgae. Various methods can be used for the adsorption of heavy metals on the outer surface of microalgae. These methods include the ionic exchange of heavy metal ions with cell wall cation. The heavy metals are moved into the cytoplasm by the process of diffusion (Ibuot et al. 2017, Pradhan et al. 2019).

Algal Bioremediation of a few metals

Arsenic (As) is a toxic heavy metal. Its pollution is caused by human activities including medical use, smelting and mining (Arora et al. 2017, Singh et al. 2016). Microalgae are used to reduce the pollution of arsenic by the complex formation with glutathione (Papry et al. 2019).

Cadmium (Cd) is poisonous heavy metal. It is released into the surroundings by human activities including color pigments, tanning industries and pesticide use (Abinandan et al. 2019). The dead microalgae can be used for bioremediation of Cd by surface binding mechanism.

Chromium (Cr) is used in ink, paint and dyes (Gupta and Rastogi 2008). Microalgae can be used for the removal of chromium by using the mechanism of biosorption. *Navicula pelliculosa* and *Phaeodactylum tricornutum* have been used for this purpose (Hedayatkhah et al. 2018).

13.3.4 Phytoremediation

The use of living plants to remove heavy metals and metalloids from water, soil and the air is called phytoremediation. The types of phytoremediation are:

- Phytoextraction
- Phytostabilization

- Phytofiltration
- Phytovolatilization
- Phytodegradation

Phytoextraction

It is also called phytoabsorption or phtyoaccumulation and phytosequestration. The process which involves the elimination of toxic contaminants such as heavy metals and metalloids from water and soil by using the roots of plants is known as phytoextraction (Ghosh and Singh 2005). These pollutants can be moved and collected in shoots (Muthusaravanan et al. 2018). The process in which contaminants can be gathered in shoots is a very important procedure as if the material is gathered in roots, then harvesting of biomass of roots is not possible (Halim et al. 2003, McIntyre 2003). If phytoextraction is done continuously then a very large number of pollutants can be collected in plants (Sarwar et al. 2017).

The following steps are involved in the phytoextraction process:

- The first step involves the mobilization of contaminants in the rhizosphere.
- The roots can absorb toxic contaminants like heavy metals and metalloids (Memon and Schröder 2009).
- The pollutants are translocated toward the airy parts.
- The contaminants are separated from the tissues of plants (Ali et al. 2013).

For the phytoextractions, the hyperaccumulator plants are used that must have the following properties: the growth rate of plants should be a high, large biomass (Khalid et al. 2017), tolerance to a wide variety of contaminants, the branched root system, adjustment to native environmental situation, immunity against pest and pathogen, high crop production and a higher ability to accumulate pollutants (Mahar et al. 2016, Sarwar et al. 2017).

The efficacy of the phytoextraction process depends upon the following three facts viz. the climatic condition (Bhargava et al. 2012), weather and the growth rate of plant roots.

This process can be improved by using various agents such as nitrilotriacetic acid, amino polycarboxylic acid, ethylenediaminetetraacetic acid, nitrilotriacetic acid and citric acid. This process is used for the elimination of contaminants from the contaminated areas.

There are also many disadvantages to this process. As it requires specific conditions that are necessary for the plant's tolerance to contaminants and their growth. It also requires a longer time to completely remediate the areas.

Phytostabilization

It is also called phytoimmobilization. The process in which plants can be used to reduce the bioavailability or mobility of contaminants either by ceasing the leaching to the underground water table (Sarwar et al. 2017) or its entrance into the food chain through a process such as the production of compounds that are not soluble in the roots zone or adsorption by the roots of plants (Khan et al. 2019). The properties of this process are:

- The reduction of contaminants in the polluted region by assimilation and adsorption on the roots and precipitation by roots.
- The deployment of the plant's roots to avoid pollutants movements by water, wind, draining and soil dispersal (USEPA 1999).

The goal of this process is the stabilization of contaminants rather than their elimination. It decreases their risk to humans and nature (Bolan et al. 2011). Phytostabilization not only reduces the concentration of contaminants, but also lessens the contaminant of the surrounding region (Khalid et al. 2017).

Phytostabilization has various advantages over other techniques of phytoremediation (Jadia and Fulekar 2009). It can be an effective technology when immobilization is required to protect the surface and groundwater (Gomes et al. 2016). It can be used for the bioremediation of various metals such as chromium and lead that are stabilized in the soil region (Mahar et al. 2016). It also decreases the interaction of pollutants with biota.

Phytofiltration

Rhizofilteration is another name for phytofiltration that requires two processes such as adsorption and precipitation of contaminant into the roots zone. Some plants contain various 'phytochelatins' which enhance the "binding capacity" of contaminants like metal ions.

The plants which are suitable for phytofiltration should have some properties such as:

- Their roots should grow rapidly with the ability to eliminate pollutants (Dhanam 2017).
- 'Terrestrial plants' are used for this process because they have a fibrous root system. This root system helps them to remove pollutants from the rhizospheric zone and the earth's water (Ali et al. 2013, Khan et al. 2019).

Phytovolatilization

The process by which plants can be used to absorb pollutants, and then convert them into compounds that are volatile is called phytovolatilization. These compounds are released into the air in a changed form or the same due to various forces such as transpiration and metabolism (Kumar et al. 2017). The process in which evaporation of 'water vapors' occurs from the leaf through the stomata into the air is called transpiration. Plants which have long root systems often have the capacity to absorb and transform pollutants through the formation of various enzymes (Muthusaravanan et al. 2018).

In the process of phytovolatilization, contaminants are taken up from the water or soil. These pollutants are then changed into vapors that are less toxic (Khalid et al. 2017). The pollutants are released into the atmosphere or air through the transpiration process. This method can be used for contaminants that are organic and heavy metals such as mercury and arsenic as these exist in gaseous form. There are very few plants that convert metals into volatile forms. Therefore, plants that are genetically modified can be used in the phytovolatilization process (Khalid et al. 2017). There are some limitations of this process, as the phytovolatilization process does not remove all contaminants from the environment because the pollutants are changed from one part of their physical environment to the other (Ali et al. 2013).

Phytodegradation

The process in which nutrients and pollutants are captured from the soil, water and sediments and chemical modification of pollutants takes place by the plant's metabolism is called phytodegradation (Gomes et al. 2016). It causes pollutant degradation and inactivation in the roots or shoots of plants (Bulak et al. 2014). It is also known as phytotransformation. The plants can destroy the pollutants into less toxic compounds. Plants' metabolic processes or enzymes can be used in this process (Muthusaravanan et al. 2018).

13.4 Conclusion

Bioremediation is an effective and environment-friendly technique to biologically degrade various pollutants. In the bioremediation process, many microorganisms play a major role such as bacteria, fungi, algae and other plants. Furthermore, fungus microorganisms can effectively degrade many toxic environmental pollutants. Phytoremediation represents an emerging technology through which plants can be used to remove pollution from the soil, water and other environments. Bioremediation requires less effort, is less labor intensive, cheap, eco-friendly, sustainable and relatively easy to implement. Various bioremediation methods are used to convert toxic heavy metals into non-toxic or environmentally friendly products. Biosorption, bioaccumulation, bioleaching, biotransformation and biomineralization are examples of these processes. Furthermore, because there is no acceptable endpoint, evaluating the performance of bioremediation may be difficult. More research is needed to develop bioremediation technologies and find more biological solutions for the bioremediation of heavy metal contamination from various environmental systems.

References

Abinandan, S., S. R. Subashchandrabose, K. Venkateswarlu, I. A. Perera and M. Megharaj. 2019. Acid-tolerant microalgae can withstand higher concentrations of invasive cadmium and produce sustainable biomass and biodiesel at pH 3.5. Bioresour. Technol. 281: 469–473.

Acosta-Rodríguez, I., J. F. Cárdenas-González, A. S. Rodríguez Pérez, J. T. Oviedo and V. M. Martínez-Juárez. 2018. Bioremoval of different heavy metals by the resistant fungal strain *Aspergillus niger*. Bioinorg. Chem. 2018. 1–7.

Ali, H., E. Khan and M. A. Sajad. 2013. Phytoremediation of heavy metals-concepts and applications. Chemosphere. 91: 869–881.

Arora, N., K. Gulati, A. Patel, P. A. Pruthi, K. M. Poluri and V. Pruthi. 2017. A hybrid approach integrating arsenic detoxification with biodiesel production using oleaginous microalgae. Algal Res. 24: 29–39.

Bai, R. S. and T. E. Abraham. 2001. Biosorption of Cr (VI) from aqueous solution by *Rhizopus nigricans*. Bioresour. Technol. 79: 73–81.

Balaji, S., T. Kalaivani, M. Shalini, M. Gopalakrishnan, M. A. Rashith and C. Rajasekaran. 2016. Sorption sites of microalgae possess metal binding ability towards Cr (VI) from tannery effluents-a kinetic and characterization study. Desalin. Water Treat. 31: 14518–14529.

Baldrian, P. 2003. Interactions of heavy metals with white-rot fungi. Enzym. Microb. Technol. 32: 78–91.

Bellion, M., M. Courbot, C. Jacob, D. Blaudez and M. Chalot. 2006. Extracellular and cellular mechanisms sustaining metal tolerance in ectomycorrhizal fungi. FEMS Microbiol. Letters. 254(2): 173–181.

Bhargava, A., F. F. Carmona, M. Bhargava and S. Srivastava. 2012. Approaches for enhanced phytoextraction of heavy metals. J. Environ. Manag. 105: 103–120.

Birungi, Z. S. and E. M. N. Chirwa. 2015. The adsorption potential and recovery of thallium using green micro-algae from eutrophic water sources. J. Hazard. Mater. 299: 67–77.

Bolan, N. S., J. H. Park, B. Robinson, R. Naidu and K. Y. Huh. 2011. Phytostabilization. A green approach to contaminant containment. Adv. Agron. 112: 145–204.

Bosecker, K. 2001. Microbial leaching in environmental clean-up program. Hydrometallurgy. 59: 245–248.

Brandl, H., R. Bosshard and M. Wegmann. 2001. Computer-munching microbes: metal leaching from electronic scrap by bacteria and fungi. Hydrometallurgy. 59: 319.

Bulak, P., A. Walkiewicz and M. Brzeziska. 2014. Plant growth regulators-assisted phytoextraction. Biol. Plant. 58: 1–8.

Cameron, H., M. T. Mata and C. Riquelme. 2018. The effect of heavy metals on the viability of *Tetraselmis marina* AC16-MESO and an evaluation of the potential use of this microalga in bioremediation. Peer J. 6: 5295.

Chaudhary, N., T. Banerjee and N. Ibrahim. 2014. Fungal leaching of iron ore: isolation, characterization and bioleaching studies of *Penicillium verruculosum*. Am. J. Microbiol. 50: 27–32.

Couto, S. R., M. A. Sanroman, D. Hoefer and G. M. Gubitz. 2004. Stainless steel: a novel career for the immobilization of the white rot fungus Trametes hirsute for decolorization of textile dyes. Bioresour. Technol. 95: 67–72.

Das, D., A. Chakraborty, S. Bhar, M. Sudarshan and S. C. Santra. 2013. Gamma irradiation in modulating cadmium bioremediation potential of *Aspergillus* sp. IOSR J. Environ. Sci. Toxicol. Food Technol. 3: 51–55.

Dhanam, S. 2017. Strategies of bioremediation of heavy metal pollutants toward sustainable agriculture. pp. 349–358. *In*: A. Dhanarajan [Ed.]. Sustainable Agriculture Towards Food Security. Springer Nature, Singapore.

Dhankhar, R. and A. Hooda. 2011. Fungal biosorption—an alternative to meet the challenges of heavy metal pollution in aqueous solutions. Environ. Technol. 32: 467–491.

Fourest, E. and J. Roux. 1992. Heavy metal biosorption by fungal mycelial by-products: mechanisms and influence of pH. Appl. Microbiol. Biotechnol. 37: 399–403.

Gadd, G. M. 2010. Metals, minerals and microbes: geomicrobiology and bioremediation. Microbiology. 156: 609.

Ghosh, M. and S. P. Singh. 2005. A review on phytoremediation of heavy metals and utilization of it's by products. Asian J. Energy Environ. 6(4): 18.

Gomes, M. A. C., R. A. Hauser-Davis, A. N. De Souza and A. P. Vitória. 2016. Metal phytoremediation: general strategies, genetically modified plants and applications in metal nanoparticle contamination. Ecotoxicol. Environ. Saf. 134: 133–147.

Gómez-Jacinto, V., T. García-Barrera, J. L. Gómez-Ariza, I. Garbayo-Nores and C. Vílchez-Lobato. 2015. Elucidation of the defence mechanism in microalgae *Chlorella sorokiniana* under mercury exposure. Identification of Hg–phytochelatins. Chem. Biol. Interact. 238: 82–90.

Gouma, S., S. Fragoeiro, A. C. Bastos and N. Magan. 2014. Bacterial and fungal bioremediation strategies. pp. 301–323. *In*: S. Das and H. R. Dash [eds.]. Microbial Biodegradation and Bioremediation. Elsevier.

Gu, T., S. O. Rastegar, S. M. Mousavi, M. Li and M. Zhou. 2018. Advances in bioleaching for recovery of metals and bioremediation of fuel ash and sewage sludge. Bioresour. Technol. 221: 428.

Guibaud, G., S. Comte, F. Bordas, S. Dupuy and M. Baudu. 2005. Comparison of the complexation potential of extracellular polymeric substances (EPS), extracted from activated sludges and produced by pure bacteria strains, for cadmium, lead and nickel. Chemosphere. 59: 629–638.

Gupta, V. K. and A. Rastogi. 2008. Biosorption of lead from aqueous solutions by green algae *Spirogyra* species: kinetics and equilibrium studies. J. Hazard. Mater. 152(1): 407–414.

Gutnick, D. L. and H. Bach. 2000. Engineering bacterial biopolymers for the biosorption of heavy metals; new products and novel formulation. Appl. Microbiol. Biotechnol. 54: 451–460.

Halim, M., P. Conte and A. Piccolo. 2003. Potential availability of heavy metals to phytoextraction from contaminated soils induced by exogenous humic substances. Chemosphere. 52: 265–275.

Hassen, A., N. Saidi, M. Cherif and A. Boudabous. 1998. Resistance of environmental bacteria to heavy metals. Bioresour. Technol. 64: 7–15.

Hedayatkhah, A., M. S. Cretoiu, G. Emtiazi, L. J. Stal and H. Bolhuis. 2018. Bioremediation of chromium contaminated water by diatoms with concomitant lipid accumulation for biofuel production. J. Environ. Manage. 227: 313–320.

Ibuot, A., A. P. Dean, O. A. McIntosh, J. K. Pittman and J. K. 2017. Metal bioremediation by CrMTP4 over-expressing *Chlamydomonas reinhardtii* in comparison to natural wastewater-tolerant microalgae strains. Algal Res. 24: 89–96.

Index, E. P. 2018. EPI report. Yale University. https://epi.envirocenter.yale.edu/2018-epi-report/ introduction. of its byproducts. Appl. Ecol. Environ. Res. 3: 1–18.

Jaafari, J. and K. Yaghmaeian. 2019. Optimization of heavy metal biosorption onto freshwater algae (*Chlorella coloniales*) using response surface methodology (RSM). Chemosphere. 217: 447–455.

Jadia, C. D. and M. H. Fulekar. 2009. Phytoremediation of heavy metals: recent techniques. Afr. J. Biotechnol. 8: 921–928.

Javaid, A., R. Bajwa and U. Shafique. 2011. Removal of heavy metals by adsorption on *Pleurotus ostreatus*. Biomass Bioenerg. 35: 1675–1682.

Kadukova, J. and E. Vircikova. 2005. Comparison of differences between copper bioaccumulation and biosorption. Environ. Int. 31: 227–332.

Kanamarlapudi, S. L. R. K., V. K. Chintalpudi and S. Muddada. 2018. Application of biosorption for removal of heavy metals from wastewater. Biosorption. 18(69): 70–116.

Kapoor, A. and T. Viraraghavan. 1995. Fungal biosorption-an alternative treatment option for heavy metal bearing wastewaters: a review. Bioresour. Technol. 53(3): 195–206.

Kapoor, A., T. Veeraraghavan and D. Roy Cullimore. 1999. Removal of heavy metals using the fungus *Aspergillus niger*. Bioresour. Technol. 70(1): 95–104.

Khalid, S., M. Shahid, N. K. Niazi, B. Murtaza, I. Bibi and C. Dumat. 2017. A comparison of technologies for remediation of heavy metal contaminated soils. J. Geochem. Explor. 182: 247–268.

Khan, I., M. Iqbal and F. Shafiq. 2019. Phytomanagement of lead-contaminated soils: critical review of new trends and future prospects. Int. J. Environ. Sci. Technol.

Kumar, S. S., A. Kadier, S. K. Malyan, A. Ahmad and N. R. Bishnoi. 2017. Phytoremediation and rhizoremediation: uptake, mobilization and sequestration of heavy metals by plants. pp. 367–394. *In*: D. Singh, H. Singh and R. Prabha [Eds.]. Plant-Microbe Interactions in Agro-Ecological Perspectives. Springer, Singapore.

Kunhikrishnan, A., G. Choppala, B. Seshadri, J. H. Park, K. Mbene, Y. Yan and N. S. Bolan. 2017. Biotransformation of heavy metal(loid)s in relation to the remediation of contaminated soils. In Handbook of Metal-Microbe Interactions and Bioremediation. 67–86. CRC Press.

Kurniati, E., N. Arfarita and T. Imai. 2014. Potential bioremediation of mercury-contaminated substrate using filamentous fungi isolated from forest soil. J. Environ. Sci. 26: 1223–123.

Lewis, D. and R. J. Kiff. 1998. The removal of heavy metals from aqueous effluents by immobilised fungal biomass. Environ. Technol. Lett. 9: 991–998.

Mahar, A., P. Wang, A. Ali, M. K. Awasthi, A. H. Lahori, Q. Wang, R. Li and Z. Zhang. 2016. Challenges and opportunities in the phytoremediation of heavy metals contaminated soils: a review. Ecotoxicol. Environ. Saf. 126: 111–121.

Maheshwari, S. and A. G. Murugesan. 2009. Remediation of arsenic in soil by *Aspergillus nidulans* isolated from an arsenic-contaminated site. Environ. Technol. 30(9): 921–926.

McIntyre, T. 2003. Phytoremediation of heavy metals from soils. Adv. Biochem. Eng. Biotechnol. 78: 97–123.

Memon, A. R. and P. Schröder. 2009. Implications of metal accumulation mechanisms to phytoremediation. Environ. Sci. Pollut. Res. 16: 162–175.

Mishra, D., D. J. Kim, J. G. Ahn and Y. H. Rhee. 2005. Bioleaching: a microbial process of metal recovery; a review. Met. Mater. Int. 11: 249.

Muthusaravanan, S., N. Sivarajasekar, J. S. Vivek, T. Paramasivan, M. U. Naushad and J. Prakashmaran. 2018. Phytoremediation of heavy metals: mechanisms, methods and enhancements. Environ. Chem. Lett. 16: 1339–1359.

Natarajan, G. and Y. P. Ting. 2014. Pretreatment of e-waste and mutation of alkali-tolerant cyanogenic bacteria promote gold biorecovery. Bioresour. Technol. 152: 80.

Ojha, S. K., S. Mishra, S., Kumar, S. S. Mohanty, B. Sarkar, M. Singh and G. R. Chaudhury. 2015. Performance evaluation of vinasse treatment plant integrated with physico-chemical methods. J. Environ. Biol. 36: 1269.

Papry, R. I., K. Ishii, M. A. A. Mamun, S. Miah, K. Naito, A. S. Mashio, T. Maki and H. Hasega. 2019. Arsenic biotransformation potential of six marine diatom species: effect of temperature and salinity. Sci. Rep. 9(1): 10226.

Pointing, S. 2001. Feasibility of bioremediation by white-rot fungi. Appl. Microbiol. Biotechnol. 57: 20–33.

Pradhan, D., L. B. Sukla, B. B. Mishra and N. Devi. 2019. Biosorption for removal of hexavalent chromium using microalgae *Scenedesmus* sp. J. Cleaner Prod. 209: 617–629.

Priatni, S., D. Ratnaningrum, S. Warya and E. Audina. 2017. Phycobiliproteins production and heavy metals reduction ability of *Porphyridium* sp. IOP Conf. Series: Earth Environ. Sci. 60.

Roane, T. M., I. L. Pepper and T. J. Gentry. 2015. Microorganisms and metal pollutants. pp. 415–439. *In*: I. L. Pepper, C. P. Gerba and T. J. Gentry [eds.]. Environmental Microbiology. Academic Press, San Diego, CA.

Rohwerder, T., T. Gehrke, K. Kinzler and W. Sand. 2003. Bioleaching review part A. Appl. Microbiol. Biotechnol. 63: 239.

Rudakiya, D., Y. Patel, U. Chhaya and A. Gupte. 2019. Carbon nanotubes in agriculture: production, potential, and prospects. pp. 121–130. *In*: D. G. Panpette and Y. K. Jhala [eds.]. Nanotechnology for Agriculture. Springer, Singapore.

Rudakiya, D. M. and K. S. Pawar. 2013. Evaluation of remediation in heavy metal tolerance and removal by *Comamonas acidovorans* MTCC 3364. IOSR J. Environ. Sci. Toxicol. Food Technol. 5: 26–32.

Rudakiya, D. M. and A. Gupte. 2017. Degradation of hardwoods by treatment of white rot fungi and its pyrolysis kinetics studies. Int Biodeterior Biodegradation. 120: 21–35.

Rudakiya, D. M. and K. Pawar. 2017. Bactericidal potential of silver nanoparticles synthesized using cellfree extract of *Comamonas acidovorans*: *in vitro* and *in silico* approaches. 3 Biotech. 7(2): 9.

Rudakiya, D. M. 2018. Metal tolerance assisted antibiotic susceptibility profiling in *Comamonas acidovorans*. Biometals. 31(1): 1–5.

Rudakiya, D. M., A. Tripathi, S. Gupte and A. Gupte. 2019. Fungal bioremediation: a step towards cleaner environment. pp. 229–249.. *In*: T. Satyanarayana, S. K. Deshmukh and M. V. Deshpande [eds.]. Advancing Frontiers in Mycology & Mycotechnology. Springer, Singapore.

Rudakiya, D. M., D. H. Patel and A. Gupte. 2020. Exploiting the potential of metal and solvent tolerant laccase from *Tricholoma giganteum* AGDR1 for the removal of pesticides. Int. J. Biol. Macromol. 144: 586–595.

Sarwar, N., M. Imran, M. R. Shaheen, W. Ishaque, M. A. Kamran, A. Matloob, A. Rehim and S. Hussain. 2017. Phytoremediation strategies for soils contaminated with heavy metals: modifications and future perspectives. Chemosphere. 171: 710–721.

Satyanarayana, T., S. K. Deshmukh and M. V. Deshpande. 2019. Advancing frontiers in mycology and mycotechnology: basic and applied aspects of fungi. Springer Nature, Singapore.

Say, R., N. Yilmaz and A. Denizli. 2004. Removal of chromium (VI) ions from synthetic solutions by the fungus *Penicillium purpurogenum*. Eng. Life Sci. 4: 276–280.

Shah, D., D. M. Rudakiya, V. Iyer and A. Gupte. 2018. Simultaneous removal of hazardous contaminants using polyvinyl alcohol coated *Phanerochaete chrysosporium*. Int. J. Agri. Environ. Biotechnol. 11(2): 235–241.

Siegel, S. M., B. Z. Siegel and K. E. Clark. 1983. Biocorrosion: solubilization and accumulation of metals by fungi. Water Air Soil Pollut. 19: 229–236.

Singh, N. K., A. S. Raghubanshi, A. K. Upadhyay and U. N. Rai. 2016. Arsenic and other heavy metal accumulation in plants and algae growing naturally in contaminated area of West Bengal, India. Ecotoxicol. Environ. Saf. 130: 224–233.

Srichandan, H., A. Pathak, S. Singh, K. Blight, D. J. Kim and S. W. Lee. 2014. Sequential leaching of metals from spent refinery catalyst in bioleaching-bioleaching and bioleaching-chemical leaching reactor: comparative study. Hydrometallurgy. 150: 130.

Srichandan, H., S. Singh, K. Blight, A. Pathak, D. J. Kim, S. Lee and S. W. Lee. 2015. An integrated sequential biological leaching process for enhanced recovery of metals from decoked spent petroleum refinery catalyst: a comparative study. International Journal of Mineral Processing 134: 66–73.

Tayang, A. and L. S. Songachan. 2021. Microbial bioremediation of heavy metals. Current Science. 120(6): 00113891.

Tobin, J. M., D. G. Cooper and R. J. Neufeld. 1984. Uptake of metal ions by *Rhizopus arrhizus* biomass. Appl. Environ. Microbial. 47: 821–824.

Townslcy, C. C., I. S. Ross and A. S. Atkins. 1986. Biorecovery of metallic residues from various industrial effluents using filamentous fungi. pp. 279–289. *In*: R. W. Lawrence, B. RMR and H. G. Enner [Eds.]. Fundamental and Applied Biohydrometallurgy. Elsevier, Amsterdam.

Tsezos, M. 1984. Recovery of uranium from biological adsorbents-desorption equilibrium. Biotechnol. Bioeng. 26: 973–981.

USEPA. 1999. Report on Bioavailability of Chemical Wastes With Respect to the Potential for Soil Remediation. T28006: QT-DC-99-003260.

Valix, M., F. Usai and R. Malik. 2001. Fungal bio-leaching of low-grade laterite ores. Miner Eng. 14: 197–203.

Vardhan, K. H., P. S. Kumar and R. C. Panda. 2019. A review on heavy metal pollution, toxicity and remedial measures: current trends and future perspectives. J. Mol. Liq. 111: 197.

Veglio, F. and F. Beolchini. 1997. Removal of metals by biosorption: a review. Hydrometallurgy. 44: 301–316.

Vidali, M. 2001. Bioremediation. An overview. Pure Appl Chem. 73(7):1163–1172.

Volesky, B. 2007. Biosorption and me. Water Res. 41(18): 4017–4029.

Wang, J. and C. Chen. 2009. Biosorbents for heavy metals removal and their future. Biotechnol. Adv. 27: 195–2.

Zaki, M. S., N. El-Battrawy, A. M. Hammam and S. I. Shalaby. 2014. Aquatic bioremediation of metals. Life Sci. J. 11(4): 1394.

Zhang, J. L. and C. L. Qiao. 2002. Novel approaches for remediation of pesticide pollutants. Int. J. Environ. Pollut. 18(5): 423–433.

CHAPTER 14

Phytoremediation of Wastewater Discharged from Paper and Pulp, Textile and Dairy Industries using Water Hyacinth (*Eichhornia crassipes*)
A Sustainable Approach

Apurba Koley,[1] Anudeb Ghosh,[1] Sandipan Banerjee,[2] Nitu Gupta,[3] Richik Ghosh Thakur,[1] Binoy Kumar Show,[1] Shibani Chaudhury,[1] Amit Kumar Hazra,[4] Andrew B. Ross,[5] Gaurav Nahar[6] and Srinivasan Balachandran[1],*

14.1 Introduction

The presence of toxic heavy metals in water bodies like rivers, lakes and reservoirs affect local inhabitants, animals and other living organisms depending on the water or source (Rai et al. 2002). The concentration of these metals above the threshold limit or optimal concentration can cause negative impacts on plant systems due to oxidative stress and inhibited cytoplasmic enzymes (Assche and Clijsters 1990). Intake of toxic metals causes serious health and growth-related issues

[1] Bioenergy Laboratory, Department of Environmental Studies, Siksha-Bhavana, Visva-Bharati.
[2] Mycology and Plant Pathology Laboratory, Department of Botany, Visva-Bharati University, Santiniketan-731235, India.
[3] Department of Environmental Science, Tezpur University, Napaam, Tezpur, Assam - 784208, India.
[4] Socio-Energy Lab, Department of Lifelong Learning and Extension, Visva-Bharati, Santiniketan.
[5] School of Chemical and Process Engineering, University of Leeds, Leeds, LS2 9JT, United Kingdom.
[6] Defiant Renewables Pvt Ltd, 1st Floor Kant Helix, Bhoir Colony, Chinchwad, Pune-411033, India.
* Corresponding author: s.balachandran@visva-bharati.ac.in
ORCID ID: 0000-0003-4247-408X

through food-chain magnification (Rai and Tripathi 2008, Pramanik et al. 2021). Different removal methods like electrolysis, ion exchange, precipitation, absorption and reverse osmosis are used to remove heavy metals. These methods are quite expensive and are comparatively ineffective, producing a large amount of waste that is difficult to dispose of. The development of cost-effective and alternative methods for the removal of toxic substances from wastewater is a necessity (Rai and Tripathi 2007). The use of living plants to remove toxic substances from the water and soils could be one of the alternatives and viable processes (Rai 2009). Plants have the capability to accumulate metals (Co, Ca, Fe, Cu, Ni, Mo, Mg, Mn, Na, Ni, Se, V and Zn) and non-essential metals (Al, As, Au, Cd, Cr, Hg, Pb, Pd, Pt, Sb, Te, Tl and U) under different concentration (Djingova and Kuleff 2000). Three different patterns have been observed during the uptake of heavy metals by plants (i). True exclusion (metals are limited during entering), (ii). Shoot exclusion (root accumulates the metals but restricts translocation to the shoot system), and, (iii). Accumulation (plant parts concentrate the metals) (Zavoda et al. 2001, Kamal et al. 2004).

Textile wastewaters are a serious polluting threat as they contaminate the environment; without proper treatment, they lead to irreversible persistent changes in the environment. Textile wastes include dyes and have high Biological Oxygen Demand (BOD), Chemical Oxygen Demand (COD), Total Suspended Solids (TSS), Total Nitrogen (TN), heavy metals, phosphates and greases, etc. Dyes are well-known toxins which can directly affect the aquatic ecosystem (Ekanayake et al. 2021).

The pulp and paper industries are one of the major causes of aquatic pollution in India (Saadia and Ashfaq 2010, Singh et al. 2022). The major pollutants in paper and pulp industries are high in TSS, Total Dissolved Solids (TDS), Total Nitrogen (TN), adsorbable organic halides (AOX), heavy metals, as well as have high COD. A large amount of this contaminated wastewater is generated and discharged during the paper-making process, which negatively impacts wild and human life. The impurities should be removed or minimized before being discharged into the environment (Singh et al. 2022, Ashrafi et al. 2015).

Similar to the textile and paper industries, the Indian dairy industry generates wastewater which is a major cause of concern. Six to seven liters of effluent is generated while processing one liter of milk (Porwal 2015). High levels of BOD, COD, hardness, along with ions like K^+, Na^+ and Cu^{2+} ions are detected in water close to the industry due to inappropriate disposal of effluents in rivers (Kaur et al. 2018). To reduce the release of these environmental pollutants from such industries, many researchers are trying to develop sustainable, cost-effective treatment practises for cleaning the industrial effluents.

Phytoremediation is one such technique being researched to clean up wastewater generated from industrial and domestic discharge for consumption of fresh water. *Eichhornia crassipes*, i.e., Water Hyacinth (WH) is one of the promising plants popularly tested for phytoremediation of various wastewater streams (Koley et al. 2022). This aquatic macrophyte is a large-leaved, noxious, free-floating aquatic weed that belongs to the family *Pontederiaceae*. WH is a free-floating angiosperm having roots, short to long stems, broad, glossy leaves and lavender flowers, usually found in shallow water. Due to its rapid reproduction rate, it covers the water surface by forming a thick mat. Owing to its fast-growing and high nutrient uptake ability, WH is used in constructed wetlands to uptake pollutants from different wastewater sources. Even though there are many sustainable uses of WH, most research has been done on how to use it to clean up nutrients and contaminants, like heavy metals, from industrial and urban wastewater (Koley et al. 2022, Sinha et al. 2021, Qin et al. 2020, Ansari et al. 2016, Elangovan et al. 2008).

This chapter aims to review and generalize the observations from different studies on wastewater remediation of lignocellulosic processing (paper and pulp and textile) industries in addition to dairy industries from the perspective of the Indian ecosphere. Additionally, the well-established capabilities of *E. crassipes* can assist in the remediation of toxic pollutants, including heavy metals and organic pollutants and other hazardous industrial pollutants, apart from its use in various documented sustainable practises.

14.2 Bibliographic Analysis

In this study, a total of 501 articles with keywords of phytoremediation, water hyacinth, *Eichhornia crassipes*, heavy metals, bioremediation, industrial wastewater, etc., were searched on the Web of Science. By using the VOS viewer software, the keywords were evaluated (Basu et al. 2021, Show et al. 2022). The keyword co-occurrences outcomes were defined to be 15, and a network map was constructed by using the most used keywords, and the results are depicted in Figure 14.1. The cluster and the prevalence of the individual keyword were indicated by color and size, and the line between the circles represents their interlinks. From this network mapping of the keywords, it can be said that these articles are mainly focused on the following five clusters or aspects:

i. Phytoremediation of wastewater (red cluster).

ii. Different types of heavy metal accumulation by the plant systems (green cluster).

iii. Aquatic plants and their responses to heavy metal toxicity (blue cluster).

iv. Relation between aquatic macrophytes and the heavy metal pollutants (yellow cluster);

v. Mode of pollutant removal by water hyacinth (purple cluster).

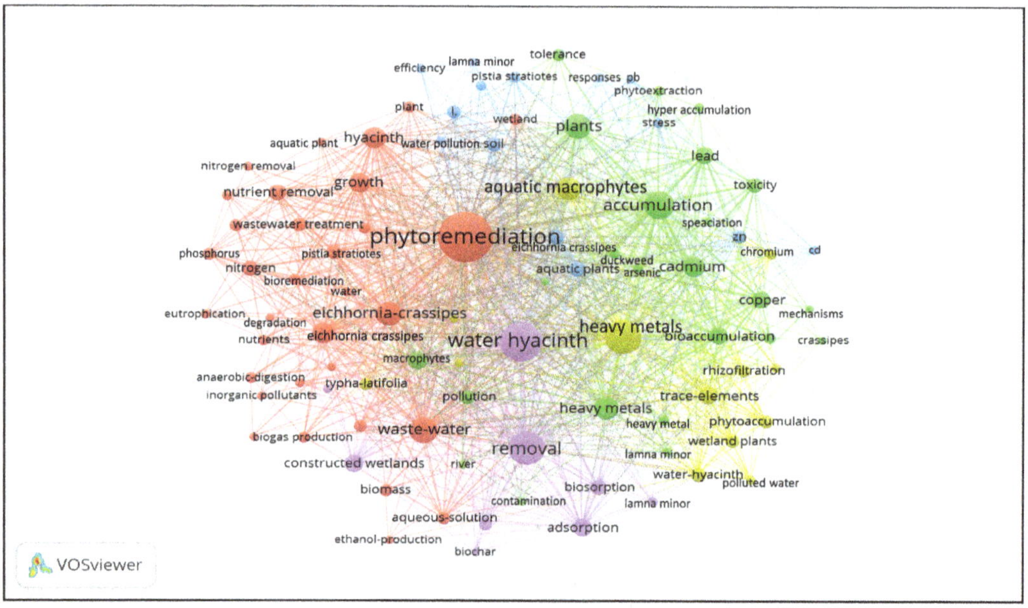

Figure 14.1. Bibliographic analysis

These keywords-oriented text mining studies explored the exploitation of aquatic plants or macrophytes as a phytoremediation agent for heavy metal contaminated wastewater resources, i.e., industrial wastewater. Therefore, interlinking these industrial pollutants and the WH can unearth a novel sustainable phenomenon, where other aquatic macrophytes can be utilized for such integrated sustainable practises (heavy metal and other organic pollutants removal or bioremediation).

14.3 Phytoremediation: A Green Technology for the Cleanup of Pollutants

Phytoremediation is an emerging and fast growing effective eco-friendly technology based on aquatic, semiaquatic and terrestrial plants for the removal of pollutants (Dhir 2013) such as metals, hydrocarbons, pesticides and chlorinated solvents (Fletcher et al. 2020, Xia et al. 2003). Plants help to degrade, assimilate, metabolize, mineralize or detoxify different pollutants (Schnoor et al. 1995) from different sources like soil, sediment, aquatic medium and atmosphere (Dhir 2013).

The phytoremediation consists of four major processes, i.e., phytoextraction, phytostabilization, phytovolatilization and rhizofiltration where the pollutants are directly accumulated and metabolized in plant tissues or transported and released as exudates by biochemical mechanisms (Schnoor et al. 1995).

In the phytoextraction process, plants and roots extract the metals and translocate them to shoots. The shoots and roots are harvested to remove the contaminants. Phytoextraction is considered a low-impact technique, having 10 times lower operational cost per hectare as compared to conventional remediation techniques (Jadia and Fulekar 2009, Salt et al. 1995). In phytostabilization, plant roots limit the bioavailability and mobility of the contaminants. As a result, it is referred to as inactivation and is used to remediate soil, sediment and sludge contaminants (Jadia and Fulekar 2009, USEPA 2000). Plants involved in phytovolatilization uptake the contaminants and transform them into volatile forms. The converted less toxic substances are released into the atmosphere through transpiration (USEPA 2000, Jadia and Fulekar 2009). In the phytofiltration or rhizofiltration process, plants are used to remediate water and wastewater with a low concentration of contaminants (Ensley 2000) (Figure 14.2). Both terrestrial and aquatic plants absorb, concentrate and precipitate contaminants in the roots. Heavy metals like Pb, Cd, Cu, Ni, Zn and Cr can be accumulated through this process and are primarily retained within the roots (USEPA 2000).

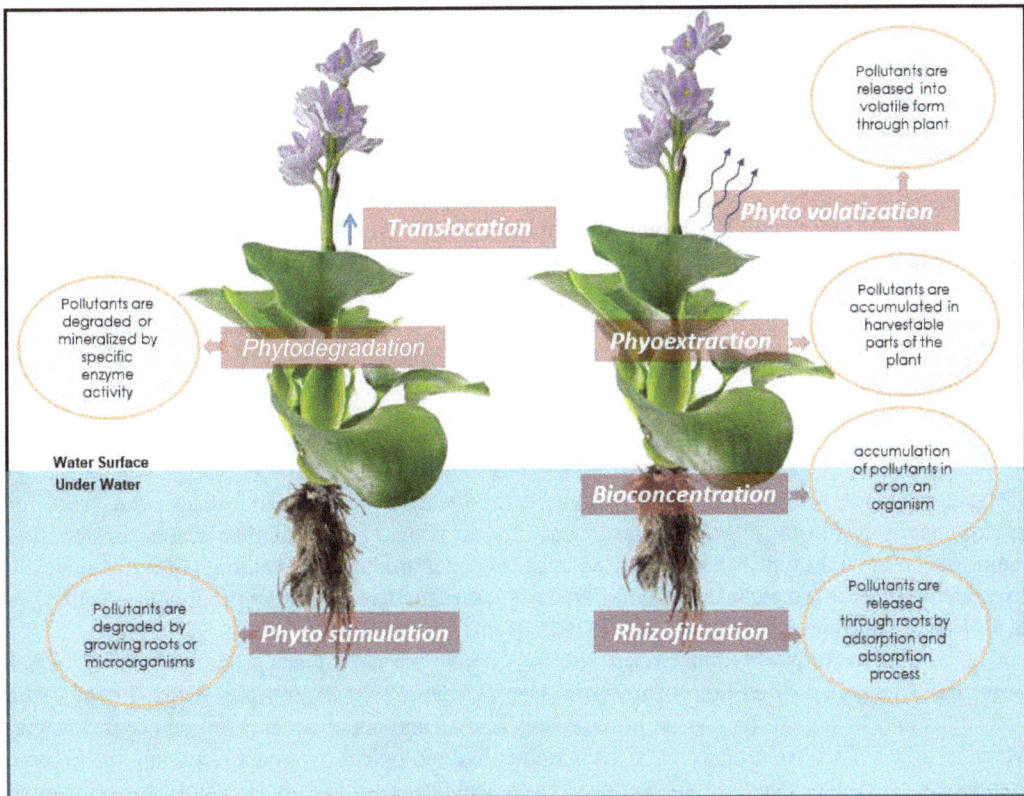

Figure 14.2. Types of phytoremediation.

14.4 WH: An Aquatic Macrophyte

WH is a free-floating, invasive aquatic macrophyte abundant in the South and northern regions of India (Kumar and Chauhan 2019). The temperature of 28°C to 30°C (Malik 2007) with pH ranging from 6.5 to 8.5, and salt concentration less than 2%, are suitable for its optimum growth.

The maximum permissible limit of N, P and K for growth are 20, 30 and 50 ppm, respectively (Kumar and Chauhan 2019).

WH's ability to cope with difficult environmental conditions makes it desirable for the remediation process. During the 1970s and 1980s, WH was generally used for wastewater treatment as the plant roots are very useful for the absorption of nutrients. The nutrients absorbed by the plant are stored for growth (Alves et al. 2003, Reddy et al. 1983, Reddy et al. 1987). It can absorb nitrogen in two stages, first from the medium and another from the sediment. It can also remove P, K and other organic pollutants, including heavy metals (Chua 1998). WH can control algal blooms and water quality (Barry 1998). It removes a large number of pollutants, such as BOD, N, COD, P, DO, heavy metals, etc. (Gupta et al. 2012, Rezania et al. 2015) from the water bodies.

The ability of WH to grow in heavily polluted water makes it favored among plants that use it in the phytoremediation process (Mahalakshmi et al. 2019, Sarkar et al. 2017, Ebel et al. 2007, Roy and Hänninen 1994). According to Malik (2007) and Smolyakov (2012) a medium sized WH plant can proliferate/spread 2 million/hectares. WH grows in water, which is enriched with nutrients, but fails to grow in saline coastal waters (Jafari 2010, Rezania et al. 2015). Based on dry weight, the growth rate of WH is $0.04 – 0.08$ kg dry weight/m^2/d, while the rate based on the surface coverage is, $1.012 – 1.077$ m^2d^{-1} (Gopal 1987, Rezania et al. 2015).

14.5 Industrial Lignocellulosic Wastewater

In recent times, lignocellulosic biomass or waste has gained popularity as a green and sustainable alternative energy resource for producing second-generation biofuels and bio-based chemicals (Menon and Rao 2012, Chandel et al. 2018). Lignocellulosic biomass comprises of cellulose, hemicelluloses and lignin. In different industrial processes, the physical and chemical structure of the lignocellulosic biomass is transformed to suit the requirements of the process (Zhao et al. 2012, Kumar et al. 2017). The two major industries in India that produce a large amount of lignocellulosic waste include paper and pulp and the textile industry. Apart from this, the dairy industry also produces a vast amount of wastewater, which can be remediated by WH. Although there are other lignocellulosic industries which produce large amounts of wastewater, like the sugar industries, oil industries, rice and other food processing industries, this chapter focuses on paper and pulp, textiles and dairy industries.

14.5.1 Paper and Pulp Industry

The paper and pulp industry (P&P) is one of the largest water-consuming industries, which requires approximately 5–100 m^3 of water to produce 1 ton of paper. The amount of water consumption mainly depends on the type of substrate used, types of paper produced, and amount of water recycled. In modern industries, the consumption of water is lesser for 1 ton of paper production (Nakamura et al. 1997, Doble and Kumar 2005, Holik 2006, Nemerow 2007).

According to the World Bank Group report, P&P industries in Canada produce 20 100 m^3 waste water per 1 ton of air-dried pulp (The World Bank Group 1999). Poor water management in the industry results in large water consumption as only a small amount of water is reused (Toczyłowska-Mamińska 2017). Conventional treatment methods like biological treatment eliminate the organic components from the water. On the contrary, biological decomposition is insufficient to reduce the refractory organics like COD and other pollutants load below the safe discharge limit (Sevimli 2005). To meet the recent environmental standards and limit water use, the conventional P&P industries are forced to switch to sophisticated technologies such as membrane separation, electrowinning, electrocoagulation and bioreactors (Toczyłowska-Mamińska 2017). Untreated P&P industry wastewater contains a brown color and persistent organics, which cause serious environmental issues if the water is discharged into the lakes or rivers directly (Hao et al. 2000). Organic and inorganic pollutants, including tannins, chloride compounds, adsorbable organic halides (AOX), COD, TS,

TDS and TSS, phenols and lignins must be minimized through different treatment processes to meet the pollution standards (Ashrafi et al. 2015).

India's first paper mill was established at Serampore, West Bengal, in 1812, and hosts the 15th largest paper and pulp industry in the world. The per capita, paper consumption in India is 5.5 kg y^{-1} which is 10 times lower than the global average consumption, i.e., 50 kg y^{-1}. The P&P industries in India are classified into three categories as per the production capacity (tons per annum, TPA), i.e., large scale (greater than 33,000 TPA), medium scale (between 10,000 and 33,000 TPA) and small scale (less than 10,000 TPA). In India mainly, four types of broad categories of paper are produced and consumed which include firstly, cultural type paper which comes in the form of cream woven, maplitho, bond paper, chromo paper. Secondly, industrial paper is manufactured in the form of kraft paper, single layer paper board, multilayer board and duplex board. Thirdly specialty paper which is security paper, grease proof paper, electrical grades of paper), and fourthly, newsprint (glazed, non-glazed) (MOEF 2010).

Wastewater generated from the P&P industries contains a high amount of organic matter, COD, BOD, TSS, Lignin, AOX, Chlorides, PO_4^{3-}, NO_3^-, phenols, etc. The study by Sharma et al. (2021) observed the high amount (693 mg/l) of phenol in the P&P industry effluent apart from a high amount of PO_4^{3-} and AOX, whereas Tarlan et al. (2002) observed 267 mg/l and 46.3 mg/l, respectively of these pollutants. Table 14.1 lists the different pollutants generated in P & P industry and Table 14.2 shows the Indian standards for Wastewater Generation from P&P industry as given by CPCB (2000).

Kraft paper, writing paper and hardboard manufacturing industries produce heavy metals like Fe, Mn, Zn, Cu, Cr, Cd, Ni and Pb, along with organic wastes (Sharma et al. 2020). Different heavy metals released from different paper-making processes are described in Table 14.3. Heavy metal tends to bind with lignocellulosic waste and forms complexes like Persistent Organic Pollutants (POPs), which are a severe threat to the environment and they tend to accumulate in the food chain (Singh et al. 2015).

14.5.2 Textile Industry

The textile industry is one of the important industries which affect the economy of the world. It is one of India's largest industrial sectors with a large unmatched raw material base, as globally, India is the 6th largest exporter of textile and apparels (T&A), and total exports stood at a significant 11.4% in 2020–21, with 4% of the global trade share in T&A. India exports textile products to more than 100 countries. Approximately 47% of Indian textile products are exported to the USA, EU and the UK. Besides the economic development, this industry contributes to many direct and indirect employment all over the country, including a large number of women and the rural population (Ministry of Textile 2022).

Textile industries are broadly classified into three types based on raw materials used, i.e., plant-based or cellulose materials, protein fabrics and synthetic materials. Plant based cellulosic materials are obtained from plants like cotton, rayon and linen. Protein fabrics are animal based, which include wool, silk and mohair and synthetic fabrics, which are nylon, polyester and acrylic, etc., are made artificially. Dry and wet processes are followed during fiber production in textile industries. A large amount of freshwater is used during the wet process and produces vast quantities of contaminated wastewater. This process consists of sizing, de-sizing, sourcing, bleaching, mercerizing, dyeing, printing and finishing techniques (Babu et al. 2007, Liu et al. 2010). The different processes produce different types of waste in each step. During the de-sizing process, enzymes, starch and waxes are used. In the process of scouring sodium hydroxide, surfactants, soaps, fats, pectin, oils and waxes are used in large amounts. The bleaching process includes different chemicals, i.e., hydrogen peroxide, sodium silicate and organic stabilizer. During the process of dyeing and printing, metals, salts, surfactants, colors, metals, urea, formaldehyde and other solvents are released into the wastewater (Holkar et al. 2016, Yaseen and Scholz 2019). Wastewater generated from the textile

Table 14.1. Characterization of paper and pulp mill waste water (mgL⁻¹).

Type of Water	pH	EC (us cm⁻¹)	Lignin (mgL⁻¹)	Colour (pcu)	COD (mgL⁻¹)	BOD (mg L⁻¹)	TDS (mg L⁻¹)	TSS (mgL⁻¹)	TS (mgL⁻¹)	Adsorbabale Organic Halides (AOX) (mg L⁻¹)	Chlorides (mgL⁻¹)	Phosphate (mg L⁻¹)	Nitrogen (mg L⁻¹)	Phenol (mgL⁻¹)	Potassium (mg L⁻¹)	References
Pulp and Paper mill effluent	7.80	-	6380.56	5205.5	5280									54		Chuphal et al. 2005
Pulp and Paper mill effluent	7 – 8.5			700 – 1200	600 – 1000	600 – 1000	1100 – 1400	120 – 180		10 – 18						Dhall 2014
Pulping process wastewater	8.07				4374											Prasongsuk et al. 2009
Pulp effluent				4018*2	1248					46.3						Tarlan et al. 2002
Paper mill water	6.86 – 7.12	0.746 – 0.791*4		1750	2000 – 2100	615 – 670*3	1760*3	40*3	2100*3		48 – 62*3					Mahesh et al. 2006
Pulp and paper wastewater	13.5	870		290	1669.7	460	1790	40	1860		0.112 – 10,121					Wang et al. 2007
Pulp and paper industry wastewater	8 ± 0/1			1002.7 ± 57*2	1500 ± 150	563 ± 30		2138 ± 50 8								Abedinzadeh et al. 2018
Corrugated board factory waste water	7.1			680*2	470											Sevimli 2005
Paper and pulp wastewater	7.44	2940			1089			345	2420		76,22					Pizzichini et al. 2005
Pulp and paper mill waste water	6.6 – 10	1650 – 1790		1000 – 1984*2	490 – 1114	810 – 1095	1310 – 1560	40 – 260								Khansorthong and Hunsom 2009
Pulp and paper wastewater	7.5	1,175	1,155	Brown	657		587									Azadi Aghdam et al. 2016

Sample	pH													Reference
Pulp and paper mill effluent			1000 ± 50*²	500 ± 25					15 ± 1					Hooda et al. 2015
Recycled paper mill wastewater	6.2 – 7.8			3380 – 4930	1650 – 2565		1900 – 3138							Gupta and Gupta 2019
Pulp and paper mill wastewater	6.5			3791	1197		1241							Gupta and Gupta 2019
Paper pulp and paper mill wastewater	9.8			1303 ± 25	148 ± 5					871.6 ± 2.3 ug/L				Rodrigues et al. 2008
Paper and pulp wastewater	9.7		16*¹	48700	15500			44*¹						Kumar et al. 2015
Paper and pulp wastewater	10.2		13.2*¹	45600	13800			42*¹						Kumar et al. 2015
Paper and pulp wastewater	8.8		14.4*¹	40000	16500			38*¹						Kumar et al. 2015
Pulp and paper wastewater	7.78 ± 0.02	4.13 ± 0.04		1480.96 ± 21.65	1027.21 ± 6.67	1612.77 ± 32.10				164.13 ± 8.97	101.87 ± 1.42		149.04 ± 3.16	Kumar et al. 2020
Pulp and paper effluent	7.80	3.87*⁵		981.7 ± 58.3	230.2 ± 22.4	2129.17	1179.17	3316.67		2.7 ± 0.38		4.93 ± 0.9		Arivoli et al. 2015
Pulp and paper effluent	8.5 ± 0.52	1934 ± 94.63	693*⁴	33685 ± 269.03	9354 ± 205	1789 ± 31.25	86 ± 1.01	2369 ± 101	5.32 ± 0.11	267 ± 6.37	209 ± 5.79	693 ± 64.87		Sharma et al. 2021

*¹gm L⁻¹, *²Pt-Co, *³mg dm⁻³, *⁴mmho cm⁻¹, *⁵ds m⁻¹, *⁶gm L⁻¹

Table 14.2. Wastewater generation standard for Paper and Pulp industry (CPCB 2000).

Industry		pH	SS mg L⁻¹	BOD₃ mg L⁻¹	COD mg L⁻¹	Particulate Matter mg Nm⁻³	H₂S mg Nm⁻³
Large Pulp and Paper Industry (Capacity 24000 Ton/annum)		6.5 – 8.5	100	30	350	250	10
Small Paper and Pulp Industry	Inland Surface Water	5.5 – 9.0	100	30	-	-	-
	Land	5.5 – 9.0	100	100			

industries has different standards as per different pollution control authorities worldwide. As per the Central Pollution Control Board of India, the standards for different textile wastewater effluents are described in Table 14.4.

Textile wastewater contains pollutants like inorganic, organic and polymeric products (Brown and Laboureur 1983). The pollutants include dyes, coloring agents, etc. The dye and coloring agents are not only harmful to the biological ecosystems, they also block sunlight by covering the surface layer of soil and water bodies (Choi et al. 2004). Table 14.5 summarizes the pollutants present and their concentrations in the textile water from different studies conducted worldwide. Wastewaters from textile industries have shown different levels of various inorganic pollutants, i.e., NH_3, PO_4^{3-} and N, along with high DO, BOD, COD, TSS, TDS, EC, etc. According to a study by Adelodun et al. (2021) and Lin and Peng (1994) large amounts of oil and grease are also present in the textile wastewater.

Along with organic pollutants, textile wastewater is a contributor to heavy metals like Al, Cd, Cu, Ni, Zn, Pb, Fe, Cr and Hg, etc. A study by Türksoy et al. (2021) has observed a high amount of Cu (643.74), Zn (398.28) and Ni (95.56). Another study by Sungur and Gülmez (2015) showed the highest amount of Al and Fe (103.13 mgL⁻¹ and 80.13 mg kg⁻¹) during the wet digestion of cotton in the textile industry. These authors also found high amounts of Cd, Pb and Cr of 19.43, 189.4 and 382.3 mg kg⁻¹, respectively, in textile sludge. The different concentrations of heavy metals in textile wastewater observed by different studies are described in Table 14.6.

14.5.3 Dairy Industry

The dairy industry generates wastewater during the packaging of the produced milk. The water also includes cleaning agents like detergents necessary for cleanliness. Overflow, leakages and improper management also contribute to more wastewater. Waste produced from the dairy industry is broadly classified into two types, i.e., effluent or wastewater and solid waste (Adesra et al. 2021). Approximately 1–3 liter of wastewater is produced during the processing of 1 liter of milk, which contains high organic pollutants, i.e., carbohydrates, protein and fat (Kushawha et al. 2011), BOD, COD, TDS, TSS (Bazrafshan et al. 2013). Characteristics and quantity of the wastewater vary based on the design of the plant, type of product processed, type of treatment, water consumption (Lawrence et al. 2005) and season (Verma and Singh 2017). Before the discharge of the wastewater from the dairy industry, the standards as per government rules of different countries should be maintained. Different organic pollutants produced by the dairy industry are described in Table 14.7.

In 2020, global dairy exports reached to 78 million tons (in milk equivalents), an upsurge of 1.5% from 2019. Between 2016 and 2019, the dairy price index increased from USD 83 to 103 per ton worldwide, and annual worldwide milk production in 2020 reached 860 million tons, a rise of 1.4% from 2019, mainly due to increased pro-duction in Asia, Europe and North America (FAO 2020). The standards set by environment (protection) rules of India, EPA and the World Bank group for the maximum permissible limits of different pollutants from dairy wastewater are described in Table 14.8.

Table 14.3. Different heavy metals released from Paper and Pulp Wastewater.

Type of Water	Cr mg kg^{-1}	Cd mg kg^{-1}	Mn mg kg^{-1}	Ni mg kg^{-1}	Pb mg kg^{-1}	Fe mg kg^{-1}	Cu mg kg^{-1}	Zn mg kg^{-1}	Hg mg kg^{-1}	References
Paper mill sludge	187.3	8.75	–	–	62.5	–	144.4	–	–	Suthar et al. 2014
Pulp and paper industrial wastes (Lime mud)	10.11	0.683	184.25	10.39	31.03	199.97	8.748	8.748	0.140	Sthiannopkao and Sreesai 2009
Pulp and paper industrial wastes (Recovery boiler ash)	1.210	0.40	42.88	0.403	20.70	37.77	3.360	37.790	0.085	Sthiannopkao and Sreesai 2009
Pulp and paper mill effluents sludge leachate	–	9.11 ± 0.01	18.27 ± 0.20	5.21 ± 0.04	–	98.30 ± 1.80	3.21 ± 0.01	51.00 ± 0.00	0.014 ± 0.80	Sharma et al. 2020
pulp and paper industry effluent	–	0.0128 ± 0.001[*]	0.21 ± 0.027[*]	0.06 ± 0.007[*]	–	1.56 ± 0.091[*]	0.24 ± 0.022[*]	0.39 ± 0.03[*]	–	Arivoli et al. 2015
Discharged pulp paper mill effluent	2.30 ± 0.06[*]	0.255 ± 0.01[*]	11.00 ± 0.3[*]	3.30 ± 0.02[*]	1.05 ± 0.01[a]	67.53 ± 2.0[*]	2.15 ± 0.06[*]	13.90 ± 0.3[*]	–	Chandra et al. 2017
Pulp Paper Industry discharged effluent	–	10.36 ± 0.02[*]	29.34 ± 0.82[*]	7.01 ± 0.05[*]	1.03 ± 0.01[a]	98.20 ± 1.55[*]	57.31 ± 0.63[*]	5.31 ± 0.05[*]	2.37 ± 0.5[*]	Sharma et al. 2021

[*]—The values provided are in mgL^{-1}

[a]—stand for mg/L

Table 14.4. Wastewater generation standard for Textile industries (mg L^{-1}).

Textile industries	pH	Suspended Solids	BOD$_3$	Oil and Grease	Bio-assay	Cr	Sulfide	Phenol	References
Cotton Textile Industry	5.5–9.0	100	150	10	90% survival of fish after 96 hr in 100% effluent	–	–	–	CPCB 2000
Composite Woolen Mill	5.5–9.0	100	100	10	90% survival of fish after 96 hr in 100% effluent	2	2	5	CPCB 2000

14.6 Application of WH in Phytoremediation of Industrial Wastewater

14.6.1 Removal of Heavy Metals from Industrial Wastewater

WH is one of the common phytoremediation plants in India, has a strong capacity for heavy metal accumulation (Bioaccumulation factor > 10,000 times), which leads to the removal of organic and inorganic pollutants from water and wastewater (Yan and Guo 2017). It removes pollutants using physical or biological treatment or a combination. WH is known for absorbing suspended solids during the physical treatment process using precipitation and absorption processes. Metals and other pollutants bound with the suspended solids are then co-precipitated (Huang and Xu 2008). The hairy fibrous root system of the plant supports the accumulation of suspended solids, microorganisms, colloids and protozoa (Zhou et al. 2005, Nawirska 2005). The absorbed pollutants are transferred to the leaves and stems after accumulation by the roots, and roots protect the plant by accumulating large amounts in the root and transfer a small number of pollutants to the leaves and stems (Cai et al. 2004). Heavy metals are accumulated by the WH in the apoplast (Cell Wall) and then transferred across the plasma membrane. The root walls of WH play an important role by blocking the pollutant uptake with the help of pectin substances such as polygalacturonic acid and cellulose molecules. This cellulose or polygalacturonic acid molecule contains carboxyl and aldehyde groups which help exchange sites for pollutants like heavy metals by chelation. In this chelation process, amino- and/ or oxygen-containing functional groups play a key role in removing heavy metals from wastewater (Zhang 2011). Among tolerance mechanisms, chelation in the cytosol attracts considerable attention: heavy metal stress could induce the formation of biomacromolecules that form chelates with heavy metal ions, thus lowering the activity of free heavy metal ions in plant cells and relieving the toxicity (Yan and Guo 2017). Two metal binding peptides, i.e., metallothionein and phytochelatins, are available in the phytoremediation plant cells; among them, metallothionein has the low-molecular-weight polypeptide containing cysteine which helps in the formation of a non-toxic or low-toxic complex with the combination of thiol (–SH) group (Margoshes and Vallee 1957).

A case study by Mokhtar et al. (2011) reported the removal of heavy metals in the textile industry using WH, where 97.3% of Cu accumulation was reported. Similarly, the 5-wk study on textile wastewater resulted in 94.87% removal of cadmium from the textile industry using WH (Ajayi and Ogunbayo 2012). Seventy to ninety percent removals of heavy metals like Fe, Pb, Cu and Cr have been observed in the study of Kolawole (2001). Another study by Mahmood et al. (2005) on textile wastewater observed 86, 88 and 83% removal of Cr, Zn and Cu, respectively. Apart from that more than 10% removal of methylene blue dye from textile wastewater has been observed in the study of Nibret et al. (2019). Water hyacinth can also remove more than 60% Mn and Pb from paper and pulp effluent (Kumar et al. 2016) and Na, K, Ca, Mg, Cd, Cr, Cu, Fe.

Table 14.5. Characterization of Textile Industry wastewater.

	pH	DO mg L⁻¹	BOD mg L⁻¹	COD mg L⁻¹	TDS mg L⁻¹	TSS mg L⁻¹	EC (Mohm cm⁻¹)	Ammonia N (Nh3-N)V mg L⁻¹	Phosphate mg L⁻¹	Copper mg L⁻¹	Zinc mg L⁻¹	Oil and Grease mg L⁻¹	Chromium (mg L⁻¹)	References
Textile Industry Wastewater	11.63	0.6	395	1926	4210		3.87	–	–	–	–	–	–	Mahajan et al. 2019
	8.1	–	48.7	178	–	5.61		12.14	–	–	–	–	3.741	Apritama et al. 2020
	13.04	–	1048	2300	4500	–	3.2*¹	–	–	–	–	–	–	Anjana and Thanga 2011
	9.87	3.8	347	2230	1450	370	–	2.5	3.78	2.45	–	–	–	Ugya 2017
	10.7	–	–	956	–	–	0.082*²	5.04	45.7	–	–	83.3	–	Adelodun et al. 2021
	10	–	500	1500	–	259	2900*¹	–	–	–	–	50	–	Lin and Peng 1994
	8.259	–	243.444	361.17	1713.17		1161.83	–	–	4.568	4.675	–	4.705	Mahmood et al. 2005

*¹μs/cm, *²S/m

Table 14.6. Different heavy metals released from Textile industry Wastewater.

Types of Water	Al	Cd	Cu	Ni	Zn	Pb	Fe	Cr	Hg	References
Textile Industry Wastewater	32.88 ± 5.99	9.00 ± 3.91	643.74 ± 27.24	95.56 ± 13.05	398.71 ± 40.59	32.28 ± 5.75	–	–	–	Türksoy et al. 2021
Dyeing effluent	–	0.01	0.17 – 0.28	0.11 – 0.22	0.11 – 0.51	0.02 – 0.10	0.39 – 0.90	0.11 – 0.21	–	Uddin 2021
Handloom dyeing effluent	–	–	1.70	–	0.73	0.17	3.81	1.35	–	Nahar et al. 2018
Textile effluent	–	24.51	–	–	17.45	11.12	–	–	–	Basha and Rajaganesh 2014
Textile effluent	–	0.002	0.078	–	1.868	0.018	1.122	0.090	3.382	Panigrahi and Santhoskumar 2020
Textile water	–	0.22	1.91	68	69	38	209	–	–	Samecka-Cymerman and Kempers 2007
Textile industry effluent	–	0.016	–	–	1.02	0.61	1.63	–	–	Odipe et al. 2018
Textile industry effluent	–	0.66	–	–	1.74	1.02	2.15	–	–	Odipe et al. 2018
Wet digestion of cotton	103.13[1]	11.86[1]	5.84[1]	2.19[1]	–	23.44[1]	80.13[1]	0.97[1]	–	Sungur and Gülmez 2015
Textile Sludge	–	19.43[1]	265[1]	66.2[1]	328[1]	189.4[1]	–	382.3[1]	0.9[1]	Sungur and Gülmez 2015

[1] mg kg^{-1}

Table 14.7. Characterization of dairy industry wastewater.

Type of Water	pH	EC (μs cm^{-1})	TSS (mg L^{-1})	TDS (mg L^{-1})	TS (mg L^{-1})	COD (mg L^{-1})	DO (mg L^{-1})	BOD (mg L^{-1})	SO$_4$ or Total S (mg L^{-1})	PO$_4$ or Total P (mg L^{-1})	NO3 (mg L^{-1})	CO$_3^{2-}$ (mg L^{-1})	Fat (g L^{-1})	Protein (g L^{-1})	Lactose (g L^{-1})	References
Dairy industry	4.6 ± 0.2	265 ± 1	420 ± 1	123 ± 05	–	118 ± 5	1.24 ± 0.4	1200 ± 11	40.80 ± 0.4	0.37 ± 0.04	0.49 ± 0.02	28 ± 0.3	–	–	–	Ondiba et al. 2022)
Dairy effluent	7.1 – 8.22	–	221 – 2,996	–	800 – 27,000	438 – 5,044	–	–	17 – 65	17 – 82	–	–	–	–	–	Tait et al. 2021
Dairy plant cleaning water	8.23 ± 0.07	–	–	–	0.50 ± 0.04*1	21.50 ± 0.71	–	4.50 ± 0.71	–	–	–	–	–	0.01 ± 0.00*1	0.33 ± 0.03*1	Alalam et al. 2021
Dairy plant cleaning water	12.45 ± 0.19	–	–	–	3.12 ± 0.24*1	354.50 ± 2.12	–	–	–	–	–	–	0.03 ± 0.02*1	0.15 ± 0.00*1	2.22 ± 0.18*1	Alalam et al. 2021
Dairy sludge	–	–	–	–	–	–	–	–	6.10 ± 3.20*4	33.04 ± 0.25*4	5.30 ± 0.12 %	–	–	–	–	Roufou et al. 2021
Dairy wastewater	–	–	3.92	–	–	260	–	130	–	531.25	–	–	–	–	–	Vinodhini and Soundhari 2019
Milk and dairy wastewater	8.34	–	5802.6	–	–	10251.2	–	4840.6	–	153.6	663	–	–	–	–	Onet 2010
Dairy fatty waste	5.3 ± 0.1	–	–	–	–	73 ± 8*1	–	–	–	–	–	–	5.85 ± 0.45*1	10.2 ± 0.3*1	–	Pascale et al. 2019
Dairy rinse water	6.72 ± 0.24	174 ± 26	–	–	–	2230 ± 141	–	–	–	6.5 ± 1.2	99.7 ± 13.2	–	–	636 ± 84*2	980 ± 59*2	Brião et al. 2019
Dairy industry effluent	6.17	1.83*3	28.37	288	316.37	71.4	4.5	36.5	–	15	18	–	–	–	–	Kaur et al. 2018
Dairy wastewater	4.67 ± 0.05	–	10.00 ± 0.01	–	–	3133 ± 438	–	2350 ± 1	–	–	75 ± 12 3133	–	–	–	2.1 ± 0.5*1	Bortoluzzi et al. 2017

*1 g L^{-1}, *2 mg L^{-1}, *3 ds m^{-1}, *4 gm kg^{-1}

Table 14.8. Wastewater generation Standard for Dairy Industries.

Industries	pH	BOD$_3$/BOD$_5$ (mgL^{-1})	SS (mg L^{-1})	Oil and Grease (mg L^{-1})	COD (mg L^{-1})	Total Nitrogen (N) (mg L^{-1})	Total Phosphorus (P) (mg L^{-1})	Total Ammonia (N) (mg L^{-1})	Waste Water Generation (m^3/Kl of milk)	References
Dairy Industries	6.5 – 8.5	100	150	10					3	The Environment (Protection) Rules, 1986
Dairy Processing Sector	6 – 9	> 90% removal, or 20 – 40	50	10 – 15	> 75% removal, or 125 – 250	> 80% removal, or 5 – 25	> 80% removal, or 2 – 5	10		U.S.EPA 2008
Dairy Industry Effluents	6 – 9	50	50	10	250	10	2			World Bank Group 1999

14.6.2 Removal of Organic Pollutants from Industrial Wastewater

WH tissues contain many enzymes that perform various biochemical processes to transfer and degrade organic compounds in joint action with rhizosphere bacteria (Lu et al. 2017, Anipeddi et al. 2022). The developed root system of water hyacinth is crucial in removing organic pollutants from water through two mechanisms:

1. Roots can absorb PCBs, PAHs as well as other organic pollutants from the aquatic environment and/or transfer pollutants to other parts of the plant, where they can be degraded into harmless molecules (Voudrias and Assaf 1996, Xia et al. 2003).

2. The root system provides a habitat for various microorganisms by releasing exudates and enzymes to stimulate microbial activity and strengthen mineralization and other forms of biotransformation by rhizosphere microorganisms (Voudrias and Assaf 1996, Xia et al. 2003).

Unlike heavy metals, the assimilation of organic pollutants by the roots of remediation plants accounts only for a small portion of the total removal (Liu et al. 2003, Zhang et al. 2011, Lu et al. 2014). Most organic pollutants, especially non-ionic ones, could enter the root system of plants via diffusion, and only some (for instance, systemic pesticides) enter plant roots via a transpiration pull of leaves (Xia 2002, Paraíba 2007, Al-Qurainy and Abdel-Megeed 2009). There are several steps in the uptake. Firstly, the organic pollutants in the water get absorbed in the "apparent free space" in the external root tissues, which account for $10 \sim 20\%$ of the volume of the plant root system (Nye and Tinker 1977); then, the organic pollutants could be transferred through the cell walls and intercellular spaces (apoplastically) all the way to endodermis. Endodermis which blocks apoplastic pathway as it contains highly-suberinized Casparian band impermeable to water; to bypass the casparian band, ions and molecules should be taken into root cells, and then continue symplastically from the cell interior to another cell interior via plasmodesmata (Wild et al. 2005).

A case study by Gamage and Yapa (2001) observed the phytoremediation capacity of WH in textile mill effluent of Sri Lanka. The study has shown a substantial reduction in VS 72.6, TDS 60, TSS 46.6, 75 BOD and 81.4% COD. Another study by Roy et al. (2010) on textile industry effluent has shown a 60% reduction in COD and the pH reduction from 11.2 to 8.6. On the other hand, Munavalli and Saler (2009) worked on dairy industry wastewater, where WH alone contributed to 30 – 45% COD removal. It has also shown a significant reduction of organic strength and has the tendency to neutralize the pH of dairy industry wastewater. Similarly, another study by Bhavsar et al. (2010) on the dairy industry wastewater observed the removal of COD levels from 810 mgL^{-1} to 200 mgL^{-1}. They also observed a pH reduction from 8.3 to 7.04 and alkalinity reduction from 600 mgL^{-1} to 480 mgL^{-1}. Safauldeen et al. (2019) conducted a phytoremediation experiment on batik textile effluent using water hyacinth, where treatment achieved 83 and 89% removal of COD and TSS, respectively (Table 14.9).

14.7 Conclusion

Water hyacinth has a high potential for removing many organic pollutants and heavy metals from wastewater under different climatic conditions. Apart from the organic pollutants, N, P, Na, Ca, N, it helps to reduce many toxic non-essential heavy metals like Al, As, Au, Cd, Cr, Hg, Pb, Pd, Pt, Sb, Te and Tl from different industrial effluents. Moreover, the study presents characteristics of the wastewater produced from the different lignocellulosic wastewater-producing industries, i.e., paper and pulp mills, dairy and textile industries. However, the dairy industry wastewater has fewer heavy metals and other pollutants than textile and paper and pulp industries. Different treatment approaches are taken for removing/minimizing the pollutants. Phytoremediation using water hyacinth would be one of the cheapest, most cost-effective and environment-friendly approaches to remediate organic and inorganic pollutants, including heavy metals. Some additional investigation on uptake efficiency, appropriate climatic conditions and post-harvest treatment processes are required to develop practical large-scale approaches to phytoremediation using water hyacinth.

Table 14.9. Phytoremediation of lignocellulosic industrial wastewater using water hyacinth.

Type of Industrial Water	Polluting Parameters	Initial Concentration (mg L⁻¹)	Final Concentration (mg L⁻¹)	Remediation Time	References
Textile Wastewater	COD	361.17	191.50	96 hrs	Mahmood et al. 2005
	pH	8.259	7.105	96 hr	
	Chromium	4.705	0.612	96 hr.	
	Zinc	4.675	0.553	96 hr	
	Copper	4.568	0.747	96 hr	
	Biological Oxygen Demand	243.44	107.33	96 hr	
	Total Solids	1713.17	955	96 hr	
	Conductivity	1884	1161.83	96 hr	
Dairy Industry Wastewater	COD	1352	280	7 d	Bhavsar et al. 2012
	pH	7.77	7.79	7 d	
	Alkalinity	580	700	7 d	
	DO	0.94	0.265	7 d	
Dairy Industry Wastewater	pH	6.69	6.50	8 d	Munavalli and Saler 2009
	Alkalinity (mg as CaCO₃)	258	326	8 d	
	EC	1.16*¹	1.29*¹	8 d	
	COD	507	77	8 d	
Dairy Industry Wastewater	Total Dissolved Solids	848	352	4 d	Trivedy and Pattanshetty 2002
	Total Suspended Solids	359	245	4 d	
	BOD	840	143	4 d	
	COD	1160	137	4 d	
	Total Nitrogen	26.60	8.9	4 d	
	Total Phosphate	8.72	1.56	4 d	
	Calcium	64.2	59.2	4 d	
	Potassium	6	10	4 d	
Textile Wastewater	MB Dye	98.6%	87.6%		Nibret et al. 2019
Batik Textile Effluent	COD			7 d	Safauldeen et al. 2019
	Color			7 d	
	Total Suspended Solids			28 d	
	pH	6	7	28 d	

Table 14.9 contd. ...

...Table 14.9 contd.

Type of Industrial Water	Polluting Parameters	Initial Concentration (mg L^{-1})	Final Concentration (mg L^{-1})	Remediation Time	References
Paper and Pulp Mill Effluent	TDS	1840	1060	60 d	Kumar et al. 2016
	EC	2.64*2	1.76*2	60 d	
	pH	7.82	7.29	60 d	
	BOD$_5$	475.10	275.68	60 d	
	TKN	192.65	82.50	60 d	
	PO$_4$	145.60	64.57	60 d	
	Na	285.44	150.33	60 d	
	K	175.50	96.37	60 d	
	Ca	435.80	305.80	60 d	
	Mg	148.35	66.40	60 d	
	Cd	2.45	1.34	60 d	
	Cr	1.38	0.69	60 d	
	Cu	5.64	2.94	60 d	
	Fe	8.95	4.86	60 d	
	Mn	3.66	1.42	60 d	
	Ni	1.74	0.73	60 d	
	Pb	1.02	0.36	60 d	
	Zn	6.90	3.10	60 d	

*1 mmho cm^{-1}, *2ds m^{-1}

14.8 Futuristic Approaches

In accordance with the phytoremediation potentiality, WH can be consider as an excellent plant for phytoremediation of different industrial wastewater. Additionally, these aquatic macrophytes can be utilized as a lignocellulolytic substrate in renewable energy production units. Moreover, to combat the ecological malaises from global environmental pollution from such different industrial resources, this aquatic weed appears as nature's green medicine for an eco-sustainable and healthy ecosphere.

Acknowledgement

Apurba Koley is thankful to the BBSRC, United Kingdom, for granting funding from the BEFWAM project: Bioenergy, Fertilizer and Clean water from Invasive Aquatic macrophytes [Grant Ref: BB/S011439/1] for financial support and research fellowship. Sandipan Banerjee and Nitu Gupta thank the Department of Biotechnology, Govt. of India, for granting DBT Twinning Project and Research Fellowship [No. BT/PR25738/NER/95/1329/2017 dated December 24, 2018]. While conducting the study, the authors are thankful for the support from, Dr. Aishiki Banerjee, Ms. Sneha Banerjee.

References

Abedinzadeh, N., M. Shariat, S. M. Monavari and A. Pendashteh. 2018. Evaluation of color and COD removal by Fenton from biologically (SBR) pre-treated pulp and paper wastewater. Process Saf. Environ. Prot. 116: 82–91.

Adelodun, A. A., O. Temitope, N. O. Afolabi, A. S. Akinwumiju, E. Akinbobola and U. O. Hassan. 2021. Phytoremediation potentials of *Eichhornia crassipes* for nutrients and organic pollutants from textile wastewater. Int. J. Phytoremediat. 23(13): 1333–1341.

Adesra, A., V. K. Srivastava and S. Varjani. 2021. Valorization of dairy wastes: integrative approaches for value added products. Indian J. Microbiol. 61(3): 270–278.

Ajayi, T. O. and A. O. Ogunbayo. 2012. Achieving environmental sustainability in wastewater treatment by phytoremediation with water hyacinth (*Eichhornia crassipes*). J. Sustain. Dev. 5(7): 80.

Alalam, S., F. Ben-Souilah, M. H. Lessard, J. Chamberland, V. Perreault, Y. Pouliot, S. Labrie and A. Doyen. 2021. Characterization of chemical and bacterial compositions of dairy wastewaters. Dairy. 2(2): 179–190.

Al-Qurainy, F. and A. Abdel-Megeed. 2009. Phytoremediation and detoxification of two organophosphorous pesticides residues in Riyadh area. World Appl. Sci. J. 6(7): 987–998.

Alves, E., L. R. Cardoso, J. Savroni, L. C. Ferreira, C. S. F. Boaro and A. C. Cataneo. 2003. Avaliações fisiológicas e bioquímicas de plantas de aguapé (*Eichhornia crassipes*) cultivadas com níveis excessivos de nutrientes. Planta Daninha. 21: 27–35.

Anipeddi, M., S. Begum and G. R. Anupoju. 2022. Integrated technologies for the treatment of and resource recovery from sewage and wastewater using water hyacinth. Biofuel Bioprod. Biorefin, pp. 293–314. Elsevier,

Anjana, S. and V. S. G. Thanga. 2011. Phytoremediation of synthetic textile dyes. Asian J. Microbiol. Biotechnol. Environ. Sci. 13: 31–34.

Ansari, A. A., R. Gill, L. Newman, S. S. Gill and G. R. Lanza. 2016. Phytoremediation: Management of Environmental Contaminants. 6: 1–576. https://doi.org/10.1007/978-3-319-99651-6.

Apritama, M. R., I. Suryawan, A. S. Afifah and I. Y. Septiariva. 2020. Phytoremediation of effluent textile wwtp for NH3-N and Cu reduction using pistia stratiotes. Plant Archives. 20(1): 2384–2388.

Arivoli, A., R. Mohanraj and R. Seenivasan. 2015. Application of vertical flow constructed wetland in treatment of heavy metals from pulp and paper industry wastewater. Environ. Sci. Pollut. Res. 22(17): 13336–13343.

Ashrafi, O., L. Yerushalmi and F. Haghighat. 2015. Wastewater treatment in the pulp-and-paper industry: a review of treatment processes and the associated greenhouse gas emission. J. Environ. Manag. 158: 146–157.

Assche, V. F. and H. Clijsters. 1990. Effects of metals on enzyme activity in plants. Plant Cell Environ. 13(3): 195–206.

Azadi Aghdam, M., H. R. Kariminia and S. Safari. 2016. Removal of lignin, COD, and color from pulp and paper wastewater using electrocoagulation. Desalin. Water Treat. 57(21): 9698–9704.

Babu, B. R., A. K. Parande, S. Raghu and T. P. Kumar. 2007. Cotton textile processing: waste generation and effluent treatment. J. Cotton Sci. 11: 141–153.

Barry, A. C. P. 1998. Preliminary investigation of an integrated aquaculture–wetland ecosystem using tertiary-treated municipal wastewater in Los Angeles County, California. Ecol. Eng. 10(4): 341–354.

Basha, S. A. and K. Rajaganesh. 2014. Microbial bioremediation of heavy metals from textile industry dye effluents using isolated bacterial strains. Int. J. Curr. Microbiol. Appl. Sci. 3: 785–794.

Basu, A., A. K. Hazra, S. Chaudhury, A. B. Ross and S. Balachandran. 2021. State of the art research on sustainable use of water hyacinth: a bibliometric and text mining analysis. In Informatics. Multidisciplinary Digital Publishing Institute. 8(2): 38.

Bazrafshan, E., H. Moein, F. KordMostafapour and S. Nakhaie. 2013. Application of electrocoagulation process for dairy wastewater treatment. J. Chem. 7–10.

Bhavsar, S. R., V. R. Pujari and V. V. Diwan. 2010. Potential of phytoremediation for dairy wastewater treatment. J. Civ. Eng. Manag. 16–23.

Bhavsar, S. R., V. R. Pujari and V. V. Diwan. 2012. Potential of phytoremediation for dairy wastewater treatment. IIOSR J. Mech. Civ. Eng. 16–23.

Bortoluzzi, A. C., J. A. Faitão, M. Di Luccio, R. M. Dallago, J. Steffens, G. L. Zabot and M. V. Tres. 2017. Dairy wastewater treatment using integrated membrane systems. J. Environ. Chem. Eng. 5(5): 4819–4827.

Brião, V. B., A. C. Vieira Salla, T. Miorando, M. Hemkemeier and D. P. CadoreFavaretto. 2019. Water recovery from dairy rinse water by reverse osmosis: giving value to water and milk solids. Resour. Conserv. Recycl. 140: 313–323.

Brown, D. and P. Laboureur. 1983. The aerobic biodegradability of primary aromatic amines. Chemosphere. 12(3): 405–414.

Cai, C., H. Wang and Z. Zhang. 2004. Removal of Cu, Pb, Cd, Zn and Fe by water hyacinth. J. Leshan Teachers College. 19(6): 69–72.

Chandel, A. K., V. K. Garlapati, A. K. Singh, F. A. F. Antunes and S. S. da Silva. 2018. The path forward for lignocellulose biorefineries: bottlenecks, solutions, and perspective on commercialization. Bioresour. Technol. 264: 370–381. doi: 10.1016/j.biortech.2018.06.00.

Chandra, R., S. Yadav and S. Yadav. 2017. Phytoextraction potential of heavy metals by native wetland plants growing on chlorolignin containing sludge of pulp and paper industry. Ecol. Eng. 98: 134–145.

Choi, J. W., H. K. Song, W. Lee, K. K. Koo, C. Han and B. K. Na. 2004. Reduction of COD and color of acid and reactive dyestuff wastewater using ozone. Korean J. Chem. Eng. 21(2): 398–403.

Chua, H. 1998. Bio-accumulation of environmental residues of rare earth elements in aquatic flora *Eichhornia crassipes* (Mart.) Solms in Guangdong Province of China. Sci. Total Environ. 214(1-3): 79–85.

Chuphal, Y., V. Kumar and I. S. Thakur. 2005. Biodegradation and decolorization of pulp and paper mill effluent by anaerobic and aerobic microorganisms in a sequential bioreactor. World J. Microbiol. Biotechnol. 21(8-9): 1439–1445.

CPCB. 2000. Environmental Standards for Ambient Air, Automobiles, Fuels, Industries and Noise, Central Pollution Control Board. Ministry of Environment and Forest.

Dhall, V. P. 2014. Biological approach for the treatment of pulp and paper industry effluent in sequence batch reactor. J. Bioremediat. Biodegrad. 5(3).

Dhir, B. 2013. Phytoremediation: role of aquatic plants in environmental clean-up. 14: 1–111. Springer, New Delhi.

Djingova, R. and I. v Kuleff. 2000. Instrumental techniques for trace analysis. In Trace metals in the Environment. 4: 137–185. Elsevier.

Doble, M. and A. Kumar. 2005. Biotreatment of industrial effluents.1st ed. Burlington: Butterworth-Heinemann. Elsevier.

Ebel, M., M. W. Evangelou and A. Schaeffer. 2007. Cyanide phytoremediation by water hyacinths (*Eichhornia crassipes*). Chemosphere. 66(5): 816–823.

Ekanayake, M. S., D. Udayanga, I. Wijesekara and P. Manage. 2021. Phytoremediation of synthetic textile dyes: biosorption and enzymatic degradation involved in efficient dye decolorization by *Eichhornia crassipes* (Mart) Solms and *Pistia stratiotes* L. Environ. Sci. Pollut. Res. 16: 20476–20486.

Elangovan, R., L. Philip and K. Chandraraj. 2008. Biosorption of chromium species by aquatic weeds: kinetics and mechanism studies. J. Hazard. Mater. 152(1): 100–112.

Ensley, B. D. 2000. Rationale for the Use of Phytoremediation. Phytoremediation of toxic metals: Using plants to Clean-up the Environment. John Wiley Publishers: New York.

FAO. 2020. The State of World Fisheries and Aquaculture 2020. Sustainability in Action.

Fletcher, J., N. Willby, D. M. Oliver and R. S. Quilliam. 2020. Phytoremediation using aquatic plants. In Phytoremediation, pp. 205–260. Springer, Cham.

Gamage, N. S. and P. A. J. Yapa. 2001. Use of water Hyacinth (*Eichhornia crassipes* (Mart) Solms) in treatment systems for textile mill effluents-a case study. J. Natl. Sci. Found. Sri Lanka. 29: 1–2.

Gopal, B. 1987. Water hyacinth. Elsevier Science Publishers.

Gupta, A. and R. Gupta. 2019. Treatment and recycling of wastewater from pulp and paper mill. In Advances in Biological Treatment of Industrial Waste Water and their Recycling for a Sustainable Future. pp. 13–49. Springer, Singapore.

Gupta, P., S. Roy and A. B. Mahindrakar. 2012. Treatment of water using water hyacinth, water lettuce and vetiver grass–a review. System. 49: 50.

Hao, O. J., H. Kim and P. C. Chiang. 2000. Decolorization of wastewater. Crit. Rev. Environ. Sci. Technol. 30(4): 449–505.

Holik, H. (Ed.). 2006. Handbook of paper and board. John Wiley and Sons.

Holkar, C. R., A. J. Jadhav, D. V. Pinjari, N. M. Mahamuni and A. B. Pandit. 2016. A critical review on textile wastewater treatments: possible approaches. J. Environ. Manag. 182: 351–366.

Hooda, R., N. K. Bhardwaj and P. Singh. 2015. Screening and identification of ligninolytic bacteria for the treatment of Pulp and Paper Mill effluent. Water Air Soil Pollut. 226(9).

Huang, B. and H. Xu. 2008. Water hyacinth on ecological damage and phytoremediation. Guangdong Water Resour. Hydropower. (3): 1–3, 11.

Jadia, C. D. and M. H. Fulekar. 2009. Phytoremediation of heavy metals: recent techniques. Afr. J. Biotechnol. 8(6).

Jafari, N. 2010. Ecological and socio-economic utilization of water hyacinth (*Eichhornia crassipes* Mart Solms). J. Appl. Sci. Environ. Manag. 14(2).

Kamal, M., A. E. Ghaly, N. Mahmoud and R. CoteCôté. 2004. Phytoaccumulation of heavy metals by aquatic plants. Environ. Int. 29(8): 1029–1039.

Kaur, V., G. Sharma and C. Kirpalani. 2018. Agro-potentiality of dairy industry effluent on the characteristics of *Oryza sativa* L. (Paddy). Environ. Technol. Innov. 12: 132–147.

Khansorthong, S. and M. Hunsom. 2009. Remediation of wastewater from pulp and paper mill industry by the electrochemical technique. Chem. Eng. J. 151(1–3): 228–234.

Kolawole, B. S. 2001. Cleaning of effluent from textile Industry by water hyacinth (B.Sc thesis). University of Agriculture, Abeokuta, Ogun state.

Koley, A., D. Bray, S. Banerjee, S. Sarhar, R. Ghosh Thahur, A. K. Hazra, N. C. Mandal et al. 2022. Water hyacinth (*Eichhornia crassipes*) a sustainable strategy for heavy metals removal from contaminated waterbodies. pp. 95–114. *In*: A. Malik, M. K. Kidwai and V. K. Garg [Eds.]. Bioremediation of Toxic Metal(loid)s. CRC Press, London

Kumar, P. and M. S. Chauhan. 2019. Adsorption of chromium (VI) from the synthetic aqueous solution using chemically modified dried water hyacinth roots. J. Environ. Chem. Eng. 7(4): 103218.

Kumar, R., K. S. Rajeev and P. S. Anirudh. 2017. Cellulose based grafted biosorbents-Journey from lignocellulose biomass to toxic metal ions sorption applications—a review. J. Mol. Liq. 232: 62–93.

Kumar, S., T. Saha and S. Sharma. 2015. Treatment of pulp and paper mill effluents using novel biodegradable polymeric flocculants based on anionic polysaccharides: a new way to treat the waste water. Int. Res. J. Eng. Technol. 2(4): 1–14.

Kumar, V., A. K. Chopra, J. Singh, R. K. Thakur, S. Srivastava and R. K. Chauhan. 2016. Comparative assessment of phytoremediation feasibility of water caltrop (*Trapa natans* L.) and water hyacinth (*Eichhornia crassipes* Solms) using pulp and paper mill effluent. Arch. Agri. Sci. J. 1(1): 13–21.

Kumar, V., J. Singh and P. Kumar. 2020. Regression models for removal of heavy metals by water hyacinth (*Eichhornia crassipes*) from wastewater of pulp and paper processing industry. Environ. Sustain. 3(1): 35–44.

Kushwaha, J. P., V. C. Srivastava and I. D. Mall. 2011. An overview of various technologies for the treatment of dairy wastewaters. Crit. Rev. Food Sci. Nutr. 51(5): 442–452.

Lawrence, W., H. Yung-Tse and S. Nazih. 2005. Physicochemical treatment processes. pp. 141–154. *In*: L. K. Wang, Y. T. Hung and N. K. Shammas [Eds.]. Handbook of Environmental Engineering. Humana Press.

Lin, S. H. and C. F. Peng. 1994. Treatment of textile wastewater by electrochemical method. Water Res. 28(2): 277–282.

Liu, J., F. Lin, Y. Wang, Z. Xu and X. Zhang. 2003. Absorption processes of macrophyte root on polycyclic aromatic hydrocarbons (naphthalene). Environ. Sci. Technol. 26(1): 32–34.

Liu, R. R., Q. Tian, B. Yang and J. H. Chen. 2010. Hybrid anaerobic baffled reactor for treatment of desizing wastewater. Int. J. Environ. Sci. Technol. 7(1): 111–118.

Lu, X., L. Liu, R. Fan, J. Luo, S. Yan, Z. Rengel and Z. Zhang. 2017. Dynamics of copper and tetracyclines during composting of water hyacinth biomass amended with peat or pig manure. Environ. Sci. Pollut. Res. 24(30): 23584–23597.

Lu, X., Y. Gao, J. Luo, S. Yan, Z. Rengel and Z. Zhang. 2014. Interaction of veterinary antibiotic tetracyclines and copper on their fates in water and water hyacinth (*Eichhornia crassipes*). J. Hazard. Mater. 280: 389–398.

Mahajan, P., J. Kaushal, A. Upmanyu and J. Bhatti. 2019. Assessment of phytoremediation potential of *Chara vulgaris* to treat toxic pollutants of textile effluent. J. Toxicol. ID 8351272 | https://doi.org/10.1155/2019/8351272.

Mahalakshmi, R., C. Sivapragasam and S. Vanitha. 2019. Comparison of BOD 5 Removal in water hyacinth and duckweed by genetic programming. Int. J. Inf. Commun. Technol. 401–408.

Mahesh, S., B. Prasad, I. D. Mall and I. M. Mishra. 2006. Electrochemical degradation of pulp and paper mill wastewater. Part 1. COD and color removal. Ind. Eng. Chem. Res. 45(8): 2830–2839.

Mahmood, Q., P. Zheng, E. Islam, Y. Hayat, M. J. Hassan, G. Jilani and R. C. Jin. 2005. Lab scale studies on water hyacinth (*Eichhornia crassipes* Marts Solms) for biotreatment of textile wastewater. Casp. J. Environ. Sci. 3(2): 83–88.

Malik, A. 2007. Environmental challenge vis a vis opportunity: the case of water hyacinth. Environ. Int. 33(1): 122–138.

Margoshes, M. and B. L. Vallee. 1957. A cadmium protein from equine kidney cortex. J. Am. Chem. Soc. 79(17): 4813–4814.

Menon, V. and M. Rao. 2012. Trends in bioconversion of lignocellulose: biofuels, platform chemicals and biorefinery concept. Prog. Energy Combust. Sci. 38: 522–550. doi: 10.1016/j.pecs.2012.02.002.

Ministry of Textiles, Government of India 21–22 Annual Report Ministry of Textiles.

MOEF. 2010. The Ministry of Environment and Forests, Government of India, Technical EIA Guidance Manual for Pulp and Paper Industries, Prepared by IL and FS Ecosmart Limited, Hyderabad.

Mokhtar, H., N. Morad and F. F. A. Fizri. 2011. Hyperaccumulation of copper by two species of aquatic plants. Int. Conf. Eng. Environ. Sci. 8: 115–118.

Munavalli, G. R. and P. S. Saler. 2009. Treatment of dairy wastewater by water hyacinth. Water Sci. Technol. 59(4): 713–722.

Nahar, K., M. Chowdhury, A. Khair, M. Chowdhury, A. Hossain, A. Rahman and K. M. Mohiuddin. 2018. Heavy metals in handloom-dyeing effluents and their biosorption by agricultural byproducts. Environ. Sci. Pollut. Res. 25(8): 7954–7967. https://doi. org/10.1007/s11356-017-1166-9.

Nakamura, Y., T. Sawada, F. Kobayashi and M. Godliving. 1997. Microbial treatment of kraft pulp wastewater pretreated with ozone. Water Sci. Technol. 35(2-3): 277–282.

Nawirska, A. 2005. Binding of heavy metals to pomace fibers. Food Chem. 90(3): 395–400.

Nemerow, N. L. 2007. Industrial waste treatment. Elsevier science and technology. Burlington, MA: Butterworth-Heinemann.

Nibret, G., S. Ahmad, D. G. Rao, I. Ahmad, M. A. M. U. Shaikh and Z. U. Rehman. 2019. Removal of methylene blue dye from textile wastewater using water hyacinth activated carbon as adsorbent: synthesis, characterization and kinetic studies. In Proceedings of International Conference on Sustainable Computing in Science, Technology and Management (SUSCOM), Amity University Rajasthan, Jaipur-India.

Nye, P. H. and P. B. Tinker. 1977. Solute movement in the soil-root system. Univ. of California Press.

Odipe, O. E., M. O. Raimi and F. Suleiman. 2018. Assessment of heavy metals in effluent water discharges from textile industry and river water at close proximity: A comparison of two textile industries from Funtua and Zaria, North Western Nigeria. Madridge J. Agric. Environ. Sci. 1(1): 1–6.

Ondiba, J. O., C. L. Kanali, B. B. Gathitu and S. N. Ondimu. 2022. Characterisation and quantification of bioprocessing effluents from coffee, dairy and tanneryplants. Int. J. Agric. Environ. Res. 08(02): 265–277.

Onet, C. 2010. Characteristics of the untreated wastewater produced by food industry. Analele Universității Din Oradea, Fascicula: Protecția Mediului, XV: 709–714.

Panigrahi, T. and A. U. Santhoskumar. 2020. Adsorption process for reducing heavy metals in Textile Industrial Effluent with low cost adsorbents. Prog. Chem. Biochem. Res. 3(2): 135–139.

Paraíba, L. C. 2007. Pesticide bioconcentration modelling for fruit trees. Chemosphere. 66(8): 1468–1475.

Pascale, N. C., J. J. Chastinet, D. M. Bila, G. L. Sant'Anna, S. L. Quitério and S. M. R. Vendramel. 2019. Enzymatic hydrolysis of floatable fatty wastes from dairy and meat food-processing industries and further anaerobic digestion. Water Sci. Technol. 79(5): 985–992.

Pizzichini, M., C. Russo and C. Di Meo. 2005. Purification of pulp and paper wastewater, with membrane technology, for water reuse in a closed loop. Desalination 178(1-3): 351–359.

Porwal, R. K. 2015. Cost control opportunities in dairy industry. Indian Dairyman. 67(2): 82–87.

Pramanik, K., S. Mandal, S. Banerjee, A. Ghosh, T. K. Maiti and N. C. Mandal. 2021. Unraveling the heavy metal resistance and biocontrol potential of *Pseudomonas* sp. K32 strain facilitating rice seedling growth under Cd stress. Chemosphere. 274: 129819.

Prasongsuk, S., P. Lotrakul, T. Imai and H. Punnapayak. 2009. Decolourization of pulp mill wastewater using thermotolerant white rot fungi. Sci. Asia. 35: 37–41.

Qin, H., M. Diao, Z. Zhang, P. M. Visser, Y. Zhang, Y. Wang and S. Yan. 2020. Responses of phytoremediation in urban wastewater with water hyacinths to extreme precipitation. J. Environ. Manag. 271: 110948.

Rai, P. K. and B. D. Tripathi. 2007. Heavy metals removal using nuisance blue green alga Microcystis in continuous culture experiment. Environ. Sci. 4(1): 53–59. doi:10.1080/15693430601164956.

Rai, P. K. and B. D. Tripathi. 2008. Heavy metals in industrial wastewater, soil and vegetables in Lohta village, India. Toxicol. Environ. Chem. 90(2): 247–257.doi:10.1080/02772240701458584.

Rai, P. K. 2009. Heavy metals in water, sediments and wetland plants in an aquatic ecosystem of tropical industrial region, India. Environ. Monit. Assess. 158(1): 433–457.

Rai, U. N., R. D. Tripathi, P. Vajpayee, V. Jha and M. B. Ali. 2002. Bioaccumulation of toxic metals (Cr, Cd, Pb and Cu) by seeds of *Euryale ferox Salisb* (Makhana). Chemosphere. 46(2): 267–272.

Reddy, G. B. and K. R. Reddy. 1987. Nitrogen transformations in ponds receiving polluted water from nonpoint sources. Am. Soc. Agron. Crop Sci. Soc. Am., and Soil Sci. Soc. Am. 16:(1):1–5.

Reddy, K. R. and J. C. Tucker. 1983. Productivity and nutrient uptake of water hyacinth, *Eichhornia crassipes* I. Effect of nitrogen source. Econ. Bot. 37(2): 237–247.

Rezania, S., M. Ponraj, A. Talaiekhozani, S. E. Mohamad, M. F. M. Din et al. 2015. Perspectives of phytoremediation using water hyacinth for removal of heavy metals, organic and inorganic pollutants in wastewater. J. Environ. Manag. 163: 125–133.

Rodrigues, A. C., M. Boroski, N. S. Shimada, J. C. Garcia, J. Nozaki and N. Hioka. 2008. Treatment of paper pulp and paper mill wastewater by coagulation-flocculation followed by heterogeneous photocatalysis. J. Photochem. Photobiol. A: Chem. 194(1): 1–10.

Roufou, S., S. Griffin, L. Katsini, M. Polańska, J. F. V. Impe and V. P. Valdramidis. 2021. The (potential) impact of seasonality and climate change on the physicochemical and microbial properties of dairy waste and its management. Trends Food Sci. Technol. 116: 1–10.

Roy, R., A. N. M. Fakhruddin, R. Khatun and M. S. Islam. 2010. Reduction of COD and pH of textile industrial effluents by aquatic macrophytes and algae. J. Bangl. Acad. Sci. 34 (1): 9–14.

Roy, S. and O. Hänninen. 1994. Pentachlorophenol: uptake/elimination kinetics and metabolism in an aquatic plant, *Eichhornia crassipes*. Environ. Toxicol. Chem. 13(5): 763–773.

Saadia, A. and A. Ashfaq. 2010. Environmental management in pulp and paper industry. J. Industrial Control Pollution. 26(1).

Safauldeen, S. H., H. A. Hasan and S. R. S. Abdullah. 2019. Phytoremediation efficiency of water hyacinth for batik textile effluent treatment. J. Ecol. Eng. 20(9): 177–187.

Salt, D. E., M. Blaylock, N. P. Kumar, V. Dushenkov, B. D. Ensley, I. Chet and I. Raskin. 1995. Phytoremediation: a novel strategy for the removal of toxic metals from the environment using plants. Biotechnol. 13(5): 468–474.

Samecka-Cymerman, A. and A. J. Kempers. 2007. Heavy metals in aquatic macrophytes from two small rivers polluted by urban, agricultural and textile industry sewages SW Poland. Arch. Environ. Contam. Toxicol. 53(2): 198–206.

Sarkar, M., A. K. M. L. Rahman and N. C. Bhoumik. 2017. Remediation of chromium and copper on water hyacinth (*E. crassipes*) shoot powder. Water Resour. Ind. 17: 1–6.

Schnoor, J. L., L. A. Light, S. C. McCutcheon, N. L. Wolfe and L. H. Carreia. 1995. Phytoremediation of organic and nutrient contaminants. Environ. Sci. Technol. 29(7): 318A–323A.

Sevimli, M. F. 2005. Post-treatment of pulp and paper industry wastewater by advanced oxidation processes. Ozone: Sci. Eng. 27(1): 37–43.

Sharma, P., S. Tripathi and R. Chandra. 2020. Phytoremediation potential of heavy metal accumulator plants for waste management in the pulp and paper industry. Heliyon. 6(7): e04559.

Sharma, P., S. Tripathi and R. Chandra. 2021. Highly efficient phytoremediation potential of metal and metalloids from the pulp paper industry waste employing *Eclipta alba* (L.) and *Alternanthera philoxeroide* (L.): Biosorption and pollution reduction. Bioresour. Technol. 319: 124147.

Show, B. K., S. Balachandran, A. Banerjee, R. GhoshThakur, A. K. Hazra, N. C. Mandal, A. B. Ross, B. Srinivasan and S. Chaudhury. 2022. Insect gut bacteria: a promising tool for enhanced biogas production. Rev. Environ. Sci. Biotechnol. 21: 1–25.

Singh, M., M. Pant, S. Diwan and V. Snasel. 2022. Genetic algorithm-enhanced rank aggregation model to measure the performance of Pulp and Paper Industries. Comput. Ind. Eng. 172: 108548.

Singh, R., S. Singh, P. Parihar, V. P. Singh and S. M. Prasad. 2015. Arsenic contamination, consequences and remediation techniques: a review. Ecotoxicol. Environ. Saf. 112: 247–270.

Sinha, D., S. Banerjee, S. Mandal, A. Basu, A. Banerjee, S. Balachandran and S. Chaudhury. 2021. Enhanced biogas production from *Lantana camara* via bioaugmentation of cellulolytic bacteria. Bioresour. Technol. 340: 125652.

Smolyakov, B. S. 2012. Uptake of Zn, Cu, Pb, and Cd by water hyacinth in the initial stage of water system remediation. Appl. Geochem. 27(6): 1214–1219.

Sthiannopkao, S. and S. Sreesai. 2009. Utilization of pulp and paper industrial wastes to remove heavy metals from metal finishing wastewater. J. Environ. Manag. 90(11): 3283–3289.

Sungur, Ş. and F. Gülmez. 2015. Determination of metal contents of various fibers used in textile industry by MP-AES. J. Spectrosc.

Suthar, S., P. Sajwan and K. Kumar. 2014. Vermiremediation of heavy metals in wastewater sludge from paper and pulp industry using earthworm *Eisenia fetida*. Ecotoxicol. Environ. Saf. 109: 177–184.

Tait, S., P. W. Harris and B. K. McCabe. 2021. Biogas recovery by anaerobic digestion of Australian agro-industry waste: A review. J. Clean. Prod. 299: 126876.

Tarlan, E., F. B. Dilek and U. Yetis. 2002. Effectiveness of algae in the treatment of a wood-based pulp and paper industry wastewater. Bioresour. Technol. 84(1): 1–5.

The Environment (Protection) Rules. 1986. Ministry of Environment and Forests, Department of Environment, Forest and Wildlife, New Delhi.

The World Bank Group. 1999. Pollution prevention and abatement handbook, 1998: toward cleaner. The International Bank for Reconstruction and Development (Washington D.C., United states).

Toczyłowska-Mamińska, R. 2017. Limits and perspectives of pulp and paper industry wastewater treatment–A review. Renew. Sustain. Energy Rev. 78: 764–772.

Trivedy, R. K. and S. M. Pattanshetty. 2002. Treatment of dairy waste by using water hyacinth. Water Sci. Technol. 45(12): 329–334.

Türksoy, R., G. Terzioğlu, İ. E. Yalçin, Ö. Türksoy and G. Demir. 2021. Removal of heavy metals from textile industry wastewater. Front. Life Sci. RT. 2(2): 44–50.

U.S. Environmental Protection Agency (EPA). 2008. EPA's 2008 Report on the Environment. National Center for Environmental Assessment, Washington, DC; EPA/600/R-07/045F.

Uddin, F. 2021. Environmental hazard in textile dyeing wastewater from local textile industry. Cellulose. 28(17): 10715–10739.

Ugya, A. Y., T. S. Imam and A. S. Hassan. 2017. Phytoremediation of textile waste water using *Azolla pinnata*; a case study. World J. Pharm. Res. 6(2).

United States Protection Agency (USPA). 2000. Introduction to Phytoremediation. EPA 600/R-99/107. U.S. Environmental Protection Agency, Office of Research and Development, Cincinnati, OH.

Verma, A. and A. Singh. 2017. Physico-chemical analysis of dairy industrial effluent. Int. J. Curr. Microbiol. Appl. Sci. 6: 1769–1775.

Vinodhini, M. and C. Soundhari. 2019. Phycoremediation of dairy effluent by using microalgal consortium. J. Pharm. Innov. 8(9): 128–133.

Voudrias, E. A. and K. S. Assaf. 1996. Theoretical evaluation of dissolution and biochemical reduction of TNT for phytoremediation of contaminated sediments. J. Contam. Hydrol. 23(3): 245–261.

Wang, B., L. Gu and H. Ma. 2007. Electrochemical oxidation of pulp and paper making wastewater assisted by transition metal modified kaolin. J. Hazard. Mater. 143(1-2): 198–205.

Wild, E., J. Dent, G. O. Thomas and K. C. Jones. 2005. Direct observation of organic contaminant uptake, storage, and metabolism within plant roots. Environ. Sci. Technol. 39(10): 3695–3702.

Xia, H. L., L. H. Wu and Q. N. Tao. 2002. Phytoremediation of some pesticides by water hyacinth (*Eichhornia crassipes* Solms). J. Zhejiang Univ. - Agric. Life Sci. 28(2): 165–168.

Xia, H., L. Wu and Q. Tao. 2003. A review on phytoremediation of organic contaminants. Ying yong sheng tai xue bao. J. Appl. Ecol. 14(3): 457–460.

Yan, S. H. and J. Y. Guo. 2017. Water hyacinth: Environmental challenges, management and utilization. In Water Hyacinth: Environmental Challenges, Management and Utilization. CRC Press, 1–327.

Yaseen, D. A. and M. Scholz. 2019. Textile dye wastewater characteristics and constituents of synthetic effluents: a critical review. Int. J. Environ. Sci. Technol. 16(2): 1193–1226.

Zavoda, J., T. Cutright, J. Szpak and E. Fallon. 2001. Uptake, selectivity, and inhibition of hydroponic treatment of contaminants. J. Environ. Eng. 127(6): 502–508.

Zhang, Z., Z. Rengel, K. Meney, L. Pantelic and R. Tomanovic. 2011. Polynuclear aromatic hydrocarbons (PAHs) mediate cadmium toxicity to an emergent wetland species. J. Hazard. Mater. 189(1-2): 119–126.

Zhao, X., L. Zhang and D. Liu. 2012. Biomass recalcitrance. Part II: Fundamentals of different pre-treatments to increase the enzymatic digestibility of lignocellulose. Biofuel Bioprod. Biorefin. 6(5): 561–579.

Zhou, W., L. Tan, D. Liu, H. Yan, M. Zhao and D. Zhu. 2005. Research advances of *Eichhornia crassipes* and it's utilization. J. Anhui Agric. Univ. 24(4): 423–428.

CHAPTER 15

Biohybrids for Environmental Remediation and Biosensing
Concept, Synthesis and Future Prospective

Archana Mishra,[1,2,]* *Ayushi Rastogi*[3] and
Avanish Singh Parmar[4]

15.1 Introduction

There is constant release of a wide range of pollutants (heavy metals, pesticides, etc.,) into water bodies which leads to pollution of water resources. Water pollution is a serious concern as it causes scarcity of drinking water worldwide. Organic as well as metal pollutants present in water bodies are harmful to all living systems and environment. Several processes such as adsorption (Douglas et al. 2016), biodegradation (Li et al. 2016), coagulation (Zhu et al. 2016) and photocatalytic oxidation (Chong et al. 2010) have been used for the removal of pollutants and efficient wastewater treatment, however; these have their own limitations, like generation of waste products, poor removal capacity, high energy demand and high cost.

Microorganisms are well known for wastewater treatment as these degrade a variety of substrates for their consumption using their metabolic diversity. A wide range of microorganism like *Aspergillus niger* (Vassilev et al. 1997), *Pseudomonas aeruginosa* (Shukla et al. 2014) and *Rhodopseudomonas sphaeroides* (Liu et al. 2015) has been used for wastewater treatment. However, limitations like slow biodegradation processes, difficulty in recovering cells, sensitivity towards the surrounding environment and poor activities of recovered cells persist. There is an urgent need to find economical cost-effective solutions for remediation of these pollutants. The combination of the microorganism with suitable support could be a low-cost, environmentally benign and efficient technique (Mishra et al. 2014, Oh et al. 2016) to achieve the desired outcomes is needed.

[1] Nuclear Agriculture and Biotechnology Division, Bhabha Atomic Research Centre, Trombay, Mumbai-400 085, India.
[2] Homi Bhabha National Institute, Anushakti Nagar, Mumbai-400 094, India.
[3] Department of Humanities and Applied Sciences, School of Management Science (SMS) Institute of Technology, College of Engineering, Lucknow – 226001, Uttar Pradesh, India.
[4] Department of Physics, Indian Institute of Technology (BHU) Varanasi,Uttar Pradesh, India.
* Corresponding author: archanam@barc.gov.in, archanamishra56@gmail.com

Biohybrid materials are composed of living cells–non-living materials or organic-inorganic or biomolecules-organic/inorganic materials. Most common biomolecules are associated/conjugated with inorganic materials to develop biohybrids which provide materials with improved features. The presence of biomolecules in biohybrid materials contributes to biomimetic properties, outstanding biocompatibility, improved functionality, etc. Due to the exceptional characteristics biohybrids are applied and hold promising applications in every field of research and development.

In this chapter, biohybrids, their two (biological and support) components, synthesis processes and their superiority over conventional materials are described. The morphology of a biohybrid material plays a very important role and contributes significantly to achieve the best suited application of developed biohybrid material. Therefore, various morphologies of biohybrid materials and their applications are discussed. The applications of biohybrids with emphasis on the bioremediation and monitoring of heavy metals and pesticides are also described. The objective of this chapter is to strengthen the interdisciplinary research between biosciences and material sciences with reasonable solutions for sustainable environment.

15.2 Biohybrids

In the field of material science, rapid development has been made in the past few decades wherein biological sciences have contributed significantly. Biohybrid materials have played an important role. A biohybrid material is composed of two components. One is the biologically active component which includes microorganisms, living cells, enzymes, etc., and the other supports (organic/inorganic nanomaterials) (Figure 15.1). Biomolecules provide a process such as their functions of synthesis, sensing, secretion, etc. However, support materials enable protection and stability to the biomolecules (Ouyang et al. 2020). As advantages of biomolecules and support materials are combined, biohybrids exhibit improved characteristics over conventional materials and offer dual functionality.

For the synthesis of biohybrid materials, many biomolecules (Mishra et al. 2017, Mishra et al. 2020a, Shukla et al. 2020, Ouyang et al. 2020) and as support material various polymers, nanoparticles, etc., have been used. A support material is present as a coating over the surface of biomolecules and gives protection against a harsh microenvironment. In order to develop biohybrids different techniques like sol-gel technique, moulding, electrospinning, spray drying, microfluidics, 3D printing have been widely applied. Using these methods, biohybrids with various morphologies

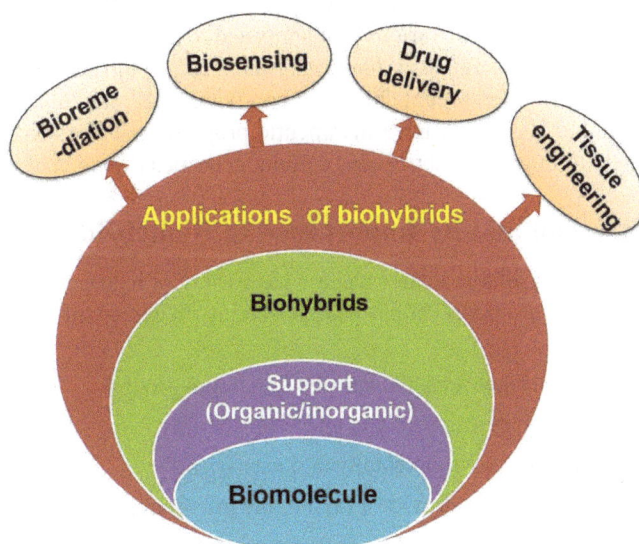

Figure 15.1. Biohybrid materials, its components and potential field of applications.

like microparticles, fibres, sheets and scaffolds can be synthesized. Biomolecules present in biohybrid materials could perform required functions and enhance the applicability of biohybrids for drug delivery, biomedical engineering, biosensing, wearable devices, etc. (Mishra et al. 2017, Mishra et al. 2021a, Rivera-Tarazona et al. 2021, Wang et al. 2022). Several reviews are available on biohybrid materials (Nguyen et al. 2018, Mishra et al. 2020b, Wang et al. 2022).

15.3 Biocomponent and its Importance in Biohybrid Material

Living cells are valuable assets. A microorganism acts as a large amount full of enzymes thus the microorganism present in the biohybrid can be used for various bioprocesses as well as for environmental monitoring. Although, living cells are very useful, however they are sensitive, vulnerable to the external environment and can be applied in relatively mild conditions. Thus, to improve its applicability it is proposed to develop a protective covering using well suited materials. Support with tuneable physicochemical properties could also be best suited as a protective covering which could improve bio-interface characteristics and cellular functions. *Sphingomonas* sp. bacterium has an organophosphorus hydrolase (OPH) enzyme which hydrolyzes methyl parathion pesticide into a coloured product. However, poor storage stability of periplasmic enzyme was a matter of concern so, *Sphingomonas* sp. cells have been associated with functionalized silica nanoparticles to develop a biohybrid and a developed biohybrid was immobilized on 96 well microplates (Mishra et al. 2017). A developed system was explored as biosensor for detecting methyl parathion pesticide. This study showed that the association of microorganisms with inorganic materials is beneficial for both the components. In a biohybrid material the limitation associated with one component can be overcome by the presence of another component. Thus, a biohybrid material offers superior features of both the components, wherein both components have their own importance.

Many living cells and microorganisms have been applied as biocomponents to develop biohybrid materials. In one study, *S. lactis* cells were combined with silica nanoparticles using an evaporation induced self-assembly process, and a biohybrid of silica NP-*S. lactis* cells were synthesized which was applied for removing uranium (Mishra et al. 2014). On a similar line of studies, *S. cerevisiae* cells were connected with silica nanoparticles and a prepared biohybrid was used for the removal of mercury (Shukla et al. 2020).

Microorganisms show distinctive features in the presence of various stimuli like temperature, light and presence of oxygen (Rivera-Tarazona et al. 2021) therefore, these features can be exploited to develop biohybrid-based robotics (Nguyen et al. 2018). Biohybrid-based robotics can be efficiently used for the detection of various pollutants and to overcome the limitations associated with conventional robotics. Microorganisms like *Escherichia coli*, could also be genetically modified to perform specific functions (Gerber et al. 2012). Even, living cells (natural killer (NK) cells, stem cells, islet cells and probiotics) which have therapeutic properties can be combined with suitable support as this will help to improve viability and cellular activities (Wang et al. 2022).

15.4 Support Material and its Importance in Biohybrid

A support is usually a solid insoluble matrix and a wide variety of matrices have been applied as a support. As reported earlier, a support should have following characteristics (Figure 15.2) (Mishra 2019):

  i. It should offer a higher surface area and various functional groups.
 ii. It should demonstrate high mechanical and chemical stability.
iii. It should be biocompatible, environmentally and user friendly.
 iv. It should show high affinity and provide a favourable microenvironment to target a molecule.
  v. It should be easily and widely available. It can also be obtained from a reliable commercial source.

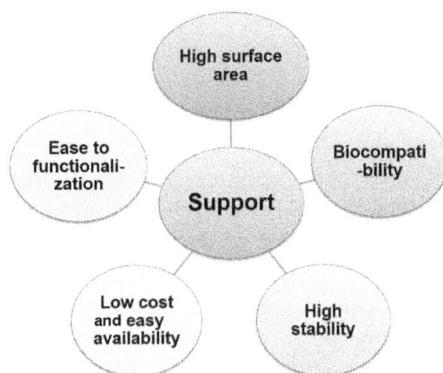

Figure 15.2. The main characteristics of support materials applied in biohybrids.

The association of living cells with support will certainly improve the application of both the components. Living cells offer functionality, however they lack stability. On the other hand, a support has high mechanical and chemical stability, while it is deficient in functionality. A biohybrid wherein both (biomolecule and support) are present, exhibit a dual and improved property. The association of living cells with a support can be achieved using various methods such as cells can be used as a template and support can be assembled on it or a biomolecule can be entrapped in support. The method was selected in such a way that it would impart a positive effect on the applicability of biomolecule (Zhu et al. 2019). A biosensor was developed using *Sphingomonas* sp. cells. The microbial cells have periplasmic enzyme which hydrolyzes methyl parathion pesticide, however the poor storage stability of enzyme was of concern. Thus, functionalized silica nanoparticles were assembled on the surface of cells and it was observed that the storage stability improved significantly (Mishra et al. 2017). In order to address stress resistance of living cells, inorganic components with tuneable properties and high stability are combined , which provides protection against stresses. A biocompatible and porous support is preferred which helps in transport of nutrients and required gases (Gerber et al. 2012). The properties of biocomponents in a biohybrid are regulated by the properties of both the biomolecule and the support. The interactions between both the components govern the biochemical and kinetic properties of biomolecule. Support materials could be classified into two major groups, i.e., organic support and inorganic support.

15.4.1 Organic Supports

Organic support materials can be further classified into two main groups: (i) synthetic polymers, and (ii) biopolymers.

15.4.1.1 Synthetic Polymers

Synthetic polymers are preferred as a support material because the monomers that help in building the polymer can be chosen as per the requirement (Zdarta et al. 2018) and the properties of the synthetic polymer are deciphered by the monomers. Various functional groups (carboxyl, hydroxyl, amine, carbonyl, epoxy and diol groups, hydrophobic alkyl groups, etc.) are present in synthetic polymers. The presence of functional groups determines the characteristics of the support and offer binding sites for biomolecules (Basso et al. 2007).

Earlier various enzymes have been associated with different synthetic polymers. *a*-amylase, tyrosinase, lipase and glucose oxidase were immobilized on polyaniline, polyamide 66 (Nylon 66), polystyrene microspheres, and amino- and carboxyl-plasma-activated polypropylene film, respectively (Jimenez Hamann and Saville 1996, Vartiainen et al. 2005, Ashly et al. 2011, Wang et al. 2015). Synthetic polymers are also commercially available and utilized for the immobilization of enzyme. Synthetic polymers have a lot of applications however, the synthesis of a synthetic polymer with the required characteristics and desired functional groups is a costly and time-consuming method which also generates chemical wastes.

15.4.1.2 Biopolymers

In comparison to synthetic polymers, biopolymers are naturally available biocompatible supports. Many biopolymers like alginate, chitosan, cellulose, and pectin, etc., are commercially available and used as supports. Biopolymers possess characters like biocompatibility, availability of various functional groups, biodegradability to harmless products, non-toxicity and affinity towards proteins. These characteristics increase the applicability of biopolymers as support for biomolecules and provide a favourable microenvironment to the attached biomolecule which helps to maintain/ improve biocatalytic aspect of biomolecule.

15.4.1.2a. Alginate: Alginate is a biopolymer which is widely applied for developing supports of various morphologies and immobilization of biomolecules. Earlier lipase was immobilized in calcium alginate beads and showed high immobilization yields (Betigeri and Neau 2002), however reusability was poor due to the leaching of the enzyme from the support. The application of alginate is limited due to its sensitivity towards divalent ions and low mechanical stability of alginate gels and substrate diffusional limitations (Coradin et al. 2003). To address the issue, a support comprising nanosilica-sodium alginate was prepared and chemotherapeutic drug doxorubicin was entrapped (Mishra et al. 2021a). It was observed that the presence of nanosilica improved the loading efficiency as well as slowed down the release of entrapped drug.

15.4.1.2b. Chitosan: Chitosan is a mainly used biopolymer as a support because of the presence of amino groups on its surface. Earlier chitosan microspheres were used for the immobilization of nuclease (Shi et al. 2011). Glucose isomerase enzyme was also immobilized on chitosan beads (Cahyaningrum et al. 2014). It was observed earlier that support-comprising chitosan in combination with alginate showed a lesser leaching effect compared to alginate alone (Betigeri and Neau 2002).

15.4.1.2c. Cellulose and Pectin: Cellulose is an abundantly available natural polymer and it has been extensively used for the immobilization of various enzymes such as, penicillin G acylase, α-amylase and tyrosinase (Mislovicova et al. 2004, Namdeo and Bajpai 2009, Labus et al. 2011). Pectin in combination with chitin and calcium alginate have improved the thermal, denaturant resistance and biocatalytic properties of entrapped enzymes due to formation of highly stable polyelectrolyte complexes (Gomez et al. 2006, Satar et al. 2008).

15.4.2 Inorganic Supports

Inorganic supports offer high thermal and mechanical stability, and are resistant towards microbial attack. These are therefore considered a suitable support for immobilization of biomolecules. In particular these are important characteristics as inorganic supports have superiority over organic supports due to their rigidity and porosity. The stiffness of the inorganic supports takes care of the invariance of pore diameter/pore volume, which further helps to maintain constant volume and shape to the support itself. It has been observed that inorganic oxides are of more importance and have wider applicability among other inorganic supports.

15.4.2.1 Inorganic Oxides as Support

Various enzymes (lipase, urease and α-amylase) have been associated with different inorganic oxides like aluminium, titanium and zirconium oxides (Yang et al. 2008, Foresti et al. 2010, Reshmi et al. 2007). It has been reported that inorganic oxides offer high mechanical resistance, stability and good sorption capacity. These are also inert in different conditions thus could be used as support for synthesizing biohybrids. Among all, silica-based materials are most preferred support due to their biocompatibility, high surface area, porous structure, wide availability, ease to functionalization, high thermal and chemical resistance and good mechanical properties. These characteristics enhance their applicability for various practical applications. Availability of silanol groups on silica's surface helps in attachment of biomolecules to develop biohybrids and it could be also utilized for functionalization using chemical agents (Zucca and Sanjust 2014).

15.4.2.2 Silica-based Supports

Silicates are the most common component on the surface of Earth. Silica is one form of the silicate (general chemical formula SiO_2 or $SiO_2.xH_2O$) which exists in two main forms crystalline and amorphous (Zucca and Sanjust 2014). Silica surfaces appear from the dehydration of hydrated silica preparations ("silicic acids") or after grinding bulky silica. Usually, two main functional groups silanols (Si–OH) and siloxanes (Si–O–Si) are available on the surface of silica. Silanols (Si–OH) are acidic and impart a negative charge to the silica surface in a wide range of pH (Davydov 2000). Due to the presence of siloxane motifs, some silica show a hydrophobic character (Rimola et al. 2013). As a result of availability of both hydrophilic and hydrophobic sites on surface and the characteristics such as weakly acidic and high tendency to take part in hydrogen bonding, silica shows high adsorption character. The property of adsorption plays a very important role as support.

15.4.2.3 Silica Nanoparticles

Silica has been extensively used throughout history because of its accessibility and ease of recovery. In recent years, colloidal silica has come up as an ideal support in the field of material science as nano- and microparticle (Bergna and Roberts 2006). The reason behind wide applicability of colloidal silica in research is the unique properties which they possess like high colloidal stability, optical transparency, chemical and thermal stability low toxicity, high surface area, biocompatibility and well-known surface chemistry (Hyde et al. 2016). Colloidal silica has a wide range of applications like catalytic supports (Chen et al. 2011, Xie et al. 2015), biosensor supports (Zhao et al. 2013, Wang et al. 2014), drug carriers (Tang et al. 2012, Rajanna et al. 2015), antifouling coatings (Zhu et al. 2014) and additives to paints/lacquers/coatings (Puig et al. 2014). As a result of many applications, colloidal silica production has become a large industry. Commercially available colloidal silica is preferred over naturally available mineral silica. As, natural silica is contaminated with various metal ions, it offers less surface area and is available in a crystalline form which is not suitable for health applications (Rahman and Padavettan 2012). Therefore chemically synthesized colloidal silica is a preferred choice for various applications.

Silica nanoparticles could be of following different types:

- Fumed (pyrogenic) silica
- Precipitated silica
- Silica gel
- Mesoporous silica nanoparticles
- Biogenic silica

Fumed silica has uses as an additive (Boldridge 2010) and used for the production of silicone rubber for medical applications (Taikum et al. 2010). However, due to its high cost and a smaller number of silanol groups, application of fumed silica is limited in research and development (Prasertsri and Rattanasom 2012). In comparison to fumed silica, precipitated silica is widely used for various applications (Kim et al. 2004, Sun et al. 2013, Pattanawanidchai et al. 2014). Silica gel offers high specific surface area, used as a desiccant (drying agent), food additive (approved by the FDA) and as a humidity indicator. Mesoporous silica is another form of silica which has gained lot of interest in nanotechnology. A wide range of mesoporous silica is available while MCM-41 and SBA-15 are the most used mesoporous silica (Katiyar et al. 2006). These have applications in the field of medicine, biosensor, imaging, etc. (Valenti et al. 2016). Biogenic silica is synthesized through the process of biosilication, wherein biomolecules are involved in the synthesis of silica. Various specialized proteins (silaffins, silacidins and cinguliums) are also known to be responsible for the formation of biogenic silica. Silaffins are polypeptides rich in lysine and serine residues with a high degree of post-translation modification (Kroger 2007). Sponges are the another major biosilica-producing organism and silicatein isoforms have been identified in siliceous sponge species which play role in synthesis of biosilica (Mishra 2019). Colloidal silica nanoparticles are commercially available and are often used for research and development purposes. One of the widely used commercially available colloidal silica is LUDOX® colloidal silica which is aqueous dispersions of silica nanoparticles in the nanometre size range. In India, various grades of silica nanoparticle in the nm size range are commercially supplied by Visa chemical industries.

1.5.5 Ways/Methods to Synthesize Biohybrids

15.5.1 Synthesis of Hybrids/Bio-hybrids using Sol-gel Method

There are other methods also available for synthesis of hybrids/bio-hybrids however, sol-gel method has been widely used (Mishra 2019, Tian et al. 1997). Due to its mild reaction conditions and compatibility with a wide variety of solvents, it could efficiently combine silica (inorganic oxide) with an organic phase/bio-component. The development of several novel polymeric-silica hybrid materials with a good performance has been reported using sol-gel process (Yang et al. 2004). There are also many reports wherein biological components, i.e., enzymes and microbes have been associated with silica through sol-gel process (Avnir et al. 1994, Kim et al. 2000, Premkumar et al. 2001).

Sol-gel method: The sol-gel process is widely used to synthesize silica, glass and ceramic materials due to its ability to form pure and homogenous products. The role of the sol-gel method has been studied since the mid 1800's. The sol-gel technique is the method in which liquid (colloidal) sol phase is transformed into a solid gel phase. This involves hydrolysis and condensation of metal alkoxides (Si $(OR)_4$) like tetraethylorthosilicate (TEOS, Si $(OC_2H_5)_4$) or inorganic salts like sodium silicate (Na_2SiO_3) in presence of acid (HCl) or base (NH_3) which act as catalyst (Nandiyanto et al. 2009). This step causes the formation of a colloidal phase which is comprised of relatively higher molecular weight intermediates and it is known as the sol phase. Intermediates formed during the reaction undergo polycondensation reactions and escort to the formation of three-dimensional (3-D) gel. Figure 15.3 shows the flow chart for the sol-gel technique.

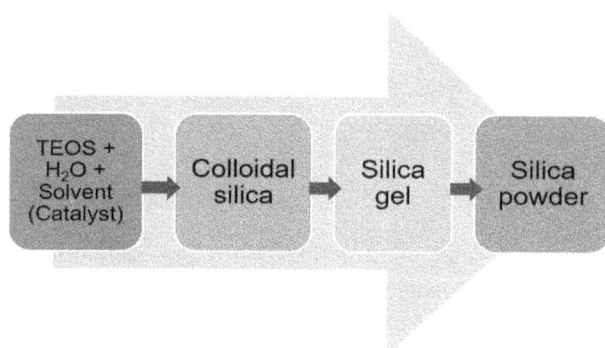

Figure 15.3. Flow chart of sol-gel process.

The following processes are involved during the sol-gel process:
  i. Hydrolysis of alkoxisilane to silanol
 ii. Alcohol condensation: reaction between silanol and alkoxisilane
iii. Water condensation: reaction between silanols

In 1968, a pioneering work was reported by Stober et al. (1968) for controlled synthesis of monodispersed silica and spherical particles. This method offers advantages over acid-catalyzed systems as using this monodispersed spherical silica particles can be prepared. However, in the sol-gel process, production of alcohol as a by-product is one of the limitations when biological components are going to be immobilized. Thus, there is a need to find suitable methods for associating bio-components with silica nanoparticles to develop biohybrids and to improve the practical applicability of associated bio-components.

15.5.2 Synthesis of Biohybrids using Spray drying Technique

Spray drying is a widely used technique for the preparation of powdered formulation using aqueous feed samples by atomization in hot drying air. It has emerged as one of the suitable techniques for synthesizing biohybrids of various morphologies like microspheres and microcapsules and holds application in food, remediation and drug delivery. Various aspects of spray drying have been explored and used as immobilizing techniques for the immobilization of proteins, enzymes and microbial whole cells (Mishra et al. 2014, Mishra et al. 2017, Mishra et al. 2021a, Mukundan et al. 2020, Shukla et al. 2020). The characteristics of the spray-dried product is regulated by various factors such as spray dryer design, feed characteristics, drying temperature and processing parameters. By using spray drying, it is even possible to prepare the product of desired characteristics by adjusting the spray drying parameters.

15.6 Different Morphologies of Biohybrids

A wide range of biohybrid materials can be synthesized depending on the applications. It could be in particles form, sheet, scaffold, fibrous, etc.

15.6.1 Biohybrid in Microparticles form

Biohybrids synthesized using spray drying techniques are mostly in the form of microparticles. Biohybrids in a microparticle shape find applications in remediation and biosensing. There are also reports when single cells were encapsulated to develop biohybrids as encapsulated single-cell offers a number of advantages (Mao et al. 2017, Kamperman et al. 2017, Lienemann et al. 2017). Different strategies are employed for encapsulation of single cells. Enzymatically crosslinked microgels were used for encapsulation of single cells (Kamperman et al. 2017). Microfluidic droplets have been widely explored to synthesize biohybrid microparticles. It also helps to retain the viability of cells (Choi et al. 2016). The flow-focusing device in combination with covalent crosslinking process was used by Headen et al. (2014) to encapsulate the cells in droplets. Water-in-oil-in-water (w/o/w) emulsion droplets with an ultrathin oil layer were also reported for immobilization of cells in the gel matrix which leads to the development of biohybrid microparticles.

15.6.2 Biohybrids in Fibrous form

Fibre-like structures have their own advantages over microparticles as these offer flexible structure, large aspect ratio and could be explored for developing complex large scale biohybrids for various applications in the biomedical engineering as these resemble nerves, blood vessels and muscle tissues. Several fibre-shaped biohybrids have been prepared (Letnik et al. 2015, Neal et al. 2014, Onoe et al. 2013). Using a sacrificial outer moulding approach, an elongated fascicle-inspired biohybrid material was prepared by Neal et al. 2014 wherein polydimethylsiloxane (PDMS) was used for moulding. A fiber-like cavity tube was constructed using a cylindrical steel pin and gelatin. Cells were seeded in the cavity tube and an increase in the temperature caused the gelatin to melt followed by solidification of cells and led to the formation of fibre-like structures.

Microfluidic spinning was also explored by Onoe et al. 2013 to develop fibre-shaped biohybrids. Using microfluidic spinning various types of cells have been applied to develop biohybrids.

Coaxial electrospinning was also applied to develop biohybrid fibres with a core-shell structure (Letnik et al. 2015). This has a water-soluble polymer containing aqueous core which maintains the survival and proliferation of yeast cells and an insoluble polymer shell to provide physical stability.

15.6.3 Biohybrids in Sheet form

Sheet shape biohybrids have potential application to be used as building blocks to fabricate living tissue models which could be used for studying the cell-cell/cell-matrix interactions, cell culture and cell therapy (Leng et al. 2012, Yang et al. 2015). These sheets could be constructed using hydrogel matrix where living cells are embedded. A mesh-like frame was used to construct cell-containing sheets by Son et al. (2016)

Microfluidics and microfabrication techniques were also used for a developing sheet-shaped biohybrid. A multilayer microfluidic system was explored to construct mosaic biohybrid sheets (Leng et al. 2012). It was observed that developed biohybrid hydrogel sheets have uniformly distributed cardiomyocytes. It is also possible to fabricate biohybrid sheets using living cells labelled with magnetic nanoparticles. This technique holds potential as it could be applied to understand the cell-cell interactions and for tissue regeneration (Ana et al. 2020).

15.6.4 Biohybrids in Scaffolds form

Scaffolds-shaped biohybrids have wide application in regenerative medicine and organ reconstruction (Capulli et al. 2017). These scaffolds are gaining attention for remediation as well as detection of various water pollutants. Scaffolds-shaped biohybrids could be prepared using various techniques (Ouyang et al. 2017, Jin et al. 2018, Compaan et al. 2020). 3D bioprinting has been also used for constructing the 3D scaffolds shaped biohybrids as it has its own advantages in creating complex structures. An "*in-situ*-crosslinking" strategy for 3D bioprinting was developed by (Ouyang et al. 2017). Here ink was used for encapsulation of fibroblasts cells and the cells present in lattice- and macro nose-structured constructs could maintain high viability even after printing. A biohybrid scaffolds with a similar multi-channel lattice structure was fabricated by Wang et al. (2021). Microorganisms have been also explored by Huang et al. (2019) to construct living material based on biofilms of *Bacillus subtilis* and resulted in biofilms demonstrating viscoelastic behaviours which could be engineered to fabricate complex structures.

15.7 Applications of Biohybrids

In comparison to conventional materials, biohybrid materials offer various advantages. Due to the presence of the biomolecule, the functionality of the biohybrid is improved and it could perform various biological processes when support provides protection against harsh conditions. These materials hold potential of application in every field of research including gene/drug delivery, biosensing, bioremediation, tissue engineering, etc. Many review articles are available on applications of biohybrids in biomedical applications, tissue engineering, biosensor, etc., where the focus is on the application of biohybrids in remediation and removal of pollutants (Xu et al. 2018, Jamal et al. 2013, Pu et al. 2020, Sankaran et al. 2019).

15.7.1 Pesticide Monitoring and Removal

Biohybrid could be applied for sensing even ultra-trace amounts of various analytes (pesticides, herbicides, etc.,) due to the presence of a biomolecule. These biomolecules have high specificity and selectivity for the analytes thus using biohybrids, it is possible to develop sensitive and stable biosensors. A nanogenerator was fabricated using *Bacillus subtilis* spores to monitor humidity (Chen et al. 2014). A *Chlorella*-based biosensor was constructed to detect metal irons in water (Roxby et al. 2020). Genetically engineered cells and microorganisms were associated with a suitable support to achieve efficient sensing of pollutant metals as well as pesticides. Liu et al. (2017) had shown that engineered living bacteria was associated with hydrogel-elastomer hybrid matrix and support has helped in growth of cells. A biohybrid comprising Silica-*Sphingomonas* sp. cells were immobilized and used as a biosensor for detection of methyl parathion pesticide (Mishra et al. 2017). This biosensor could detect methyl parathion in the linear range of 0.1 – 1 ppm concentration.

As reported, *Rhodobacter* sphaeroides reaction centre is a robust and tractable membrane protein which has potential for technological applications and lead to develop biohybrid devices for various applications like biosensing (Swainsbury et al. 2014). This study evaluated the viability of using a tiny working electrode attached to a reaction centre to create a photocurrent as the basis for a biosensor for classes of herbicides that are widely used to control weeds in important agricultural crops. The triazides atrazine and terbutryn suppressed photocurrent production in a concentration-dependent manner, but nitrile or phenylurea herbicides did not. The kinetics of charge recombination in photo-oxidized reaction centres in solution were measured, and the results revealed the same selectivity of response. The limits of detection were calculated to be about 8 nM and 50 nM, respectively, as indicated in the Table 15.1. Titrations of the reaction centre photocurrents produced half maximum inhibitory doses of 2.1 mM and 208 nM for atrazine and terbutryn, respectively. Without the need for model-dependent kinetic analysis of the signal used for detection or the use of unreasonably complex instrumentation, photocurrent attenuation provided a direct measure of herbicide concentration. It also offered the possibility of using protein-engineering to increase the selectivity and sensitivity of herbicide against the Rbasphaeroides reaction centre.

In the urban and rural regions of Campo Verde and Lucas do Rio Verde Cities, Mato Grosso State, Brazil, Table 15.1 offers information on the prevalence of the pesticides atrazine, chlorpyrifos, a-endosulfan, β-endosulfan, flutriafol, malathion and metolachlor in water bodies (Nogueira et al. 2012). In these significant grain-producing regions, samples of the surface, rain and groundwater were taken throughout the wet and dry seasons of 2007 and 2008. The results showed that rainwater had a greater variety of chemicals and a higher frequency of detection than surface and groundwater samples. Some surface and groundwater samples contained concentrations of atrazine, endosulfan and malathion that were higher than those prescribed by Brazilian regulations, and some samples contained a higher amount of degraded product of deisopropylatrazine and endosulfansulfate in comparison to their parent compounds. The results highlight the fragility of the local water infrastructure and the danger of pesticide contamination in significant headwater streams.

15.7.2 Heavy Metal Monitoring and Remediation

The scarcity of drinking water due to increase of pollutants in water bodies is a matter of concern (Mishra et al. 2021b, Mishra et al. 2022). Thus, remediation of pollutants from waste water is of great significance for environment and human health. Among various techniques, bioremediation and adsorption have wider applicability (Shukla et al. 2022). A large range of sorbents have been applied for the removal of heavy metals. Among all nanosize silica have been widely applied for sorption of radionuclide like uranium (Gabriel et al. 2001, Metilda et al. 2005). Nanoparticles have many advantages however, their application as a sorbent has certain limitations such as harmful impact on human health, the environmental fate of nanoparticles, and technical limitations like difficulty in separation.

Several microbial systems have been used for bioremediation of heavy metals and radionuclide like uranium (Sar et al. 2004, Kazy et al. 2009, Wang et al. 2010, Melo et al. 2020). However, poor mechanical strength, a small size with low density and less rigidity of microorganism biomass are the limiting factors (Wang et al. 2010). Association of biomass with a suitable support could be an efficient way for improving its application for bioremediation. With this motive, microorganisms and other biomolecules have been conjugated with suitable supports to develop biohybrids and used as sorbents.

There has been a need for suitable methods for the preparation of biohybrids in large quantities. Thus, the spray drying technique was applied to prepare a wide range of biohybrids using nanosilica and various microorganisms, in our laboratory. Various biohybrids in microstructures morphologies using silica nanoparticles and biomolecules through the process of evaporation induced self-assembly (Mishra et al. 2014, Mukundan et al. 2020). In one of the studies, microstructures of silica NPs and *Streptococcus lactis* (*S. lactis*) cells were prepared, utilized for removal of uranium and adsorbed uranium was recovered using ultra sonication (Mishra et al. 2014, Lahiri et al. 2021). It is possible to get different morphologies of microstructures by tuning the physico-chemical parameters of evaporation induced by self-assembly process. Synthesized products were characterized and it has shown that morphology depended on removal of uranium. Among spherical and doughnut-shaped microstructures, doughnut-shaped microstructures have shown higher uranium uptake. The adsorbed uranium was also recovered using ultrasonic treatment however, after treatment the binding capacity of microstructures were reduced during reusability studies. The plausible cause for reduced sorption may be changes in morphology of microstructures and decrease in number of adsorption sites after acid and ultrasonic treatment. On a similar line of studies, a bio-hybrid of silica NPs and *S. cerevisiae* cells were developed and employed for mercury (Hg^{2+}) removal. It has shown 98% mercury sorption with maximum sorption 185.19 mgg^{-1} (Shukla et al. 2020). It was observed that these microstructures could be used as a sorbent for remediation of mercury. In summary, spray drying has emerged as a simple, efficient and cost-effective process for synthesizing functionalized silica-based sorbents using microbial cells. Prepared biohybrids are of higher length scale thus they could be easily separated once the sorption process is completed. Physico-chemical characterization has also shown that microbial cell surfaces are available for binding to metal ions in the biohybrids. These characteristics make synthesized biohybrids a suitable and efficient sorbent. *S. lactis* cells express β-galactosidase activity and hydrolyzes lactose, thus the enzyme activity was studied in the microstructures comprising *S. lactis* cells and silica NPs (Mukundan et al. 2020). The result was encouraging as the cells entrapped in microstructures have shown higher enzymatic activity than the free cells.

The environment's growing concentration of heavy metal ions creates a major hazard to living beings. For the quick and highly effective removal of diverse contaminants, adsorption methods based on a range of micro/nanomaterials and micro/nanomanipulation are being used. Magnetic Fe_3O_4 nanoparticles were grown *in-situ* on hydrothermally treated fungus spores to create a particular type of biohybrid adsorbent (Zhang et al. 2018). Due to their porous structure and high adsorbing components, such organic/inorganic Porous Spore@Fe_3O_4 Biohybrid Adsorbents (PSFBAs) could efficiently adsorb and remove heavy metal ions. In comparison to their non-motile counterparts, magnetic PSFBAs in controllably collective motion exhibited improved adsorption capacity and quicker removal times for a number of heavy metal ions. It was shown how the collective magnetic actuation of PSFBAs crowded into a small fluidic channel worked. In comparison to untreated and static equivalents, using these adsorbents and a magnetically propelled swarming micro robotic method, lead ions in polluted water were swiftly removed from 5 ppm down to 0.9 ppm. Four successive cycles show that such magnetically propelled PSFBAs may be reused following simple separation and post-treatment (Zhang et al. 2018). A possible approach for the decontamination of toxins in environmental restoration would be the combination of biological entities with swarming microrobotic systems.

15.8 Conclusion and Future Prospective

In this chapter, biohybrids, their different components and the importance of the components and methods of synthesizing biohybrids were described. The association of biomolecules with suitable supports enhances the applicability of biohybrids. Biomolecules have enzymatic activity and other biological processes thus biohybrids could be used as smart sensing materials against various analytes. On the other hand, the support acts as a protective covering and protects the biomolecules from the surrounding harsh environment.

Table 15.1. List of various biohybrids applied for remediation and biosensing.

Note over I_{C50}/K_i/LOD columns: (PPDB: The Pesticide Properties Database, http://sitem.herts.ac.uk/aeru/footprint/en/index.htm accessed in June 2012)

Biohybrid	Pollutants	Parameters	I_{C50} (nM)	K_i (nM)	LOD (nM)	Linear range/ (μg/ml)	SIM ions[a]	LOD (μg/L)	LOQ (μg/L)	P_{vap}[c]/mPa (25°C)	K_H[d]/(Pa m³/mol)	K_{OC}[e]/ (mL/g)	$t_{1/2}$[f]/day	S_w[g]/(mg/L) (20°C)
Photoelectrochemical cell employing purple bacterial reaction centre	**Herbicides** 1. Stigmatellin		280 60	165	10									
	2. Terbutryn		208 10	123	8.3									
	3. Atrazine		2100 100	1200	49									
Swarming biohybrid adsorbents/magnetic biohybrid adsorbents	**Heavy Metals** 1. Ni (II) 2. CO (II) 3. Mn (II) 4. Hg (II) 5. Cd (II) 6. Pb (II)	Removal efficiency 27.3 to 48.3% (static) 29.2 to 85.2% (swarming)												
	Herbicides 1. Metolachlor					0.02 – 24.28	162.1, 238.0, 146.0	0.02	0.02	1.7	2.40×10^{-3}	200	90	530
	2. Phenanthrene[b]					0.20 – 21.24	188.1, 189.0, 184.0	-	-	-	-	-	-	-
	3. Atrazine					-	200.0, 215.0, 202.0	0.03	0.20	0.039	1.50×10^{-4}	100	75	35
	4. DEA					0.82 – 28.88	172.0, 187.0, 174.0	0.06	0.82	12.4	1.55×10^{-4}	72	45	3200
	5. DIA					0.84 – 36.94	173.0, 158.0, 144.9	0.17	0.84	-	1.52×10^{-5}	142	-	670
	Insecticides 1. Methyl Parathion					0.22 – 24.50	262.9, 108.9, 124.9	0.12	0.22	2.0	8.57×10^{-3}	240	12	55
	2. Malathion					0.20 – 28.50	173.0, 124.9, 157.9	0.03	0.20	3.1	1.00×10^{-3}	217	0.17	148
	3. α – Endosulfan					0.22 – 17.44	240.8, 194.9, 236.8	0.06	0.22	0.83	1.48	11500	50	0.32
	4. β – Endosulfan					0.10 – 14.46	194.9, 236.8, 240.8	0.03	0.10	0.83	1.48	11500	50	0.32
	5. Endosulfansulfate					0.22 – 24.42	271.8, 273.8, 228.8	0.08	0.22	0.83	1.48	5194	-	0.48
	6. Chlorpyrifos					0.28 – 18.48	196.9, 198.9, 313.9	0.03	0.22	1.43	4.78×10^{-1}	8151	50	1.05
	Fungicides 1. Flutriafol					0.20 – 20.58	123.0, 164.0, 219.0	0.06	0.20	0.44	1.27×10^{-6}	205	1358	95

Biohybrid	Metal ion	Removal efficiency	References
S. lactis-silica NP	Uranium (VI)	92%	Mishra et al. 2014
S. cerevisiae-silica NP	Mercury (Hg²⁺)	98%	Shukla et al. 2020

[a]Ions used for substance quantification (first ion) and identification (second and third ions) during the Selected Ion Monitoring (SIM); [b]internal standard; [c]Pvap: vapour pressure; [d]KH: Henry's Law constant; [e]KOC: soil organic carbon sorption coefficient; [f]t1/2: half-live; [g]Sw: water solubility; -: metabolite; *Pesticide Properties Database* (PPDB). (Agriculture & Environment Research Unit, University of Hertfordshire, PPDB: *The Pesticide Properties Database*, http://sitem.herts.ac.uk/aeru/footprint/en/index.htm accessed in June 2012.)

Along with sol-gel, spray drying and other advanced techniques like 3D bioprinting and microfluidics could be applied for developing efficient biohybrids. Using these techniques, biohybrids in various morphologies such as microparticles, sheet, fibre and scaffold could be understood. Depending on their morphology, biohybrids could be used for various applications like drug/gene delivery, biosensing, biomedical and environmental remediation.

Though a lot of advances have been made in the field of synthesis and application of biohybrids, there are yet some challenges involved in synthesizing biohybrids in large quantities with accurate precision and design, and understanding applications of biohybrids for constant monitoring and applying them for continuous reactors. For this purpose, there is need to work in an interdisciplinary way, where scientists from different areas like material science, biological sciences, chemical sciences could come together. Regarding biocomponents, there is need for developing engineered biomolecules with high biological activity, which could be associated with suitable supports. Till today, as support organic and nanoparticle materials have been used , now there is need to develop biogenic supports with extraordinary characteristics like intricate design of the surface structure and chemical moieties. There is also a need to develop suitable fabricating process/techniques which have precise control over operational parameters and in which the potential of both components could be fully utilized.

In summary, the combination of biomolecules with supports enables biohybrid materials to act as biomimetic material which results in a dynamic smart material. One can clearly see that advanced characteristics (biocompatibility, flexibility, sensitivity towards analytes) associated with biohybrid materials will enable them to achieve the desired results not only in biosensing and remediation but also in several research areas.

References

Agriculture & Environment Research Unit, University of Hertfordshire, PPDB: The Pesticide Properties Database, 2012. http://sitem.herts.ac.uk/aeru/footprint/en/index.htm accessed in June 2012.

Ana, S. S., L. F. Santos, M. C. Mendes and J. F. Mano. 2020. Multi-layer pre-vascularized magnetic cell sheets for bone regeneration. Biomater. 231: 119664.

Ashly, P. C., M. J. Joseph and P. V. Mohanan. 2011. Activity of diastase *a*-amylase immobilized on polyanilines (PANIs). Food Chem. 127: 1808–1813.

Avnir, D., S. Braun, O. Lev and M. Ottolenghi. 1994. Enzymes and other proteins entrapped in sol-gel Materials. Chem. Mater. 6: 1605.

Basso, A., P. Braiuca, S. Cantone, C. Ebert, P. Linda, P. Spizzo, P. Caimi, U. Hanefeld, G. Degrassi and L. Gardossi. 2007. *In silico* analysis of enzyme surface and glycosylation effect as a tool for efficient covalent immobilisation of CalB and PGA on Sepabeads. Adv. Synth. Catal. 349: 877–886.

Betigeri, S. S. and S. H. Neau. 2002. Immobilization of lipase using hydrophilic polymers in the form of hydrogel beads. Biomater. 51: 3627.

Bergna, H. E. and W. O. Roberts. 2006. Colloidal science in colloidal silica fundamentals and applications. pp. 575–588. CRC Press: Boca Raton.

Boldridge, D. 2010. Morphological characterization of fumed silica aggregates. Aerosol. Sci. Technol. 44: 182.

Cahyaningrum, S. E., N. Herdyastusi and D. K. Maharani. 2014. Immobilization of glucose isomerase in surface-modified chitosan gel beads. Res. J. Pharm. Biol. Chem. Sci. 5: 104.

Capulli, A. K., M. Y. Emmert, F. S. Pasqualini, D. Kehl, E. Caliskan, J. U. Lind, S. P. Sheehy, S. J. Park, S. Ahn, B. Weber, J. A. Goss, S. P. Hoerstrup, K. K. Parker. 2017. JetValve: rapid manufacturing of biohybrid scaffolds for biomimetic heart valve replacement. Biomaterials 133: 229–241.

Chen, L., J. Hu, Z. Fang, Y. Qi and R. Richards. 2011. Gold nanoparticles intercalated into the walls of mesoporous silica as a versatile redox catalyst. Ind. Eng. Chem. Res. 50: 13642.

Chen, Xi, L. Mahadevan, A. Driks and O. Sahin. 2014. *Bacillus spores* as building blocks for stimuli-responsive materials and nanogenerators. Nat. Nanotechnol. 9(2): 137.

Chang-Hyung., C., H. Wang, H. Lee, J. H. Kim, L. Zhang, A. Mao, D. J. Mooney and D. A. Weitz. 2016. One-step generation of cell-laden microgels using double emulsion drops with a sacrificial ultra-thin oil shell. Lab Chip. 16(9): 1549.

Choi, C.-H., H. Wang, H. Lee, J. H. Kim, L. Zhang, A. Mao, D. J. Mooney and D. A. Weitz. 2016. One-step generation of cell-laden microgels using double emulsion drops with a sacrificial ultra-thin oil shell. Lab Chip 16 (9): 1549–1555.

Chong, M. N., B. Jin, C. W. K. Chow and C. Saint. 2010. Recent developments in photocatalytic water treatment technology: a review. Water Res. 44: 2997–3027. doi: 10.1016/j.watres.2010.02.039.

Compaan, A. M.., K. Song, W. Chai and Y. Huang. 2020. Cross-linkable microgel composite matrix bath for embedded bioprinting of perfusable tissue constructs and sculpting of solid objects. ACS Appl. Mater. Interfaces. 12(7): 7855.

Coradin, T., N. Nassif and J. Livage. 2003. Silica-alginate composites for microencapsulation. Appl. Microbiol. Biotechnol. 61: 429.

Davydov, V. Y. 2000. Adsorption on silica surfaces; E. Papirer [Ed.]. CRC Press, Taylor & Francis Group: Santa Barbara, CA, USA, 90: 63.

Douglas, G. B., M. Lurling and B. M. Spears. 2016. Assessment of changes in potential nutrient limitation in an impounded river after application of lanthanum-modified bentonite. Water Res. 97: 47–54. doi: 10.1016/j. watres. 2016.02.005.

Foresti, M. L., G. Valle, R. Bonetto, M. L. Ferreira and L. E. Briand. 2010. FTIR, SEM and fractal dimension characterization of lipase B from *Candida antarctica* immobilized onto titania at selected conditions. Appl. Surf. Sci. 256: 1624–1635.

Gabriel, U., L. Charlet, C. W. Schlapfer, J. C. Vial, A. Brachmann and G. Geipel. 2001. Uranyl surface speciation on silica particles studied by time-resolved laser-induced fluorescence spectroscopy. J. Colloid. Interf. Sci. 239: 358.

Gerber, L. C., F. M. Kbehler, R. N. Grass and J. S. Wendelin. 2012. Incorporating microorganisms into polymer layers provides bioinspired functional living materials. Proc. Natl. Acad. Sci. U. S. A 109(1): 90.

Gomez, L., H. L. Ramırez, A. Neira-Carrillo and R. Villalonga. 2006. Polyelectrolyte complex formation mediated immobilization of chitosan–invertase neoglycoconjugate on pectin-coated chitin. BioprocBiosyst. Eng. 28: 387.

Headen, D. M., G. Aubry, H. Lu and A. J. Garcia. 2014. Microfluidic-based generation of size-controlled, biofunctionalized synthetic polymer microgels for cell encapsulation. Adv. Mater. 26(19): 3003.

Huang, J., S. Liu, C. Zhang, X. Wang, J. Pu, F. Ba, S. Xue, H. Ye, T. Zhao, K. Li, Y. Wang, J. Zhang, L. Wang, C. Fan, T. K. Lu and C. Zhong. 2019. Programmable and printable *Bacillus subtilis* biofilms as engineered living materials. Nat. Chem. Biol. 15(1): 34.

Hyde, E. D. E. R., A. Seyfaee, F. Neville and R. Moreno-Atanasio. 2016. Colloidal silica particle synthesis and future industrial manufacturing pathways: a review. Ind. Eng. Chem. Res. 55: 8891.

Jamal, M., S. S. Kadam, R. Xiao, F. Jivan, T. M. Onn, R. Fernandes, T. D. Nguyen and D. H. Gracias. 2013. Bio-origami hydrogel scaffolds composed of photocrosslinked PEG bilayers. Adv. Health. Mater. 2(8): 1142.

Jimenez-Hamann, M. C. and B. A. Saville. 1996. Enhancement of tyrosinase stability by immobilization on Nylon 66. Food Bioprod. Process Trans. Inst. Chem. Eng. C. 74: 47–52.

Jin, Y., W. Chai and Y. Huang. 2018. Fabrication of stand-alone cell-laden collagen vascular network scaffolds using fugitive pattern-based printing-then-casting approach, ACS Appl. Mater. Interfaces. 10(34): 28361.

Kamperman, T., S. Henke, A. van den Berg, S. R. Shin, A. Tamayol, A. Khademhosseini, M. Karperien and J. Leijten. 2017. Single cell microgel based modular bioinks for uncoupled cellular micro- and macroenvironments. Adv. Health. Mater. 6(3): 1600913.

Kamperman, T., S. Henke, W. C. Visser, M. Karperian and J. Leijten. 2017. Centering single cells in microgels via delayed crosslinking supports long-term 3D culture by preventing cell escape. Small. 13(22): 1603711.

Katiyar, A., S. Yadav, P. G. Smirniotis and N. G. Pinto. 2006. Synthesis of ordered large pore SBA-15 spherical particles for adsorption of biomolecules. J. Chromato. A. 1122: 1-2(13).

Kazy, S. K., S. F. D'Souza and P. Sar. 2009. Uranium and thorium sequestration by a *Pseudomonas* sp., Mechanism and chemical characterization. J. Hazard. Mater. 163: 65.

Kim, S. F., R. B. John and L. W. James. 2000. Encapsulation of sulfate-reducing bacteria in a silica host. J. Mater. Chem. 10: 1099.

Kim, J. K., J. K. Park and H. K. Kim. 2004. Synthesis and characterization of nanoporous silica support for enzyme immobilization. Colloids Surf., A. 241: 113.

Kroger, N. 2007. Prescribing diatom morphology: toward genetic engineering of biological nanomaterials. Curr. Opin. Chem. Biol. 11: 662.

Labus, K., A. Turek, J. Liesiene and J. Bryjak. 2011. Efficient *Agaricusbisporus* tyrosinase immobilization on cellulose-based carriers. Biochem. Eng. J. 56: 232.

Lahiri, S., A. Mishra, D. Mandal, R. L. Bhardwaj and P. R. Gogate. 2021. Sonochemical recovery of uranium from nanosilica-based sorbent and its biohybrid. Ultrasonics Sonochemistry. 76: 105667.

Leng, L., A. McAllister, B. Zhang, M. Radisic and A. Günther. 2012. Mosaic hydrogels: one-step formation of multiscale soft materials. Adv. Mater. 24(27): 3650.

Letnik, I., R. Avrahami, J. S. Rokem, A. Greiner, E. Zussman and C. Greenblatt. 2015. Living composites of electrospun yeast cells for bioremediation and ethanol production. Biomacromolecules. 16(10): 3322.

Li, R., L. Morrison, G. Collins, A. Li and X. Zhan. 2016. Simultaneous nitrate and phosphate removal from wastewater lacking organic matter through microbial oxidation of pyrrhotite coupled to nitrate reduction. Water Res. 96: 32–41. doi: 10.1016/j.watres.2016.03.034.

Lienemann, P. S., T. Rossow, A. S. Mao, Q. Vallmajo-Martin, M. Ehrbar and D. J. Mooney. 2017. Single cell-laden protease-sensitive microniches for long term culture in 3D. Lab Chip. 17(4): 727.

Liu, X., T. Tzu-Chieh, E. Tham, H. Yuk, S. Lin, T. K. Lu and X. Zhao. 2017. Stretchable living materials and devices with hydrogel–elastomer hybrids hosting programmed cells. Proc. Natl. Acad. Sci. U. S. A. 114(9): 2200.

Liu, B. F., Y. R. Jin, Q. F. Cui, G. J. Xie, Y. N. Wu and N. Q. Ren. 2015. Photofermentation hydrogen production by *Rhodopseudomonasspnov* strain A7 isolated from the sludge in a bioreactor. Int. J. Hydrogen Energy. 40: 8661–8668.

Mao, A. S., J. W. Shin, S. Utech, H. Wang, O. Uzun, W. Li, M. Cooper, Y. Hu, L. Zhang, D. A. Weitz and D. J. Mooney. 2017. Deterministic encapsulation of single cells in thin tunable microgels for niche modelling and therapeutic delivery. Nat. Mater. 16(2): 236.

Melo, J. S., A. Tripathi, J. Kumar, A. Mishra, S. Bhanu and K. Bhainsa. 2020. Immobilization: then and now. pp. 1–84. *In*: A. Tripathi and J. S. Melo [Eds.]. I Immobilization Strategies. Gels Horizons: From Science to Smart Materials. Singapore: Springer.

Metilda, J. P., M. Gladis and T. P. Rao. 2005. Catechol functionalized aminopropyl silica gel: synthesis, characterization and preconcentrative separation of uranium (VI) from thorium (IV), Radiochim. Acta. 93: 219.

Mishra, A., J. S. Melo, D. Sen and S. F. D'Souza. 2014. Evaporation induced self assembled microstructures of silica nanoparticles and *Streptococcus lactis* cells as sorbent for uranium (VI). J. Colloid Interface Sci. 414: 33–40.

Mishra, A., J. Kumar and J. S. Melo. 2017. An optical microplate biosensor for the detection of methyl parathion pesticide using a bio-hybrid of *Sphingomonas* sp. cells-silica nanoparticles. Biosensor and Bioelectronics. 87: 332–338.

Mishra, A. 2019. Synthesis, characterization and applications of silica based biohybrid materials, Ph.D. Thesis, Homi Bhabha National Institute, Mumbai, India.

Mishra, A., J. S. Melo, A. Agrawal, Y. Kashyap and D. Sen. 2020a. Preparation and application of silica nanoparticles-*Ocimumbasilicum* seeds bio-hybrid for the efficient immobilization of invertase enzyme. Colloids Surface B. Biointerfaces. 188: 110796.

Mishra, A., J. Kumar and J. S. Melo. 2020b. Silica based bio-hybrid materials and their relevance to bionanotechnology. Austin J. Plant Biol. 6(1): 1024–1028.

Mishra, A., V. K. Pandey, B. S. Shankar and J. S. Melo. 2021a. Spray drying as an efficient route for synthesis of silica nanoparticles-sodium alginate biohybrid drug carrier of doxorubicin. Colloids Surface B: Biointerfaces. 197: 111445.

Mishra, A., J. Kumar, J. S. Melo and B. P. Sandaka. 2021b. Progressive development in biosensors for detection of dichlorvos pesticide: a review. J. Environ. Chem. Eng. 9(2): 105067.

Mishra, A., S. Mukundan and J. Kumar. 2022. An overview of metal-organic frameworks for detection of pesticides. *In*: Ram K. Gupta, Tahir Rasheed, Tuan Anh Nguyen and Muhammad Bilal [Eds.]. I Metal-Organic Frameworks-Based Hybrid Materials for Environmental Sensing and Monitoring. CRC Press Taylor & Francis, USA, P. 7, eBook ISBN 9781003188148. doi.10.1201/9781003188148-21.

Mislovicova, D., J. Masarova, A. Vikartovska, P. Germeiner and E. Michalkova. 2004. Biospecific immobilization of mannan–penicillin G acylase neoglycoenzyme on Concanavalin A-bead cellulose. J. Biotechnol. 110: 11.

Mukundan, S., J. S. Melo, D. Sen and J. Bahadur. 2020. Enhancement in β-galactosidase activity of *Streptococcus lactis* cells by entrapping in microcapsules comprising of correlated silica nanoparticles. Colloids Surface B: Biointerfaces. 195: 111245.

Namdeo, M. and S. K. Bajpai. 2009. Immobilization of a-amylase onto cellulose-coated magnetite (CCM) nanoparticles and preliminary starch degradation study. J. Mol. Catal. B-Enzym. 59: 134.

Nandiyanto, A. B. D., S.-G. Kim, F. Iskandar and K. Okuyama. 2009. Synthesis of silica nanoparticles with nanometer-size controllable mesopores and outer diameters. Microporous and Mesoporous Mater. 20(3): 447.

Neal, D., M. S. Sakar, L.-L. S. Ong and H. H. Asada. 2014. Formation of elongated fascicle-inspired 3D tissues consisting of high-density, aligned cells using sacrificial outer molding. Lab Chip. 14(11): 1907.

Nguyen, P. Q., N.-M. D. Courchesne, A. Duraj-Thatte, P. Praveschotinunt and N. S. Joshi. 2018. Engineered living materials: prospects and challenges for using biological systems to direct the assembly of smart materials. Adv. Mater. 30(19): e1704847.

Nogueira, E. N., E. F. G. C. Dores, A. A. Pinto, R. S. Amorim, M. L. Ribeiro and C. Lourencetti. 2012. Currently used pesticides in water matrices in Central-Western Brazil. J. Braz. Chem. Soc. 23(8): 1476–1487.

Oh, S. Y., Y. D. Seo, B. Kim, I. Y. Kim and D. K. Cha. 2016. Microbial reduction of nitrate in the presence of zero-valent iron and biochar. Bioresour. Technol. 200: 891–896. doi: 10.1016/j.biortech.2015.11.021.

Onoe, H., T. Okitsu, A. Itou, M. Kato-Negishi, R. Gojo, D. Kiriya, K. Sato, S. Miura, S. Iwanaga, K. Kuribayashi-Shigetomi, Y. T. Matsunaga, Y. Shimoyama and S. Takeuchi. 2013. Metre-long cell-laden microfibres exhibit tissue morphologies and Functions. Nat. Mater. 12(6): 584.

Ouyang, L. H., B. Christopher, W. Sun and J. A. Burdick. 2017. A generalizable strategy for the 3D bioprinting of hydrogels from nonviscous photo-crosslinkable inks. Adv. Mater. 29(8): 1604983.

Ouyang, L., J. P. K. Armstrong, M. Salmeron-Sanchez and M. M. Stevens. 2020. Assembling living building blocks to engineer complex tissues. Adv. Funct. Mater. 30(26): 1909009.

Pattanawanidchai, S., S. Loykulnant, P. Sae-oui, N. Maneevas and C. Sirisinha. 2014. Development of eco-friendly coupling agent for precipitated silica filled natural rubber compounds. Polym. Test. 34: 58.

Prasertsri, S. and N. Rattanasom. 2012. Fumed and precipitated silica reinforced natural rubber composites prepared from latex system: mechanical and dynamic properties. Polym. Test. 31: 593.

Premkumar, J. R., O. Lev, R. Rosen and S. Belkin. 2001. Encapsulation of luminous recombinant *E. coli* in sol-gel silicate films. Adv. Mater. 13(23): 1773.

Pu, J., Y. Liu, J. Zhang, B. An, Y. Li, X. Wang, K. Din, C. Qin, K. Li, M. Cui, S. Liu, Y. Huang, Y. Wang, Y. Lv, J. Huang, Z. Cui, Z. Suwen and C. Zhong. 2020. Virus disinfection from environmental water sources using living engineered biofilm materials. Adv. Sci. 7(14): 1903558.

Puig, M., L. Cabedo, J. J. Gracenea, A. Jiménez-Morales, J. Gámez-Pérez and J. J. Suay. 2014. Adhesion enhancement of powder coatings on galvanised steel by addition of organo-modified silica particles. Prog. Org. Coat. 77: 1309.

Rahman, I. A. and V. Padavettan. 2012. Synthesis of silica nanoparticles by sol-gel: size-dependent properties, surface modifications and applications in silica-polymer nanocomposites—a review. J. Nanomater. 1.

Rajanna, S. K., D. Kumar, M. Vinjamur and M. Mukhopadhyay. 2015. Silica aerogel microparticles from rice husk ash for drug delivery. Ind. Eng. Chem. Res. 54: 949.

Reshmi, R., G. Sanjay and S. Sugunan. 2007. Immobilization of *a*-amylase on zirconia: a heterogeneous biocatalyst for starch hydrolysis. Catal. Commun. 8: 393.

Rimola, A., D. Costa, M. Sodupe, J.-F. Lambert and P. Ugliengo. 2013. Silica surface features and their role in the adsorption of biomolecules: computational modelling and experiments. Chem. Rev. 113: 4216.

Rivera-Tarazona, L. K., Z. T. Campbell and T. H. Ware. 2021. Stimuli-responsive engineered living materials. Soft Matter. 17(4): 785.

Roxby, D. N., H. Rivy, C. Gong, X. Gong, Z. Yuan, G. E. Chang and Y. C. Chen. 2020. Microalgae living sensor for metal ion detection with nanocavity enhanced photoelectrochemistry. Biosens. Bioelectron. 165: 112420.

Sankaran, S., J. Becker, C. Wittman and A. dCampo. 2019. Optoregulated drug release from an engineered living material: self-replenishing drug depots for long-term, light-regulated delivery. Small. 15(5): e1804717.

Sar, P., S. K. Kazy and S. F. D'Souza. 2004. Radionuclide remediation using a bacterial biosorbent. Int. Biodeter. Biodegr. 54: 193.

Satar, R., M. Matto and Q. Husain. 2008. Studies on calcium alginate—pectin gel entrapped concanavalin A–bitter gourd (*Momordica charantia*) peroxidase complex. J. Sci. Ind. Res. India. 67: 609.

Shi, L. E., Z. X. Tang, Y. Yi, J. S. Chen, W. Y. Xiong and G. Q. Ying. 2011. Immobilization of nuclease p1 on chitosan micro-spheres. Chem. Biochem. Eng. Q. 25: 83.

Shukla, P., A. Mishra, S. Manivannan, J. S. Melo and D. Mandal. 2020. Parametric optimization for adsorption of mercury (II) using self-assembled bio-hybrid. J. Environ. Chem. Eng. 8(3): 103725.

Shukla, P., A. Mishra, S. Manivanna and D. Mandal. 2022. Metal-Organic-Frames (MOFs) based electrochemical sensors for sensing heavy metal contaminated liquid effluents: a review. Nanoarchitectonics. 3(2): 46–60.

Shukla, V. Y., D. R. Tipre and S. R. Dave. 2014. Optimization of chromium (VI) detoxification by *Pseudomonas aeruginosa* and its application for treatment of industrial waste and contaminated soil. Bioremediat. J. 18: 128–135. doi: 10.1080/10889868.2013.834872.

Son, J., C. Y. Bae and J.-K. Park. 2016. Freestanding stacked mesh-like hydrogel sheets enable the creation of complex macroscale cellular scaffolds. Biotechnol. J. 11(4): 585.

Sun, Z., C. Bai, S. Zheng, X. Yang and R. L. Frost. 2013. A comparative study of different porous silica minerals supported TiO_2 catalysts. Appl. Catal. A. 458: 103.

Stöber, W., A. Fink and E. Bohn. 1968. Controlled growth of monodisperse silica spheres in the micron size range. J. Colloid. Inter. Sci. 26(1): 62.

Swainsbury, D. J. K., V. M. Friebe, R. N. Frese and M. R. Jones. 2014. Evaluation of a biohybrid photoelectrochemical cell employing the purple bacterial reaction centre as a biosensor for herbicides. Biosensors and Bioelectronics. 58: 172–178.

Taikum, O., R. Friehmelt and M. Scholz. 2010. The last 100 years of fumed silica in rubber reinforcement. Rubber World. 242: 35.

Tang, F., L. Li and D. Chen. 2012. Mesoporous silica nanoparticles: synthesis, biocompatibility and drug delivery. Adv. Mater. 24: 1504.

Tian, D., P. Dubois and R. Jérôme. 1997. Biodegradable and biocompatible inorganic–organic hybrid materials. I. Synthesis and characterization. J. Poly. Sci. Part A: Poly. Chem. 35(11): 2295.

Valenti, G., R. Rampazzo, S. Bonacchi, L. Petrizza, M. Marcaccio, M. Montalti, L. Prodi and F. Paolucci. 2016. Variable doping induces mechanism swapping in electrogenerated chemiluminescence of Ru(bpy)32$^+$ core–shell silica nanoparticles. J. Am. Chem. Soc. 138(49): 15935.

Vassilev, N., M. Fenice, F. Federici and R. Azcon. 1997. Olive mill waste water treatment by immobilized cells of *Aspergillus niger* and its enrichment with soluble phosphate. Process. Biochem. 32: 617–620. doi: 10.1016/S0032-9592(97) 00024-1.

Vartiainen, J., M. Rättö and S. Paulussen. 2005. Antimicrobial activity of glucose oxidase-immobilized plasma-activated polypropylene films. Packag. Technol. Sci. 18: 243–251.

Wang, J.-S., X.-J. Hu, Y.-G. Liu, S.-B. Xie and Z.-L. Bao. 2010. Biosorption of uranium (VI) by immobilized *Aspergillus fumigatus* beads. J. Environ. Radioactiv. 101: 504.

Wang, K., P. Liu, Y. Ye, J. Li, W. Zhao and X. Huang. 2014. Fabrication of a novel laccase biosensor based on silica nanoparticles modified with phytic acid for sensitive detection of dopamine. Sens. Actuators B. 197: 292.

Wang, W., W. Zhou, J. Li, Hao, D. Z. Su and G. Ma. 2015. Comparison of covalent and physical immobilization of lipase in gigaporous polymeric microspheres. Bioprocess Biosys. Eng. 38: 2107.

Wang, X., Y. Yu, C. Yang, C. Shao, K. L. Shi, Shang, F. Ye and Y. Zhao. 2021. Microfluidic 3D printing responsive scaffolds with biomimetic enrichment channels for bone regeneration. Adv. Funct. Mater. 31(40): 2105190.

Wang, C., Z. Zhang, J. Wang, Q. Wang and L. Shang. 2022. Biohybrid materials: structure design and biomedical applications. Materials Today Bio. 16: 100352.

Xie W., L. Hu and X. Yang. 2015. Basic ionic liquid supported on mesoporous SBA-15 silica as an efficient heterogeneous catalyst for biodiesel production. Ind. Eng. Chem. Res. 54: 1505.

Xu, H., M. Medina-Sánchez, V. Magdanz, L. Schwarz, F. Hebenstreit and O. G. Schmidt. 2018. Sperm-hybrid micromotor for targeted drug delivery. ACS Nano. 12(1): 327.

Yang, S. H., J. Choi, L. Palanikumar, E. S. Choi, J. Lee, J. Kim, I. S. Choi and J.-H. Ryu. 2015. Cytocompatible in situ cross-linking of degradable LbL films based on thiol–exchange reaction. Chem. Sci. 6(8): 4698.

Yang, Y., H. Yang, M. Yang, Y. Liu, G. Shen and R. Yu. 2004. Amperometric glucose biosensor based on a surface treated nanoporous ZrO₂/Chitosan composite film as immobilization matrix. Analytica. Chimica. Acta. 525(2): 213.

Yang, Z., S. Si and C. Zhang. 2008. Study on the activity and stability of urease immobilized ontonanoporous alumina membranes. Microporous Mesoporous Mater. 111: 359.

Zdarta, J., A. S. Meyer, T. Jesionowski and M. Pinelo. 2018. A general overview of support materials for enzyme immobilization: characteristics, properties, practical utility. Catalysts. 8: 92.

Zhang, Y., K. Yan, F. Ji and L. Zhang. 2018. Enhanced removal of toxic heavy metals using swarming biohybrid adsorbents. Adv. Funct. Mater. 1806340.

Zhao, W., Y. Fang, Q. Zhu, K. Wang, M. Liu, X. Huang and J. Shen. 2013. A novel glucose biosensor based on phosphonic acid-functionalized silica nanoparticles for sensitive detection of glucose in real samples. Electrochim. Acta. 89: 278.

Zhu, L. J., L. P. Zhu, J. H. Jiang, Z. Yi, Y. F. Zhao, B. K. Zhu and Y. Y. Xu. 2014. Hydrophilic and anti-fouling polyethersulfone ultrafiltration membranes with poly(2-hydroxyethyl methacrylate) grafted silica nanoparticles as additive. J. Membr. Sci. 451: 157.

Zhu G., Q. Wang, J. Yin, Z. Li, P. Zhang, B. Ren, G. Fan and P. Wan. 2016. Toward a better understanding of coagulation for dissolved organic nitrogen using polymeric zinc-iron-phosphate coagulant. Water Res. 100: 201–210. doi: 10.1016/j.watres. 2016.05.035.

Zhu, W., J. Guo, S. Amini, Y. Ju, J. O. Agola, A. Zimpel, J. Shang, A. Noureddine, F. Caruso, S. Wuttke, J. G. Croissant and C. J. Brinker. 2019. SupraCells: living mammalian cells protected within functional modular nanoparticle-based exoskeletons. Adv. Mater. 31(25): e1900545.

Zucca, P. and E. Sanjust. 2014. Inorganic materials as supports for covalent enzyme immobilization: methods and mechanisms. Mol. 19: 14139.

CHAPTER 16

Remediation for Heavy Metal Contamination
A Nanotechnological Approach

*Rubina Khanam, Amaresh Kumar Nayak** and
Dibyendu Chatterjee

16.1 Introduction

Pollution is defined as the presence of undesirable chemical objects that obstruct natural processes or have negative consequences for living beings and the environment. Pollution is increasing at an alarming rate as a result of industrialization and the massive rise in population that leads to rising urbanization. Environmental pollution identification, treatment and prevention is a critical step toward long-term environmental sustainability. Environment sustainability refers to the responsible and justifiable relationship between humans and the environment, as well as the intelligent use of resources, ensuring environmental safety for current and future generations. Economic and environmental sustainability are inextricably linked (Fajardo et al. 2020). A lot of work is being done right now to discover and create persuasive and dependable ways for degrading or transforming environmental contaminants of concern. Nanoremediation is a groundbreaking remediation technique that employs nanomaterials having high surface: volume ratio, low reduction potential and quantum confinement making them efficient for the detoxification and alteration of hazardous recalcitrant pollutants in the system (Fajardo et al. 2020). In particular, when compared to standard remediation procedures (viz., chemical oxidation, thermal decomposition and solvent co-flushing), the use of nanomaterials in environmental remediation has gained a lot of attention. Nanoremediation methods have drawn a lot of interest because of their unique qualities, such as cost-effectiveness, sensitivity, superior electrical properties, high surface area and improved catalytic properties (Shafi et al. 2021). These techniques have the potential to give long-term solutions to environmental pollution issues, while also reducing the cost of cleaning (Shafi et al. 2021). Nanostructure-based technologies have the potential to reduce not only the overall costs of cleaning up large-scale contaminated areas but also to reduce clean-up time, minimize the need for polluted material treatment and disposal, and decrease pollutant concentrations to near zero-all *in-situ* (Corsi et al. 2018). Nanoremediation involves the applications of nano-sized metal and bimetallic

ICAR-National Rice Research Institute, Cuttack 753006, Odisha, India.
* Corresponding author: aknayak20@yahoo.com

particles, nanodots, carbon nanotubes and nanocomposites, etc., for breaking down contaminants (Shafi et al. 2021). These nanomaterials in the form of sensors, catalysts and adsorbents ensure rapid detection and immediate detoxification of pollutants such as heavy metals and metalloids from contaminated land sites (Figure 16.1). This chapter summarizes the application of nanotechnology (1) to detect and quantify trace pollutants in the environment. (2) to decontaminate the environment using nano adsorbents, nanocatalysts, nano clay composites, etc.

16.2 Human Health and Heavy Metal Toxicity

Large quantities of metals (Cu, Cd, Pb, Zn, etc.) and metalloids (particularly As) have been released into the environment as a result of industrial operations, causing significant ecological harm. Refineries, mining activities, sludge disposal and manufacturing sectors such as paints, electronic and electrical gadgets, batteries, fertilizers, pesticides and other industries all contribute to the contamination of water bodies. Furthermore, metals and metalloids cause a risk to human health because many of them are hazardous even at low concentrations, and some are even carcinogenic (e.g., As). Unlike organic contaminants, which break down into nontoxic smaller molecules, metal(loid)s are resistant to many biological processes and hazardous to humans and other living things. Lead, Cd, Cr, Hg and As metals are well-known hazardous metals. These metals have no beneficial biological role, but they can mimic other beneficial elements and hinder biological activities. As a result, metal toxicity has emerged as a serious concern, with several health risks, such as lung cancer, skin cancer, bladder cancer, etc. (Table 16.1).

Table 16.1. Effects of metal(loid)s on human health.

Metal(loid)s	Effect on Human	References
Arsenic (As)	Inflammation, chronic respiratory illness, liver fibrosis, cardiovascular disorders, skin cancer, lung cancer may be caused by As, which is classified as a class I carcinogen.	Mawia et al. 2021
Mercury (Hg)	The human neurological system is extremely susceptible to mercury, which causes acrodynia, or pink sickness. Mercury is a highly carcinogenic element that can disrupt brain functioning and cause timidity, difficulties in memory, tremors, irritability and visual or hearing abnormalities.	Khanam et al. 2020
Cadmium (Cd)	Cadmium is Group 1 carcinogen for humans. Most common occurring toxicity symptoms are kidney diseases, bone disorders, anemia and cancer.	Khanam et al. 2020
Lead (Pb)	Long-term exposure can induce birth abnormalities, mental retardation, brain damage, psychosis, autism, weight loss, hyperactivity, paralysis, muscle weakness, kidney damage and death.	Khanam et al. 2020
Chromium (Cr)	Chronic exposure can lead to the respiratory tract and stomach cancer and liver damage	Khanam et al. 2020

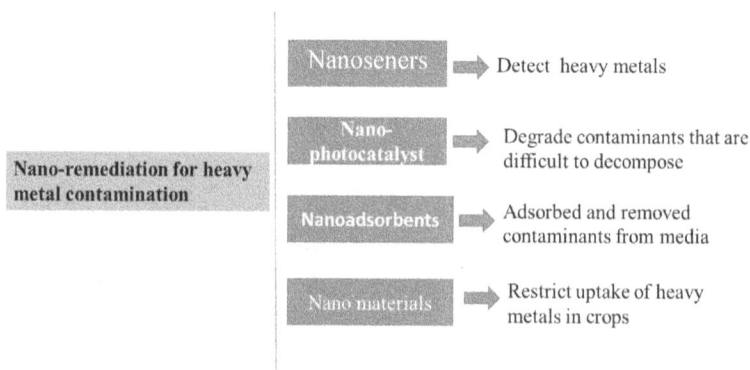

Figure 16.1. Application nanoremediation techniques for reducing heavy metal contamination in the environment.

16.3 Application of Nanomaterials for Detection of Heavy Metal Contamination

16.3.1 Nano-sensors for Detection of Heavy Metals

Nanotechnology is a rapidly developing area that has the potential to provide a new generation of technically advanced environmental sensing instruments and devices. Researchers are currently focusing their attention on creating integrated sensing devices with a low-cost technology that can identify susceptible contaminants even at trace levels. Depending on various signal transduction methods, nanosensors are divided into optical, nanobiosensers, electrochemical and Field-Effect Transistor (FET) sensors. The introduction of nanostructures and nanomaterials into sensors significantly improves the selectivity, sensitivity, multiplexed detection capabilities and mobility of the devices. In addition, inorganic materials have been combined with proteins, DNA, microbes and small molecules to selectively link heavy metals as a probe for recognition at the molecular level (Maghsoudi et al. 2021).

16.3.1.1 Optical Sensors

16.3.1.1.1 Fluorescent Sensors

The principle of fluorescence sensing is related to alterations in physicochemical characteristics of fluorophores induced by analytes, measured through changes in the intensity of fluorescence, anisotropy and lifetime, all of which are associated with transferring of energy mechanisms (Sauer 2003). Organic dyes, toolboxes and Graphene Oxide (GO) are commonly used as fluorophores in Foerster Resonance Energy Transfer sensors (FRET) because of their commercial availability. They are cheap and can be manufactured in large quantities. This battery-powered sensor had high sensitivity (2 ppt) and selectivity for mercury (Hg^{2+}) over lead (Pb^{2+}) and copper (Cu^{2+}) (Darbha et al. 2007) (Table 16.2).

Table 16.2. Fluorescent nanosensors for detecting heavy metal ions.

Nanosensor	Strategy of sensing	Metals	References
CNPs	The fluorescence was reduced by successful chelation of functions with metal ions	Pd^{2+} and Hg^{2+}	Sharma et al. 2016
GSH-Mn-ZnS QDs LDH NCs	Intensity of luminescence is reduced by nonradiative recombination	Pb^{2+}, Cd^{3+}, and Hg^{2+}	Liu et al. 2017
N-CQDs	Nonradiative transfer of electrons causes quenching of fluorescence	Hg^{2+}	Zhang and Chen 2014
S-GQDs	The fluorescence intensity was reduced when Fe^{+3} and S-GQDs functioned coordinately	Fe^{3+}	Li et al. 2014
PPVs@MSN @CdTe NCs	The chelation of CdTe QDs with Cu^{+2} causes a change in luminescence from red to green	Cu^{2+}	Sha et al. 2015

16.3.1.1.2 Plasmonic Sensors

Surface plasmon resonance is caused by a group of free conduction electrons in a noble metal resonating with incoming electromagnetic radiation (Li et al. 2012). This kind of sensor detects analytes in a direct, visible and rapid manner by reducing expenses. Different heavy metals, like Cu^{2+}, As^{3+}, Hg^{2+} and Pb^{2+}, have been monitored using the colorimetric detection technique (Chai et al. 2010).

16.3.1.1.3 Surface-enhanced Raman Scattering (SERS) Sensors

Although SERS sensors are widely utilized for chemical and biological sensing as well as medical diagnostics (Li et al. 2013a), only a few studies confirmed detection of heavy metal (Mulvihill et al. 2008, Han et al. 2010) (Table 16.3). A glutathione-based SERS sensor was developed for As^{3+}

Table 16.3. SERS nanosensors for detecting heavy metal ions.

Nanosensor	Sensing strategy	Metals	References
AuNPs/rGO colloid	The Pb^{2+} accelerated breakdown of AuNPs causes a reduction in the Raman signal of graphene.	Pb^{2+}	Zhao et al. 2016
AuNRs	Hg–Au production causes compositional and morphological alterations in the NRs, resulting in blue shifts in the SPR peak.	Hg^{2+}	Schopf et al. 2017
AgNPs conjugates	Aggregation of AgNPs caused by glycine coordination with Hg^{2+} and Cu^{2+} resulted in higher Raman band intensity.	Cu^{2+} and Hg^{2+}	Li et al. 2013b
Flu-PBADiol-Au@AgNPs	Raman signals are quenched attributable to the boronate ester being cleaved by F.	F^-	Zhang et al. 2017
AuNPs	Anti aggregation related to autocatalytic action results in OPDA oxidation to poly-OPDA or OPDAox by metal NPs.	Hg^{2+}, Pb^{2+} and Cr^{4+}	Patil et al. 2017

detection that can selectively bind to As^{3+} ions via the As-O link (Han et al. 2010). SERS provides information on the existence of a chemical element as well as its chemical composition, which is significant in metal ion toxicological studies since different heavy metals complexes have different degrees of toxicity in humans and animals.

16.3.1.2 Nano Biosensing

Due to its vast potential for developing and fabricating future multifunctional nano sensing systems, nanobiosensing is an emerging multidisciplinary research area that goes hand in hand with biotechnology and molecular biology. This has made it easier to equip sensors with multiplexing capabilities so that they can detect diverse analytes simultaneously.

(i) *Aptasensors:* Aptamer-based DNA sensors or aptasensors are a stable and extremely sensitive technology because of their strong affinity to a wide range of binding sites. Single-stranded DNA (ssDNA) or RNA molecules are known as aptamers. When compared to antibodies and other biorecognition elements, they represent the next category of potential biorecognition elements. These are usually utilized in manufacturing low-cost Hg^{2+} ion detection nanosensors with an attached battery-powered visual reader connected to a smartphone's existing camera module.

(ii) *DNA-nano sensors:* The use of particular, pre-defined DNA sequences has been proposed as a potential synthesis method for metal nanoparticles with a regulated shape and size. The development of fluorescent nanoclusters is aided by both single-stranded DNA (ssDNA) (Petty et al. 2004) and double-stranded DNA (dsDNA) (Rotaru et al. 2010) sequences. Mercury (Hg^{2+}) and Silver (Ag^+) ions are detected using DNA nanosensors. Silver ions (Ag^+) have a strong affinity for cytosine-rich DNA; for this reason DNA nanosensors have been developed for visual detection of Ag in tap water and river samples (Kumar and Guleria 2020). The affinity of the thiamine rich oligonucleotide for Hg, which inhibits elongation of DNA, is applied in the detection procedure of mercury (Hg^{2+}) (Kumar and Guleria 2020).

16.3.1.3 Chemically Synthesized Nanomaterial-based Sensors

Electrochemical sensors have received a lot attention recently because of their characteristic ability to provide label-free detection and are widely recognized for their sensitivity and mobility (Table 16.4). Conducting wires are used to capture the electrochemical sensor signals. While heavy metals have different redox potentials, bare electrodes without chemical detection probes are used to select specific heavy metal ions. Voltammetry, conductometry, potentiometry, amperometry and impedemetry are some of the transducing methods utilized in electrochemical sensing (Kimmel et al. 2012). Mercury-based electrodes were first to be known for electrochemical detection of heavy metals (Varun and Daniel 2018).

Table 16.4. Chemically synthesized nanomaterials used for detecting heavy metals.

Nanomaterials	Method of detection	Metals	References
BiNPs	Electrochemical	Ni^{2+}, Pb^{2+} and Cd^{2+}	Niu et al. 2015
Graphene/CeO$_2$	Electrochemical	Cd^{2+}, Pb^{2+}, Cu^{2+}, and H^{g2+}	Xie et al. 2015
Core-shell SiO@Ag NPs	Colorimetric	Hg^{2+}	Boken and Kumar 2014
Carbon xerogel BiNPs	Electrochemical	Pb^{2+} and Cd^{2+}	Gich et al. 2013
GQDs functionalized AuNPS	Electrochemical	Hg^{2+} and Cu^{2+}	Ting et al. 2015

16.3.1.4 Field-effect Transistor (FET) Sensors

Heavy metals have been detected using FET sensors that take advantage of the interaction between the analyte and the semiconductor resistance (Chen et al. 2011). The conductivity variations caused by the selective redox interaction between single-walled carbon nanotubes (SWCNTs) and Hg^{2+} have been used to create a FET sensor (Kim et al. 2009). In both drinking water and aqueous solution, this sensor demonstrated a broad detection limit from 10 nM to 1 mM, as well as high selectivity for Hg^{2+} over the other metal ions.

16.4 Application of Nanomaterials for Decontamination of Heavy Metals

Heavy metals and metalloids (viz., Pb, Cd, Fe, Se, Ni, Cu, Co, As, etc.) are hazardous and persistent. Recent advances in nanotechnology have prepared the path for novel approaches to heavy metal remediation, particularly in wastewater treatment, where nanomaterials are used in water purification via mechanisms such as adsorption and degradation of toxic materials. According to research, nanoparticles can assist in reducing the adverse effects or absorption/uptake of heavy metals. A study was conducted to evaluate how nano-TiO$_2$ (2–6 nm) can reduce the effect of wastewater (containing heavy metals) on the development of maize seedlings. Analysis of wastewater revealed that it was unfit for irrigation as it contained high levels of heavy metals (Zn, Cu, Fe, Mn, Cr and Cd) beyond the permissible levels for irrigation. During *in vitro* studies, the nano-TiO$_2$ suspension was administered at different concentrations, either autoclaved wastewater or deionized water. Nano-TiO$_2$ at a concentration of 25 mg L^{-1} significantly reduced the negative effects of wastewater on maize development indices (Yaqoob et al. 2018). Further, the effects of nano-Fe$_3$O$_4$ in reducing heavy metal toxicity in wheat seedlings were investigated by Konate et al. (2017). The use of nano-Fe$_3$O$_4$ (@2000 mg L^{-1}) in a 1 Mm solution dramatically reduced the negative impact of heavy metals on the growth in wheat seedlings. The beneficial impacts of nano-Fe$_3$O$_4$ under heavy metal stress might be due to increased antioxidant enzyme activity (Konate et al. 2017). Praveen et al. (2018) reported the effectiveness of iron oxide nanoparticles (Fe$_3$O$_4$ NP) in dropping detrimental effects of As in *Brassica juncea*. They also reported that the application of Fe$_3$O$_4$ NP further decreased the stress in plants induced by As contamination. Under cadmium stress, zinc oxide nanoparticles were found to improve wheat chlorophyll content, gas exchange characteristics, antioxidant enzymes, zinc absorption and yield while lowering Cd concentration. Siddiqui and Al-Whaibi (2014) discovered that 8 g L^{-1} of 12 nm nano-silicon oxide increased seed germination in *Lycopersicum esculentum*. It has also been observed that nanostructured silicon dioxide aids in the reduction of plant transpiration rates, as well as the improvement of the plant's green color and shoot growth. Application of nano SiO$_2$ reduced As uptake in rice seedling through improving pectin synthesis and the mechanical force of the cell wall. Recently, various iron nano-materials have been employed to minimize heavy metal uptake in plants. In the modern era, the application of nanoparticles not only reduces the uptake of heavy metals in plants but also decontaminates the environment through different mechanisms. Next the different applications of nanomaterials (viz., photocatalyst, nano absorbents, nano clay polymers, etc.) to reduce the heavy metal burden in the environment are described.

16.4.1 Nano-photocatalyst

Photocatalysis is a photochemical reaction based on electron/hole pair redox interactions when exposed to light (Danish et al. 2021). Photocatalysis has the potential to degrade contaminants that are difficult to decompose. The photocatalytic process is regarded as an effective remediation method for converting harmful chemicals into environmentally friendly products. In the presence of abundant solar radiation, photocatalysts accelerate chemical reactions. In general, photocatalysis is a redox process in which holes are created in the Valence Band (VB) of a photocatalyst and electrons are generated in the Conduction Band (CB), resulting in the emergence of highly energetic and reactive species such as hydroxyl (OH^-) and superoxide radicals (O^{2-}). Due to the production of extremely energetic radicals that function as potent oxidizing agents, photocatalysis serves as a fruitful technique for the removal of hazardous chemical compounds (Ajmal et al. 2014). Nanomaterial photocatalysis for environmental cleanup has been the subject of significant research for the past decade. To cleanse soil, purify the air and detoxify wastewater, a wide range of nanostructured photocatalysts have been synthesized (viz., oxides and sulfide of metals, composite oxides, carbon derivatives, graphene-based photocatalysts, dendrimers and polymeric nanocomposites (Danish et al. 2021). In comparison to conventional photocatalysts, graphene-based photocatalysts have been found to enhance the activity due to their large surface area, nanosize and greater electronic motions. Oxide-based nanomaterials, such as Fe_3O_4, TiO_2, ZnO and their composites, are also excellent catalysts for the removal of heavy metals. Several researches have reported the use of these oxides in environmental remediation, notably in photocatalytic degradation of organic contaminants (Czech et al. 2020, Sabzehei et al. 2020, Masudy-Panah et al. 2019). By reducing the distance between photon absorption sites and limiting electron-hole (e--$h+$) recombination, these nanoparticles with a large surface area and porosity have greater photocatalytic activity. Transition metal oxides and their composites have strong photocatalytic activity for organic pollutant photodegradation. TiO_2 has demonstrated outstanding photocatalytic destruction of contaminants due to its high resistance against photochemical, exceptional surface qualities, microstructural features and large surface area; nanosized metal oxides are employed in adsorption processes (Hitam et al. 2018, Rahman et al. 2011). The surface energy of metal absorbents is improved by decreasing their size and generating more active sites on their surface for the adsorption of pollutant molecules (Gusain et al. 2019). Immobilizing ZnO NPs on polymer substrates is another agreeable technique in the production of modified-ZnO photocatalysts. According to research, ZnO/polymer nanocomposite achieves the requisite photocatalytic activity (Shirdel and Behnajady 2020). Hydrothermal synthesis and homogeneous precipitation are the most appropriate procedures to create ZnO photocatalytic materials. A large number of studies have demonstrated CuO's remarkable performance in the photocatalytic breakdown of contaminants. Chen et al. (2020) reported a 97% photodegradation of crystal violet dye utilizing monoclinic crystalline CuO nanoparticles when exposed to visible light. Carbon nanotubes (CNTs) are also used as innovative materials for photocatalyst due to their higher quantum efficiency and excellent chemical stability. Gupta (2017) successfully developed ultrathin photocatalyst (SWCNTs-TiO_2) for wastewater treatment.

16.4.2 Nanoadsorbents

Heavy metal discharge from industrial, municipal, agricultural and household wastewater has become a significant ecological concern. Over the last few decades, a new class of nano-adsorbents has been developed to combat this rising menace. Adsorption is commonly used for heavy metal removal due to its low cost, efficiency and simplicity. They have acquired recognition as a result of their particular characteristics, and have demonstrated outstanding promise in the treatment of wastewater and industrial effluents for reuse in a wide range of applications for the long-term sustainability of the environment. Adsorption has long been known as a phenomenon in water treatment. It is a common experience in the gaseous phase, but is used effectively in the treatment of pollutants from soil and water. The specificity for a particular hazardous chemical, the adsorption efficiency and the benefit-cost ratio are important factors in selecting adsorbents. In the recent past, different nanomaterials viz. carbon analogues, carbonaceous nanostructured composites, nano-magnetic materials, microporous glasses, adsorbent sieves ceramics, clay polymers, etc. have been synthesized (Kumar and Guleria 2020). In comparison to the materials utilized historically and commercially as adsorbents, they have been found to have a high adsorption capacity for heavy metals from wastewater. Among nanoabsorbents, carbon nanomaterials are most extensively used for the removal of heavy metals. The carbon nanomaterials can be classified as (i) Zero-dimensional (having all the three dimensions less than 100 nm); for example, are fullerene and quantum dots. (ii) One-dimensional (1 D) (having only one dimension larger than 100 nm and two dimensions smaller than 100 nm), e.g., nanotubes of carbon and titanium. (iii) Two-dimensional nanomaterials (with two dimensions greater than 100 nm), e.g., graphene. (iv) Three dimensional (3-D) (all dimensions are greater than 100 nm), e.g., graphite and some nono-composites. Highly efficient nano absorbents are described next.

16.4.2.1 Carbon Nano Tubes (CNTs)

Engineered materials with unique characteristics such as electrical conductivity, optical activity, mechanical strength and surface morphologies are known as carbon nanotubes (CNTs). Due to their high porosity, light mass density, large specific area, hollow structure and strong interactions with pollutant molecules, they are used as effective adsorbents. Internal sites, interstitial channels, grooves and the outer surface of CNT bundles are all adsorption sites. When compared to internal sites, exterior sites attain equilibrium faster. The adsorption process is largely driven by chemical interactions between the surface functional groups of CNTs and metal ions. CNTs often include –OH, –C=O and –COOH groups, depending on the synthesis and purifying procedure. However, some other functional groups can also be added to CNTs through oxidation using Pd, Ni or Pt or removed from CNTs through heat treatment at 2200°C. Cu, Zn and Pb were removed using CNTs manufactured using the microwave irradiation technique. Kaushal and Singh (2017) reported removal efficiency of 111 mg g^{-1}, 71 mg g^{-1}, and 85 mg g^{-1} for Cu (II), Pb(II) and Zn(II), respectively by these types CNTs. Carbon nanotubes can be classified based on their dimensions into zero-dimensional (0-D, e.g., fullerene and quantum dots), one dimensional (1-D, e.g., nanotubes and titanium), two dimensional (2-D, e.g., graphene), or multi-dimensional. The one-dimensional single wall CNTs (SWCNTs) are promising for their nano-sized porous structure, high surface area and easy surface functionalization. Thus, SWCNTs are widely used for the removal of metal pollutants. Researchers demonstrated SWCNT-based nanocomposite could efficiently adsorb Hg (more than 99.56% within 7 min). In an experiment conducted by Anitha et al. (2015) to evaluate the adsorption capacity of unadorned SWCNTs and their functionalized SWCNTs (viz., SWCNTs-OH, SWCNTs-NH$_2$ and SWCNTs-COOH) to remove meta(loids). It has been observed that SWCNTs-COOH have the highest adsorption capacities (150–230%) compared to unadorned SWCNTs. However, SWCNTs-OH and SWCNTs-NH have resulted in poor adsorption capacity. SWCNTs-COOH have also shown the adsorption capacity of 96, 77, and 56 mg g^{-1} for Pb^{2+}, Cu^{2+}, and Cd^{2+}, respectively. There is another type of CNTs developed by researchers called multiwall carbon nanotubes (MWCNTs). Zhang et al. (2012) created MWCNTs-TiO$_2$ and used them to photodegrade methylene blue. Zhang et al. (2012) reported MWCNTs based nano-adsorbent can efficiently removal Pb (II) and Hg (II) to the extent of 65.40 mg g^{-1} and 65.52 mg g^{-1}, respectively. Moreover, sulfonated multi-walled CNTs (s-MWCNTs) were synthesized by Ge et al. (2014) through processing p-MWCNTs with strong sulfuric acid at high temperatures to adsorb Cu (II). However, since their discovery in the early 1990s, CNT toxicity has been a significant issue. Another disadvantage of employing CNTs is their low removal effectiveness and limited selectivity. Furthermore, because the walls of carbon nanotubes are not reactive, the sorption capacities for metal ions are relatively low (Mukhopadhyay et al. 2020).

16.4.2.2 Metal based Nano-adsorbents

Metal oxides, such as Fe, Cu, Ti, Mn, Mg, Zn, Si and Al, can be used to synthesize nanoparticles (Table 16.5). Due to metal-ligand precipitation or the generation of ternary ligands, metal oxide nanoparticles have a greater degree of adsorption than regular sized oxide (Stietiya and Wang 2014). Metal oxide-based nanomaterials with very well structural, crystalline and surface properties act as semiconductors with large band gaps. They also have advantageous properties such as non-toxicity and excellent water stability. Metal oxides include ferric oxides, Ti, Ce, and Zn, provide low-cost adsorbents. The adsorption capacity of metal-based nanoparticles is pH deepened. Heavy metals adsorption from the media enhanced with pH due to an increase in electrostatic interactions and formation of ionic or covalent bonds. In addition, improvement in pH favors the release of protons of nanosorbent, enhancing the negatively charged sites. Nassar (2012) showed 36 mg g^{-1} absorption capacity of Fe_3O_4 nanoadsorbents for the removal of Pb (II) ions. Moreover, ultrafine magnesium ferrite ($MgO.27Fe_2.5O_4$) nanoadsorbent synthesized by Tang et al. (2013) functioned as a good adsorbent for As from contaminated water. Alumina nanoadsorbent produced through the solution combination synthesis method was reported to be highly efficient in removing Zn (II) from wastewaters. It is notable that nanosized iron oxide particles have superparamagnetism, which distinguishes them from other oxide nanoparticles. Super paramagnetic nanoparticles have a higher surface area, are biocompatible, less hazardous, chemically inert, have a low diffusion resistance and their surface may be changed with organic molecules, inorganic ions or functional groups, making them ideal surfaces for absorbing heavy metals. Fe_3O_4 superparamagnetic nanoadsorbents with a diameter less than

Table 16.5. Removal efficiency of different metal based nano absorbents and carbon nanotube-based adsorbents.

Metal based nano-adsorbent	Metal removal	Efficiency (mg g^{-1})	References
Fe_3O_4 nanoadsorbents	Pb (II)	36	Nassar 2012
Anatase oxide nanoadsorbent	Pb, Cu, As	31.25, 23.74, 16.98	Kocabas-Atkali and Yurum 2013
Magnesium ferrite nanoadsorbent	As	127.4 (As^{+3}) and 83.2 (As^{+5})	Tang et al. 2013
Manganese feroxyhyte nanoadsorbent	As	11.7 μg mg^{-1} (As^{+3}), 6.7 μg/mg (As^{+5})	Tresintsi et al. 2013
Fe-La composite oxide	As	58.2	Zhang et al. 2012
Alumina nanoadsorbent	Zn	1047.83	Bhargavi et al. 2015
Fe_3O_4 magnetic nanoparticles	Hg	-	Nassar 2012
Mercapto-functionalized nano-Fe_3O_4 magnetic nanoadsorbent (SH-Fe_3O_4-NMPs)	Hg	-	Chalasani and Vasudevan 2012
Fe_3O_4 superparamagnetic nanoadsorbent coated with ascorbic acid	As	16.56 (As^{+3}) and 46.06 (As^{+5})	Nassar 2012
$MnFe_2O_4$ nanoadsorbent	Pb, Cu and Cd	481.2 for Pb^{+2}, 386.2 for Cu^{+2} and 345.5 for Cd^{+2}	Ma et al. 2015
SWCNTs	Hg	41.66	Alijani and Shariatinia 2018
SWCNTs-Fe_3O_4-CoS	Hg	1666	Alijani and Shariatinia 2018
MWCNTs	Cr	1.26	Dehghani et al. 2015
Al_2O_3-MWCNTs	Pb	-	Gupta 2017
Porous graphene	As	-	Tabish et al. 2018
Activated carbon	Pb, Cd, Cu, Ni	238.1, 96.2, 87.7, 52.4	Li et al. 2018

10 nm and coated with ascorbic acid were highly efficient in removing As. These nanoadsorbentions had the maximum adsorption capacity of 16.56 mg g^{-1} and 46.06 mg g^{-1} for As(III) and As(V), respectively (Feng et al. 2012). Further, Fe_2O_4 MNPs coated with oxide shells of Mn-Co is an excellent adsorber of Pb (II) (481.2 mg g^{-1}), Cu (II) (386.2 mg g^{-1}) and Cd (II) (345.5 mg g^{-1}) (Ma et al. 2013). Silica nanoparticles are also considered to be highly efficient in removing heavy metals. Silica nanoparticles along with graphite oxide were used to remove Zn, Ni, Cr, Pb and Cd by Sheet et al. (2014). When metal oxides are used as adsorbents without any supporting elements (viz., clays or zeolites), they may have limited adsorption capacity, smaller surface area and release hazardous metals into the environment.

16.4.4.3 Clay-polymer Nanocomposites

Describing the removal of metalloids from polluted media and many types of adsorbents (organic and inorganic materials) have already been addressed. The clay polymer adsorbents include zeolites (activated and natural), natural clay minerals, modified clay minerals, etc. (Mukhopadhyay et al. 2020). Available clay minerals are of low cost, readily available and can be exclusively used. However, their low surface area, less effectiveness for micro pollutants and lack of standard protocols for regeneration are major limitations. In the recent past, researchers have begun to utilize Clay-Polymer Nanocomposites (CPNs) obtained by clay minerals and resins conjointly as a single adsorbent. The nanocomposites have shown immense potential for removal of major pollutants in water such as Cu, Cr, As, Pb, Ni and Cd (Table 16.6). With the development of nanoscience, several clay polymer nanocomposites are synthesized, such as, epoxy-clay nanocomposites, polystyrene-clay nanocomposites, polyurethane-clay nanocomposites, magnetic clay polymer nanocomposites, polyimide-clay hybrids, etc. Recently, magnetized-CPN was developed using bentonite clay, iron oxide nanoparticles and monomer methyl methacrylate for the removal of Cr (VI) (Sundaram et al. 2018). The application of these CPN had an adsorption capacity of 113 mg g^{-1}. Whereas chitosan-Al-pillared-montmorillonite nanocomposite can adsorb Cr (VI) 15.67 mg g^{-1}

Table 16.6. Efficiency of clay-polymer nanocomposites in metal removal.

Clay-polymer nanocomposites	Metal removed	Efficiency (mg g^{-1})	References
Alginate–montmorillonite nanocomposite	Pb	238.1	Shawky 2011
Polyacrylic acid/bentonite	Ni	270.27	Mukhopadhyay et al. 2020
Polyacrylic acid/bentonite	Pb	1666.67	Bulut and Tez 2009
Chitosan-Al-pillared-montmorillonite nanocomposite	Pb	-	Mukhopadhyay et al. 2020
Chitosan immobilized on bentonite	Ni	15.82	Mukhopadhyay et al. 2020
Polyacrylic acid/bentonite	Cd	416.67	Mukhopadhyay et al. 2020
Modifed bentonite clay composits	Zn	-	Mukhopadhyay et al. 2020
Polyetherimide/porous activated bentonite	Cd	-	
Chitosan-Al-pillared-montmorillonite nanocomposite	Cu	-	Pereira et al. 2013
Chitosan:clay (0.45:1)	Cu	-	Pereira et al. 2013
Montmorillonite and chitosan	Se	18.4	
Pillared bentonite by MnCl2	Pb	12.6	Mukhopadhyay et al. 2020
Chitosan-bentonite clay	Ni, Cd	-	Mukhopadhyay et al. 2020
Alginate–montmorillonite nanocomposite	Fe		Mukhopadhyay et al. 2020
Bentonite/thiourea-formaldehyde composite	Mn	4.81	Mukhopadhyay et al. 2020
Na-montmorillonite/cellulose	Cr	22.2	Mukhopadhyay et al. 2020
Chitosan–clay composites and oxides	Se		Bleiman and Mishael 2010

(Sakulthaew et al. 2017). Recently, chitosan-based clay composite showed high adsorption of Cu (II) in the tune of 96.0 and 99.5%, respectively, at a pH of 6.5 (Sakulthaew et al. 2017). In addition, intercalation polymerization generated organo-bentonite and polyacrylonitrile composite showed the removal of Cd, Zn and Cu at the tune of 52.6, 65.4 and 77.4 mg g^{-1}, respectively (Mukhopadhyay et al. 2020).

16.5 Future Research Need

In this chapter, nanotechnology-based remediation techniques for removing metal(loid)s pollutants from the environment have been described thoroughly. These nano-structured materials have shown great potential in the decontamination of heavy metal-contaminated media. Applications of nanotechnology have immense potential for detecting polluted sites and their remediation sustainably. All the nanoadsorbents discussed earlier have tremendous potential for the removal of pollutants from the environment. Incorporating clay at appropriate doses with a variety of inorganic polymers and biopolymers (or nanoclay) for the development of nanocomposites enhanced the effectiveness of nanoabsorbents for the removal of pollution. Furthermore, the majority of research demonstrated the efficiency of nanomaterials at the laboratory scale, with only a few attempts to evaluate at the actual polluted locations. To continue the economic feasibility and sustainability of nanotechnology for the removal of metal pollutants is the need of the hour. Furthermore, to boost the sustainability of these remediation techniques, proper safety protocols and standardized methodology should be developed. In addition, awareness should be generated for nanotechnology-based remediation through more field-level applications. Sustainable agriculture, food availability and nutritional security are among the century's core sustainable development goals. Therefore, it is critical to use the benefits of nanotechnology in accomplishing the feat by boosting plant nutrient availability and lowering plant losses on agricultural soils. Approaches are needed to the synthesis of nanomaterials that are more environmentally friendly and sustainable. Further, advanced tools for ecotoxicological evaluation and safety measures should be developed. The establishment of regulatory agencies at national and international levels is essential to monitor the diffusion of nanoparticles in the environment and to implement nanotechnology-based approaches for improved results, with the goal of minimizing negative consequences.

References

Ajmal, A., I. Majeed, R. N. Malik, H. Idriss and M. A. Nadeem. 2014. Principles and mechanisms of photocatalytic dye degradation on TiO$_2$ based photocatalysts: a comparative overview. Rsc. Adv. 4(70): 37003–37026.
Alijani, H. and Z. Shariatinia. 2018. Synthesis of high growth rate SWCNTs and their magnetite cobalt sulfide nanohybrid as super-adsorbent for mercury removal. Chem. Eng. Res. Des. 129: 132–149.
Anitha, K., S. Namsani and J. K. Singh. 2015. Removal of heavy metal ions using a functionalized single-walled carbon nanotube: a molecular dynamics study. J. Phys. Chem. A. 119(30): 8349–8358.
Bhargavi, R. J., U. Maheshwari and S. Gupta. 2015. Synthesis and use of alumina nanoparticles as an adsorbent for the removal of Zn(II) and CBG dye from wastewater. Int. J. Ind. Chem. 6: 31–41.
Bleiman, N. and Y. G. Mishael. 2010. Selenium removal from drinking water by adsorption to chitosan–clay composites and oxides: batch and columns tests. J. Hazard. Mater. 183(1-3): 590–595.
Boken, J. and D. Kumar. 2014. Detection of toxic metal ions in water using SiO 2@ Ag Core-Shell nanoparticles. Int. J. Environ. Res. Dev. 4: 303–308.
Bulut, Y. and Z. Tez. 2009. Adsorption of heavy metal ions from aqueous solutions by bentonite. J. Colloid Interface Sci. 332(1): 46–53.
Chai, F., C. Wang, T. Wang, L. Li and Z. Su. 2010. Colorimetric detection of Pb^{2+} using glutathione functionalized gold nanoparticles. ACS Appl. Mater. Interface. 2(5): 1466–1470.
Chalasani, R. and S. Vasudevan. 2012. Cyclodextrin functionalized magnetic iron oxide nanocrystal: a host-carrier for magnetic separation of non-polar molecules and arsenic from aqueous media. J. Mater. Chem. 22: 14925–14931.
Chen, K. I., B. R. Li and Y. T. Chen. 2011. Silicon nanowire field-effect transistor-based biosensors for biomedical diagnosis and cellular recording investigation. Nano Today. 6(2): 131–154.

Chen, K., X. Wang, P. Xia and J. Xie. 2020. Efficient removal of tetrabromodiphenyl ether with a Z-scheme Cu2O (rGOTiO2) photocatalyst under sunlight irradiation. Chemosphere. 254: 126806.

Corsi, I., M. Winther-Nielsen, R. Sethi, C. Punta, C. Della Torre, G. Libralato, G. Lofrano, L. Sabatini, M. Aiello, L. Fiordi and F. Cinuzzi. 2018. Ecofriendly nanotechnologies and nanomaterials for environmental applications: key issue and consensus recommendations for sustainable and ecosafenanoremediation. Ecotoxicol. Environ. Saf. 154: 237–244.

Czech, B, P. Zygmunt, Z. C. Kadirova, K. Yubuta and M. Hojamberdiev. 2020. Effective photocatalytic removal of selected pharmaceuticals and personal care products by Elsmoreite/Tungsten Oxide@Zns Photocatalyst. J. Environ. Manag. 270: 110870.

Danish, M. S. S., L. L. Estrella, I. M. A. Alemaida, A. Lisin, N. Moiseev, M. Ahmadi, M. Nazari, M. Wali, H. Zaheb and T. Senjyu. 2021. Photocatalytic applications of metal oxides for sustainable environmental remediation. Metals. 11(1): 80.

Darbha, G. K., A. Ray and P. C. Ray. 2007. Gold nanoparticle-based miniaturized nanomaterial surface energy transfer probe for rapid and ultrasensitive detection of mercury in soil, water, and fish. Acs Nano. 31(3): 208–14.

Dehghani, M. H., M. M. Taher, A. K. Bajpai, B. Heibati, I. Tyagi, M. Asif, S. Agarwal and V. K. Gupta. 2015. Removal of noxious Cr (VI) ions using single-walled carbon nanotubes and multi-walled carbon nanotubes. Chem. Eng. J. 279: 344–352.

Fajardo, C., S. Sánchez-Fortún, G. Costa, M. Nande, P. Botías, J. García-Cantalejo, G. Mengs and M. Martín. 2020. Evaluation of nanoremediation strategy in a Pb, Zn and Cd contaminated soil. Sci. Total Environ. 706: 136041.

Feng, L., M. Cao, X. Ma, Y. Zhu and C. Hu. 2012. Superparamagnetic high-surface-area Fe_3O_4 nanoparticles as adsorbents for arsenic removal. J. Hazard. Mater. 217: 439–446.

Ge, Y., Z. Li , D. Xiao, P. Xiong and N. Ye. 2014. Sulfonated multi-walled carbon nanotubes for the removal of copper (II) from aqueous solutions. J. Ind. Eng. Chem. 20(4): 1765–71.

Gich, M., C. Fernández-Sánchez, L. C. Cotet, P. Niu and A. Roig. 2013. Facile synthesis of porous bismuth–carbon nanocomposites for the sensitive detection of heavy metals. J. Mater. Chem. A 1(37): 11410–8.

Gupta, R. K. 2017. Oil/water separation techniques: a review of recent progresses and future directions. J. Mater. Chem. A. 5(31): 16025–16058.

Gusain, R., K. Gupta, P. Joshi and O. P. Khatri. 2019. Adsorptive removal and photocatalytic degradation of organic pollutants using metal oxides and their composites: a comprehensive review. Adv. Colloid Interface Sci. 272: 102009.

Han, D., S. Y. Lim, B. J. Kim, L. Piao and T. D. Chung. 2010. Mercury (II) detection by SERS based on a single gold microshell. Chem. Commun. 46(30): 5587–5589.

Hitam, C. N. C., A. A. Jalil, S. Triwahyono, A. F. A. Rahman, N. S. Hassan, N. F. Khusnun, S. F. Jamian, C. R. Mamat, W. Nabgan and A. Ahmad. 2018. Effect of carbon-interaction on structure-photoactivity of Cu doped amorphous TiO_2 catalysts for visible-light- oriented oxidative desulphurization of dibenzothiophene. Fuel. 216: 407–417.

Kaushal, A. and S. K. Singh. 2017. Removal of heavy metals by nanoadsorbents: a review. J. Environ. Biotechnol. Res. 6(1): 96–104.

Khanam, R., A. Kumar, A. K. Nayak, M. Shahid, R. Tripathi, S. Vijayakumar, D. Bhaduri, U. Kumar, S. Mohanty, P. Panneerselvam and D. Chatterjee. 2020. Metal(loid)s (As, Hg, Se, Pb and Cd) in paddy soil: bioavailability and potential risk to human health. Sci. Total Environ. 699: 134330.

Kim, T. H., J. Lee and S. Hong. 2009. Highly selective environmental nanosensors based on anomalous response of carbon nanotube conductance to mercury ions. J. Phys. Chem. C. 113(45): 19393–19396.

Kimmel, D. W., G. LeBlanc, M. E. Meschievitz and D. E. Cliffel. 2012. Electrochemical sensors and biosensors. Anal. Chem. 84(2): 685–707.

Kocabas-Atakli, Z. Ö. and Y. Yürüm. 2013. Synthesis and characterization of anatase nanoadsorbent and application in removal of lead, copper and arsenic from water. Chem. Eng. J. 225: 625–635. http://dx.doi.org/10.1016/j.cej.2013.03.106.

Konate, A., X. He, Z. Zhang, Y. Ma, P. Zhang, G. M. Alugongo and Y. Rui. 2017. Magnetic (Fe3O4) nanoparticles reduce heavy metals uptake and mitigate their toxicity in wheat seedling. Sustain. 9: 790.

Kumar, V. and P. Guleria. 2020. Application of DNA-Nanosensor for environmental monitoring: recent advances and perspectives. Curr. Pollut. Rep. 12: 1–21.

Li, F., J. Wang, Y. Lai, C. Wu, S. Sun, Y. He and H. Ma. 2013a. Ultrasensitive and selective detection of copper (II) and mercury (II) ions by dye-coded silver nanoparticle-based SERS probes. Biosens. Bioelectron. 39(1): 82–87.

Li, M., S. K. Cushing, J. Zhang, J. Lankford, Z. P. Aguilar, D. Ma and N. Wu. 2012. Shape-dependent surface-enhanced Raman scattering in gold–Raman-probe–silica sandwiched nanoparticles for biocompatible applications. Nanotechnol. 23(11): 115501.

Li, M., S. K. Cushing, H. Liang, S. Suri, D. Ma and N. Wu. 2013b. Plasmonic nanorice antenna on triangle nanoarray for surface-enhanced Raman scattering detection of hepatitis B virus DNA. Anal. Chem. 85(4): 2072–2078.

Li, S., Y. Li, J. Cao, J. Zhu, L. Fan and X. Li. 2014. Sulfur-doped graphene quantum dots as a novel fluorescent probe for highly selective and sensitive detection of Fe^{3+}. Anal. Chem. 86(20): 10201–7.

Li, J., X. Xing, J. Li, M. Shi, A. Lin, C. Xu and R. Li. 2018. Preparation of thiol-functionalized activated carbon from sewage sludge with coal blending for heavy metal removal from contaminated water. Environl. pollut. 234: 677–683.

Liu, J., G. Lv, W. Gu, Z. Li, A. Tang and L. Mei. 2017. A novel luminescence probe based on layered double hydroxides loaded with quantum dots for simultaneous detection of heavy metal ions in water. J. Mater. Chem. 5(20): 5024–5030.

Ma, H., H. Wang and C. Na. 2015. Microwave-assisted optimization of platinum-nickel nanoalloys for catalytic water treatment Appl. Catal. B Environ. 163: 198–204.

Ma, Z., D. Zhao, Y. Chang, S. Xing, Y. Wu and Y. Gao. 2013. Synthesis of MnFe2 O4 and Mn–Co oxide core–shell nanoparticles and their excellent performance for heavy metal removal. Dalt. Trans. 42(39): 14261–14267.

Maghsoudi, A. S., S. Hassani, K. Mirnia and M. Abdollahi. 2021. Recent advances in nanotechnology-based biosensors development for detection of arsenic, lead, mercury, and cadmium. Int. J. Nanomed. 16: 803.

Masudy-Panah, S., R. Katal, N. D. Khiavi, E. Shekarian, J. Hu and X. Gong. 2019. A high-performance cupric oxide photocatalyst with palladium light trapping nanostructures and a hole transporting layer for photoelectrochemical hydrogen evolution. J. Mater. Chem. A 7: 22332–22345.

Mawia, A. M., S. Hui, L. Zhou, H. Li, J. Tabassum, C. Lai, J. Wang, G. Shao, X. Wei, S. Tang, J. Luo, S. Hu and P. Hu. 2021. Inorganic arsenic toxicity and alleviation strategies in rice. J. Hazard. Mater. 408: 124751. https://doi.org/10.1016/j.jhazmat.2020.124751.

Mukhopadhyay, R., D. Bhaduri, B. Sarkar, R. Rusmin, D. Hou, R. Khanam et al. 2020. Clay–polymer nanocomposites: progress and challenges for use in sustainable water treatment. J. Hazard. Mater. 383: 121125.

Mulvihill, M., A. Tao, K. Benjauthrit, J. Arnold and P. Yang. 2008. Surface-enhanced Raman spectroscopy for trace arsenic detection in contaminated water. Angewandte Chemie International Edition. 47(34): 6456–60.

Nassar, N. N. 2012. Iron oxide nanoadsorbents for removal of various pollutants from wastewater: an overview. pp. 81–118. *In*: A. Bhatnagar [Ed.]. Application of Adsorbents for Water Pollution Control. Bentham Science Publishers, India.

Niu, P., C. Fernández-Sánchez, M. Gich, C. Ayora and A. Roig. 2015. Electroanalytical assessment of heavy metals in waters with bismuth nanoparticle-porous carbon paste electrodes. Electrochimica Acta. 165: 155–161.

Patil, P. O., P. V. Bhandari, P. K. Deshmukh, S. S. Mahale, A. G. Patil, H. R. Bafna, K. V. Patel and S. B. Bari. 2017. Green fabrication of graphene-based silver nanocomposites using agro-waste for sensing of heavy metals. Res. Chem. Intermed. 43(7): 3757–3773.

Pereira, F. A., K. S. Sousa, G. R. Cavalcanti, M. G. Fonseca, A. G. de Souza and A. P. Alves. 2013. Chitosan-montmorillonite biocomposite as an adsorbent for copper (II) cations from aqueous solutions. Int. J. Biol. Macromol. 61: 471–478.

Petty, J. T., J. Zheng, N. V. Hud and R. M. Dickson. 2004. DNA-templated Ag nanocluster formation. J. Am. Chem. Soc. 126(16): 5207–5212.

Praveen, A., E. Khan, D. S. Ngiimei, M. Perwez, M. Sardar and M. Gupta. 2018. Iron oxide nanoparticles as nano-adsorbents: a possible way to reduce arsenic phytotoxicity in Indian mustard plant (*Brassica juncea* L.). J. Plant Growth Regul. 37(2): 612–624.

Rahman, A. F. A., A. A. Jalil, S. Triwahyono, A. Ripin, F. F. A. Aziz, N. A. A. Fatah, N. F. Jaafar, C. N. C. Hitam, N. F. M. Salleh and N. S. Hassan. 2011. Strategies for introducing titania onto mesostructured silica nanoparticles targeting enhanced photocatalytic activity of visible-light-responsive Ti-MSN catalysts. J. Clean Prod. 143: 1–12.

Rotaru, A., S. Dutta, E. Jentzsch, K. Gothelf and A. Mokhir 2010. Selective dsDNA-templated formation of copper nanoparticles in solution. Angewandte Chemie International Edition. 49(33): 5665–7.

Sabzehei, K., S. H. Hadavi, M. G. Bajestani and S. Sheibani. 2020. Comparative evaluation of copper oxide nano-photocatalyst characteristics by formation of composite with TiO_2 and Zno. Solid State Sci. 10: 106362.

Sakulthaew, C., C. Chokejaroenrat, A. Poapolathep, T. Satapanajaru and S. Poapolathep. 2017. Hexavalent chromium adsorption from aqueous solution using carbon nano-onions (CNOs). Chemosphere. 184: 1168–1174.

Sauer, M. 2003. Single-molecule-sensitive fluorescent sensors based on photoinduced intramolecular charge transfer. Angewandte Chemie International Edition. 42(16): 1790–3.

Schopf, C., A. Martín and D. Iacopino. 2017. Plasmonic detection of mercury via amalgam formation on surface-immobilized single Au nanorods. Sci. Technol. Adv. Mater. 18(1): 60–67.

Sha, J. C., C. Tong, H. Zhang, L. Feng and B. Liu. 2015. CdTe QDs functionalized mesoporous silica nanoparticles loaded with conjugated polymers: a facile sensing platform for cupric (II) ion detection in water through FRET. Dyes Pigments. 113: 102–9.

Shafi, A., S. Bano, N. Khan, S. Sultana, Z. Rehman, M. M. Rahman, M. M. Sabir, F. Coulon and M. Z. Khan. 2021. Nanoremediation technologies for sustainable remediation of contaminated environments: recent advances and challenges. Chemosphere. 130065.

Sharma, V., A. K. Saini and S. M. Mobin. 2016. Multicolour fluorescent carbon nanoparticle probes for live cell imaging and dual palladium and mercury sensors. J. Mater. Chem. B. 4(14): 2466–2476.

Shawky, H. A. 2011. Improvement of water quality using alginate/montmorillonite composite beads. J. Appl. Polymer Sci. 119(4): 2371–2378.

Sheet, I., A. Kabbani and H. Holail. 2014. Removal of heavy metals using nanostructured graphite oxide, silica nanoparticles and silica/graphite oxide composite. Energy procedia. 50: 130–138.

Shirdel, B. and M. A. Behnajady. 2020. Visible-light induced degradation of rhodamine B by Ba-doped ZnO Nanoparticles. J. Mol. Liq. 315: 113633.

Siddiqui, M. H. and M. H. Al-Whaibi. 2014. Role of nano-SiO2 in germination of tomato (*Lycopersicum esculentum* seeds Mill.). Saudi J. Biol. Sci. 21: 13–17.

Stietiya, M. H. and J. J. Wang. 2014. Zinc and cadmium adsorption to aluminum oxide nanoparticles affected by naturally occurring ligands. J. Environ. Qual. 43(2): 498–506.

Sundaram, R. M., A. Sekiguchi, M. Sekiya, T. Yamada and K. Hata. 2018. Copper/carbon nanotube composites: research trends and outlook. Royal Soc. Open Sci. 5(11): 180814.

Tabish, T. A., F. A. Memon, D. E. Gomez, D. W. Horsell and S. Zhang. 2018. A facile synthesis of porous graphene for efficient water and wastewater treatment. Sci. Rep. 8(1): 1–14.

Tang, W., Y. Su, Q. Li, S. Gao and J. K. Shang. 2013. Superparamagnetic magnesium ferrite nanoadsorbent for effective arsenic (III, V) removal and easy magnetic separation. Water Res. 47(11): 3624–3634.

Ting, S. L., S. J. Ee, A. Ananthanarayanan, K. C. Leong and P. Chen. 2015. Graphene quantum dots functionalized gold nanoparticles for sensitive electrochemical detection of heavy metal ions. Electrochimica Acta. 172: 7–11.

Tresintsi, S., K. Simeonidis, S. Estradé, C. Martinez-Boubeta, G. Vourlias, F. Pinakidou, M. Katsikini, E. C. Paloura, G. Stavropoulos and M. Mitrakas. 2013. Tetravalent manganese feroxyhyte: a novel nanoadsorbent equally selective for As (III) and As (V) removal from drinking water. Environ. Sci. Technol. 47(17): 9699–9705.

Varun, S. and S. C. Daniel. 2018. Emerging nanosensing strategies for heavy metal detection. Nanotechnol. Sustain. Water Resour. 5: 199–255.

Xie, Y. L., S. Q. Zhao, H. L. Ye, J. Yuan, P. Song and S. Q. Hu. 2015. Graphene/CeO$_2$ hybrid materials for the simultaneous electrochemical detection of cadmium (II), lead (II), copper (II), and mercury (II). J. Electroanal. Chem. 15: 757: 235–242.

Yaqoob, S., F. Ullah, S. Mehmood, T. Mahmood, M. Ullah, A. Khattak and M. A. Zeb. 2018. Effect of waste water treated with TiO2 nanoparticles on early seedling growth of *Zea mays* L. J. Water Reuse Desalin. 8(3): 424–431.

Zhang, C., J. Sui, Y. Li, Tang and W. Cai. 2012. Efficient removal of heavy metal ions by thiol-functionalized superparamagnetic carbon nanotubes. Chem. Eng. J. 210: 45–52.

Zhang, J., L. He, P. Chen, C. Tian, J. Wang, B. Liu, C. Jiang and Z. Zhang. 2017. A silica-based SERS chip for rapid and ultrasensitive detection of fluoride ions triggered by a cyclic boronate ester cleavage reaction. Nanoscale. 9(4): 1599–606.

Zhang, R. and W. Chen. 2014. Nitrogen-doped carbon quantum dots: facile synthesis and application as a "turn-off" fluorescent probe for detection of Hg^{2+} ions. Biosens. Bioelectron. 55: 83–90.

Zhao, L., W. Gu, C. Zhang, X. Shi and Y. Xian. 2016. *In situ* regulation nanoarchitecture of Au nanoparticles/reduced graphene oxide colloid for sensitive and selective SERS detection of lead ions. J. Colloid Interface Sci. 465: 279–285.

CHAPTER 17

Modification Strategies of g-C$_3$N$_4$ for Potential Applications in Photocatalysis
A Sustainable Approach towards the Environment

Sachin Shoran,[1] *Sweety Dahiya,*[1] *Anshu Sharma*[2]
and *Sudesh Chaudhary*[1],*

17.1 Introduction

Graphitic carbon nitride (g-C$_3$N$_4$) has been used as a visible-light-active polymeric photocatalyst (460 nm) with a bandgap of 2.7 eV. g-C$_3$N$_4$ has become useful in engineering and physics because of its low-cost and eco-friendly synthesis techniques. It is also a stable catalyst with promising physicochemical properties (Ong et al. 2016). In comparison with other semiconductors, g-C$_3$N$_4$ can be successfully prepared by multiple processes and has the desired electrical structures and morphologies, as well as good thermal stability in the air up to 600°C (Liu et al. 2021, Taha et al. 2021). The most frequently used precursors of g-C$_3$N$_4$ are urea, thiourea, dicyandiamide, cyanamide, melamine and ammonium thiocyanate (Figure 17.2). Among several forms of carbon nitrides, such as cubic C$_3$N$_4$, pseudocubic C$_3$N$_4$, α-C$_3$N$_4$ and β-C$_3$N$_4$; g-C$_3$N$_4$ is the most stable phase under ambient circumstances (Ong et al. 2016). Researchers have proposed many strategies for improving the performance and modulating the characteristics of g-C$_3$N$_4$, such as heterojunction with other materials and doping with metal sulfides, noble metals, nonmetals and metal oxides nanoparticles are examples of these materials (Ren et al. 2019, Jiang et al. 2020, Wei et al. 2022). Metal oxides enhance the efficiency of g-C$_3$N$_4$, such as boosting light absorption and lowering the electron and hole's recombination by encouraging charge carrier separation. This is primarily because of their appropriate band structure (Li et al. 2016, Fu et al. 2018). The g-C$_3$N$_4$ structure has been extensively

[1] Center of Excellence for Energy and Environmental Studies, Deenbandhu Chhotu Ram University of Science and Technology, Murthal-131039 (Haryana), India.
[2] Department of Physics under School of Engineering and Technology, Central University of Haryana, Mahendergarh-123031 (Haryana), India.
* Corresponding author: sudesh.energy@dcrustm.org

employed in several applications, particularly those connected to energy. The energy needed to make electricity and heat will double by 2050, primarily because of industrialization, urbanization and population growth (Dai et al. 2012)). Oil, coal and other fossil fuels should be used less often since they negatively influence the environment (Ong et al. 2016). Solar power and photocatalysis are two solutions (Hasanvandian et al. 2022). Both need appropriate semiconductors, such as g-C$_3$N$_4$, that have high activity for various catalytic processes, including water splitting, hydrogen production, the degradation of organic pollutants and the conversion of CO$_2$ (Zou et al. 2018, Arumugam et al. 2022, Yang et al. 2022). Additionally, g-C$_3$N$_4$ can be utilized to clean wastewater and control microorganisms (Zhang et al. 2019).

This chapter has focused on several broad details on the characterization and structure of bare g-C$_3$N$_4$. Some possible adjustments, including doping to improve the characteristics of g-C$_3$N$_4$ are also discussed. The studies on altering the structure and characteristics of g-C$_3$N$_4$ to increase its effectiveness for various applications by mixing it with metals, metal oxides and nonmetals have been described. In the final section of this chapter, the authors have made several recommendations for prospective future studies in this area which, to our knowledge, have not yet been done.

17.2 Structure and Properties of g-C$_3$N$_4$

Carbon Nitride (CN) has been studied since 1834 when Berzelius made the first linear CN polymer, which he called "melon" (Liebig 1834). Since then, the CN has been studied. In 1922, Franklin discovered a specific form of graphitic carbon nitride by letting mercuric thiocyanate break down in heat (Franklin 1922). In 1989, it was hypothesized that a substance known as -C$_3$N$_4$ could be created if carbon replaced silica in the structure of Silicon Nitride (Si$_3$N$_4$). Liu and Cohen (1989), Teter and Hemley (1996) predicted that there would be five different phases of CN: α-C$_3$N$_4$, β-C$_3$N$_4$, c-C$_3$N$_4$, p-C$_3$N$_4$ and g-C$_3$N$_4$. Except for g-C$_3$N$_4$, all CNs are rigid materials based on their crystal structures. So, g-C$_3$N$_4$ is much easier to change in shape and structure. As a result, the research of g-C$_3$N$_4$ gained popularity, and other varieties of g-C$_3$N$_4$ were created.

In g-C$_3$N$_4$, both the N and C atoms are sp^2 hybridized. They are linked together by bonds, which make a hexagonal shape. This six-atom ring is the triazine ring (Figure 17.1). A small unit connects the three triazine rings to a C-N bond. The N atom at the end of each triazine ring in g-C$_3$N$_4$ connects it to the following ring. This makes a planar grid structure that can grow indefinitely. The C and N atoms in g-C$_3$N$_4$ are both sp^2 hybridized, allowing them to form strongly conjugated bonds with lone

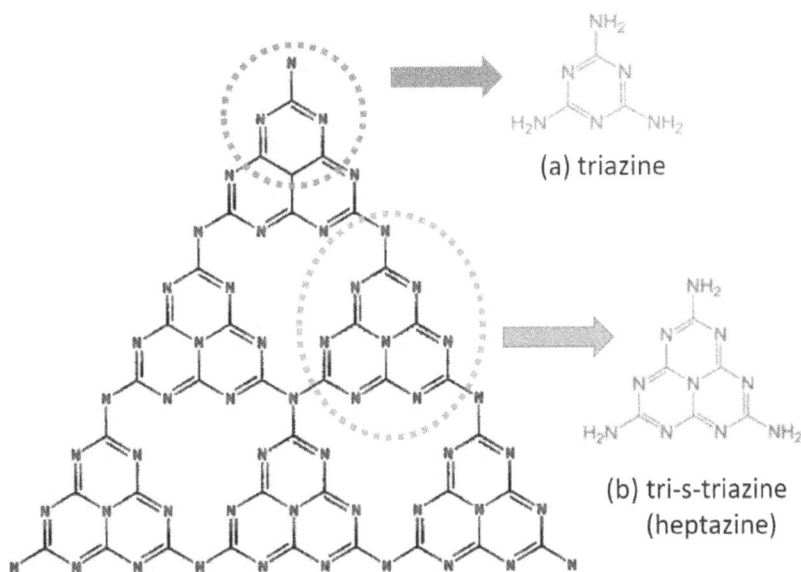

Figure 17.1. Triazine and heptazine structures of g-C$_3$N$_4$.

(a)

(b)

Graphitic Cabon Nitride (g-C₃N₄)

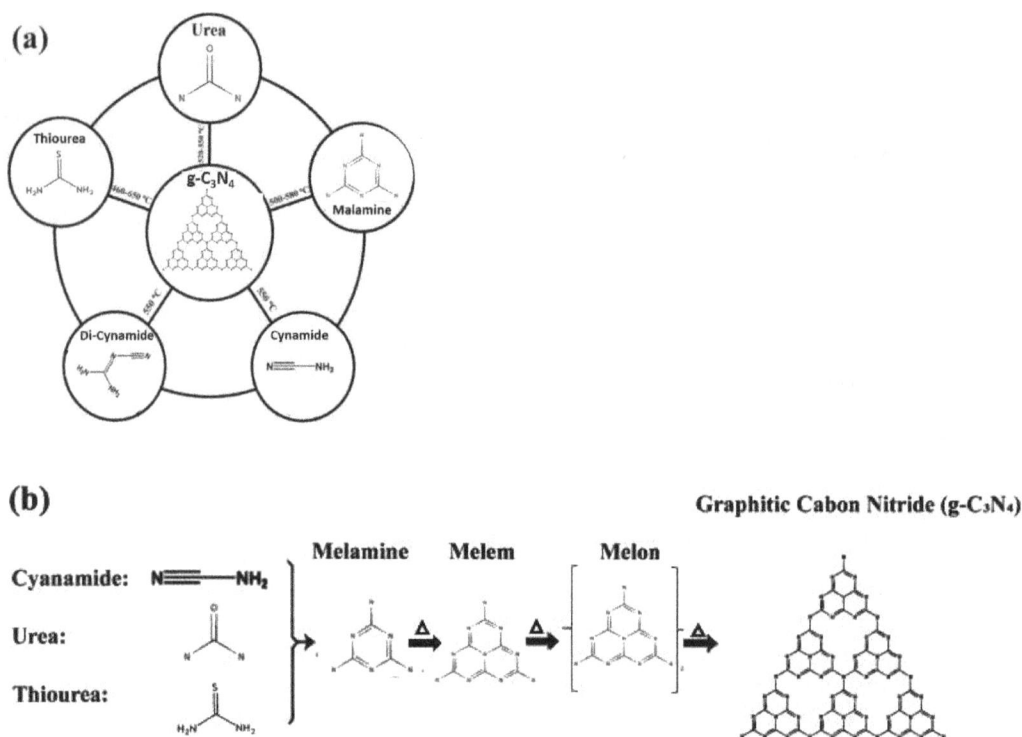

Figure 17.2(a). Synthesis precursors for g-C$_3$N$_4$ preparation; (b) synthesis procedure using urea, thiourea, cyanamide and melamine for g-C$_3$N$_4$.

electron pairs on pz orbitals. This results in the formation of a highly delocalized conjugated system (Hao et al. 2020, Li et al. 2019). The electron-hole recombination rate of bulk g-C$_3$N$_4$ is relatively high because of the presence of low-coordinated N atoms in both the Conduction band (CB) and the valence band (VB) that forms low delocalization bonds.

Most researchers are interested in how to modify g-C$_3$N$_4$ to improve its photocatalytic performance (Mishra et al. 2019, Zhang et al. 2019), such as converting energy, breaking down pollutants (Ren et al. 2019) and controlling microbes, etc. (Zhang et al. 2019).

Even though some studies offer suggestions for how to alter g-C$_3$N$_4$, a comprehensive analysis of the fabrication of g-C$_3$N$_4$ with varied dimensionalities and their implications on diverse applications connected to the environment and energy has not yet been done. This chapter provides an up-to-date review of how to make g-C$_3$N$_4$ with different dimensions and how it can be used for energy and the environment. Additionally, an overview of the research's state and suggestions on how g-C$_3$N$_4$ will develop in the future and what it could be used for has been provided.

17.3 g-C$_3$N$_4$ Preparation

The production of g-C$_3$N$_4$ has been done using a variety of synthetic methods (Figure 17.3). Here the numerous ways by which g-C$_3$N$_4$ may be made synthetically, most notably Chemical Vapor Deposition (CVD), electrochemical deposition, thermal decomposition, solid-state reactions and solvothermal reactions are summarized (Figure 17.4).

17.3.1 Chemical Vapor Deposition (CVD)

"A vacuum deposition technique called chemical vapor deposition (CVD) is used to create high-quality, high-performance solid materials." Ye et al. 2016 carried out chemical vapor deposition (CVD) to develop a layer of graphitic carbon nitride on Indium-Tin-Oxide (ITO). In this procedure,

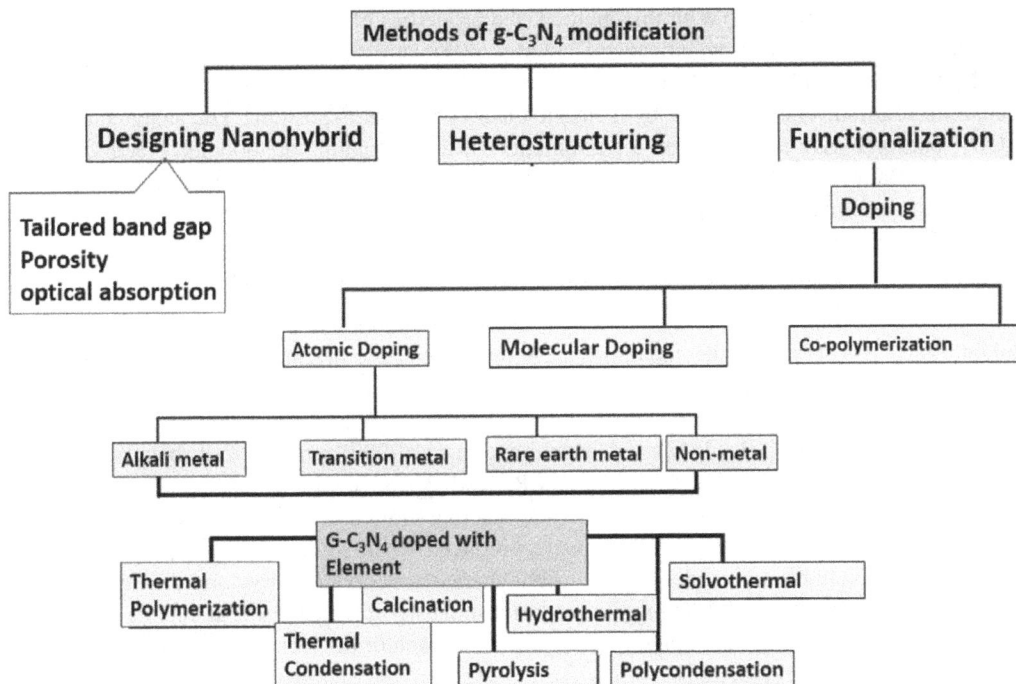

Figure 17.3. Modification methods for g-C₃N₄ and various doping strategies.

the ITO substrate was placed above the thiourea and melamine mixture in a crucible before it was moved to the muffle furnace. There are some benefits and drawbacks of CVD. CVD does not require a high vacuum to manufacture high-purity composites and may deposit various materials.

In contrast, a high vacuum environment is needed for Physical Vapor Deposition (PVD), which includes sputtering. The disadvantage of CVD is that several CVD precursors, such as Ni $(CO)_4$, B_2H_6 and $SiCl_4$, are expensive, extremely poisonous, explosive or corrosive. This process can also produce dangerous byproducts such as CO or HF. The substrates are restricted since they need to withstand high temperatures.

17.3.2 Solid-state Reaction

"The solid-state reaction is generated by contact, reaction, nucleation and crystal growth between solid interfaces at high temperature" (Baughman 1974, Wang et al. 2001). Reaction parameters such as pressures and temperature of synthesis are varied to control the crystallinity, morphology and properties of the g-C₃N₄. Lu et al. 2007 used high pressure to make g-C₃N₄ material whose crystallinity could be changed. Under 40 MPa and 220°C, sodium azide and 1,3,5-trichlorotriazine were mixed for this method. It is important to note that as the pressure in the vessel went up, g-C₃N₄ became more crystalline. When Zn particles were added to the reactor, the shape changed into nanowires, which was the most crucial change.

17.3.3 Electrochemical Deposition

"Electrochemical deposition is a method in which current moves through positive and negative ions in an electrolyte solution under the influence of an external electric field." This causes electrons to be oxidized and reduced on the electrodes to form coatings. This method has received attention because it can make products better in terms of their quality and properties. Fu et al. (1999), Li et al. (2017), Shaikh et al. (2017) were the first to use the electrodeposition method to make g-C₃N₄ from a solution of dicyandiamide. With this method, the properties of g-C$_3$N$_4$ can be changed by using different precursors in different solvents. Li et al. (2004) were able to show that highly crystalline g-C$_3$N$_4$ thin films could be made at room temperature using cyanuric chloride and an acetonitrile melamine solution. This was done by a simple electrochemical deposition. The shape and size of g-C$_3$N$_4$ particles were controlled by a combination of the templating method and electrochemical deposition. Bai et al. (2010) reported hollow g-C$_3$N$_4$ microspheres were produced using different-sized silica nanospheres. The spheres have an average diameter of about 1 m and are made up of nanoparticles that range in size from 5 to 30 nm.

17.3.4 The Solvothermal Process

Solvothermal is a way to make something new by reacting the original mixture with organic or non-aqueous solvents at a certain temperature and pressure in a closed system (like an autoclave). This method makes materials that have good crystallinity, a shape that can be controlled and good dispersity. Chen et al. (2018) and Montigaud et al. (2000) reported that cyanuric chloride and melamine in triethylamine under 130 MPa and 250°C could be used to make g-C$_3$N$_4$. Guo et al. (2003) made g-C$_3$N$_4$ nanocrystallites by reacting sodium amide and cyanuric chloride with benzene. Interestingly, g-C$_3$N$_4$ nanotubes were produced when sodium azide substituted sodium amide. This is likely because sodium azide might change how crystals grow and line up. Bai et al. (2003) made g-C$_3$N$_4$ by heating ammonium chloride and carbon tetrachloride at 400°C. g-C$_3$N$_4$ can also be made by a solvothermal reaction using carbon tetrachloride, melamine or dicyandiamide at 4.5 MPa and 290°C (Li et al. 2007).

17.3.5 Thermal Decomposition

Thermal decomposition is a common way to make a lot of g-C$_3$N$_4$ because it uses many resources, is easy to carry out and does not cost much. Overall, different ways to make g-C$_3$N$_4$ have different benefits, so researchers should consider their goals when choosing an excellent way to make g-C$_3$N$_4$. The thermal decomposition of nitrogen-rich precursors like urea, dicyandiamide, cyanamide, etc., can also be used to produce g-C$_3$N$_4$ (Lan et al. 2016, Shan et al. 2016).

Overall, g-C$_3$N$_4$ can be created by solvothermal reaction, solid-state reaction, electrochemical deposition and thermal decomposition. The solid-state reaction is not used very often because it needs high temperatures and pressures, is expensive and harmful to the environment. Electrochemical deposition can be used to make g-C$_3$N$_4$ film or spheres at room temperature. These materials could be used to make sensors and photo electrocatalysts, even though this method might not work for mass production. g-C$_3$N$_4$ with a unique shape can also be made using the solvothermal reaction, but the process is more complicated and needs high pressure. The most common way to make g-C$_3$N$_4$ is through thermal decomposition, which is easy and does not cost much. However, more heat is needed for the process to work, and a lot of ammonia will be produced. The g-C$_3$N$_4$ made by the thermal decomposition process usually has a bulky structure that restricts its use. Overall, different ways to make g-C$_3$N$_4$ have multiple benefits, so researchers must consider their goals when choosing the best way to do it.

17.4 Modifications to Improve Efficiency

To improve the photocatalytic and other applications of g-C$_3$N$_4$, several morphologies with enhanced properties have been created by heterojunction building and doping with metals, metal oxides and nonmetals. The modification strategy of g-C$_3$N$_4$, including the creation of composites are covered here.

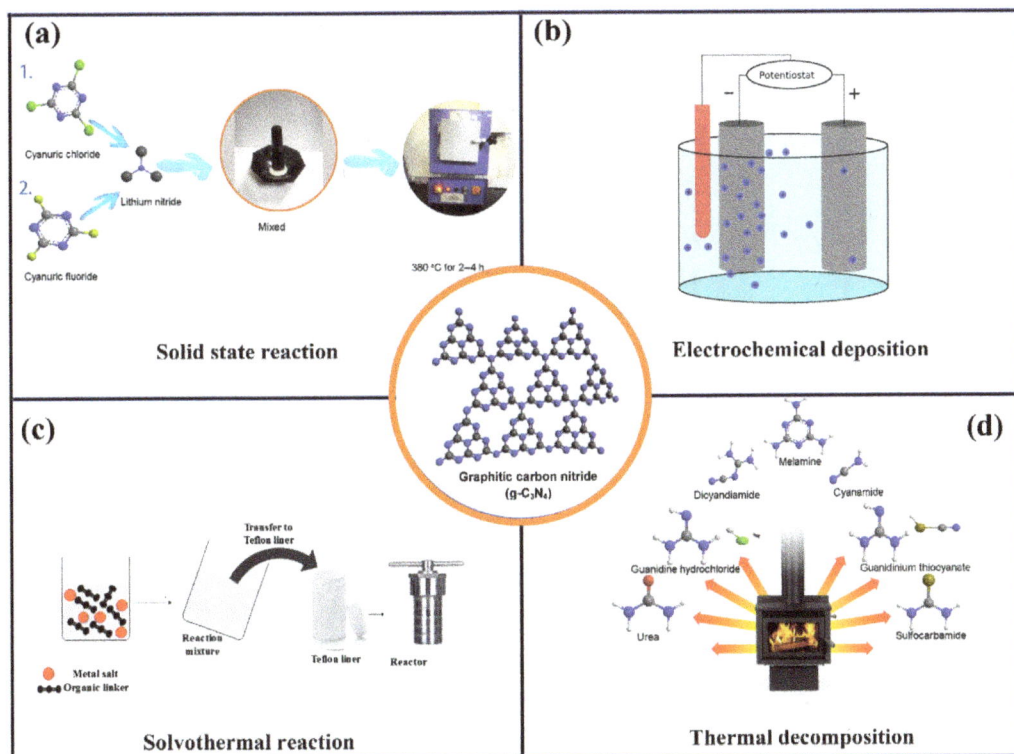

Figure 17.4. Synthetic approaches of g-C3N4 including (a) solid-state reaction, (b) electrochemical deposition, (c) solvothermal reaction and (d) thermal decomposition.

17.4.1 *g-C₃N₄ Modifications by Constructing Heterojunctions*

By promoting the charge carrier's separation and lowering e^--h^+ hole recombination, various metal oxides, including tin oxide, zinc oxide, iron oxide, etc., can increase the photocatalytic activity of g-C₃N₄. As a result, modified g-C₃N₄ nanocomposites have improved electric, magnetic and photocatalytic properties and apply to a wide range of applications such as CO_2 reduction H_2 generation, degradation of organic and inorganic dyes, NO oxidation and sensing, etc. (Chen et al. 2020, Rabani et al. 2021).

A few g-C₃N₄ heterojunction topologies based on metal oxides to g-C₃N₄ are compared here. For g-C₃N₄-metal oxide photocatalysts, charge carrier separation can take five different forms:

  i. Type I heterojunction
 ii. Type II heterojunction
iii. Z-scheme heterojunction
 iv. p-n heterojunction
  v. Schottky junction

Most of g-C₃N₄ metal oxide photocatalysts exhibit type II and Z-scheme charge carrier separation processes. Here these two types of heterojunctions will be discussed. "In type II heterojunctions, two semiconductors are bonded together to produce stable heterojunctions." The semiconductor A's valence band (VB) is positioned higher than semiconductor B's. The photoinduced hole, in this instance, traveled from the VB of semiconductor B to semiconductor A because of the disparity in voltages (Figure 17.5a). On the opposite side, electrons are moved from semiconductor B's conduction band (CB) to semiconductor A's. The improved separation of the electrons and holes

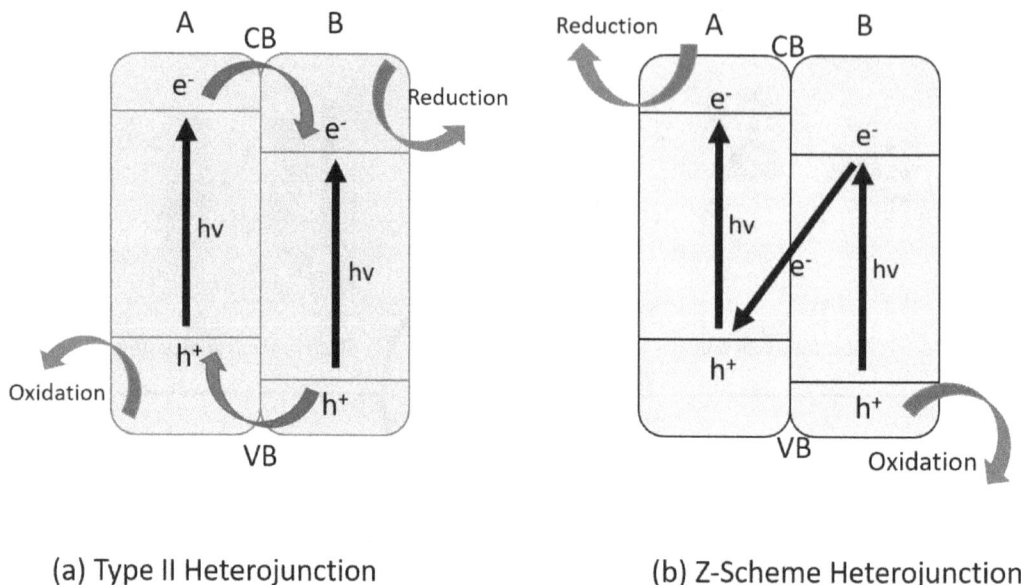

(a) Type II Heterojunction   (b) Z-Scheme Heterojunction

Figure 17.5. (**a**) Type II heterojunction, (**b**). Z-scheme heterojunction.

will lengthen the lifespan of the electrons and slow down recombination. Building type II systems for photocatalysis in various applications is highly desirable.

The direct Z-scheme heterojunction was first proposed by (Bard and Fox 1995). The produced electrons on the CB of semiconductor B move to the VB of semiconductor A, as illustrated in Figure 17.5b, where they mix with the photogenerated holes. By increasing the separation of electrons and holes, these photocatalysts can reduce recombination while enhancing redox capability. Different electron and hole migration patterns are caused by the electric field's driving force, which is created by the band edge positions of various semiconductors. Charge carriers begin to move in the electrical field direction to decrease the system's energy. As a result, various systems are identified based on electron and hole migration in the heterojunction. The flow of electrons and holes is primarily responsible for the differences between each type of heterojunction.

17.4.2 g-C₃N₄ Modifications by Doping

17.4.2.1 Metal-based Doping

The use of g-C_3N_4 in wastewater remediation and water filtration has just evolved to a nascent level. The application of noble metals (i.e., Pd, Ru, Ag, Au, Pt, Ir and Os) as dopants is a costly process that has limited their use on a broad scale due to their rarity and corrosive character. Therefore, it is crucial to design a metal-doped g-C_3N_4 based photocatalyst to produce real earth-abundant photocatalytic systems for water and wastewater treatment. By reducing the band gap, improving visible light absorption and increasing surface area, the addition of metals as dopants boosts the photocatalytic capabilities of g-C_3N_4.

The most used synthesis method is thermal condensation, which involves mixing the appropriate soluble metal salt with the g-C_3N_4 precursor in distilled water while ensuring a constant heat source to achieve the goal of a modulated band gap (Jiang et al. 2017, Zhu et al. 2015). The modification of g-C_3N_4 has also been done using this method. By employing ferric chloride as the Fe-precursor (Tonda et al. 2014), synthesized Fe-Doped g-C_3N_4 nanosheets. To create a more effective and recyclable photocatalyst for wastewater disinfection, several research groups have synthesized metal-doped g-C_3N_4.

For extended visible-light photocatalysis, metallic impurities like Na^+ and K^+ were inserted into the N sites of the g-C_3N_4, taking advantage of the improved electronic structures, adjustable band gaps and improved surface area for efficient charge transfer. The adsorption efficiency of Na-doped $g-C_3N_4$ produced from cyanamide and NaCl as precursors were reported by (Fronczak et al. 2017). Textural characteristics of Na-doped $g-C_3N_4$ were studied using N_2-adsorption/desorption isotherms. The findings showed that Na concentration was increased. The photocatalytic activity and visible light absorption of $g-C_3N_4$ were significantly enhanced by deprotonation with Na^+.

Similarly (Hu et al. 2014), created potassium (K) doped $g-C_3N_4$ with a band gap optimized for removing RhB dye under visible light irradiation using dicyandiamide (DCDA) monomer and potassium hydrate as precursors. Due to its enhanced electronic structure and redox potentials for adequate consumption of photogenerated electrons, it was inferred from several investigations that Na-doped $g-C_3N_4$ demonstrated higher photocatalytic capacity. Strong interactions between metal dopants and lone pair electrons on $g-C_3N_4$ nitrogen pots make it easy for metal cations to get into the framework. Alkali metal addition showed stable chemical activity despite being very reactive, which justifies the conclusion that it will increase the photodegradation of organic contaminants.

17.4.2.2 Transition Metal Ion Doping

In addition to alkali-metal doping, other metals, including Fe, W, Zr, Pd, Cu and so on, have also been widely used to change the optical and electrical properties of $g-C_3N_4$ (Sudrajat 2018, Gonçalves et al. 2018). For significant photocatalytic activity, metal doping can effectively minimize the band gap, boost light absorption, speed up charge movement and extend the lifetime of charge carriers (Shandilya et al. 2019, Zhang et al. 2015). The strong interactions between the negatively charged nitrogen atoms and the cations are attributed to lone pairs. Using conventional precursors and a basic pyrolysis procedure, molybdenum-doped $g-C_3N_4$ catalysts were synthesized by (Zhang et al. 2015). They discovered that adding Mo species can significantly lower the rate at which photogenerated charges recombine, expand the response to visible light, increase surface area, produce mesoporous structure and provide narrow band gap energy. As a result, the CO_2 reduction activity of $g-C_3N_4$ catalysts doped with Mo was considerably higher for electrons in the nitrogen pots of $g-C_3N_4$, causing the compound to capture the ions readily. The $g-C_3N_4$ was functionalized with better carrier mobility, improved electron-hole separation and a reduced band gap using noble metals like Pt and Pd (Hu et al. 2014). Noble metals like Pt and Pd doping can enhance the g-photocatalytic C_3N_4's activity, but its expensive price makes it impractical for everyday use.

17.4.2.3 Rare Earth Metal Doping

By thermally condensing urea and cerium sulfate precursors, the rare earth metal Cerium (Ce) was integrated into pure $g-C_3N_4$, which led to 90% Rhodamine B (RhB) degradation (Jin et al. 2015). The interstitial occupancy of Ce^{+3} coupled to a lone pair of N atoms was revealed by XPS spectrum analysis. Ce^{+3}'s ionic radius, which is substantially bigger than that of N and C, was also discovered to be 0.103 nm. As a result, the Ce dopant is encouraged to have a higher electron density and interstitial attachment. Xu et al. (2013) doped Europium (Eu) in $g-C_3N_4$ and thoroughly examined the photocatalytic mechanism and influence of impurity concentration. The optimized composite showed a high rate of degradation of MB dye (81.57%) compared to bulk $g-C_3N_4$'s efficiency of just 53%.

To create a heterogeneous photocatalyst, Ruthenium (Ru) was used in complex formation with $g-C_3N_4$. The Ru-doped $g-C_3N_4$ increased apparent quantum yield to 5.7% and the efficiency of CO_2 reduction to formic acid (Kuriki et al. 2015). Recently, synthetic Samarium (Sm^{+3}) doped $g-C_3N_4$ has been used to remove Methylene Blue (MB), Rhodamine B (RhB) and eosin yellow dyes with remarkable success. It was discovered that Sm^{+3}'s empty 4f orbital effectively captured photogenerated e^-, which decreased the recombination rate and enhanced photocatalytic activity (Thomas et al. 2018).

By integrating NH_4F into the $g-C_3N_4$ framework, fluorine-doped $g-C_3N_4$ was synthesized (Das et al. 2018). Due to nitrogen's and fluorine's electronegativity, the doped fluorine will attach to carbon rather than nitrogen, partially converting C-sp2 to C-sp3. Due to this, fluorine doping caused the C–F bonding, which reduced the band gap of g–C3N4 from 2.69 eV to 2.63 eV. Additionally, DFT studies showed that adding F to the carbon in the bay pushes the VB and CB to higher energy values. F-doped g-C_3N_4 displayed around 2.7 folds more activity than untreated g-C_3N_4 in photocatalytic hydrogen evolution.

17.4.2.4 Metal Oxide Doping

One of the most promising semiconductors photocatalysts under investigation is TiO_2, which also has high stability, a cost-effective synthesis method and an appropriate conduction band location. In 1972, the pioneers (Fujishima and Honda 1972) investigated the photocatalytic efficiency of TiO_2. The researchers wanted to minimize the material's bandgap by doping it with other elements and composing it with other substances to absorb visible light energy to increase TiO_2's photocatalytic efficiency. The g-C_3N_4 -TiO_2 heterojunction may be generated using various synthetic techniques, including solvothermal treatment, hydrothermal treatment, co-calcination and microwave assistance (Acharya and Parida 2020). The hydrothermal and calcination methods are often used to fabricate g-C_3N_4 - TiO_2 heterojunction (Li et al. 2015). Kočí et al. (2017) successfully used a hydrothermal method, followed by a calcination approach, to deposit TiO_2 on the surface of g-C_3N_4 (Rathi et al. 2018). The CuNi@g-C_3N_4 - TiO_2 nanocatalyst exhibited a three and five times greater photocatalytic activity RhB degradation than bare g-C_3N_4 and TiO_2 nanorods. The produced heterojunction exhibits greater photocatalytic effectiveness than g-C_3N_4 and TiO_2 alone, according to research by (Alcudia-Ramos et al. 2020).

ZnO nanostructures have attracted researchers due to their inexpensive fabrication cost, low toxicity, good stability, high aspect ratio and appropriate bandgap energy. The g-C_3N_4-ZnO structure has been created using various techniques, including hydrothermal, solvothermal, atomic layer deposition and more (Mohammadi et al. 2020, Tan et al. 2019). It has been demonstrated that adding ZnO to g-C_3N_4 may improve photocatalytic activity, including charge transfer and separation and reduce the recombination of the photogenerated carriers (Paul et al. 2020, Ramachandra et al. 2020).

To improve the activity of the bulk g-C_3N_4, other types of metal oxides, including V_2O_5, NiO, MoO_3, Cu_2O, Co_3O_4, CeO_2, Bi_2O_3, Al_2O_3, etc., are also utilized (Jin et al. 2020, Shi et al. 2019, Sumathi et al. 2019). By combining with other metals or doping with different substances, the g-C_3N_4-based heterojunctions can be altered (Chaudhary and Ingole 2020). Cu is a standard metal that is used to enhance photocatalytic activity. Other metal photocatalysts for enhanced photocatalytic activity include Bi, Pd, Pt, Ni and Cd (Karimi et al. 2020, Zhao et al. 2021). In addition to the composites based on metal oxide g-C_3N_4, some semiconductors are also used. Excellent stability, economic synthesis, improved photogenerated electron reservoirs, photocatalysis and electrocatalysis are all possible with carbon-based nanomaterials (Fan et al. 2019).

17.4.2.5 Nonmetal Oxide-Based g-C_3N_4 Nanocomposite

Regarding widespread practical applications, there are considerable obstacles due to the dearth of metal-based photocatalysts in nature and their high price. In addition, these metal-doped semiconductor photocatalysts self-degrade due to thermal instability. As a result, research into nonmetal-modified g-C_3N_4 has grown. Nonmetals can acquire electrons and establish powerful covalent bonds in the parent lattice structure due to their inherent characteristics, such as high electronegativity and ionization energies. To increase g-C_3N_4's photodegradation activity, phosphorus, sulfur, oxygen, carbon, boron nitrogen and halogens have been used as nonmetal dopants.

Zhou et al. (2015) synthesized P-doped g-C_3N_4 via an economic copolymerization process. P-doped g-C_3N_4 shows a rapid 100% RhB dye degradation activity. In a different work, CeO_2 was added via calcination after being heated with melamine and diammonium hydrogen phosphate to couple it to P-doped g-C_3N_4. At an optimal concentration of 13.8% CeO_2, the synthesized CeO_2-P doped g-C_3N_4 reduced the band gap of pure g-C_3N_4 from 2.71 eV to 2.32 eV. This behavior is attributed to synergistic interactions between CeO_2 and P-doped g-C_3N_4 and was further evaluated to be 7.4 and 4.9 fold better than CeO_2 and g-C_3N_4, respectively (Luo et al. 2015).

For the fabrication of S doped g-C$_3$N$_4$ photocatalyst (Wang et al. 2015), used melamine and thiourea precursors calcined at 520°C. The findings of photocatalytic reduction of CO$_2$ showed that the CH$_3$OH yield with pure g-C$_3$N$_4$ was 0.81 mol g^{-1}, whereas it was 1.12 mol g^{-1} for g-C$_3$N$_4$ doped with S. Using melamine and thiourea as common precursors (Liang et al. 2016), created a series of S doped g-C$_3$N$_4$ grafted with zinc phthalocyanines (ZnTNPc). S-doped g-C$_3$N$_4$ and ZnTNPc showed a synergistic relationship for the photocatalytic elimination of Methylene blue dye, which was 4.5-fold more than that of zinc phthalocyanines (ZnTNPc). According to Mott-Schottky Relationship, adding S atoms narrows the band gap and lowers the Conduction Band (CB) from 1.04V to 0.83V, which facilitates the photocatalytic activity of MB.

By condensing oxalic acid and urea at a high temperature of 550°C (Qiu et al. 2017), created porous O-doped g-C$_3$N$_4$. The band gap was reduced from 2.91 eV to 2.07 eV due to the inclusion of the O atom into the g-C$_3$N$_4$ lattice. To enable the oxidation of benzene to phenol and other non-toxic chemicals under the influence of visible light, a fluorinated g-C$_3$N$_4$ heterogeneous photocatalyst was produced. The preparation was initiated by adding NH$_4$F by thermal precursor in a consistent, easy one-pot facile thermal polymerization technique to design and create B-doped g-C$_3$N$_4$ nanosheets.

17.5 Application of Metal Oxide-Based g-C$_3$N$_4$ Nanocomposites

17.5.1 Photocatalysts

17.5.1.1 H$_2$ Generation via Water Splitting

More and more individuals are calling for the use of reliable, cost-effective and renewable energy sources instead of finite fossil fuel supplies. This substitution is a successful treatment for greenhouse gas emissions and global warming. The energy content of hydrogen is higher than that of hydrocarbon fuels, ranging from 120 to 142 MJ kg^{-1}. Thus, it is predicted that by 2080, hydrogen will produce 90% of all energy. As a result, several researchers (Paul et al. 2020, Shi et al. 2021) considered H$_2$ creation a unique and ecologically favorable study issue. The photocatalytic water splitting approach employing metal oxide-g-C$_3$N$_4$ heterojunctions and abundant light sources is one of the most current methods for producing hydrogen (Ji et al. 2018). The photocatalysts' band positions should be changed to make the CB position more favorable than the H$_2$O oxidation potential for O$_2$ production and more harmful than the H$_2$O reduction potential for H$_2$ production. Several heterojunctions are more anodic than H$_2$O reduction potential to demonstrate good activity under visible light, when considering the band edges position of some metal oxide g-C$_3$N$_4$ composites identified earlier. TiO$_2$-g-C$_3$N$_4$ heterojunctions are a prime example of how well they work as a photocatalytic for H$_2$ evolution. According to (Marchal et al. 2018) analysis of Au/(TiO$_2$-g-C$_3$N$_4$), optimum component ratios and contact quality improved visible light absorption with the appropriate band locations for photogenerated charge carriers. Table 17.1 lists other studies on applying metal oxide composites based on g-C$_3$N$_4$ for water splitting.

17.5.1.2 Reduction of CO$_2$

One of the leading environmental issues brought on by fossil fuels is CO$_2$ emission, that raises the surface temperature of the globe. For two reasons, photocatalytic CO$_2$ reduction offers a green solution to this issue. The production of energy fuels like CH$_4$, CH$_3$OH and other fuels helps to meet future energy demands and reduce CO$_2$ emissions. Different metal oxide g-C$_3$N$_4$-based systems convert CO$_2$ via photocatalysis significantly as they have advantageous band edge locations. As g-C$_3$N$_4$-based composites for CO$_2$ conversion, ZnO and TiO$_2$ are frequently employed (Mulik et al. 2021). TiO$_2$ and g-C$_3$N$_4$ were combined to create a photocatalyst by (Wang et al. 2020) utilizing ball milling and calcination. High CH$_4$ and CO evolution yields of 72.2 and 56.2 mol g^{-1} are achieved due to the heterostructure between TiO$_2$ and C3N4's low charge recombination rate and high separation. As shown in Table 17.2, other scientists have also studied the CO$_2$ reduction of photocatalysts based on g-C$_3$N$_4$-metal oxide.

Table 17.1. Earlier reported works on g-C_3N_4-based photocatalytic water splitting.

Photocatalyst	Light Source	Photocatalysis Rate	References
TiO_2-g-C_3N_4	Xenon lamp (320nm)	76.25 µmol·h^{-1}	Qu et al. 2016
TiO_2 nanodots/g-C_3N_4	Xenon lamp (300W)	H_2 evolution rate=1318.3 µmol g^{-1} O_2 evolution rate=638.7 µmol g^{-1}	Jiang et al. 2022
N-doped ZnO-g-C_3N_4	PLS-SXE-300C UV lamp	152.7 µmol·h^{-1}	Liu et al. 2019
S-$Cu2O$/g-C_3N_4	Xenon lamp (300W)	24.83 µmol·h^{-1}	Gu et al. 2021
BiO_2/g-C_3N_4	Xenon lamp (500nm)	8,542 µmol·g^{-1}	Alhaddad et al. 2020
Mn_3O_4/g-C_3N_4	Xenon lamp (300W)	2700 mmol g^{-1} h^{-1}	Li et al. 2021
g-C_3N_4/Nitrogen-Doped Carbon Dots/WO_3	Xenon lamp (300W)	3.27 mmol g^{-1} h^{-1}	Song et al. 2021
$MoO3$-x-g-C_3N_4	Xenon lamp (300W)	22.8 mmol h^{-1}	Guo et al. 2020
TiO_2/Ti_3C_2/g-C_3N_4	Xenon lamp (300W)	2592 mmol·g^{-1}	Hieu et al. 2021
ZnO/Au/g-C_3N_4	Xenon lamp (150W)	3.69 µmol h^{-1} cm^{-2}	Wen et al. 2020

Table 17.2. A list of CO_2 reduction applications of the metal oxide-based g-C_3N_4 photocatalysts.

Photocatalyst	Light Source	Photocatalysis Rate	References
NiO-g-C_3N_4	Xenon lamp (300 W)	4.17	Tang et al. 2018
g-C_3N_4 foam-Cu_2O	350 W Lamp	8.182	Sun et al. 2019
ZnO/Au/g-C_3N_4	UV-Vi's lamp (300W)	689.7 µmol/m^2 (CO evolution)	Li et al. 2021
TiO_2/g-C_3N_4	UV-Vi's lamp (8W)	CH_4 and CO yields of 72.2 and 56.2 µmol g^{-1}	Wang et al. 2020
ZnO/g-C_3N_4	Xenon lamp (300 W)	The CH_3OH production rate was 1.32 µmol h^{-1} g^{-1}	Nie et al. 2018
g-C_3N_4/3D ordered microporous (3DOM)-WO_3	visible light	48.7 µmol h^{-1} g^{-1}	Tang et al. 2022
$NiTO_3$/g-C_3N_4	Xenon lamp (300 W)	CH_3OH production is 13.74 molg^{-1} h^{-1}	Guo et al. 2021

17.5.1.3 Photodegradation of Organic Pollutants

Several environmental issues are brought on by the growth of an increasing number of dye-related businesses, including textile, food and furniture manufacturers. Additionally, having an unfavorable visual effect on water sources, organic dyes also cause wastewater to have a higher COD. Diverse techniques such as adsorption, membrane separation and coagulation have been explored to remove organic pollutants from effluents, but these only move organic dyes from the liquid phase of wastewater to the solid phase. This creates secondary pollution in the environment. Most metal oxides may degrade and transform organic colors into particles during photocatalysis, which uses solar energy to start the reaction. The most stable dyes in water at room temperature are Rhodamine B (RhB), Methylene Orange (MO) and Methylene Blue (MB) (Si et al. 2020). Since MB and RhB are toxic dyes, the health of people and aquatic animals may be negatively impacted by their high concentration. Wastewater treatment is a pressing issue because of the high resistivity of the RhB and MB under various environmental conditions.

Therefore, providing a reliable and affordable technique to remove MB and RhB from sewage is imperative. According to reports, the optimal composition for g-C_3N_4-ZnO nanocomposites is 30% weight, which results in the maximum MB degradation efficiency (Liu et al. 2018). The more g-C_3N_4, the more likely it is that electrons and holes will recombine, decreasing the photocatalytic activity. Additionally, it should be noted that this composite's RhB degrading efficiency is approximately 2.1 times greater than that of pure ZnO (Chen et al. 2018). g-C_3N_4-TiO_2 heterojunction, in addition

Table 17.3. List of the studies carried out on organic pollutant degradation by g-C_3N_4-based heterojunctions.

Photocatalyst	Light Source	Organic Pollutant	Degradation Efficiency %	Time (hr)	References
CeO_2/g-C_3N_4	Xe lamp (400 W)	Rose Bengal Crystal violet	79.2 76.7	1.5	Shoran et al. 2022
TiO_2@g-C_3N_4	Xe lamp (100 W)	RhB	95.68	--	Hao et al. 2017
MoS_2-g-C_3N_4@TiO_2	Xe lamp (350 W)	Methylene Blue	97.55	1	Karpuraranjith et al. 2022
ZnO-g-C_3N_4	Xe lamp (500 W)	Methylene Blue	75	3	Ngullie et al. 2020
ZnO-g-C_3N_4	Xe lamp (300 W)	cephalexin oxidation	98.9	1	Li et al. 2018
WO_3-g-C_3N_4	Xe lamp (300 W)	tetracycline	90.54	1	Pan et al. 2020
Ag-WO_3/g-C_3N_4	Xe lamp (500 W)	oxytetracycline hydrochloride	97.74	1	Ouyang et al. 2022
Fe_3O_4/CeO_2/g-C_3N_4	Xe lamp (300 W)	tetracycline hydrochloride	96.63	3	Wang et al. 2022
Co_3O_4-g-C_3N_4	Xe lamp (250 W)	Methyl Orange	100	3	Han et al. 2014
MoO_3-g-C_3N_4	Xe lamp (150 W)	Rhodamine B	93	3	Adhikari and Kim 2020
g-C_3N_4 -NiO	LED-light (30 W)	Methyl Orange (MO)	96.8	2	Chen et al. 2019
WO_3/g-C_3N_4	Xe lamp (300 W)	sulfamethoxazole	91.7	4	Zhu et al. 2017
g-C_3N_4 /Bi_2O_3/TiO_2	Xe lamp	Methylene Blue (MB)	77.5	3	Zhang et al. 2015

to g-C_3N_4-ZnO heterojunction, has a strong capacity in the degradation of RhB (Li et al. 2019, Xia et al. 2019). Table 17.3 lists many ongoing research projects on the photodegradation applications of g-C_3N_4-based heterojunctions.

17.5.2 Sensors

A g-C_3N_4 nanosheet is an excellent option for a modified electrode for sensors that can detect analytes like dopamine, hydrogen peroxide, glucose, etc. These advantages include outstanding fluorescence quenching abilities, quick response to external stimulations, high sensitivity to analytes, high level of stability light and electricity conversion properties and biocompatibility (Zou et al. 2018, Wang et al. 2019). As gas sensors, metal oxide semiconductor/g-C_3N_4 composites are frequently utilized (Rahman et al. 2021). Consequently, the metal oxide-loaded g-C_3N_4 has also disclosed new sensors to identify various materials. g-C_3N_4-TiO_2-based structures are one of the most popular composites for sensing and other applications. Due to this, the composite exhibited exceptional stability, repeatability and excellent selectivity, another heterojunction utilized in UV-assisted gas sensors is ZnO-g-C_3N_4. It is demonstrated that ZnO-g-C_3N_4 has significantly greater ethanol (C_2H_5OH) detecting capacity than bare ZnO and g-C_3N_4. The best sensing performance was demonstrated by the ZnO containing 8% g-C_3N_4, which is attributed to the efficient separation of electrons and holes between g-C_3N_4 and ZnO and the catalytic impact of UV light at room temperature (Zhai et al. 2018). The applications of g-C_3N_4-metal oxide heterojunctions as a sensor material are listed in Table 17.4.

Table 17.4. The list of sensing-related applications for g-C_3N_4-metal oxide heterojunction.

Photocatalyst	Application	Limit of detection (LOD)	References
MoO_3-g-C_3N_4	Detection of Furazolidone	1.4 nM	Balasubramanian et al. 2019
Co_3O_4-g-C_3N_4	Detection of environmental phenolic hormones	$10-9$ mol L^{-1}	Sun et al. 2018
g-C_3N_4-NiO	Detection of quercetin	0.002 µM	Selvarajan et al. 2018
NiO-Co_3O_4-g-C_3N_4	Detection of tetra bromobisphenol-A	~ 0.1 mmol L^{-1}	Liu et al. 2017
g-C_3N_4-Mn_3O_4	H_2S sensor	0.13 µg mL^{-1}	Huan et al. 2010
WO_3/g-C_3N_4	Detection of phosmet	3.6 nM	Bilal and Hassan 2021
CuO-g-C_3N_4	Aflatoxin B1 sensing	6.8 pg mL^{-1}	Mao et al. 2021

17.5.3 Anti-Bacterial Study

Under variations in ultraviolet or visible light irradiation, the g-C_3N_4 can eradicate a number of sizes, forms and architectures of bacteria, viruses, microorganisms and microalgae. Effective composites to control microbes and water disinfection rely on fabricating g-C_3N_4 with various metal oxides. Bio-hazards causing human health problems are frequently present in wastewater and polluted water and contain different viruses, fungi, bacteria, etc. The groundbreaking research by (Matsunaga et al. 1985) demonstrated that TiO_2 could aid in the UV-light-induced inactivation of bacteria like *Escherichia coli*, *Saccharomyces cerevisiae* and *Lactobacillus acidophilus*. Additionally, g-C_3N_4/Cr-ZnO nanocomposites with superior antibacterial activity against Gram-positive (*Staphylococcus aureus*, *Bacillus subtilis*) and the 60% g-C_3N_4/5%Cr-ZnO nanocomposite exhibited the most potent antibacterial activity in this study. *Escherichia coli* and *Staphylococcus aureus* showed excellent sterilizing performance for Fe-SnO_2/g-C_3N_4 under sunlight, near-ultraviolet light and daylight bulbs (Chen et al. 2020). Under daytime lighting, this structure's sterilizing efficacy is largely justified. Graphitic carbon nitride nanosheets coated with magnetic silver-iron oxide nanoparticles showed antibacterial activity against *E. coli* germs (Pant et al. 2017). Additional research on the disinfectant properties of g-C_3N_4 metal oxide-based nanoparticles is described in Table 17.5.

Table 17.5. Studies on antibacterial properties of g-C_3N_4 metal oxide-based nanomaterials.

Photocatalyst	Light Source	Bacteria	Inhibition Amount	References
CeO_2/g-C_3N_4	Xe lamp	*E. coli* *S. aureus* *B. cerrous* *S. abony*	*E. coli* = 19.9 mm *S. aureus* = 18.9 mm *B. cerrous* = 16.04 mm *S. abony* = 18.05 mm	Shoran et al. 2022
g-C_3N_4 /TiO_2/Ag	Xe lamp	*E. coli*	84%	Sun et al. 2018
g-C_3N_4-m-Bi_2O_4	halogen lamp (500 W)	*E. coli* *S. aureus*	*E. coli* = ~11 ± 0.5 mm *S. aureus* = 12–13 ± 0.5 mm	Shanmugam et al. 2020
Cu_2O-g-C_3N_4	fluorescent lamp (36 W)	*P. aeruginosa,* *B. subtilis,* *S. aureus,* *E. coli*	*P. aeruginosa* = 6 ± 0.09 mm *B. subtilis* = 22 ± 1.67 mm *S. aureus* = 11 ± 1.22 mm *E. coli* = 22 ± 1.67 mm	Meenakshisundaram et al. 2019
TiO_2/g-C_3N_4	--	*E. coli*	16%	Xu et al. 2016
α-Fe_2O_3/ CeO_2 decorated g-C_3N_4	Xe light (500W)	*S. aureus,* *E. coli*	*S. aureus* = 11 ± 0.5 mm *E. coli* = 12 mm	Vignesh et al. 2019
Ag/AgO-g-C_3N_4	Tungsten lamp (100 W)	*E. coli*	99%	Meenakshisundaram et al. 2019

17.5.4 Other Applications

These composites have a wide range of different uses because of the distinct characteristics of g-C$_3$N$_4$ based heterojunctions (Vignesh et al. 2019). Rechargeable Lithium-Ion Batteries (LIBs) have drawn a great deal of academic interest due to the rising energy demand for electronic gadgets and automobiles (Li et al. 2015). Anode materials, which comprise a sizeable portion of LIBs, should have higher stability and increased specific capacity. The hydrothermal approach was used by (Tran et al. 2019) to develop SnO$_2$ on the graphite oxide/g-C$_3$N$_4$. SnO$_2$ is a novel lipophilic material with a high specific capacity, low Li$^+$ insertion potential, greater number of sources and other characteristics. This composite demonstrated high cycling and reversible capacity for lithium storage, possibly due to the presence of graphite oxide-g-C$_3$N$_4$ or g-C$_3$N$_4$. SnO$_2$@g-C$_3$N$_4$ based nanocomposites can replace next-generation high-power and low-cost LIBs. Another use for g-C$_3$N$_4$ based materials is in producing NH$_3$ by photocatalytic nitrogen fixation, which is an environmentally friendly and long-lasting process. However, g-effectiveness C$_3$N$_4$'s for nitrogen fixation activities has been constrained by low stability, negligible surface-active sites and a high carrier recombination rate. The carrier separation should be improved by doping with additional elements or composing with other metal oxides. For the reduction processes, nitrogen gas should finally totally adsorb on the photocatalyst. According to (Kong et al. 2020), the cyano group in g-C$_3$N$_4$ modified by the cyano group increases the photocatalytic activities for N$_2$ fixation by up to 128 times. As explained earlier, sulfur and iodine can enhance the carriers' activities in the bulk of g-C$_3$N$_4$. The fact that g-C$_3$N$_4$ -metal oxide-based heterojunction may be widely employed to detect various heavy metals, gases, biological materials and organic and inorganic compounds (Ahmad et al. 2020). Supercapacitors and desulfurization are two further applications for g-C$_3$N$_4$-metal oxide-based composites that could be utilized extensively (Ma et al. 2019).

17.6 Conclusion

g-C$_3$N$_4$ is a metal-free semiconductor with an adjustable band gap, excellent chemical and thermal stability and appealing electrical characteristics. Graphitic carbon nitride absorbs UV light and has a band gap of 2.7 eV. Modifying g-C$_3$N$_4$ via doping and blending with other materials to produce composites can result in improved optoelectronic characteristics, and the composites can exhibit synergistic properties. Many studies used doping elements to increase the effectiveness of bare g-C$_3$N$_4$ light harvesting, while mixing g-C$_3$N$_4$ with other materials, such as metals, metal oxides and nonmetals is another strategy to deal with this issue. The examination of g-C$_3$N$_4$ heterojunctions with designed bandgaps and optimized surfaces to improve the absorption spectrum towards the visible-light region, reduced charge carrier recombination, and increased surface adsorption and reaction are among the critical areas of g-C$_3$N$_4$ based material research. The highlighted applications of modified g-C$_3$N$_4$ included water splitting, bacterial disinfection, energy storage, photodegradation of organic pollutants, CO$_2$ reduction, sensing, etc. Even though most recent research has shown that g-C$_3$N$_4$-based photocatalysts work very well, their full potential has not yet been fully realized. The main problems are in finding a green way to make a photocatalyst with a high surface area and good photostability, testing how well a g-C$_3$N$_4$ based photocatalyst works with real industrial wastewater and improving reactor design to get the best photocatalytic activity.

References

Acharya, R. and P. Kulamani. 2020. A review on TiO$_2$/g-C$_3$N$_4$ visible-light-responsive photocatalysts for sustainable energy generation and environmental remediation. J. Environ. Chem. Eng. 8(4).

Adhikari, S. and Do H. Kim. 2020. Heterojunction C$_3$N$_4$/MoO$_3$ microcomposite for highly efficient photocatalytic oxidation of Rhodamine B. Appl. Surf. Sci. 511.

Ahmad, R., N. Tripathy, A. Khosla, M. Khan, P. Mishra, W. A. Ansari, M. A. Syed and Y.-B. Hahn. 2020. Review—recent advances in nanostructured graphitic carbon nitride as a sensing material for heavy metal ions. J. Electrochem. Soc. 167(3): 037519.

Alcudia-Ramos, M. A., M. O. Fuentez-Torres, F. Ortiz-Chi, C. G. Espinosa-González, N. Hernández-Como, D. S. García-Zaleta, M. K. Kesarla, J. G. Torres-Torres, V. Collins-Martínez and S. Godavarthi. 2020. Fabrication of G-C$_3$N$_4$/TiO$_2$ heterojunction composite for enhanced photocatalytic hydrogen production. Ceramics Int. 46(1): 38–45.

Alhaddad, M., R. M. Navarro, M. A. Hussein and R. M. Mohamed. 2020. Bi$_2$O$_3$/g-C$_3$N$_4$ nanocomposites as proficient photocatalysts for hydrogen generation from aqueous glycerol solutions beneath visible light. Ceramics Int. 46(16): 24873–24881.

Arumugam, M., M. Tahir and P. Praserthdam. 2022. Effect of nonmetals (B, O, P, and S) doped with porous g-C$_3$N$_4$ for improved electron transfer towards photocatalytic CO$_2$ reduction with water into CH$_4$. Chemosphere. 286 (Pt 2).

Bai, X., J. Li and C. Cao. 2010. Synthesis of hollow carbon nitride microspheres by an electrodeposition method. Appl. Surf. Sci. 256(8): 2327–2331.

Bai, Y. J., B. Lü, Z. G. Liu, L. Li, De L. Cui, X. G. Xu and Q. L. Wang. 2003. Solvothermal preparation of graphite-like C$_3$N$_4$ nanocrystals. J. Crystal Growth. 247(3-4): 505–508.

Balasubramanian, P., A. Muthaiah, S. M. Chen and T. W. Chen. 2019. Sonochemical synthesis of molybdenum oxide (MoO$_3$) microspheres anchored graphitic carbon nitride (g-C$_3$N$_4$) ultrathin sheets for enhanced electrochemical sensing of furazolidone. Ultrason. Sonochem. 50: 96–104.

Bard, A. J. and M. A. Fox. 1995. Artificial photosynthesis: solar splitting of water to hydrogen and oxygen. Acc. Chem. Res. 28(3): 141–145.

Baughman, R. H. 1974. Solid-state synthesis of large polymer single crystals. J. Polym. Sci.: Polym. Phys. Ed. 12(8): 1511–1535.

Bilal, S., Mudassir M. Hassan, Fayyaz M. Ur Rehman, M. Nasir, Jamil A. Sami and A. Hayat. 2021. An Insect Acetylcholinesterase Biosensor Utilizing WO 346: 12889.

Chaudhary, P. and P. P. Ingole. 2020. *In-situ* solid-state synthesis of 2D/2D interface between Ni/NiO hexagonal nanosheets supported on g-C$_3$N$_4$ for enhanced photo-electrochemical water splitting. Int. J. Hydrog. Energy. 45(32): 16060–16070.

Chen, C., M. Xie, L. Kong, W. Lu, Z. Feng and J. Zhan. 2020. Mn3O4 nanodots Loaded G-C$_3$N$_4$ nanosheets for catalytic membrane degradation of organic contaminants. J. Hazard. Mater. 390.

Chen, G., S. Bian, C. Y. Guo and X. Wu. 2019. Insight into the Z-scheme heterostructure WO3/g-C$_3$N$_4$ for enhanced photocatalytic degradation of methyl orange. Mater. Lett. 236: 596–599.

Chen, Q., H. Huijie, D. Zhang, S. Hu, T. Min, B. Liu, C. Yang, W. Pu, J. Hu and J. Yang. 2018. Enhanced visible-light driven photocatalytic activity of hybrid ZnO/g-C$_3$N$_4$ by high performance ball milling. J. Photochem. Photobiol. A: Chem. 350: 1–9.

Chen, T., Q. Hao, W. Yang, C. Xie, D. Chen, C. Ma, W. Ya and Y. Zhu. 2018. A honeycomb multilevel structure Bi$_2$O$_3$ with highly efficient catalytic activity driven by bias voltage and oxygen defect. Appl. Catal. B: Environ. 237: 442–448.

Chen, X., Q. Wang, J. Tian, Y. Liu, Y. Wang and C. Yang. 2020. A study on the photocatalytic sterilization performance and mechanism of Fe-SnO$_2$/g-C$_3$N$_4$ heterojunction materials. New J. Chem. 44(22): 9456–9465.

Dai, L., D. W. Chang, J. B. Baek and W. Lu. 2012. Carbon nanomaterials for advanced energy conversion and storage. Small. 8(8): 1130–1166.

Das, D., D. Banerjee, M. Mondal, A. Shett, B. Das, N. S. Das, U. K. Ghorai and K. K. Chattopadhyay. 2018. Nickel doped graphitic carbon nitride nanosheets and its application for dye degradation by chemical catalysis. Mater. Res. Bull. 101: 291–304.

Fan, W., Z. Wei, Z. Zhang, F. Qiu, C. Hu, Z. Li, M. Xu and J. Qi. 2019. High-performance carbon-based perovskite solar cells through the dual role of PC61BM. Inorg. Chem. Front. 6(10): 2767–2775.

Franklin, E. C. 1922. The ammono carbonic acids. J. Am. Chem. Soc. 44.

Fronczak, M., M. Krajewska, K. Demby and M. Bystrzejewski. 2017. Extraordinary adsorption of methyl blue onto sodium-doped graphitic carbon nitride. J. Phys. Chem. C 121(29): 15756–15766.

Fu, J., J. Yu, C. Jiang and B. Cheng. 2018. G-C$_3$N$_4$-based heterostructured photocatalysts. Adv. Energy Mater. 8(3).

Fu, Q., C. B. Cao and H. E. S. Zhu. 1999. Preparation of carbon nitride films with high nitrogen content by electrodeposition from an organic solution. J. Mater. Sci. Lett. 18(18): 1485–1488.

Fujishima, A. and K. Honda. 1972. Electrochemical photolysis of water at a semiconductor electrode. Nature. 238: 37–38.

Gonçalves, D. A. F., R. P. R. Alvim, H. A. Bicalho, A. M. Peres, I. Binatti, P. F. R. Batista, L. S. Teixeira, R. R. Resende and E. Lorençon. 2018. Highly dispersed Mo-doped graphite carbon nitride: potential application as oxidation catalyst with hydrogen peroxide. New J. Chem. 42(8): 5720–5727.

Gu, Y., A. Bao, X. Zhang, J. Yan, Q. Du, M. Zhang and X. Qi. 2021. Facile fabrication of sulfur-doped Cu_2O and g-C_3N_4 with Z-Scheme structure for enhanced photocatalytic water splitting performance. Mater. Chem. Phys. 266: 124542.

Guo, H., S. Wan, Y. Wang, W. Ma, Q. Zhong and J. Ding. 2021. Enhanced photocatalytic CO_2 reduction over direct Z-Scheme $NiTiO_3$/g-C_3N_4 nanocomposite promoted by efficient interfacial charge transfer. Chem. Eng. J. 412: 12864.

Guo, Q., Y. Xie, X. Wang, S. Lv, T. Hou and X. Liu. 2003. Characterization of well-crystallized graphitic carbon nitride nanocrystallites via a benzene-thermal route at low temperatures. Chem. Phys. Lett. 380(1-2): 84–87.

Guo, Y., B. Chang, T. Wen, S. Zhang, M. Zeng, N. Hu, Y. Su, Z. Yang and B. Yang. 2020. A Z-scheme photocatalyst for enhanced photocatalytic H2 evolution, constructed by growth of 2D plasmonic MoO_3-x nanoplates onto 2D g-C_3N_4 nanosheets. J. Colloid Interface Sci. 567: 213–223.

Han, C., L. Ge, C. Chen, Y. Li, X. Xiao, Y. Zhang and L. Guo. 2014. Novel visible light induced Co_3O_4-g-C_3N_4 heterojunction photocatalysts for efficient degradation of methyl orange. Appl. Catal. B: Environ. 147: 546–553.

Hao, J., S. Zhang, F. Ren, Z. Wang, J. Lei, X. Wang, T. C. and L. Li. 2017. Synthesis of TiO_2@g-C_3N_4 core-shell nanorod arrays with Z-scheme enhanced photocatalytic activity under visible light. J. Colloid Interface Sci. 508: 419–425.

Hao, Q., C. Xie, Y. Huang, D. Chen, Y. Liu, W. Wei and B.-J. Ni. 2020. Accelerated Separation of Photogenerated Charge Carriers and Enhanced Photocatalytic Performance of g-C 3: 249–258.

Hasanvandian, F., M. Moradi, S. A. Samani, B. Kakavandi, S. R. Setayesh and M. Noorisepehr. 2022. Effective promotion of g–C_3N_4 photocatalytic performance via surface oxygen vacancy and coupling with bismuth-based semiconductors towards antibiotics degradation. Chemosphere. 287 (Pt 3).

Hieu, V. Q., T. C. Lam, A. Khan, T. T. T. Vo, T. Q. Nguyen, V. D. Doan, D. L. Tran, V. T. Le and V. A. Tran. 2021. TiO_2/Ti_3C_2/g-C_3N_4 ternary heterojunction for photocatalytic hydrogen evolution. Chemosphere. 285: 13142.

Hu, S., F. Li, Z. Fan, F. Wang, Y. Zhao and Z. Lv. 2014. Band gap-tunable potassium doped graphitic carbon nitride with enhanced mineralization ability. Dalton Transact. 44(3): 1084–1092.

Hu, S., L. Ma, J. You, F. Li, Z. Fan, F. Wang, D. Liu and J. Gui. 2014. A simple and efficient method to prepare a phosphorus modified G-C 3N4 visible light photocatalyst. RSC Adv. 4(41): 21657–21663.

Huan, Z., J. Chang and J. Zhou. 2010. Low-temperature fabrication of macroporous scaffolds through foaming and hydration of tricalcium silicate paste and their bioactivity. J. Mater. Sci. 45(4): 961–968.

Ji, C., S. N. Yin, S. Sun and S. Yang. 2018. An *in situ* mediator-free route to fabricate Cu_2O/g-C_3N_4 Type-II heterojunctions for enhanced visible-light photocatalytic H2 generation. Appl. Surf. Sci. 434: 1224–1231.

Jiang, H., Y. Li, D. Wang, X. Hong and B. Liang. 2020. Recent advances in heteroatom doped graphitic carbon nitride (g-C_3N_4) and g-C_3N_4/Metal oxide composite photocatalysts. Curr. Organ. Chem. 24(6): 673–693.

Jiang, L., X. Yuan, Y. Pan, J. Liang, G. Zeng, Z. Wu and H. Wang. 2017. Doping of graphitic carbon nitride for photocatalysis: a reveiw. Appl. Catal. B: Environ. 217: 388–406.

Jiang, Y., Z. Sun, Q. Chen, C. Cao, Y. Zhao, W. Yang, L. Zeng and L. Huang. 2022. Fabrication of 0D/2D TiO_2 Nanodots/g-C_3N_4 S-Scheme heterojunction photocatalyst for efficient photocatalytic overall water splitting. Appl. Surf. Sci. 571: 15128.

Jin, C., M. Wang, Z. Li, J. Kang, Y. Zhao, J. Han and Z. Wu. 2020. Two dimensional Co_3O_4/g-C_3N_4 Z-scheme heterojunction: mechanism insight into enhanced peroxymonosulfate-mediated visible light photocatalytic performance. Chem. Eng. J. 398: 125569.

Jin, R., S. Hu, J. Gui and D. Liu. 2015. A convenient method to prepare novel rare earth metal ce-doped carbon nitride with enhanced photocatalytic activity under visible light. Bull. Korean Chem. Soc. 36(1): 17–23.

Karimi, M. A., M. Atashkadi, M. Ranjbar and A. Habibi-Yangjeh. 2020. Novel visible-light-driven photocatalyst of NiO/Cd/g-C_3N_4 for enhanced degradation of methylene blue. Arab. J. Chem. 13(6): 5810–5820.

Karpuraranjith, M., Y. Chen, S. Rajaboopathi, M. Ramadoss, K. Srinivas, D. Yang and B. Wang. 2022. Three-dimensional porous $MoS2$ nanobox embedded g-C_3N_4@TiO_2 architecture for highly efficient photocatalytic degradation of organic pollutant. J. Colloid Interface Sci. 605: 613–623.

Koči, K., M. Reli, I. Troppová, M. Šihor, J. Kupková, P. Kustrowski and P. Praus. 2017. Photocatalytic decomposition of N_2O over TiO_2/g-C_3N_4 Photocatalysts Heterojunction. Appl. Surf. Sci. 396: 1685–1695.

Kong, Y., C. Lv, C. Zhang and G. Chen. 2020. Cyano group modified G-C_3N_4: molten salt method achievement and promoted photocatalytic nitrogen fixation activity. Appl. Surf. Sci. 515: 14600.

Kuriki, R., K. Sekizawa, O. Ishitani and K. Maeda. 2015. Visible-light-driven CO_2 reduction with carbon nitride: enhancing the activity of ruthenium catalysts. In Angewandte Chemie - Int. Ed. 54: 2406–2409. 1002/anie.201411170: 54.

Lan, Z. A., G. Zhang and X. Wang. 2016. A facile synthesis of Br-modified g-C_3N_4 semiconductors for photoredox water splitting. Appl. Catal. B: Environ. 192: 116–125.

Li, C., C. B. Cao and H. S. Zhu. 2004. Graphitic carbon nitride thin films deposited by electrodeposition. Mater. Lett. 58(12-13): 1903–1906.

Li, G., X. Nie, J. Chen, Q. Jiang, T. An, P. K. Wong, H. Zhang, H. Zhao and H. Yamashita. 2015. Enhanced visible-light-driven photocatalytic inactivation of *Escherichia coli* using g-C3N4/TiO2 hybrid photocatalyst synthesized using a hydrothermal-calcination approach. Water Res. 86: 17–24.

Li, J., C. Cao and H. Zhu. 2007. Synthesis and characterization of graphite-like carbon nitride nanobelts and nanotubes. Nanotechnol. 18(11): 5.

Li, J., Y. Huan and Z. Zhu. 2016. Improved photoelectrochemical performance of Z-scheme g-C$_3$N$_4$/Bi$_2$O$_3$/BiPO$_4$ heterostructure and degradation property. Appl. Surf. Sci. 385: 34–41.

Li, J., X. Zhang, F. Raziq, J. Wang, C. Liu, Y. Liu et al. 2017. Improved photocatalytic activities of G-C$_3$N$_4$ nanosheets by effectively trapping holes with halogen-induced surface polarization and 2,4-dichlorophenol decomposition mechanism. Appl. Catal. B: Environ. 218: 60–67.

Li, N., Y. Tian, J. Zhao, J. Zhang, W. Zuo, L. Kong and H. Cui. 2018. Z-scheme 2D/3D g-C$_3$N$_4$@ZnO with enhanced photocatalytic activity for cephalexin oxidation under solar light. Chem. Eng. J. 352: 412–422.

Li, X., Y. Feng, M. Li, W. Li, H. Wei and D. Song. 2015. Smart hybrids of Zn2GeO4 nanoparticles and ultrathin G-C$_3$N$_4$ layers: synergistic lithium storage and excellent electrochemical performance. Adv. Funct. Mater. 25(44): 6858–6866.

Li, X., J. Xiong, Y. Xu, Z. Feng and J. Huang. 2019. Defect-assisted surface modification enhances the visible light photocatalytic performance of g-C$_3$N$_4$@C-TiO$_2$ direct Z-scheme heterojunctions. Cuihua Xuebao/Chinese J. Catal. 40(3): 424–433.

Li, Y., Z. Ruan, Z. He, J. Li, K. Li, Y. Yang, D. Xia, K. Lin and Y. Yuan. 2019. Enhanced photocatalytic H2 evolution and phenol degradation over sulfur doped meso/macroporous g-C$_3$N$_4$ spheres with continuous channels. Int. J. Hydrogen Energy. 44(2): 707–719.

Li, X., H. Jiang, C. Ma, Z. Zhu, X. Song, H. Wang, P. Huo and X. Li. 2021. Local surface plasma resonance effect enhanced Z-scheme ZnO/Au/g-C$_3$N$_4$ film photocatalyst for reduction of CO2 to CO. Appl. Catal. B: Environ. 283: 11963.

Li, Y., S. Zhu, X. Kong, X. Liang, Z. Li, S. Wu, C. Chang, S. Luo and Z. Cui. 2021. *In situ* synthesis of a novel Mn3O4/g-C$_3$N$_4$ p-n heterostructure photocatalyst for water splitting. J. Colloid Int. Sci. 586: 778–784.

Liang, Q., M. Zhang, C. Liu, S. Xu and Z. Li. 2016. Sulfur-doped graphitic carbon nitride decorated with zinc phthalocyanines towards highly stable and efficient photocatalysis. Appl. Catal. A: General. 519: 107–115.

Liebig, J. 1834. About some nitrogen compounds. Ann. Pharm. 10: 10.

Liu, A. Y. and M. L. Cohen. 1989. Prediction of new low compressibility solids. Sci. 245(4920): 841–842.

Liu, L., Xi L., Y. Li, F. Xu, Z. Gao, X. Zhang, Y. Song, H. Xu and H. Li. 2018. Facile synthesis of few-layer g-C$_3$N$_4$/ZnO composite photocatalyst for enhancing visible light photocatalytic performance of pollutants removal. Colloids Surf. A: Physicochem. Eng. Asp. 537: 516–523.

Liu, X., R. Ma, L. Zhuang, B. Hu, J. Chen, X. Liu and X. Wang. 2021. Recent developments of doped G-C$_3$N$_4$ photocatalysts for the degradation of organic pollutants. Crit. Rev. Environ. Sci. Technol. 51(8): 751–790.

Liu, Y., J. Jiang, Y. Sun, S. Wu, Y. Cao, W. Gong and J. Zou. 2017. NiO and Co$_3$O$_4$ Co-Doped g-C$_3$N$_4$ nanocomposites with excellent photoelectrochemical properties under visible light for detection of tetrabromobisphenol-A. RSC Adv. 7(57): 36015–36020.

Liu, Y., H. Liu, H. Zhou, T. Li and L. Zhang. 2019. A Z-scheme mechanism of N-ZnO/g-C$_3$N$_4$ for enhanced H2 evolution and photocatalytic degradation. Appl. Surf. Sci. 466: 133–140.

Lu, X., L. Gai, D. Cui, Q. Wang, X. Zhao and X. Tao. 2007. Synthesis and characterization of C$_3$N$_4$ nanowires and Pseudocubic C$_3$N$_4$ polycrystalline nanoparticles. Mater. Lett. 61(21): 4255–4258.

Luo, J., X. Zhou, L. Ma and X. Xu. 2015. Enhancing visible-light photocatalytic activity of g-C$_3$N$_4$ by doping phosphorus and coupling with CeO$_2$ for the degradation of methyl orange under visible light irradiation. RSC Adv. 5(84): 68728–68735.

Ma, R., J. Guo, D. Wang, M. He, S. Xun, J. Gu, W. Zhu and H. Li. 2019. Preparation of highly dispersed WO3/few layer g-C$_3$N$_4$ and its enhancement of catalytic oxidative desulfurization activity. Colloids Surf. A: Physicochem. Eng. Asp. 572: 250–258.

Mao, L., X. Xue, X. Xu, W. Wen, M. M. Chen, X. Zhang and S. Wang. 2021. Heterostructured CuO-g-C$_3$N$_4$ nanocomposites as a highly efficient photocathode for photoelectrochemical Aflatoxin B1 sensing. Sens. Actuators B: Chem. 329: 12914.

Marchal, C., T. Cottineau, M. G. Méndez-Medrano, C. Colbeau-Justin, V. Caps and V. Keller. 2018. Au/TiO$_2$–GC$_3$N$_4$ nanocomposites for enhanced photocatalytic H2 production from water under visible light irradiation with very low quantities of sacrificial agents. Adv. Energy Mater. 8(14): 1702142.

Matsunaga, T., R. Tomoda, T. Nakajima and H. Wake. 1985. Photoelectrochemical sterilization of microbial cells by semiconductor powders. FEMS Microbiol. Lett. 29 (1-2): 211–214.

Meenakshisundaram, I., S. Kalimuthu, P. P. Gomathi and S. Karthikeyan. 2019. Facile green synthesis and antimicrobial performance of Cu_2O nanospheres decorated G-C_3N_4 nanocomposite. Mater. Res. Bull. 112: 331–335.

Mishra, A., A. Mehta, S. Basu, N. P. Shetti, K. R. Reddy and T. M. Aminabhavi. 2019. Graphitic carbon nitride (g-C_3N_4)-based metal-free photocatalysts for water splitting: a review. Carbon. 149: 693–721.

Mohammadi, I., F. Zeraatpisheh, E. Ashiri and K. Abdi. 2020. Solvothermal synthesis of G-C_3N_4 and ZnO nanoparticles on TiO_2 nanotube as photoanode in DSSC. Int. J. Hydrogen Energy. 45(38): 18831–18839.

Montigaud, H., B. Tanguy, G. Demazeau, I. Alves and S. Courjault. 2000. Dream or Reality? Solvothermal Synthesis as Macroscopic Samples of the C. 3: 2547–2552.

Mulik, B. B., B. D. Bankar, A. V. Munde, P. P. Chavan, A. V. Biradar and B. R. Sathe. 2021. Electrocatalytic and catalytic CO_2 hydrogenation on ZnO/g-C_3N_4 hybrid nanoelectrodes. Appl. Surf. Sci. 538: 14812.

Ngullie, R. C., O. A. Saleh, B. Kandasamy, P. Shanmugam, T. Pazhanivel and P. Arunachalam. 2020. Synthesis and characterization of efficient ZnO/g-C_3N_4 nanocomposites photocatalyst for photocatalytic degradation of methylene blue. Coatings 10(5): 500.

Nie, N., L. Zhang, J. Fu, B. Cheng and J. Yu. 2018. Self-assembled hierarchical direct Z-Scheme g-C_3N_4/ZnO microspheres with enhanced photocatalytic CO_2 reduction performance. Appl. Surf. Sci. 441: 12–22.

Ong, W. J., L. L. Tan, Y. H. Ng, S. T. Yong and S. P. Chai. 2016. Graphitic carbon nitride (g-C_3N_4)-based photocatalysts for artificial photosynthesis and environmental remediation: are we a step closer to achieving sustainability? Chem. Rev. 116(12): 7159–7329.

Ouyang, K., B. Xu, C. Yang, H. Wang, P. Zhan and S. Xie. 2022. Synthesis of a novel Z-scheme Ag/WO3/g-C_3N_4 nanophotocatalyst for degradation of oxytetracycline hydrochloride under visible light. Mater. Sci. Semiconductor Process. 137.

Pan, T., D. Chen, W. Xu, J. Fang, S. Wu, Z. Liu, K. Wu and Z. Fang. 2020. Anionic polyacrylamide-assisted construction of thin 2D-2D WO3/g-C_3N_4 step-scheme heterojunction for enhanced tetracycline degradation under visible light irradiation. J. Hazard. Mater. 393: 12236.

Pant, B., M. Park, J. H. Lee, H. Y. Kim and S. J. Park. 2017. Novel magnetically separable silver-iron oxide nanoparticles decorated graphitic carbon nitride nano-sheets: a multifunctional photocatalyst via one-step hydrothermal process. J. Colloid Interface Sci. 496: 343–352.

Paul, A. M., A. Sajeev, R. Nivetha, K. Gothandapani, K. Bhardwaj, K. Govardhan et al. 2020. Cuprous oxide (Cu_2O)/ Graphitic Carbon Nitride (g-C_3N_4) nanocomposites for electrocatalytic hydrogen evolution reaction. Diamond Relat. Mater. 107: 107899.

Paul, D. R., S. Gautam, P. Panchal, S. P. Nehra, P. Choudhary and A. Sharma. 2020. ZnO-modified g-C_3N_4: a potential photocatalyst for environmental application. ACS Omega. 5(8): 3828–3838.

Qiu, P., C. Xu, H. Chen, F. Jiang, X. Wang, R. Lu and X. Zhang. 2017. One step synthesis of oxygen doped porous graphitic carbon nitride with remarkable improvement of photo-oxidation activity: role of oxygen on visible light photocatalytic activity. Appl. Catal. B: Environ. 206(1016): 319–327.

Qu, A., X. Xu, H. Xie, Y. Zhang, Y. Li and J. Wang. 2016. Effects of calcining temperature on photocatalysis of G-C_3N_4/TiO_2 composites for hydrogen evolution from water. Mater. Res. Bull. 80: 167–176.

Rabani, I., R. Zafar, K. Subalakshmi, H. S. Kim, C. Bathula and Y. S. Seo. 2021. A facile mechanochemical preparation of Co_3O_4@g-C_3N_4 for application in supercapacitors and degradation of pollutants in water. J. Hazard. Mater. 407.

Rahman, N., J. Yang, Zulfiqar, M. Sohail, R. Khan, A. Iqbal, C. Maouche et al. 2021. Insight into metallic oxide semiconductor (SnO2, ZnO, CuO, α-Fe2O3, WO3)-Carbon Nitride (g-C_3N_4) heterojunction for gas sensing application. Sens. Actuators A: Phys. 332: 11312.

Ramachandra, M., S. D. Kalathiparambil, R. Pai, J. R. Jaleel and D. Pinheiro. 2020. Improved photocatalytic activity of G-C_3N_4/ZnO: A PotentialDirect Z-Scheme Nanocomposite. Chem. Sel. 5(38): 11986–11995.

Rathi, A. K., H. Kmentová, A. Naldoni, A. Goswami, M. B. Gawande, R. S. Varma, S. Kment and R. Zboril. 2018. Significant Enhancement of photoactivity in hybrid TiO_2/g-C_3N_4 nanorod catalysts modified with Cu-Ni-based nanostructures. ACS Appl. Nano Mater. 1(6): 2526–2535.

Ren, Y., D. Zeng and W. J. Ong. 2019. Interfacial engineering of graphitic Carbon Nitride (g-C_3N_4)-based metal sulfide heterojunction photocatalysts for energy conversion: a review. Cuihua Xuebao/Chinese J. Catal. 40(3): 289–319.

Selvarajan, S., A. Suganthi and M. Rajarajan. 2018. Fabrication of G-C_3N_4/NiO heterostructured nanocomposite modified glassy carbon electrode for quercetin biosensor. Ultrason. Sonochem. 41: 651–660.

Shaikh, A.V., R. S. Mane, O. S. Joo, S. H. Han and H. M. Pathan. 2017. Electrochemical deposition of cadmium selenide films and their properties: a review. J. Solid State Electrochem. 21(9): 2517–2530.

Shan, W., Y. Hu, Z. Bai, M. Zheng and C. Wei. 2016. *In situ* preparation of G-C_3N_4/bismuth-based oxide nanocomposites with enhanced photocatalytic activity. Appl. Catal. B: Environ. 188: 1–12.

Shandilya, P., D. Mittal, A. Sudhaik, M. Soni, P. Raizada, A. K. Saini and P. Singh. 2019. GdVO4 modified fluorine doped graphene nanosheets as dispersed photocatalyst for mitigation of phenolic compounds in aqueous environment and bacterial disinfection. Sep. Purif. Technol. 210: 804–816.

Shanmugam, V., K. S. Jeyaperumal, P. Mariappan and A. L. Muppudathi. 2020. Fabrication of novel G-C_3N_4 based MoS_2 and Bi_2O_3 nanorod embedded ternary nanocomposites for superior photocatalytic performance and destruction of bacteria. New J. Chem. 44(30): 13182–13194.

Shi, J. W., Y. Zou, L. Cheng, D. Ma, D. Sun, S. Mao, L. Sun, C. He and Z. Wang. 2019. *In-situ* phosphating to synthesize Ni2P decorated NiO/g-C_3N_4 p-n junction for enhanced photocatalytic hydrogen production. Chem. Eng. J. 378.

Shi, J., B. Zheng, L. Mao, C. Cheng, Y. Hu, H. Wang, G. Li, D. Jing and X. Liang. 2021. MoO_3/g-C_3N_4 Z-scheme (S-Scheme) system derived from MoS2/melamine dual precursors for enhanced photocatalytic H2 evolution driven by visible light. Int. J. Hydrogen Energy. 46(3): 2927–2935.

Shoran, S., S. Chaudhary and A. Sharma. 2022. Photocatalytic dye degradation and antibacterial activities of CeO_2/g-C_3N_4 nanomaterials for environmental applications. Environ. Sci. Pollut. Res.

Si, Y., X. Zhang, T. Liang, X. Xu, L. Qiu, P. Li and S. Duo. 2020. Facile *in-situ* synthesis of 2D/3D g-C_3N_4/Cu_2O heterojunction for high-performance photocatalytic dye degradation. Mater. Res. Express. 7(1): 15524.

Song, T., X. Zhang and P. Yang. 2021. Bifunctional nitrogen-doped carbon dots in g-C_3N_4/WOx heterojunction for enhanced photocatalytic water-splitting performance. Langmuir. 37(14): 4236–4247.

Sudrajat, H. 2018. A one-pot, solid-state route for realizing highly visible light active Na-Doped GC_3N_4 photocatalysts. J. Solid State Chem. 257: 26–33.

Sumathi, M., A. Prakasam and P. M. Anbarasan. 2019. Fabrication of hexagonal disc shaped nanoparticles G-C_3N_4/NiO heterostructured nanocomposites for efficient visible light photocatalytic performance. J. Cluster Sci. 3(30): 757–766.

Sun, Y., J. Jiang, Y. Liu, S. Wu and J. Zou. 2018. A facile one-pot preparation of Co_3O_4/g-C_3N_4 heterojunctions with excellent electrocatalytic activity for the detection of environmental phenolic hormones. Appl. Surf. Sci. 430: 362–370.

Sun, Z., W. Fang, L. Zhao, H. Chen, X. He, W. Li, P. Tian and Z. Huang. 2019. G-C_3N_4 foam/Cu_2O QDs with excellent CO_2 adsorption and synergistic catalytic effect for photocatalytic CO_2 reduction. Environ. Int. 130: 10489.

Taha, M. M., L. G. Ghanem, M. A. Hamza and N. K. Allam. 2021. Highly stable supercapacitor devices based on three-dimensional bioderived carbon encapsulated g-C_3N_4 nanosheets. ACS Appl. Energy Mater. 4(9): 10344–10355.

Tan, X., X. Wang, H. Hang, D. Zhang, N. Zhang, Z. Xiao and H. Tao. 2019. Self-assembly method assisted synthesis of g-C_3N_4/ZnO heterostructure nanocomposites with enhanced photocatalytic performance. Opt. Mater. 96.

Tang, J. Y., R. T. Guo, W. G. Zhou, C. Y. Huang and W. G. Pan. 2018. Ball-flower like NiO/g-C_3N_4 heterojunction for efficient visible light photocatalytic CO_2 reduction. Appl. Catal. B: Environ. 237: 802–810.

Tang, Z., C. Wang, W. He, Y. Wei, Z. Zhao and J. Liu. 2022. The Z-Scheme g-C_3N_4/3DOM-WO3 photocatalysts with enhanced activity for CO_2 photoreduction into CO. Chinese Chem. Lett. 33 (in press).

Teter, D. M. and R. J. Hemley. 1996. Low-compressibility carbon nitrides. Science. 271(5245): 53–55.

Thomas, J., K. S. Ambili and S. Radhika. 2018. Synthesis of Sm3+-doped graphitic carbon nitride nanosheets for the photocatalytic degradation of organic pollutants under sunlight. Catal. Today. 310: 11–18.

Tonda, S., S. Kumar, S. Kandula and V. Shanker. 2014. Fe-doped and -mediated graphitic carbon nitride nanosheets for enhanced photocatalytic performance under natural sunlight. J. Mater. Chem. A 2(19): 6772–6780.

Tran, H. H., P. H. Nguyen, M. L. P. Le, S.-J. Kim and V. Vo. 2019. SnO_2 Nanosheets/Graphite Oxide/g-C_3N_4 composite as enhanced performance anode material for lithium ion batteries. Chem. 715: 284–292.

Vignesh, S., S. Suganthi, J. K. Sundar and V. Raj. 2019. Construction of α-Fe2O3/CeO2 decorated g-C_3N_4 nanosheets for magnetically separable efficient photocatalytic performance under visible light exposure and bacterial disinfection. Appl. Surf. Sci. 488: 763–777.

Wang, H., H. Li, Z. Chen, J. Li, X. Li, P. Huo and Q. Wang. 2020. TiO_2 modified G-C_3N_4 with enhanced photocatalytic CO_2 reduction performance. Solid State Sci. 100: 10609.

Wang, K., Q. Li, B. Liu, B. Cheng, W. Ho and J. Yu. 2015. Sulfur-doped g-C_3N_4 with enhanced photocatalytic CO_2-Reduction Performance. Appl. Catal. B: Environ. 176-177: 44–52.

Wang, S., J. Long, T. Jiang, S. Li, D. Li, X. Xie and F. Xu. 2022. Magnetic Fe3O4/CeO2/g-C_3N_4 composites with a visible-light response as a high efficiency fenton photocatalyst to synergistically degrade tetracycline. Sep. Purif. Technol. 278: 11960.

Wang, W., Y. Zhan and G. Wang. 2001. One-step, solid-state reaction to the synthesis of copper oxide nanorods in the presence of a suitable surfactant. Chem. Commun. 8(8): 727–728.

Wang, Y., R. Zhang, Z. Zhang, J. Cao and T. Ma. 2019. Host–guest recognition on 2D graphitic carbon nitride for nanosensing. Adv Mater Interfaces. 6(23): 19014.

Wei, T., J. Xu, C. Kan, L. Zhang and X. Zhu. 2022. Au tailored on G-C₃N₄/TiO₂ heterostructure for enhanced photocatalytic performance. J. Alloys Compd. 894.

Wen, P., Y. Sun, H. Li, Z. Liang, H. Wu, J. Zhang, H. Zeng, S. M. Geyer and L. Jiang. 2020. A highly active three-dimensional Z-Scheme ZnO/Au/g-C₃N₄ photocathode for efficient photoelectrochemical water splitting. Appl. Catal. B: Environ. 263.

Xia, Y., L. Xu, J. Peng, J. Han, S. Guo, L. Zhang, Z. Han and S. Komarneni. 2019. TiO₂@g-C₃N₄ core/shell spheres with uniform mesoporous structures for high performance visible-light photocatalytic application. Ceramics Int. 45(15): 18844–18851.

Xu, D., X. Li, J. Liu and L. Huang. 2013. Synthesis and photocatalytic performance of europium-doped graphitic carbon nitride. J. Rare Earths 31(11): 1085–1091.

Xu, J., Y. Li, X. Zhou, Y. Li, Z. D. Gao, Y. Y. Song and P. Schmuki. 2016. Graphitic C₃N₄-sensitized TiO₂ nanotube layers: a visible-light activated efficient metal-free antimicrobial platform. Chem. - A European J. 22(12): 3947–3951.

Yang, S., Q. Sun, W. Han, Y. Shen, Z. Ni, S. Zhang, L. Chen, L. Zhang, J. Cao and H. Zheng. 2022. A simple and highly efficient composite based on G-C₃N₄ for super rapid removal of multiple organic dyes from water under sunlight. Catal. Sci. Technol. 12(3): 786–798.

Ye, L., D. Wang and S. Chen. 2016. Fabrication and enhanced photoelectrochemical performance of MoS2/S-doped g-C₃N₄ heterojunction film. ACS Appl. Mater. Interfaces. 8(8): 5280–5289.

Zhai, J., T. Wang, C. Wang and D. Liu. 2018. UV-light-assisted ethanol sensing characteristics of g-C₃N₄/ZnO composites at room temperature. Appl. Surface Sci. 441: 317–323.

Zhang, C., Y. Li, D. Shuai, Y. Shen, W. Xiong and L. Wang. 2019. Graphitic carbon nitride (g-C₃N₄)-based photocatalysts for water disinfection and microbial control: a review. Chemosphere. 214: 462–479.

Zhang, M., X. Bai, D. Liu, J. Wang and Y. Zhu. 2015. Enhanced catalytic activity of potassium-doped graphitic carbon nitride induced by lower valence position. Appl. Catal. B: Environ. 164: 77–81.

Zhang, P., X. Li, C. Shao and Y. Liu. 2015. Hydrothermal synthesis of carbon-rich graphitic carbon nitride nanosheets for photoredox catalysis. J. Mater. Chem. A 3(7): 3281–3284.

Zhang, S., P. Gu, R. Ma, C. Luo, T. Wen, G. Zhao, W. Cheng and X. Wang. 2019. Recent developments in fabrication and structure regulation of visible-light-driven g-C₃N₄-based photocatalysts towards water purification: a critical review. Catal. Today. 335: 65–77.

Zhang, Y., J. Lu, M. R. Hoffmann, Q. Wang, Y. Cong, Q. Wang and H. Jin. 2015. Synthesis of G-C₃N₄/Bi₂O₃/TiO₂ composite nanotubes: enhanced activity under visible light irradiation and improved photoelectrochemical activity. RSC Adv. 5(60): 48983–48991.

Zhao, X., J. Guan, J. Li, X. Li, H. Wang, P. Huo and Y. Yan. 2021. CeO₂/3D g-C₃N₄ heterojunction deposited with Pt cocatalyst for enhanced photocatalytic CO₂ reduction. Appl. Surf. Sci. 537.

Zhou, Y., L. Zhang, J. Liu, X. Fan, B. Wang, M. Wang, W. Ren, J. Wang, M. Li and J. Shi. 2015. Brand new P-doped g-C₃N₄: enhanced photocatalytic activity for H2 evolution and Rhodamine B degradation under visible light. J. Mater. Chem. A 3(7): 3862–3867.

Zhu, W., F. Sun, R. Goei and Y. Zhou. 2017. Construction of WO₃-g-C₃N₄ composites as efficient photocatalysts for pharmaceutical degradation under visible light. Catal. Sci. Technol. 7(12): 2591–2600.

Zhu, Y. P., T. Z. Ren and Z. Y. Yuan. 2015. Mesoporous phosphorus-doped g-C₃N₄ nanostructured flowers with superior photocatalytic hydrogen evolution performance. ACS Appl. Mater. Interfaces. 7(30): 16850–16856.

Zou, J., S. Wu, Y. Liu, Y. Sun, Y. Cao, J. P. Hsu, A. T. S. Wee and J. Jiang. 2018. An ultra-sensitive electrochemical sensor based on 2D g-C₃N₄/CuO nanocomposites for dopamine detection. Carbon. 130: 652–663.

Zou, W., B. Deng, X. Hu, Y. Zhou, Y. Pu, S. Yu, K. Ma, J. Sun, H. Wan and L. Dong. 2018. Crystal-plane-dependent metal oxide-support interaction in CeO₂/g-C₃N₄ for photocatalytic hydrogen evolution. Appl. Catal. B: Environ. 238: 111–118.

Index

P

PAH degradation 91, 99, 100
PAHs 126, 132, 133
Parathion 136, 139, 131, 137
Pesticide pollution 110, 112, 113
Pesticides 125, 126, 129, 133, 137, 262, 263, 270, 271
Phenylurea 126, 130
Photocatalytic degradation 21, 74, 284
Phytoremediation 2, 10–12, 67, 70, 74–80, 188, 193, 195–197, 199, 231, 233
Pollutants 1–6, 12, 13
Polycyclic Aromatic Hydrocarbons 67, 78
Pseudomonas putida 125–128, 130, 133

R

reactor's design 45
Remediation 70, 74, 75, 77–80, 262, 269–274, 279, 283, 284, 288

S

Silica NPs 272
Soil 67–80
Sustainable 1
Sustainable cleanup 1
Sustainable environment 263

T

TCP 131
Technologies 148, 156
Toxic metals 167, 169–180
Toxicity 187–195, 198

W

waste reuse 41
Wastewater 109
Water hyacinth 238–240, 248, 253, 254

About the Editors

Dr. Anju Malik is presently working as Associate Professor and Chairperson at Department of Energy and Environmental Sciences, Chaudhary Devi Lal University, Sirsa, India. She has completed her M.Phil. and Ph.D. from the School of Environmental Sciences, Jawaharlal Nehru University, New Delhi. She was awarded with a fellowship from the Belgian Government, under the bilateral cultural exchange programme at University of Ghent, Belgium. Dr. Anju has also received research scholarships from University Grant Commission (UGC) and Council of Scientific and Industrial Research (CSIR), New Delhi, India. She has worked in the past as a scientist under the Indian Council of Forestry Research and Education, Dehradun, India. With more than 18 years of research and teaching experience, her research interest is focused on trace elemental contamination, speciation and remediation. She is a life member of The Indian Science Congress Association, Kolkata, India. She has guided several M.Sc., M.Phil. and Ph.D. students. She has published research papers in various peer-reviewed national and international journals of repute. She has also published one book and 10 book chapters in edited books published by reputed international publishers. She is a reviewer of several international journals.

Prof (Dr.) Vinod Kumar Garg is presently working as Professor at the Department of Environmental Science and Technology, Central University of Punjab, Punjab, India. He is an experienced researcher with more than 30 years of involvement in leading, supervising, and undertaking research. He along with his research group are working on water and wastewater pollution monitoring and abatement, solid waste management, pesticide degradation, radioecology, green chemistry and heavy metal detoxification. He has published more than 200 research and review articles, and six editorials in peer-reviewed international and national journals of repute with more than 18000 citations and h-index 73. In addition, he has published eight books and 15 book chapters and completed 10 sponsored research projects as principal investigator funded by various agencies and departments. He was awarded "Thomson Reuters Research Excellence—India Citation Awards 2012". He is an active member of various scientific societies and organizations including, the Biotech Research Society of India, the Indian Nuclear Society, etc.

For Product Safety Concerns and Information please contact our EU
representative GPSR@taylorandfrancis.com
Taylor & Francis Verlag GmbH, Kaufingerstraße 24, 80331 München, Germany

www.ingramcontent.com/pod-product-compliance
Lightning Source LLC
Chambersburg PA
CBHW080929220326
41598CB00034B/5724

9 781032 234922